Multifunctional Polymer Composite Materials

Multifunctional Polymer Composite Materials

Guest Editors

Md Najib Alam
Vineet Kumar

MDPI

Basel • Beijing • Wuhan • Barcelona • Belgrade • Novi Sad • Cluj • Manchester

Guest Editors
Md Najib Alam
School of Mechanical Engineering
Yeungnam University
Gyeongsan
Republic of Korea

Vineet Kumar
School of Mechanical Engineering
Yeungnam University
Gyeongbuk
Republic of Korea

Editorial Office
MDPI AG
Grosspeteranlage 5
4052 Basel, Switzerland

This is a reprint of the Special Issue, published open access by the journal *Polymers* (ISSN 2073-4360), freely accessible at: https://www.mdpi.com/journal/polymers/special_issues/924C3B9473.

For citation purposes, cite each article independently as indicated on the article page online and as indicated below:

Lastname, A.A.; Lastname, B.B. Article Title. *Journal Name* **Year**, *Volume Number*, Page Range.

ISBN 978-3-7258-4985-7 (Hbk)
ISBN 978-3-7258-4986-4 (PDF)
https://doi.org/10.3390/books978-3-7258-4986-4

Contents

About the Editors

Md Najib Alam

Md Najib Alam obtained his B.Sc. and M.Sc. degrees from the University of Kalyani, India, in 2006 and 2008, respectively. He completed his Ph.D. at the same university in 2014, where he addressed numerous challenges related to rubber vulcanization chemistry. Following his Ph.D., he pursued postdoctoral research at Chulalongkorn University, Thailand, from 2015 to 2019. From 2019 to 2022, he served as a research fellow at the School of Mechanical Engineering, Yeungnam University, Republic of Korea. In 2022, he transitioned to the role of Assistant Professor at the same institution. Throughout his research career, he has published over 50 original research articles in internationally recognized, peer-reviewed journals, including the *Journal of Industrial and Engineering Chemistry, Composites Science and Technology, and Composites Part B: Engineering*. Currently, he is serving as a Guest Editor for various Special Issues in MDPI's journal *Polymers*. His research focuses on rubber vulcanization, novel rubber composites, stretchable sensors, and smart materials.

Vineet Kumar

Vineet Kumar has been an Assistant Professor since 2016 at the Department of Mechanical Engineering, Yeungnam University, Gyeongsan-si, Republic of Korea. He completed his Ph.D. in the Department of Material Science at the University of Milan-Bicocca, Milan, Italy, in 2014. He finished his Master of Science degree in Environment Management from the Forest Research Institute, Dehradun, India, in 2008. He has co-authored more than 100 peer-reviewed articles in reputed SCI journals. He has also served as a Topic Editor for the *Polymers* journal of MDPI since 2021. Moreover, he has edited as a Guest Editor for nine Special Issues in the *Polymers* journal of MDPI and *Frontiers in Materials*.

Preface

Polymer composites are composed of a polymer matrix, reinforcing fillers, and other additives. Here, the polymer matrix used can be thermoplastic, thermoset, or elastomer. An elastomeric matrix is frequently used due to its ability to stretch, be lightweight, and be flexible. Functional fillers, such as carbon nanotubes, graphene, and carbon black, are added to impart reinforcement and enhance their properties. These improved properties are mechanical, electrical, and thermal properties. Unlike traditional composites, these multifunctional composites exhibit robust sensing capability, self-healing, energy storage, or environmental responsiveness. The design of these multifunctional composites needs an interdisciplinary approach in order to achieve robust performance.

Keeping these ideas in mind, the present Reprint provides an in-depth study of multifunctional aspects in polymer composites. It further offers the integration of various functionalities into a single lightweight and durable system, making these polymer composite materials promising for next-generation technologies. Finally, each chapter in the Reprint will cover the innovations, novel results, and the discussion of these multifunctional polymer composites. We are confident that readers of this Reprint will gain meaningful insights into multifunctional composites.

Md Najib Alam and Vineet Kumar

Guest Editors

polymers MDPI

Editorial

Multifunctional Polymer Composite Materials

Vineet Kumar and Md Najib Alam *

School of Mechanical Engineering, Yeungnam University, 280, Daehak-ro, Gyeongsan 38541, Republic of Korea; vineetfri@gmail.com

* Correspondence: mdnajib.alam3@gmail.com

1. Introduction

Recently, polymer composites have evolved from simple polymeric materials into advanced engineering systems capable of delivering multiple functionalities simultaneously [1,2]. This evolution has given rise to the concept of multifunctionality in polymer composite materials. They refer to the integration of diverse properties within a single system, making them suitable for a wide range of applications [3]. Multifunctionality is primarily enabled by the incorporation of functional fillers, such as carbon nanotubes [4], metal oxides [5], and various ceramic materials [6]. While many types of polymer matrices are available, elastomeric matrices are particularly favored. Among them, polyurethane [7], silicone rubber [8], and polydimethylsiloxane (PDMS) [9] are especially popular in energy-harvesting and -sensing applications. This choice is due to their inherent flexibility, stretchability, light weight, and ease of processing [10]. These characteristics facilitate the development of smart polymeric materials suitable for various applications. These applications are wearable sensors, soft robotics, energy-harvesting systems, vibration dampers, and advanced sensing platforms [11,12]. However, realizing true multifunctionality in polymer composites requires a careful balance of several critical properties. These include mechanical stiffness, uniform filler dispersion, and strong interfacial bonding between the filler and polymer matrix. Furthermore, maintaining mechanical integrity is essential, as it directly impacts the durability and intrinsic flexibility of the composite material [13].

However, achieving an optimal balance among these properties remains a significant challenge. To address this, filler functionalization is often employed to improve both filler dispersion and interfacial interactions between the filler and the rubber matrix [14]. Multifunctional polymer composites thus represent a transformative class of materials, like soft, adaptive, and intelligent, that are key enablers for next-generation technologies [15]. These advanced composites are specifically engineered for emerging applications such as human–machine interfaces, healthcare monitoring devices, and energy-harvesting systems. In light of these developments, this Special Issue aims to explore the multifunctionality of polymer composites in depth. It will highlight strategies for tailoring material properties to meet specific performance requirements. Therefore, they provide valuable insights and pathways for the practical implementation of these composites. Finally, the ability to control and optimize a wide range of mechanical, electrical, and structural properties enables the development of application-specific solutions. The following sections will review relevant literature and discuss how multifunctional polymer composites are being utilized across various domains. Finally, portable electronics, energy-harvesting technologies, and advanced sensing platforms are some important and targeted applications for this Special Issue.

2. Overview of Published Articles

Kumar et al. [16] report using silicone rubber-based composites and their multifunctionality as energy-harvesting systems. The authors use titanium carbide (TiC) and molybdenum disulfide (MoS_2) to reinforce the matrix and improve the electrical conductivity of the composites. The results show that the output voltages were 3.5 mV (6 phr of TiC) and 6.7 mV (6 phr of MoS_2). Similarly, Iyer et al. [17] designed and fabricated composites based on poly (ethylene oxide), and their multifunctional prospects for sodium ion batteries were demonstrated. Here, the authors develop a novel approach to designing the batteries by making them capable of operating at room temperature with high performance. The properties were fantastic, such as tensile strength of 32.1 MPa and an ionic conductivity of 1.01×10^{-4} S cm^{-1}. Moreover, the battery shows high electrochemical stability and an energy density of 14.2 Wh kg^{-1}. Therefore, this multifunctionality of the composites demonstrates a promising potential for developing high-performance batteries. Assadakorn et al. [18] use the Dunlop process to fabricate composites based on high-surface-area silica and natural rubber latex. The results show that the four types of silica particles were used, some with hydrophobicity and some with hydrophilicity. The hardness was tested, and it was 45 for non-silica foam rubber and 48 for hydrophilic fumed silica-based samples. Overall, the study concludes that the hydrophobic fumed silica was better than the hydrophilic ones in both fumed and precipitated silica particles in rubber composites. Another interesting review study by Malashin et al. [19] shows multifunctionalities in polymer composites. This review provides a special focus on new generation processing techniques like 3D and 4D printing. The review study further highlights that these emerging printing technologies are very promising in obtaining consistent composite samples. Finally, the paper summarizes the key takeaways, challenges, and future prospects of these techniques.

Sriani et al. [20] demonstrate ultrafiltration in whey protein separations. Here, the authors modify the polysulfone-based flat membrane with hydroxyapatite (HA) during fabrication using a wet-phase method. The results are interesting, as the permeability flux increases to 38% after adding 0.3 wt% of HA. Therefore, a simple and economical modification of the polysulfone membrane with HA resulted in improving permeability without sacrificing the separation efficiency. Therefore, this work provides an efficient route to obtain scalable membrane production for the high whey protein separation industry. Another study by Bril' et al. [21] discusses the use of polymer composites for multifunctional applications like flexible electronics and electromagnetic compatibility. These composites are fabricated by mixing silver nanowires with a PET polymer matrix. As multifunctional applications, authors study the shielding efficiency for a 500–600 nm thick porous sample. The results show that the samples were 40 dB, while as the thickness increased to 3.1–4.1 μm, the efficiency increased to 85–90 dB. Finally, the results show that the samples were able to show their application for portable electronics and electromagnetic compatibility. Inna et al. [22] present an interesting study on possible staining and surface roughness of restored dental applications. This in vitro study provides a breakthrough in teeth restoration by micro-abrasion and resin infiltration (RIT). The study shows that RIT proves to be promising for higher durability with lower roughness. Fatkullin et al. [23] studied the switching of the polymer composites based on laser-induced graphene for multifunctional applications like tunable electronics. Moreover, the fabricated composites show strong potential for their use in energy storage, sensing, and bioelectronics. The new strategy proposed by the authors concludes that the tailoring of the polymer composite's electrochemical properties with dynamic switching is possible. This aspect opens new pathways for flexible electronics with tunable properties and performance.

Repin et al. [24] study the multifunctionality of cellulose-based composite materials for water extraction from atmospheric air. A strong and porous composite system was developed that acts like a carrier for the precipitation of hygroscopic agents. The hygroscopic agents developed in this work were $CaCl_2$ and 1-butyl-3-methylimidazolium chloride. Here, the use of 1-butyl-3-methylimidazolium chloride was found most promising for water absorption efficiency and recyclable potential. Plamadiala et al. [25] present a review article summarizing enhanced performance of polylactic acid (PLA) filaments in FDM. The polylactic acid was proffered because of its biodegradability and scalability due to easy processing. Finally, this study highlights the multifunctional use of PLA, including the potential challenges and processing conditions. Janmanee et al. [26] developed a biosensor based on polymer composites for the detection of dopamine (DA). The study shows that the DA imbalance is also associated with brain malfunction and results in diseases like Parkinson's. The developed biosensor shows high-sensitivity selectivity in detecting DA content from 5 to 180 μM. Thus, the study supports that biosensing technology can be used to detect DA early and can avoid neurological disorders. Finally, Kumar et al. [27] provide a comprehensive review for studying physical activity monitoring through polymer composite sensors. This review study helps in understanding the early trends on the use of composite-based sensors for detecting physical activity from year 2020–2025. The composites are based on carbon nanomaterials as a conducting material and silicone rubber as the polymer matrix. The physical activities monitored in this work involve sensing running, cycling, and swimming. The electrical properties reviewed in this work include resistance, response time, and gauge factors of the strain sensor. Finally, the key challenges and future prospects of this work are summarized and reported.

3. Summary and Future Outlook

Although significant advancements have been made in the development of polymer composite materials, several critical challenges remain. These include difficulties in scalability, ensuring performance reliability, and achieving mechanical integration for long-term durability [28]. Addressing these issues requires focused research, particularly in the areas of filler dispersion, filler–polymer interfacial interactions, and material durability [29]. For instance, achieving uniform filler dispersion is essential for ensuring balanced and consistent performance in polymer composites. Poor dispersion often leads to filler aggregation, which reduces the effective interfacial area and hampers stress transfer during mechanical deformation [30]. Additionally, engineering the filler–matrix interface can substantially enhance stress distribution, charge transport, and overall durability. In contrast, weak interfacial bonding can cause performance degradation under mechanical, electrical, or thermal stress. Embedding multifunctional filler particles into high-performance functional devices also requires careful consideration of interface design, packaging, and interconnect architecture [31]. Fortunately, many of these challenges can be addressed through strategic approaches. For example, functionalizing filler particles can greatly improve dispersion and interfacial bonding, thereby unlocking the full performance potential of multifunctional composites [32]. Another promising route involves using hybrid filler systems, which can exhibit synergistic effects and deliver robust performance even at low filler loadings. Furthermore, there is a growing interest in bio-based polymers and biodegradable fillers, aimed at creating environmentally friendly multifunctional composites [33]. These sustainable materials are particularly well-suited for applications in transient electronics and green manufacturing. Looking ahead, multifunctional polymer composites are expected to play a pivotal role in numerous fields, including wearable electronics, soft robotics, biomedical devices, energy-harvesting systems, and structural health monitoring [34]. Overall, the future of composite-based sensors and devices appears highly promising, offering a sustainable, resilient, and versatile pathway toward green energy and next-generation sensing technologies.

Acknowledgments: The authors thank all the contributors and reviewers for their valuable contributions and support from section editors of this Special Issue.

Conflicts of Interest: The authors declare no conflicts of interest.

References

1. Al-Amiery, A.A.; Fayad, M.A.; Wahhab, H.A.A.; Al-Azzawi, W.K.; Mohammed, J.K.; Majdi, H.S. Interfacial engineering for advanced functional materials: Surfaces, interfaces, and applications. *Results Eng.* **2024**, *22*, 102125. [CrossRef]
2. Ganeshkumar, S.; Rahman, H.A.; Gowtham, T.M.; Adithya, T.; Suyambulinagm, I.; Maniraj, J. Multifunctional Polymer Composites: Design, Properties, and Emerging Applications—A Critical Review. In *International Conference on Eco-Friendly Fibers and Polymeric Materials*; Springer Nature: Singapore, 2024; pp. 637–649.
3. Luo, L.; Zhang, F.; Wang, L.; Liu, Y.; Leng, J. Recent advances in shape memory polymers: Multifunctional materials, multiscale structures, and applications. *Adv. Funct. Mater.* **2024**, *34*, 2312036. [CrossRef]
4. Vidakis, N.; Petousis, M.; Velidakis, E.; Tzounis, L.; Mountakis, N.; Boura, O.; Grammatikos, S.A. Multi-functional polyamide 12 (PA12)/multiwall carbon nanotube 3D printed nanocomposites with enhanced mechanical and electrical properties. *Adv. Compos. Mater.* **2022**, *31*, 630–654. [CrossRef]
5. Wawrzyńczak, A.; Chudzińska, J.; Feliczak-Guzik, A. Metal and metal oxides nanoparticles as nanofillers for biodegradable polymers. *ChemPhysChem* **2024**, *25*, e202300823. [CrossRef] [PubMed]
6. Kumar, R.; Pandey, R.K.; Shukla, S.S.; Gidwani, B. Ceramic fillers, fibers, and acrylics. In *Magnetic Polymer Composites and Their Emerging Applications*; CRC Press: Boca Raton, FL, USA, 2024; pp. 289–313.
7. Jiang, R.; Zheng, X.; Zhu, S.; Li, W.; Zhang, H.; Liu, Z.; Zhou, X. Recent advances in functional polyurethane chemistry: From structural design to applications. *Chem. Sel.* **2023**, *8*, e202204132. [CrossRef]
8. Kumar, V.; Parvin, N.; Joo, S.W.; Mandal, T.K.; Park, S.S. Great carbon nano materials based composites for electronic skin: Intelligent sensing, and self-powered nano generators. *Nano Energy* **2025**, *137*, 110805. [CrossRef]
9. Liu, J.; Yao, Y.; Li, X.; Zhang, Z. Fabrication of advanced polydimethylsiloxane-based functional materials: Bulk modifications and surface functionalizations. *Chem. Eng. J.* **2021**, *408*, 127262. [CrossRef]
10. Qi, D.; Zhang, K.; Tian, G.; Jiang, B.; Huang, Y. Stretchable electronics based on PDMS substrates. *Adv. Mater.* **2021**, *33*, 2003155. [CrossRef]
11. Zhu, M.; Biswas, S.; Dinulescu, S.I.; Kastor, N.; Hawkes, E.W.; Visell, Y. Soft, wearable robotics and haptics: Technologies, trends, and emerging applications. *Proc. IEEE* **2022**, *110*, 246–272. [CrossRef]
12. Lee, J.H.; Cho, K.H.; Cho, K. Emerging trends in soft electronics: Integrating machine intelligence with soft acoustic/vibration sensors. *Adv. Mater.* **2023**, *35*, 2209673. [CrossRef]
13. Zhao, Z.; Liu, K.; Liu, Y.; Guo, Y.; Liu, Y. Intrinsically flexible displays: Key materials and devices. *Natl. Sci. Rev.* **2022**, *9*, nwac090. [CrossRef] [PubMed]
14. Chandrashekar, A.; Hegde, M.; Krishna, S.; Gopi, J.A.; Kotresh, T.M.; Prabhu, T.N. Non-covalent surface functionalization of nanofillers towards the enhancement of thermal conductivity of polymer nanocomposites: A mini review. *Eur. Polym. J.* **2023**, *198*, 112379. [CrossRef]
15. Luan, C.; Movva, S.; Wang, K.; Yao, X.; Zhang, C.; Wang, B. Towards next-generation fiber-reinforced polymer composites: A perspective on multifunctionality. *Funct. Compos. Struct.* **2019**, *1*, 042002. [CrossRef]
16. Kumar, V.; Alam, M.N.; Yewale, M.A.; Park, S.S. Multifunctional Aspects of Mechanical and Electromechanical Properties of Composites Based on Silicone Rubber for Piezoelectric Energy Harvesting Systems. *Polymers* **2024**, *16*, 2058. [CrossRef]
17. Iyer, V.; Petersen, J.; Geier, S.; Wierach, P. Design and Characterization of Poly (ethylene oxide)-Based Multifunctional Composites with Succinonitrile Fillers for Ambient-Temperature Structural Sodium-Ion Batteries. *Polymers* **2024**, *16*, 2806. [CrossRef] [PubMed]
18. Assadakorn, D.; Liu, G.; Hao, K.; Bai, L.; Liu, F.; Xu, Y.; Guo, L.; Liu, H. Effects of BET Surface Area and Silica Hydrophobicity on Natural Rubber Latex Foam Using the Dunlop Process. *Polymers* **2024**, *16*, 3076. [CrossRef]
19. Malashin, I.; Masich, I.; Tynchenko, V.; Gantimurov, A.; Nelyub, V.; Borodulin, A.; Martysyuk, D.; Galinovsky, A. Machine Learning in 3D and 4D Printing of Polymer Composites: A Review. *Polymers* **2024**, *16*, 3125. [CrossRef]
20. Sriani, T.; Mahardika, M.; Arifvianto, B.; Yusof, F.; Whulanza, Y.; Prihandana, G.S.; Baskoro, A.S. Study of Polysulfone-Impregnated Hydroxyapatite for Ultrafiltration in Whey Protein Separation. *Polymers* **2024**, *16*, 3079. [CrossRef]

21. Bril', I.Y.; Voronin, A.; Fadeev, Y.; Pavlikov, A.; Govorun, I.; Podshivalov, I.; Parshin, B.; Makeev, M.; Mikhalev, P.; Afanasova, K.; et al. Laser-Induced Silver Nanowires/Polymer Composites for Flexible Electronics and Electromagnetic Compatibility Application. *Polymers* **2024**, *16*, 3174. [CrossRef]
22. Inna, T.; Krajangta, N.; Rakmanee, T. The Staining Susceptibility and Surface Roughness of Teeth Restored by Microabrasion and Resin Infiltration: An In Vitro Study. *Polymers* **2024**, *16*, 3523. [CrossRef]
23. Fatkullin, M.; Petrov, I.; Dogadina, E.; Kogolev, D.; Vorobiev, A.; Postnikov, P.; Chen, J.-J.; de Oliveira, R.F.; Kanoun, O.; Rodriguez, R.D.; et al. Electrochemical Switching of Laser-Induced Graphene/Polymer Composites for Tunable Electronics. *Polymers* **2025**, *17*, 192. [CrossRef] [PubMed]
24. Repin, D.; Gablina, M.; Repina, N.; Cherednichenko, K.; Li, W.; Gushchina, Y.; Ivanov, E.; Melnikov, V.; Fakhrullin, R.; Vinokurov, V. Cellulose-Based Composite Materials for Fresh Water Extraction from Atmospheric Air. *Polymers* **2025**, *17*, 328. [CrossRef]
25. Plamadiala, I.; Croitoru, C.; Pop, M.A.; Roata, I.C. Enhancing Polylactic Acid (PLA) Performance: A Review of Additives in Fused Deposition Modelling (FDM) Filaments. *Polymers* **2025**, *17*, 191. [CrossRef]
26. Janmanee, R.; Sriwichai, S. Development of an Electrochemical Biosensor Based on Polypyrrole-3-carboxylic Acid/Polypyrrole/Au Nanoparticle Composites for Detection of Dopamine. *Polymers* **2025**, *17*, 754. [CrossRef]
27. Kumar, V.; Alam, M.N.; Manik, G.; Park, S.S. Recent Advancements in Rubber Composites for Physical Activity Monitoring Sensors: A Critical Review. *Polymers* **2025**, *17*, 1085. [CrossRef] [PubMed]
28. Oladele, I.O.; Omotosho, T.F.; Adediran, A.A. Polymer-based composites: An indispensable material for present and future applications. *Int. J. Polym. Sci.* **2020**, *2020*, 8834518. [CrossRef]
29. Zhang, X.; Li, B.W.; Dong, L.; Liu, H.; Chen, W.; Shen, Y.; Nan, C.W. Superior energy storage performances of polymer nanocomposites via modification of filler/polymer interfaces. *Adv. Mater. Interfaces* **2018**, *5*, 1800096. [CrossRef]
30. Šupová, M.; Martynková, G.S.; Barabaszová, K. Effect of nanofillers dispersion in polymer matrices: A review. *Sci. Adv. Mater.* **2011**, *3*, 1–25. [CrossRef]
31. Garner, L.; Sane, S.; Suh, D.; Byrne, T.; Dani, A.; Martin, T.; Mello, M.; Patel, M.; Williams, R. Finding Solutions to the Challenges in Package Interconnect Reliability. *Intel Technol. J.* **2005**, *9*, 297. [CrossRef]
32. Maalihan, R.D.; Domalanta, M.R.B.; Corrales, A.C.C.; Caldona, E.B. Advances in interfacially engineered surface-functionalized fillers for multifunctional polymer composite coatings. *Polym. Compos.* **2025**, *46*, 5857–5881. [CrossRef]
33. Garrison, T.F.; Murawski, A.; Quirino, R.L. Bio-based polymers with potential for biodegradability. *Polymers* **2016**, *8*, 262. [CrossRef] [PubMed]
34. Biswas, S.; Lee, S.W.; Lee, Y.; Choi, H.J.; Chen, J.; Yang, X.; Du, Y.; Falcone, N.; Barros, N.R.D.; Lee, S.M.; et al. Emerging energy harvesters in flexible bioelectronics: From wearable devices to biomedical innovations. *Small Sci.* **2024**, *4*, 2300148. [CrossRef] [PubMed]

polymers

MDPI

Article

Development of an Electrochemical Biosensor Based on Polypyrrole-3-carboxylic Acid/Polypyrrole/Au Nanoparticle Composites for Detection of Dopamine

Rapiphun Janmanee [1,*] and Saengrawee Sriwichai [2,3,*]

[1] Chemistry Program, Faculty of Science and Technology, Pibulsongkram Rajabhat University, Phitsanulok 65000, Thailand
[2] Department of Chemistry, Faculty of Science, Chiang Mai University, Chiang Mai 50200, Thailand
[3] Center of Excellence in Materials Science and Technology, Chiang Mai University, Chiang Mai 50200, Thailand
* Correspondence: rapiphun16@psru.ac.th (R.J.); saengrawee.s@cmu.ac.th (S.S.)

Abstract: Dopamine (DA) is a neurotransmitter that works in the brain. It plays several important roles in executive functions, including motor control, memory, mood, motivation, and reward. DA imbalances are associated with diseases in the nervous system such as Parkinson's disease, schizophrenia, Alzheimer's disease, and attention deficit hyperactivity disorder (ADHD). Therefore, the development of a biosensor for the detection of precise amounts of DA is of great interest. In this research, polypyrrole-3-carboxylic acid/polypyrrole/gold nanoparticle (PP3C/PPy/AuNPs) composites were developed for the electrochemical detection of DA. Firstly, a PP3C/PPy/AuNPs composite thin film was synthesized by electropolymerization on a fluorine-doped tin oxide (FTO)-coated glass substrate. Subsequently, cyclic voltammetry (CV), scanning electron microscopy (SEM), and differential pulse voltammetry (DPV) were used for the characterization and study of the efficiency of the obtained conducting polymer–gold nanoparticle composite thin film for the detection of DA. The proposed electrochemical sensor showed good sensitivity and selectivity for the detection of DA with a wide detection linear range from 5 to 180 μM (R^2 = 0.9913). The limit of detection (LOD) and limit of quantitation (LOQ) values were 9.72 nM and 0.032 μM, respectively. Therefore, it can be concluded that the electrochemically fabricated PP3C/PPy/AuNPs composite thin film can be applied as an electrochemical biosensor for the detection of dopamine for the early diagnosis of various neurological disorders in the future.

Keywords: polypyrrole; gold nanoparticle; dopamine; electrochemical; biosensor

1. Introduction

Dopamine (DA) is an important neurotransmitter in biological systems [1,2]. DA significantly influences various cognitive and physiological processes, including mood regulation [3], motivation [4], endocrine regulation [5], and movement [6]. The normal DA concentration in the nervous and peripheral systems can range from 0.01 to 1000 μM [7–10]. DA dysfunction causes several neurodegenerative diseases and psychiatric disorders, such as Parkinson's disease, Huntington's disease, Alzheimer's disease, and depression [4,5,10,11]. Consequently, it is essential to accurately and sensitively determine DA levels in biological fluids for both research and clinical applications. Various analytical methods have been utilized for monitoring DA levels including calorimetric methods [12], chromatographic methods [13], spectrophotometric methods [14], and immunoassay [15]. These methods have unique advantages but are greatly limited due to their operational complexity, high costs, time-consuming processes, and tedious sample pre-treatments [16–18]. Hence, to address this issue, the development of an electrochemical biosensor for DA determination has been extensively considered.

Currently, electrochemical biosensors using several techniques, including cyclic voltammetry (CV) [19], amperometry [20], differential pulse voltammetry (DPV) [21], or square-wave voltammetry [22], have been actively explored for DA level detection due to their simple operation, cost-effectiveness, rapid response, and high sensitivity and selectivity [23–25]. However, in biological matrices, the high concentration of ascorbic acid (AA) and uric acid (UA) is a huge problem in DA detection. The overlapping oxidation peaks of DA, AA, and UA pose challenges to the accurate detection of DA [22,26,27]. Consequently, most researchers are focused on the enhancement of the sensitivity and selectivity of electrochemical biosensors for DA detection in biological samples containing various interferences.

To date, the surface modification of working electrodes with nanomaterials and nanocomposites, including conducting polymers (CPs), noble metal nanomaterials, metal oxide nanomaterials, and carbon-based materials, has been focused on enhancing their selectivity and sensitivity [24,26,28,29]. Among the various materials, CPs have been extensively used in electrochemical biosensors due to their excellent electrical properties, biocompatibility, and environmental stability [30–32]. Polypyrrole (PPy) and its derivatives are among the most widely used CPs for electrochemical biosensor fabrications due to their good environmental stability in air and aqueous media, facile synthesis via both chemical and electrochemical methods, and higher electronic conductivity compared to many other CPs [25,33–36]. Moreover, to improve the performance of PPy-based electrochemical biosensors, the synthesis of composite materials by combining PPy with nanomaterials, such as metal oxides, carbon nanotubes (CNTs), graphene, or metal nanoparticles (NPs), has attracted much interest [1,37–42].

Recently, composite nanomaterials based on CPs and NPs of various metals, including Ag, Pt, Pd, and Au, have been increasingly utilized in the fabrication of electrochemical sensors to improve their analytical performance [11,43–47]. Gold nanoparticles (AuNPs) are one of the metal nanoparticles that serve as an advantageous material for PPy-based electrochemical biosensors due to their great conductivity, good electrocatalytic activity, and enhanced electron transfer with a large surface area [25,29]. Therefore, an electrochemical dopamine biosensor was developed using PPy-decorated AuNPs as a composite to enhance the CPs' properties in this study.

In this present study, we have attempted to integrate the advantages of nanomaterials based on CPs and NPs, utilizing a composite material composed of polypyrrole-3-carboxylic acid, polypyrrole, and gold nanoparticles (PP3C/PPy/AuNPs) to prepare an electrochemical biosensor for detecting DA with high sensitivity and selectivity. From this interesting perspective, the presented composite of this PP3C/PPy/AuNPs for electrochemical dopamine biosensors has never been reported. First, the PP3C/PPy/AuNPs composite was synthesized by electropolymerization under optimal conditions. The fabrication of the proposed electrochemical DA biosensor based on PP3C/PPy/AuNPs composites is presented in Scheme 1. The fabricated PP3C/PPy/AuNPs composite thin film exhibited excellent electrochemical performance for label-free DA detection using the DPV technique. The developed electrochemical biosensor based on PP3C/PPy/AuNPs demonstrated high selectivity for detecting DA with no interference effects from AA and UA. The proposed PP3C/PPy/AuNPs thin film showed excellent efficiency for application as an electrochemical biosensor for DA detection in real sample analysis, which has the potential to be applied in the future of human health diagnoses.

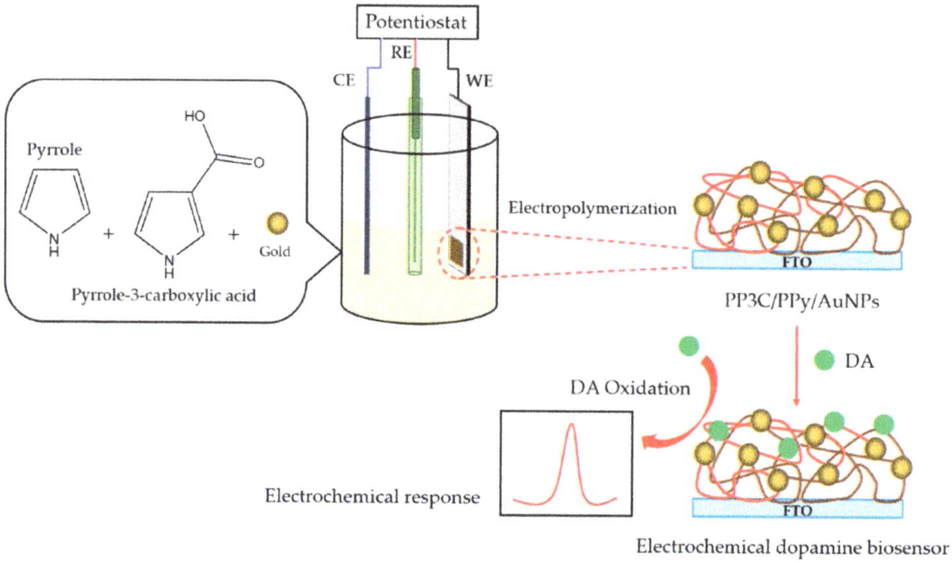

Scheme 1. Schematic diagram of the fabrication of PP3C/PPy/AuNPs composites for use as electrochemical DA biosensors.

2. Materials and Methods

2.1. Chemicals and Materials

Dopamine (DA), gold solution colloidal (18–20 nm), pyrrole (Py), pyrrole-3-carboxylic acid (P3C), ascorbic acid (AA), uric acid (UA), and phosphate-buffered saline (PBS) were acquired from Sigma-Aldrich (Darmstadt, Germany). Potassium chloride (KCl) and sulfuric acid (H_2SO_4) were purchased from RCI Labscan (Bangkok, Thailand). Potassium hexacyanoferrate ($K_3Fe(CN)_6$) was purchased from Fisher Scientific (Mumbai, India). Potassium hexacyanoferrate (II) trihydrate ($K_4Fe(CN)_6$) was acquired from Loba Chemie™ (Mumbai, India). Deionized water (DI) was applied for all aqueous solutions. All chemicals were of analytical grade and used as received.

2.2. Electropolymerization of PP3C/PPy/AuNPs

The PP3C/PPy/AuNPs composite thin film was fabricated by electropolymerization with the following procedure: Firstly, 10 mM of P3C/Py monomer solution was prepared by dissolving 0.120% w/v P3C and 0.065% w/v Py in 0.5 M H_2SO_4. Then, 40 µL of 0.0065wt% colloidal solution of gold was added to 10 mM of P3C/Py precursor solution and sonicated for 5 min. The electropolymerization of PP3C/PPy/AuNPs was performed using CV using a standard three-electrode electrochemical cell setup. A fluorine-doped tin oxide (FTO)-coated glass substrate (Sigma-Aldrich, Darmstadt, Germany) was used as a working electrode (area of 0.25 cm^2). The counter electrode was Pt wire, and the reference electrode was Ag/AgCl (3 M KCl). The potential was scanned from −1.0 to 1.0 V for 5 cycles. The scan rate was set at 20 mV/s. The individual PP3C and PPy were both produced under the same electrochemical polymerization conditions of the PP3C/PPy/AuNPs composite thin film.

2.3. Structural Characterizations of PP3C/PPy/AuNPs

The electrochemical characteristics of the fabricated PP3C/PPy/AuNPs in 10.0 mM of freshly prepared $K_3Fe(CN)_6/K_4Fe(CN)_6$ (1:1, containing PBS solution, pH

7.4) ($[Fe(CN)_6]^{3-/4-}$) were monitored using CV between −0.5 to 1.0 V under different scan rates (5–100 mV/s) on a potentiostat Autolab204 (Metrohm Autolab B.V., Utrecht, The Netherlands) with a three-electrode system. The electrochemical cell consisted of the Ag/AgCl reference electrode in 3 M KCl, Pt counter electrode, and FTO working electrode. The NOVA 2.1 software was employed to simulate all the electrochemical data reported in this research. Morphological studies were examined using scanning electron microscopy (SEM, Thermo Fisher Scientific, Apreo S, Waltham, MA, USA). The functional groups of the composite were identified by attenuated total reflectance–Fourier transform infrared spectroscopy (ATR-FTIR, Bruker Tensor 27, Billerica, MA, USA). X-ray photoelectron spectroscopy (XPS, AXIS ultra DLD spectrometer, Manchester, UK) was used to characterize the chemical constituents of the prepared composite. All specimens intended for XPS analysis were prepared on an ITO glass substrate.

2.4. Determination of DA

The electrochemical detection of DA based on the obtained PP3C/PPy/AuNPs-modified FTO electrode was monitored using the CV and DPV techniques. The CV measurement for 20 μM DA in PBS solution was recorded at 20 mV/s with a potential between −1.0 and 1.0 V. In addition, DPV measurements were performed to study the sensitivity performance. The potential ranged between −0.2 to 1.0 V with a scan rate of 20 mV/s, while the DA concentration was varied from 5 to 180 μM. The DPV responses were investigated at 20 μM of DA containing 200 μM AA and 200 μM UA as common interferences in the evaluation of selectivity. For stability determination, the biosensor electrodes were tested at two-day intervals for 10 days. Three similar independent electrodes were employed for the detection of 100 μM DA to evaluate reproducibility.

3. Results and Discussion

3.1. Fabrication of PP3C/PPy/AuNPs

The electrochemical property of the PP3C/PPy/AuNPs composite thin film was investigated using the CV technique. Figure 1 shows the CV curves of the PP3C/PPy/AuNPs composite thin film. The current in the oxidation onset began at about 0.6 V, and the current during the cathodic scan slightly decreased at about 0.3 V, which is attributed to the beginning of the formation of the copolymer film and the dedoping process of the copolymer films on the FTO electrode, respectively [37]. The peak current appeared at about −0.3 V, corresponding to the presence of gold nanoparticles on the modified working electrode. The CV responses of the PP3C/PPy/AuNPs composite thin film and the PPy electrode are presented in Figure S1. It was found that only the CV response of the PP3C/PPy/AuNPs electrode exhibited notable current peaks at −0.3 V, which were associated with the presence of AuNPs in the PP3C/PPy/AuNPs composite thin film. Moreover, it was found that the oxidation/reduction currents of the PP3C/PPy/AuNPs composite thin film increased with the number of cycles, as shown in the insets of Figure 1. This indicates an increase in the thickness of PP3C/PPy/AuNPs [48]. Furthermore, the detailed structural morphologies and chemical compositions of the PP3C/PPy/AuNPs composite thin film were investigated using SEM, ATR-FTIR, and XPS techniques.

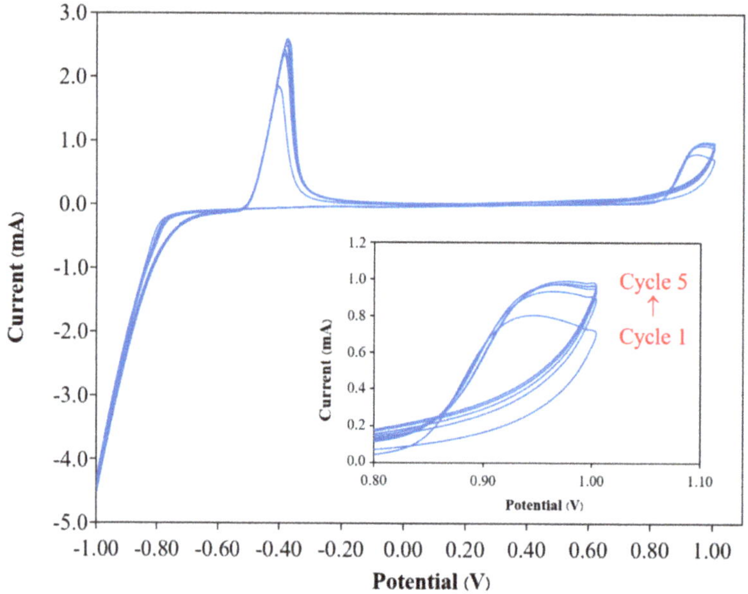

Figure 1. CV responses of the fabricated PP3C/PPy/AuNPs composite thin film.

3.2. Characterization of PP3C/PPy/AuNPs Composite Thin Film

The SEM was applied to characterize the surface morphological features of the PP3C PPy and PP3C/PPy/AuNPs, as illustrated in Figure 2. The morphology of the PP3C was a spherical morphology with nanoscale structures in random clusters, as seen in Figure 2a. Figure 2b shows the granular morphology of PPy, which is densely packed, with granules smaller than those in PP3C. As for the PP3C/PPy/AuNPs composite, the SEM image in Figure 2c reveals a cauliflower-like morphology of PP3C/PPy/AuNPs, and the granules are larger than those in PP3C and PPy. It suggests a higher reaction area and larger amounts of catalytic sites for electrochemical reactions [1,34,49]. The SEM-EDX spectrum illustrated in Figure S2 clearly demonstrates the presence of AuNPs within the fabricated PP3C/PPy/AuNPs.

The functional groups of the prepared PP3C, PPy, and PP3C/PPy/AuNPs were identified by ATR-FTIR spectroscopy. The ATR-FTIR spectra of PP3C, PPy, and PP3C/PPy/AuNPs in the range of 500–3500 cm^{-1} are presented in Figure 3. In the ATR-FTIR spectra of the PP3C, a very broad band peak at around 2400–3400 cm^{-1} is ascribed to the stretching vibration of O–H and –C=O of the carboxylic acid from P3C. The C–O of the carboxylic acid group showed a peak at 1289 cm^{-1} [50]. The bipolaron band of aromatic rings for PP3C appeared at 1104 cm^{-1} [34]. The C–H out-of-plane and the C–C out-of-plane ring deformation in the PP3C structure showed peaks at 808 cm^{-1} and 602 cm^{-1}, respectively [48]. In the ATR-FTIR spectra of PPy, the peak observed at 3123 cm^{-1} is due to the C–H stretching of the pyrrole ring [51]. Peaks appeared at 1558 1188 cm^{-1} and 1044 cm^{-1}, indicating the stretching vibrations of C=C and C–N bonds and the C–H in-plane vibration, respectively. The N–H in-plane vibrations are presented at 923 cm^{-1}. The bending vibrations of C–H out of plane are represented by the peak at 795 cm^{-1} [51,52]. The peak at 596 cm^{-1} indicates ring deformation [53]. The ATR-FTIR spectrum of the PP3C/PPy/AuNPs composite represents all characteristic peaks of both PP3C and PPy. Moreover, a red shift for PP3C/PPy/AuNPs (1561 cm^{-1}) indicates the doping of AuNPs through the PP3C/PPy/AuNPs composite [48]. The results demonstrate that the

PP3C/PPy/AuNPs composite could be successfully synthesized through electrochemical polymerization.

Figure 2. Surface morphology of (**a**) PP3C, (**b**) PPy, and (**c**) PP3C/PPy/AuNPs by SEM analysis.

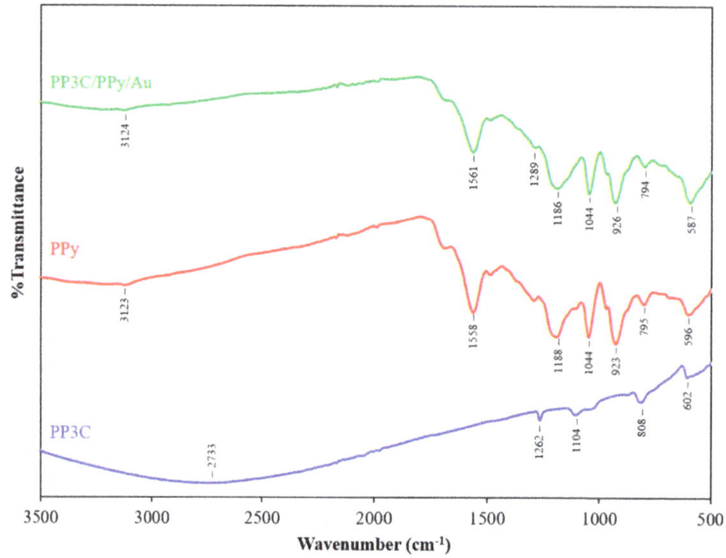

Figure 3. ATR-FTIR spectra of PP3C, PPy, and PP3C/PPy/AuNPs.

The illustrative XPS data of PP3C, PPy, and PP3C/PPy/AuNPs are displayed in Figure 4. The C1s, N1s, and O1s peaks appeared at a binding energy of about 285, 402 and 533 eV, respectively, as shown in the survey spectra (Figure 4a). The peak is observed within the energy range of 440–460 eV, indicating the presence of the In spectrum associated with the ITO glass substrate. The C1s, N1s, and O1s spectra of PPy, PP3C, and PP3C/PPy/AuNPs are presented in Figure 4b, c, and d, respectively. For PPy, the C1s spectra showed binding energy at 284.9 (green line), 285.9 (blue line), 286.9 (red line), 288.3 (black line), and 289.3 eV (purple line), corresponding to C–C, C–N, C–O, C=O, and $\pi = \pi^*$ [54,55]. The C1s spectra of PP3C showed binding energy at 284.9 (green line), 285.9 (blue line), 286.9 (red line), 288.1 (purple line), and 289.4 eV (black line). The C1s regions of the XPS spectrum of PP3C/PPy/AuNPs are similarly found at 284.9 (green line), 286.1 (blue line), 286.9 (red line), 287.8 (black line), and 289.2 eV (purple line), respectively. The deconvoluted N1s spectra of PPy can be divided into four component peaks at 400.4 (green line), 401.3 (blue line), 402.2 (red line), and 403.1 eV (black line), corresponding to –NH–, $-N^+-$, NH^+-, and $-NH_2^+$, respectively [7,56]. The N1s spectra of PP3C and PP3C/PPy/AuNPs exhibited the identical –NH–, $-N^+-$, and $=NH^+-$ at 400.2 (blue line), 401.3 (red line), and 402.3 eV (black line), respectively. The O1s spectra of PP3C showed four peaks at 530.6 (green line), 532.2 (blue line), 533.4 (red line), and 534.6 eV (black line), respectively. The peak at 530.6 is due to O–C, 532.2 eV can be associated with the carbonyl (O=C) group, and the 533.6–534.6 eV is due to the carboxyl group (–COOH). All the peaks that appeared in the O1s region of PP3C were also present in the PP3C/PPy/AuNPs composite. The XPS results indicate that the presented PP3C/PPy/AuNPs composite could be successfully formed in this study.

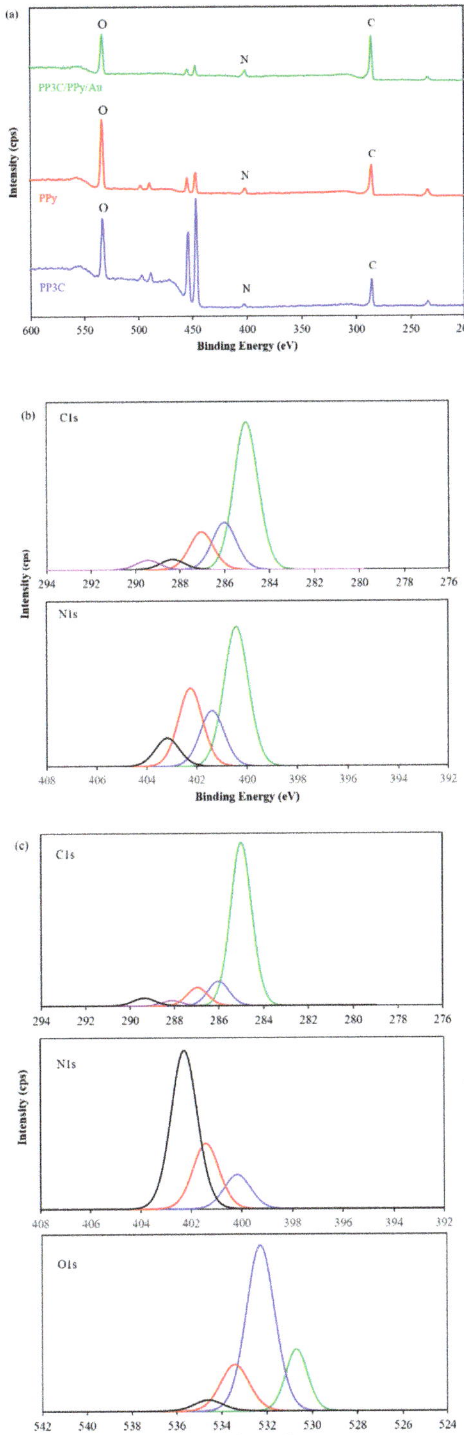

Figure 4. *Cont.*

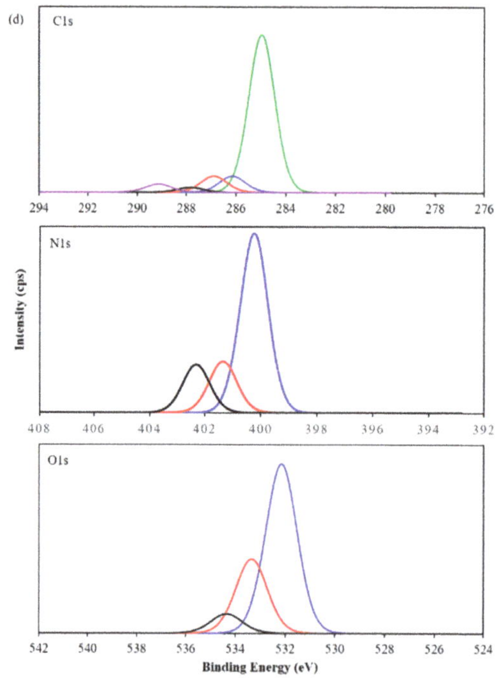

Figure 4. (**a**) Survey scan of PPy, PP3C, and PP3C/PPy/AuNPs and XPS spectra of (**b**) PPy, (**c**) PP3C, and (**d**) PP3C/PPy/AuNPs.

Furthermore, the electrochemical property of the PP3C/PPy/AuNPs-modified electrode in $[Fe(CN)_6]^{3-/4-}$ solution was studied. The CV responses of the PP3C/PPy/AuNPs at various scan rates from 5 to 100 mV/s are presented in Figure 5a, and the redox peak current increased with the increase in the scan rate. Figure 5b shows a good linear relationship in the plot of the oxidation and reduction peak current with the square root of scan rates, which indicated the diffusion-controlled redox process. It could be concluded that the PP3C/PPy/AuNPs composite thin film had sufficient electrochemical reactivity for device development [57].

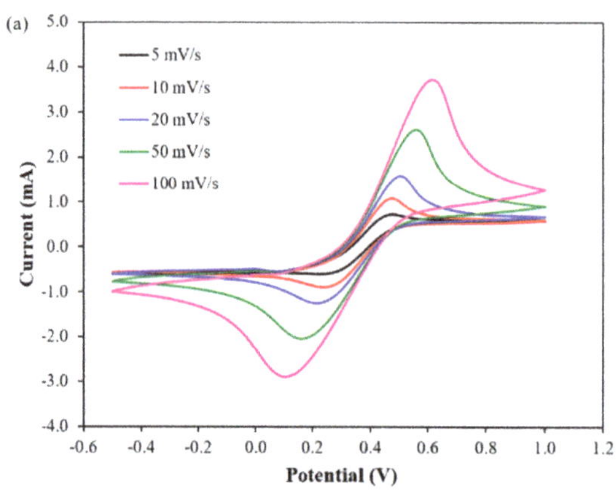

Figure 5. *Cont.*

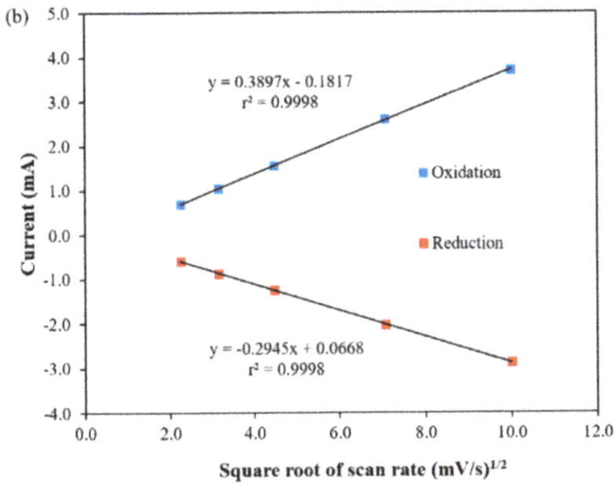

Figure 5. (**a**) CVs and (**b**) plot of peak current of PP3C/PPy/AuNPs at various scan rates.

3.3. Electrochemical Detection of DA at PP3C/PPy/AuNPs

The electrochemical biosensor based on the obtained PP3C/PPy/AuNPs-modified electrode was compared to the bare FTO electrode for the detection of DA using the CV technique. Figure 6 presents the current responses of the PP3C/PPy/AuNPs composite thin film-modified electrode and the bare FTO electrode to 20 μM DA in PBS solution. The CV response of the bare FTO electrode shows no significant current peaks during the electrochemical processes. On the other hand, the proposed electrochemical DA biosensor demonstrated a high current response. These results are attributed to the interactions between the redox species of dopamine and the electrode. [4]. Figure 7 represents the schematic of the electrochemical process for detecting DA based on the PP3C/PPy/AuNPs composite thin film-modified electrode. When an electrode potential is applied, dopamine is easily oxidized to dopamine-O-quinone (DAQ), leading to the two electrons and protons being transferred between the electrode and dopamine. The changes in currents due to the dopamine electrons released during the oxidation of DA are measured [29,58]. The dopamine oxidation is permanent due to the irreversibility of this reaction under an applied potential; electrons flowing from dopamine to the electrode cannot diffuse out from the electrode surface [59].

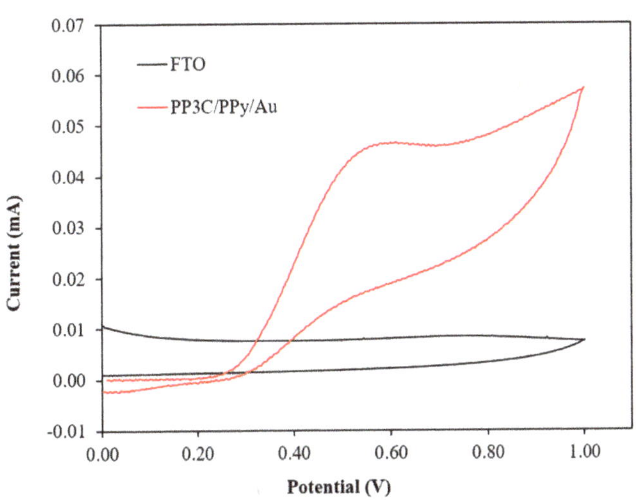

Figure 6. CV of FTO and PP3C/PPy/AuNPs in the presence of 20 μM dopamine.

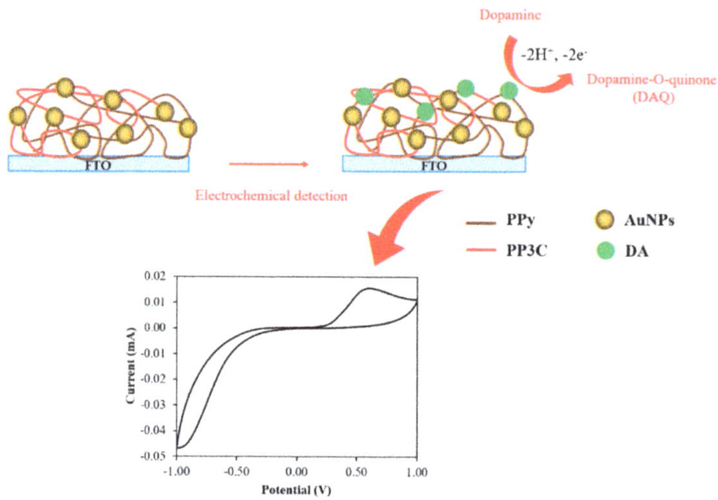

Figure 7. The electrochemical mechanism of PP3C/PPy/AuNPs for the detection of DA.

3.4. Performance of Electrochemical DA Biosensor

The electrochemical DA biosensor based on PP3C/PPy/AuNPs was used to detect various concentrations of DA from 5 to 180 μM using the DPV technique. The DPV study demonstrated its precise detection capability with a wide linear range for the DA detection of PP3C/PPy/AuNPs. The DPV responses for various DA concentrations ranging from 5 to 180 μM are illustrated in Figure 8. The oxidation current responses of dopamine were observed at approximately 0.4 V, which gradually increased with increasing DA concentration, as indicated in Figure 8a. A corresponding linear relationship of the current response is shown in Figure 8b. It is expressed in the linear regression equations of I (mA) = $0.0005C_{DA}$ + 0.0033, with correlation coefficients (r^2) of 0.9913. The limit of detection (LOD) value was 9.72 nM, and the limit of quantitation (LOQ) value was 0.032 μM, based on the observed linear equation (S/N = 3). The DPV response of PP3C/PPy/AuNPs obtained in PBS solution is presented in Figure S3 of the Supplementary Information. The sensitivity of the electrochemical DA biosensor based on PP3C/PPy/AuNPs was found to be 2 μA μM^{-1} cm^{-2}. The DA detection performance of the developed PP3C/PPy/AuNPs-modified electrode was compared with other recent reports as presented in Table 1. The proposed electrochemical biosensor based on PP3C/PPy/AuNPs exhibits comparable or better analytical performance compared to previously reported studies. The primary strength of this study is its broad linear range compared to the method based on the DPV technique. However, the detection limits achieved in this study are less favorable compared to the findings of other research endeavors. Additionally, the fabrication of the biosensor in this study was efficient, involving only a few steps and requiring a short time, which was a significant advantage. Therefore, it could be concluded that the fabricated PP3C/PPy/AuNPs electrochemical biosensor has the potential for detecting DA in real samples.

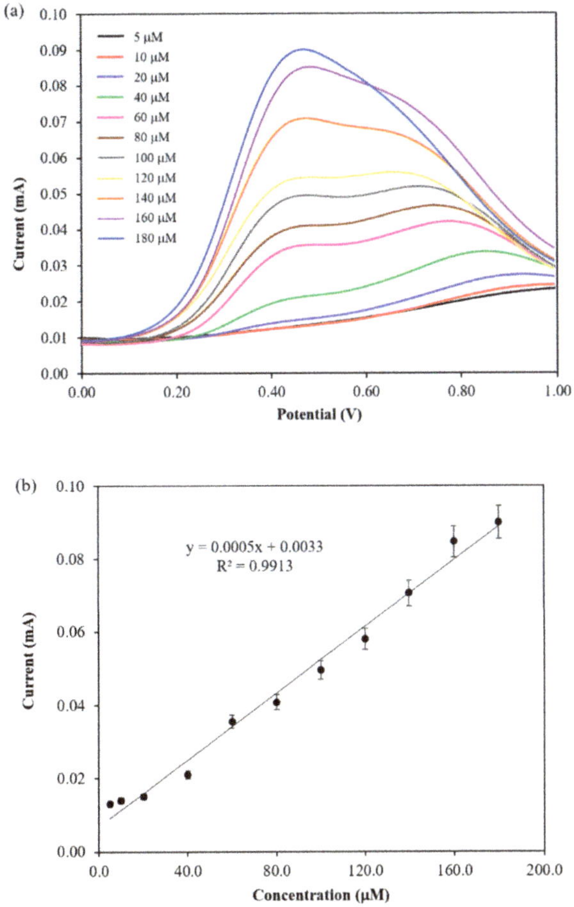

Figure 8. (**a**) DPV responses and (**b**) the calibration curve of PP3C/PPy/AuNPs for 5–180 μM DA.

Table 1. Comparison of the PP3C/PPy/AuNPs performance with other electrochemical DA biosensors.

Modified Electrode	Method	Linear Range (μM)	LOD (μM)	Sensitivity	Ref.
PPy–MCM-48	SWV	2–250	0.70	0.0058 (μA μM^{-1})	[19]
PPy/rGO/NiO/GCE	CV	10–1000	0.195	25.89 (A mM^{-1} cm^{-2})	[6]
OONfPPy/PGE	DPV	1–1000	0.00695	-	[60]
ZnO-Cu_xO-PPy/GCE	DPV	0.1–130	0.04	0.28 (μA μM^{-1})	[61]
PP3C/EOPGE	AdSDPV	0.01–7.5	0.0025	17.23 (μA μM^{-1})	[5]
PP3C/PPy/AuNPs	DPV	5–180	0.00972	2 (μA μM^{-1} cm^{-2})	Present work

Abbreviations: polypyrrole–mesoporous silica molecular (PPy–MCM-48); square-wave voltammetry (SWV); cyclic voltammetry (CV); glassy carbon electrode (GCE); nickel oxide (NiO); polypyrrole (PPy); reduced graphene oxide (rGO), overoxidized nanofiber polypyrrole (OONfPPy); pencil graphite electrode (PGE); differential pulse voltammetry (DPV); zinc oxide (ZnO); copper oxide (Cu_xO); adsorptive stripping differential pulse voltammetry (AdSDPV); overoxidized pencil graphite electrode (EOPGE); poly(pyrrole-3-carboxylic acid) (PP3C); gold nanoparticles (AuNPs).

3.5. Determination of Selectivity, Stability, and Reproducibility

The modified PP3C/PPy/AuNPs electrode for detecting DA was evaluated for selectivity, stability, and reproducibility, which are crucial for its practical applications. The DPV technique was used to evaluate the selectivity of the PP3C/PPy/AuNPs electrochemical biosensor for detecting DA with the addition of AA and UA as interferences. To simulate the coexistence of dopamine bio-molecules, a mixture of 200 μM AA and 200 UA μM was added with 20 μM of DA. The DPV responses of PP3C/PPy/AuNPs to the combination of 20 μM DA with 200 μM AA and UA interferences are shown in Figure 9. The DPV responses demonstrated that the developed sensor exhibits remarkable selectivity for DA detection with negligible current responses observed with the addition of other bio-molecule interferences.

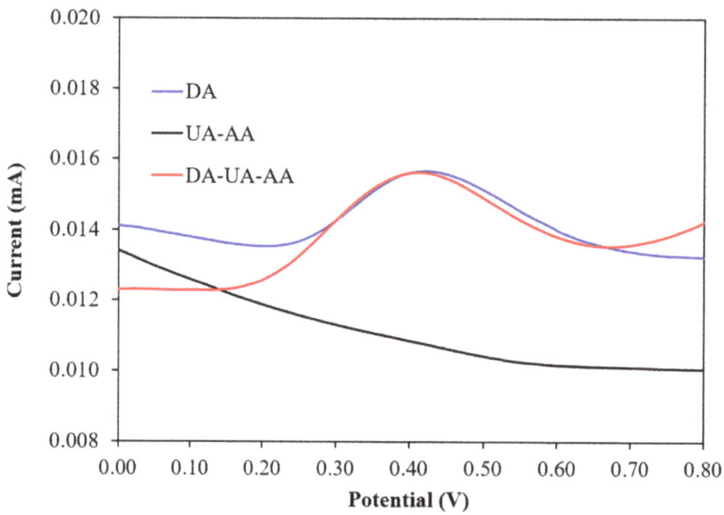

Figure 9. DPV responses of PP3C/PPy/AuNPs for the detection of 20 μM DA, 200 μM AA, and UA, as well as 20 μM DA, with the addition of 200 μM AA and UA interference.

The stability and reproducibility of the prepared PP3C/PPy/AuNPs electrochemical biosensor were monitored by the DPV technique. To minimize the impact of temperature and humidity on the biosensor's performance, which can affect binding kinetics, material properties, and electrochemical signals, the biosensor electrodes were stored in a dry environment at room temperature prior to use. The DPV responses, monitored at two-day intervals for 10 days, were analyzed to assess the durability of the biosensor electrodes. As seen in Figure 10, the DPV results indicate that the PP3C/PPy/AuNPs preserved 91.30% of its initial response, highlighting the considerable stability of the prepared PP3C/PPy/AuNPs. Furthermore, the fabrication reproducibility of the prepared PP3C/PPy/AuNPs electrochemical biosensor was investigated. The three electrodes were applied to detect a concentration of 100 μM DA. It was determined that the electrochemical biosensor had a relative standard deviation (RSD) of 4.21%. This result indicates a high level of reproducibility in the fabrication of the PP3C/PPy/AuNPs composite. Thus, it could be concluded that the proposed electrochemical biosensor shows excellent selectivity, high stability, and reproducibility for the detection of DA.

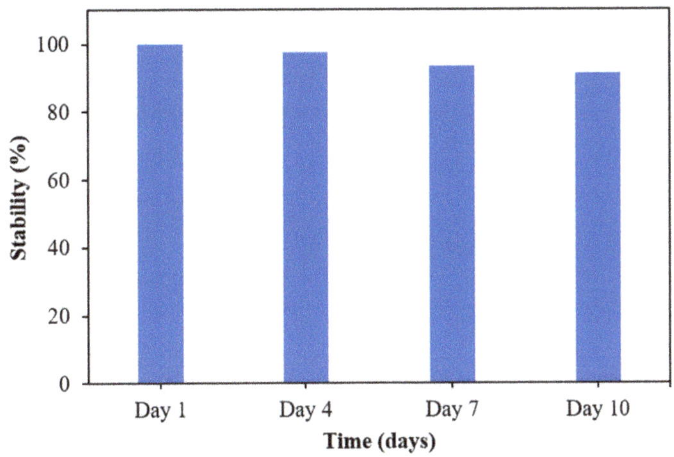

Figure 10. Stability histograms of PP3C/PPy/AuNPs for 100 μM DA.

4. Conclusions

The PP3C/PPy/AuNPs composite thin film was successfully fabricated by electropolymerization. The presented PP3C/PPy/AuNPs composite thin film exhibited a good electroactivity property in neutral PBS solution (pH 7.4), suggesting potential application for use as an electrochemical DA biosensor. The fabricated PP3C/PPy/AuNPs-modified electrode proved to be sensitive and selective in DA detection through DPV measurements. The fabricated PP3C/PPy/AuNPs exhibited a wide linear detection range, good sensitivity, and high selectivity. The modified electrochemical DA biosensor showed improved efficiency in detecting DA, achieving an LOD of 9.72 nM and a sensitivity of 2 $\mu A \mu M^{-1} cm^{-2}$. The obtained results revealed that the electrochemically fabricated PP3C/PPy/AuNPs composite thin film can be applied as a label-free electrochemical biosensor for detecting DA in real samples for the early diagnosis of various neurological disorders in the future.

Supplementary Materials: The following supporting information can be downloaded at: https://www.mdpi.com/article/10.3390/polym17060754/s1, Figure S1. CV responses of the fabricated PPy and PP3C/PPy/AuNPs composite thin films. Figure S2. The EDX spectrum of PP3C/PPy/AuNPs composite thin film. Figure S3. DPVs of PP3C/PPy/AuNPs obtained in PBS buffer and the presence of 5 μM DA.

Author Contributions: Conceptualization, R.J. and S.S.; Methodology, R.J. and S.S.; Validation, R.J. and S.S.; Formal analysis, R.J.; Resources, R.J.; Data curation, R.J.; Writing—original draft, R.J.; Writing—review & editing, S.S.; Supervision, S.S.; Funding acquisition, R.J. and S.S. All authors have read and agreed to the published version of the manuscript.

Funding: This research received no external funding.

Institutional Review Board Statement: Not applicable.

Data Availability Statement: The original contributions presented in this study are included in the article/Supplementary Material. Further inquiries can be directed to the corresponding author.

Acknowledgments: The authors gratefully acknowledge the Research and Development Institute of Pibulsongkram Rajabhat University (RDI-2-64-34, RDI-1-65-12) for financial support. Special thanks should be given to the Center of Excellence in Materials Science and Technology, Chiang Mai University.

Conflicts of Interest: The authors declare no conflict of interest.

References

1. Lin, M. A dopamine electrochemical sensor based on gold nanoparticles/over-oxidized polypyrrole nanotube composite arrays. *RSC Adv.* **2015**, *5*, 9848–9851. [CrossRef]
2. Qian, T.; Yu, C.; Zhou, X.; Ma, P.; Wu, S.; Xu, L.; Shen, J. Ultrasensitive dopamine sensor based on novel molecularly imprinted polypyrrole coated carbon nanotubes. *Biosens. Bioelectron.* **2014**, *58*, 237–241. [CrossRef]
3. Panapimonlawat, T.; Phanichphant, S.; Sriwichai, S. Electrochemical Dopamine Biosensor Based on Poly(3-aminobenzylamine) Layer-by-Layer Self-Assembled Multilayer Thin Film. *Polymers* **2021**, *13*, 1488. [CrossRef] [PubMed]
4. Anbumannan, V.; Kumar, R.T.R.; Suresh, K. Enhanced electrochemical detection of dopamine by graphene oxide/tungsten trioxide nanocomposite. *Mater. Sci. Semicond. Process.* **2021**, *127*, 105696. [CrossRef]
5. Özcan, A.; İlkbaş, S.; Atılır Özcan, A. Development of a disposable and low-cost electrochemical sensor for dopamine detection based on poly(pyrrole-3-carboxylic acid)-modified electrochemically over-oxidized pencil graphite electrode. *Talanta* **2017**, *165*, 489–495. [CrossRef] [PubMed]
6. Priyanto, S.A.N.; Yulianti, E.S.; Zakiyuddin, A.; Rahman, S.F. Amperometric Biosensor Detecting Dopamine Based on Polypyrrole/Reduced Graphene Oxide/Nickel Oxide/Glassy Carbon Electrode. *J. Electr. Comput. Eng.* **2024**, *2024*, 7453474. [CrossRef]
7. Kaewda, C.; Sriwichai, S. Label-Free Electrochemical Dopamine Biosensor Based on Electrospun Nanofibers of Polyaniline/Carbon Nanotube Composites. *Biosensors* **2024**, *14*, 349. [CrossRef]
8. Li, J.; Zhao, J.; Wei, X. A sensitive and selective sensor for dopamine determination based on a molecularly imprinted electropolymer of o-aminophenol. *Sens. Actuators B Chem.* **2009**, *140*, 663–669. [CrossRef]
9. Matt, S.M.; Gaskill, P.J. Where Is Dopamine and how do Immune Cells See it?: Dopamine-Mediated Immune Cell Function in Health and Disease. *J. Neuroimmune Pharmacol.* **2020**, *15*, 114–164. [CrossRef]
10. Liu, X.; Liu, J. Biosensors and sensors for dopamine detection. *View* **2021**, *2*, 20200102. [CrossRef]
11. Mahmoudian, M.R.; Basirun, W.J.; Binti Alias, Y. Sensitive Dopamine Biosensor Based on Polypyrrole-Coated Palladium Silver Nanospherical Composites. *Ind. Eng. Chem. Res.* **2016**, *55*, 6943–6951. [CrossRef]
12. Liang, L.; Zhao, Z.; Ye, F.; Zhao, S. Rapid and sensitive colorimetric detection of dopamine based on the enhanced-oxidase mimicking activity of cerium(iv). *N. J. Chem.* **2021**, *45*, 6780–6786. [CrossRef]
13. Gu, H.; Varner, E.L.; Groskreutz, S.R.; Michael, A.C.; Weber, S.G. In Vivo Monitoring of Dopamine by Microdialysis with 1 min Temporal Resolution Using Online Capillary Liquid Chromatography with Electrochemical Detection. *Anal. Chem.* **2015**, *87*, 6088–6094. [CrossRef]
14. Ramadan, M.A.A.; Almasri, I.; Khayal, G. Spectrophotometric Determination of Dopamine in Bulk and Dosage Forms Using 2,4-Dinitrophenylhydrazine. *Turk. J. Pharm. Sci.* **2020**, *17*, 679–685. [CrossRef]
15. Moghzi, F.; Soleimannejad, J.; Sañudo, E.C.; Janczak, J. Dopamine Sensing Based on Ultrathin Fluorescent Metal–Organic Nanosheets. *ACS Appl. Mater. Interfaces* **2020**, *12*, 44499–44507. [CrossRef] [PubMed]
16. Ma, C.; Wen, Y.; Qiao, Y.; Shen, K.Z.; Yuan, H. A Dopamine Detection Sensor Based on Au-Decorated NiS2 and Its Medical Application. *Molecules* **2024**, *29*, 2925. [CrossRef] [PubMed]
17. Anuar, N.S.; Basirun, W.J.; Shalauddin, M.; Akhter, S. A dopamine electrochemical sensor based on a platinum–silver graphene nanocomposite modified electrode. *RSC Adv.* **2020**, *10*, 17336–17344. [CrossRef] [PubMed]
18. Zhang, J.; Zhou, J.; Yu, Y.; Chen, J.; Xu, W. Electrochemical Dopamine Biosensor Using Over-Oxidized Polypyrrole with Assistance of Graphene. In Proceedings of the 2016 13th IEEE International Conference on Solid-State and Integrated Circuit Technology (ICSICT), Hangzhou, China, 25–28 October 2016; pp. 671–673.
19. Zablocka, I.; Wysocka-Zolopa, M.; Winkler, K. Electrochemical Detection of Dopamine at a Gold Electrode Modified with a Polypyrrole–Mesoporous Silica Molecular Sieves (MCM-48) Film. *Int. J. Mol. Sci.* **2019**, *20*, 111. [CrossRef]
20. Mani, V.; Devasenathipathy, R.; Chen, S.-M.; Kohilarani, K.; Ramachandran, R. A Sensitive Amperometric Sensor for the Determination of Dopamine at Graphene and Bismuth Nanocomposite Film Modified Electrode. *Int. J. Electrochem. Sci.* **2015**, *10*, 1199–1207. [CrossRef]
21. Demirkan, B.; Bozkurt, S.; Cellat, K.; Arıkan, K.; Yılmaz, M.; Şavk, A.; Çalımlı, M.H.; Nas, M.S.; Atalar, M.N.; Alma, M.H.; et al. Palladium supported on polypyrrole/reduced graphene oxidenanoparticles for simultaneous biosensing application of ascorbic acid, dopamine, anduric acid. *Sci. Rep.* **2020**, *10*, 2946. [CrossRef]
22. Alahmadi, N.; El-Said, W.A. Electrochemical Sensing of Dopamine Using Polypyrrole/Molybdenum Oxide Bilayer-Modified ITO Electrode. *Biosensors* **2023**, *13*, 578. [CrossRef]
23. Ghadimi, H.; Mahmoudian, M.R.; Basirun, W.J. A sensitive dopamine biosensor based on ultra-thin polypyrrole nanosheets decorated with Pt nanoparticles. *RSC Adv.* **2015**, *5*, 39366–39374. [CrossRef]
24. Zhang, X.; Tan, X.; Wang, P.; Qin, J. Application of Polypyrrole-Based Electrochemical Biosensor for the Early Diagnosis of Colorectal Cancer. *Nanomaterials* **2023**, *13*, 674. [CrossRef]
25. Jain, R.; Jadon, N.; Pawaiya, A. Polypyrrole based next generation electrochemical sensors and biosensors: A review. *TrAC Trends Anal. Chem.* **2017**, *97*, 363–373. [CrossRef]

26. Chang, Y.H.; Woi, P.M.; Alias, Y. The selective electrochemical detection of dopamine in the presence of ascorbic acid and uric acid using electro-polymerised-β-cyclodextrin incorporated f-MWCNTs/polyaniline modified glassy carbon electrode. *Microchem. J.* **2019**, *148*, 322–330. [CrossRef]
27. Granero, A.M.; Pierini, G.D.; Robledo, S.N.; Di Nezio, M.S.; Fernández, H.; Zon, M.A. Simultaneous determination of ascorbic and uric acids and dopamine in human serum samples using three-way calibration with data from square wave voltammetry. *Microchem. J.* **2016**, *129*, 205–212. [CrossRef]
28. Khan, R.; Anjum, S.; Fatima, N.; Farooq, N.; Shaheen, A.; Fernandez Garcia, J.; Khan, M.I.; Shanableh, A. Development of Electrochemical and Colorimetric Biosensors for Detection of Dopamine. *Chemosensors* **2024**, *12*, 126. [CrossRef]
29. Lakard, S.; Pavel, I.A.; Lakard, B. Electrochemical Biosensing of Dopamine Neurotransmitter: A Review. *Biosensors* **2021**, *11*, 179. [CrossRef]
30. Iordănescu, A.; Tertis, M.; Cernat, A.; Suciu, M.; Săndulescu, R.; Cristea, C. Poly-(pyrrole-3-carboxylic acid) Based Nanostructured Platform for the Detection of Carcinoembryonic Antigen. *Electroanalysis* **2018**, *30*, 1100–1106. [CrossRef]
31. Thunyakontirakun, W.; Sriwichai, S.; Phanichphant, S.; Janmanee, R. Fabrication of poly(pyrrole-3-carboxylic acid)/graphene oxide composite thin film for glucose biosensor. *Mater. Today Proc.* **2019**, *17*, 2070–2077. [CrossRef]
32. Gupta, S.A.; Singh, J. A Study of conducting electrochemical sensors based on molecularly imprinted polymer on carbon nanostructure using polypyrrole film: A review. *J. Sci. Res.* **2021**, *65*, 110–115. [CrossRef]
33. Sriwichai, S.; Netsuwan, P.; Phanichphant, S.; Baba, A. Electropolymerization and properties of poly (3-anilinethiophene) thin film. *Chiang Mai J. Sci.* **2016**, *43*, 863–869.
34. Luceño-Sánchez, J.A.; Díez-Pascual, A.M. Grafting of Polypyrrole-3-carboxylic Acid to the Surface of Hexamethylene Diisocyanate-Functionalized Graphene Oxide. *Nanomater.* **2019**, *9*, 1095. [CrossRef] [PubMed]
35. Nagarajan, A.; Sethuraman, V.; Sasikumar, R. Non-enzymatic electrochemical detection of creatinine based on a glassy carbon electrode modified with a Pd/Cu_2O decorated polypyrrole (PPy) nanocomposite: An analytical approach. *Anal. Methods* **2023**, *15*, 1410–1421. [CrossRef] [PubMed]
36. Sierra-Padilla, A.; García-Guzmán, J.J.; López-Iglesias, D.; Palacios-Santander, J.M.; Cubillana-Aguilera, L. E-tongues/noses based on conducting polymers and composite materials: Expanding the possibilities in complex analytical sensing. *Sensors* **2021**, *21*, 4976. [CrossRef]
37. Janmanee, R.; Pinwattana, K. Electrochemical Fabrication of Polypyrrole/Poly(pyrrole-3-carboxylic acid)/Graphene Oxide Composite Thin Film for Biosensor Application. *NU. Int. J. Sci.* **2015**, *1*, 52–61.
38. Kannan, A.; Radhakrishnan, S. Fabrication of an electrochemical sensor based on gold nanoparticles functionalized polypyrrole nanotubes for the highly sensitive detection of l-dopa. *Mater. Today Commun.* **2020**, *25*, 101330. [CrossRef]
39. Kawakami, H.; Ito, Y.; Chien, Y.-A.; Chen, C.-Y.; Chiu, W.-T.; Chakraborty, P.; Nakamoto, T.; Sone, M.; Chang, T.-F.M. Development of polypyrrole/nano-gold composite for non-enzymatic glucose sensors. *Micro Nano Eng.* **2022**, *14*, 100109. [CrossRef]
40. Namsheer, K.; Rout, C.S. Conducting polymers: A comprehensive review on recent advances in synthesis, properties and applications. *RSC Adv.* **2021**, *11*, 5659–5697. [CrossRef]
41. Ferrier, D.C.; Honeychurch, K.C. Carbon Nanotube (CNT)-Based Biosensors. *Biosensors* **2021**, *11*, 486. [CrossRef]
42. Esmaeili, C.; Heng, L.Y.; Chiang, C.P.; Rashid, Z.A.; Safitri, E.; Malon Marugan, R.S.P. A DNA biosensor based on kappa-carrageenan-polypyrrole-gold nanoparticles composite for gender determination of Arowana fish (Scleropages formosus). *Sens. Actuators B Chem.* **2017**, *242*, 616–624. [CrossRef]
43. Nowicka, A.M.; Fau, M.; Rapecki, T.; Donten, M. Polypyrrole-Au Nanoparticles Composite as Suitable Platform for DNA Biosensor with Electrochemical Impedance Spectroscopy Detection. *Electrochim. Acta* **2014**, *140*, 65–71. [CrossRef]
44. Vellaichamy, B.; Periakaruppan, P.; Paulmony, T. Evaluation of a New Biosensor Based on in Situ Synthesized PPy-Ag-PVP Nanohybrid for Selective Detection of Dopamine. *J. Phys. Chem. B* **2017**, *121*, 1118–1127. [CrossRef] [PubMed]
45. Dube, A.; Malode, S.J.; Alodhayb, A.N.; Mondal, K.; Shetti, N.P. Conducting polymer-based electrochemical sensors: Progress, challenges, and future perspectives. *Talanta Open* **2025**, *11*, 100395. [CrossRef]
46. Lakard, B. Electrochemical Biosensors Based on Conducting Polymers: A Review. *Appl. Sci.* **2020**, *10*, 6614. [CrossRef]
47. Naveen, M.H.; Gurudatt, N.G.; Shim, Y.-B. Applications of conducting polymer composites to electrochemical sensors: A review. *Appl. Mater. Today* **2017**, *9*, 419–433. [CrossRef]
48. Milikić, J.; Tapia, A.; Stamenović, U.; Vodnik, V.; Otoničar, M.; Škapin, S.; Santos, D.M.F.; Šljukić, B. High-performance metal (Au,Cu)–polypyrrole nanocomposites for electrochemical borohydride oxidation in fuel cell applications. *Int. J. Hydrogen Energy* **2022**, *47*, 36990–37001. [CrossRef]
49. Yuan, Y.; Li, T.; Ye, Z.; Feng, Y.; Chen, Z.; Wang, Y.; Sun, Y.; Wu, H.; Yang, Z.; Wang, Y.; et al. A One-Step Electropolymerized Biomimetic Polypyrrole Membrane-Based Electrochemical Sensor for Selective Detection of Valproate. *Front. Bioeng. Biotechnol.* **2022**, *10*, 851692. [CrossRef]
50. Rozi, N.; Hanifah, S.A.; Zaid, M.H.M.; Abd Karim, N.H.; Ikeda, M. Feasible study on poly(Pyrrole-co-Pyrrole-3-Carboxylic Acid)-modified electrode for detection of 17β-Estradiol. *Chem. Pap.* **2021**, *75*, 3493–3503. [CrossRef]

51. Sowmiya, G.; Velraj, G. Design of hollow nanosphere structured polypyrrole/Sn and SnO_2 nanoparticles by COP approach for enhanced electron transport behavior. *J. Inorg. Organomet. Polym. Mater.* **2020**, *30*, 5217–5223. [CrossRef]
52. Husain, A.; Mahajan, D.K. Effect of multi-walled carbon nanotubes on DC electrical conductivity and acetone vapour sensing properties of polypyrrole. *Carbon Trends* **2022**, *9*, 100193. [CrossRef]
53. Bachhav, S.; Patil, D. Study of ethanol vapour sensing behaviour by polypyrrole-multiwall carbon nanotubes nanocomposites. *J. Phys. Sci.* **2018**, *29*, 137–152. [CrossRef]
54. Tang, X.; Raskin, J.-P.; Kryvutsa, N.; Hermans, S.; Slobodian, O.; Nazarov, A.N.; Debliquy, M. An ammonia sensor composed of polypyrrole synthesized on reduced graphene oxide by electropolymerization. *Sens. Actuators B Chem.* **2020**, *305*, 127423. [CrossRef]
55. Rashed, M.A.; Faisal, M.; Alsaiari, M.; Alsareii, S.A.; Harraz, F.A. MWCNT-doped polypyrrole-carbon black modified glassy carbon electrode for efficient electrochemical sensing of nitrite ions. *Electrocatalysis* **2021**, *12*, 650–666. [CrossRef]
56. Sun, J.; Shu, X.; Tian, Y.; Tong, Z.; Bai, S.; Luo, R.; Li, D.; Liu, C.C. Facile preparation of polypyrrole-reduced graphene oxide hybrid for enhancing NH3 sensing at room temperature. *Sens. Actuators B Chem.* **2017**, *241*, 658–664. [CrossRef]
57. Supakiet, C.; Chammari, P.; Nawee, K.; Kontad, O. A Poly(pyrrole-3-carboxylic acid) Thin Film Modified Screen Printed Carbon Electrode as Highly Sensitive and Selective Label-free Electrochemical Immunosensing Platform. *Chiang Mai J. Sci.* **2020**, *47*, 530–541.
58. Choo, S.-S.; Kang, E.-S.; Song, I.; Lee, D.; Choi, J.-W.; Kim, T.-H. Electrochemical Detection of Dopamine Using 3D Porous Graphene Oxide/Gold Nanoparticle Composites. *Sensors* **2017**, *17*, 861. [CrossRef]
59. Keene, S.T.; Lubrano, C.; Kazemzadeh, S.; Melianas, A.; Tuchman, Y.; Polino, G.; Scognamiglio, P.; Cinà, L.; Salleo, A.; van de Burgt, Y.; et al. A biohybrid synapse with neurotransmitter-mediated plasticity. *Nat. Mater.* **2020**, *19*, 969–973. [CrossRef] [PubMed]
60. Koyun, O.; Gursu, H.; Gorduk, S.; Sahin, Y. Highly Sensitive Electrochemical Determination of Dopamine with an Overoxidized Polypyrrole Nanofiber Pencil Graphite Electrode. *Int. J. Electrochem. Sci.* **2017**, *12*, 6428–6444. [CrossRef]
61. Ghanbari, K.; Hajheidari, N. ZnO-CuxO/polypyrrole nanocomposite modified electrode for simultaneous determination of ascorbic acid, dopamine, and uric acid. *Anal. Biochem.* **2015**, *473*, 53–62. [CrossRef]

polymers

MDPI

Article

Cellulose-Based Composite Materials for Fresh Water Extraction from Atmospheric Air

Dmitry Repin [1], Mariia Gablina [1], Natalya Repina [1], Kirill Cherednichenko [1,2,*], Wenpeng Li [1], Yuliiya Gushchina [1], Evgenii Ivanov [1], Vyacheslav Melnikov [1], Rawil Fakhrullin [3,4] and Vladimir Vinokurov [1]

1 Department of Physical and Colloidal Chemistry, National University of Oil and Gas «Gubkin University», Leninsky Prospekt 65, Moscow 119991, Russia; vladimir@vinokurov.me (V.V.)
2 Faculty of Chemistry, Lomonosov Moscow State University, Leninskie Gory 1, Moscow 119991, Russia
3 Institute of Fundamental Medicine and Biology, Kazan Federal University, Kreml uramı 18, Kazan 420008, Republic of Tatarstan, Russia
4 Institute for Regenerative Medicine, Sechenov First Moscow State Medical University (Sechenov University), Moscow 119991, Russia
* Correspondence: cherednichenko.k@gubkin.ru

Abstract: The fibrous hybrid material was synthesized by suspension radical styrene polymerization on the surface of cellulose microfibers. The resulting material was used to prepare a thermally stable and mechanically strong porous composite matrix that was employed as a carrier for further precipitation of the hygroscopic agents: $CaCl_2$ and 1-butyl-3-methylimidazolium chloride. The obtained composite materials were used to capture atmospheric water at different relative humidity levels and extract fresh water. A composite material containing an ionic liquid (1-butyl-3-methylimidazolium chloride) as a hygroscopic agent demonstrated the best water absorption efficiency and reusability potential.

Keywords: atmospheric water harvesting; BMIMCl; hybrid material; ionic liquid; microfibrillated cellulose; MFC; polystyrene; PS

1. Introduction

Continuous supply of fresh water is a critical requirement for the sustainable development of modern civilization. Fresh water accounts for only 2.5% of the total 1.4×10^{18} tons of water on the planet, while the majority of this water is frozen in glaciers, ice caps, and permafrost [1]. The natural sources of fresh water are presented by lakes, rivers, and soil waters. Nevertheless, due to their uneven geographical distribution and the ever-growing demand on fresh water, the elaboration of alternative methods of fresh-water extraction has become an issue during the last decades. The purification/recycling of wastewater is a widely applied alternative approach to retrieve fresh water. Unfortunately, a number of the technological solutions related to water purification (mechanical, chemical, biological, and physicochemical methods) require large capital investment and high material and energy consumption, and often do not meet high environmental standards [2].

Extraction of fresh water from atmospheric water vapor is a promising way to supply fresh water since it does not strongly depend on the geography of the region. Devices for the collection of water from air by sorption are generally categorized into two systems: passive and active [3–5]. Passive systems rely on natural energy sources and do not require an additional power supply. In these systems, the cycle lasts 24 h: moisture is absorbed at night, and during the day, solar heat drives desorption, followed by distillation. Active systems, in contrast, can perform multiple absorption and desorption cycles within a single day, thereby increasing water production. However, this increased efficiency comes at the cost of higher energy consumption due to the use of electricity [6]. During recent decades, atmospheric water harvesting (AWH) materials, including hygroscopic inorganic compounds, functional hydrogels, ionic liquids (ILs), and metal–organic frameworks (MOFs) have attracted increasing attention [1,7,8]. Despite their excellent water absorbency, each of the above classes of AWH materials has application limitations and none can be considered ideal.

Thus, for example, hygroscopic salts have a strong dipole–dipole force that results in a great water adsorption capacity (5–6 $g \cdot g^{-1}$) [9]. G.E. William et al. applied $CaCl_2$, which has the capacity to absorb up to 97% of its weight in moisture, as a desiccant by employing a solution of salt [10]. Hygroscopic salts absorb moisture through hydration; however, they are prone to agglomeration and the formation of inactive layers on the surface of particles [11]. This can reduce permeability to moisture and, consequently, decrease the amount of water absorbed. Another issue is the liquefication of hygroscopic salts upon moisture absorption leading to equipment corrosion and environmental concerns [11,12]. Despite zeolites and silica gel being first-generation water sorption materials, they are characterized by high regeneration temperatures (above 100 °C) and low moisture absorption rates. Unlike zeolites, silica gel and hygroscopic salts of monovalent/divalent metals, MOFs can capture more water from the ambient air with a relative humidity of only 20% and do not require high temperatures for moisture desorption. A number of effective MOF-based AWH systems have been proposed recently [13–16]. However, their high cost, poor resistance to water and toxicity, which can present a danger to human health, are the bottlenecks restricting their AWH application [17,18]. Polymer hydrogels possess an excellent water absorption and can take any shape, but they have low capacity, poor physical strength, and may be subject to chemical leaching [19]. ILs are viscous fluids composed of anions and cations. They possess high hygroscopicity due to their ionic nature. The ions in the liquid create a strong electrostatic field that attracts polar water molecules. Besides electrostatic interaction, many ILs are capable of forming hydrogen bonds with water molecules, thus, enhancing water sorption [20]. However, ILs have several disadvantages, including strict requirements for their storage and transportation conditions (e.g., precise temperature and humidity control) to prevent decomposition or alteration of their properties [21]. Rather high cost of ILs can also considerably impede their practical application [20]. The leakage of IL is the last but not the least issue of this class of AWH materials.

Various approaches have been elaborated on to overcome these limitations of AWH materials [22]. For example, the employment of various supports/carriers of AWH materials can considerably enhance the atmospheric water absorption and exclude their leakage. For instance, a hydrogel layer was deposited onto an MOF/solar absorber layer to suppress convective heat loss [23]. In another research study, impregnation of the porous supports by hydroscopic salts or ILs helped to avoid their leakage [20]. Moreover, deposition of rather dense IL on the support with a high specific surface significantly enhanced the moisture absorption of this AWH material. Integration of hydroscopic $CaCl_2$ into an alginate-derived matrix enhanced AWH performance and provided cyclic stability of the obtained composite [24]. However, the current ecological concerns place strict demands on the environmental impact of moisture absorbents and supports for AWH materials, which has pushed researchers to use various eco-friendly/biodegradable materials [25].

Cellulose is a biopolymer that has been gaining interest worldwide for its biodegradability, non-toxicity, water affinity and large surface area [26]. Cellulose is one of the most abundant biopolymers, which is the reason for its low price. Owing to the high inter- and intramolecular hydrogen bonds and van der Waals forces, this biopolymer possesses high hydrophilicity and mechanical stability, which makes it an ideal component for AWH materials [25]. For instance, cellulose and its derivatives (in particular nanocellulose) have been frequently employed to prepare various hydrogels for agricultural uses (collection and sustained release of water), collection of oil spills, and drug delivery [25,27–29]. To enhance water absorption, nanocellulose-based hydrogels were modified with hydroscopic salts [30,31]. Indeed, cellulose is not the only biopolymer that can be used to create hydroscopic materials and carriers for AWH materials. Chitosan (made by treating the chitin shells of shrimp and other crustaceans with alkaline substances) can also be employed to synthesize water-absorbing hydrogels. For instance, chitosan-based hydrogels demonstrated effective water adsorption and considerable enhancement of structural durability upon repeated use [32]. Molybdenum carbide/carbon-based chitosan hydrogel (MoCC-CH) was recently designed for effectively absorbing atmospheric water and also efficiently accelerating water evaporation by tuning the biopolymer content in the hydrogel [33]. However, despite the high water sorption capacity of biopolymer-based hydrogels, their synthesis often involves expensive raw materials (e.g., nanocellulose), complex procedures, and large-scale challenges [34].

However, the wide application of nanocellulose is currently limited by its sophisticated and expensive production [35]. In this regard, employment of cheaper cellulose microfibrils can be a possible solution to apply biopolymer-based materials in industry. To the best of our knowledge, there are no reports on the application of the cellulose microfibrils as AWH material or as possible porous supports for commercial hydroscopic agents. Unlike nanocellulose, distinguished by shorter and more rigid structures, the fibrillar form exhibits an enhanced ability to form three-dimensional networks due to the efficient interactions between fibrils [36]. This makes it more suitable for the development of reinforcing composites and biomaterials while maintaining lightness and porosity.

This work aims to describe the synthesis of the composite materials based on a porous hybrid microfibrillated cellulose/polystyrene (MFC/PS) matrix and moisture-absorbing agents: $CaCl_2$ and 1-butyl-3-methylimidazolium chloride. Variation in biopolymer/polymer mass ratio allowed us to achieve optimal mechanical characteristics and shape the stability and hydrophilicity of the porous matrix. The obtained composites have been applied as new AWH materials for the first time. MFC/PS blends exhibited passive water vapor absorption from the atmosphere requiring minimal energy input. The MFC/PS-based composites represent a simple and cost-efficient alternative to nanocellulose-based AWH materials, while maintaining high water absorption and reliable performance across a wide range of humidity conditions.

2. Materials and Methods

2.1. Materials

Styrene ($\rho = 0.906$ g/cm^3 at 20 °C, purity $\geq$ 99.0%, stabilized with 4-tert-butylpyrocatechol), ammonium persulfate ($(NH_4)_2S_2O_8$, purity $\geq$ 97%), acetone ($(CH_3)_2O$, purity $\geq$ 99.5%), 1-butyl-3-methylimidazolium chloride (BMIMCl, purity $\geq$ 95%) and calcium chloride ($CaCl_2$, CC, purity $\geq$ 97%) were purchased from Sigma-Aldrich (St. Louis, MO, USA). Softwood pulp in the form of bleached sulfate pulp was supplied by Arkhangelsk PPM (Arkhangelsk, Russia).

Microfibrillar cellulose (MFC) was obtained from bleached paper pulp according to the procedure described previously [37]. Pulp water suspension (3 wt.%) of the paper pulp was treated with a supermasscolloider MKCA 6-5 grinder (Masuko, Sangyo Co., Ltd., Kawaguchi, Japan) in 7 working cycles. The resulting suspension in water was used without modification for the emulsion polymerization of styrene.

2.2. Synthesis of Hybrid Fibrils

The deposition of polystyrene (PS) onto cellulose fibrils was provided in accordance with a procedure described previously [37,38]. Styrene was added to 3 wt.% MFC water suspension; the mass ratio of MFC:styrene was 70:30. The obtained solution was heated up to 60 °C and ammonium persulfate (0.5 wt.% with respect to styrene) was added to it at constant stirring. The synthesis lasted 240 min, the obtained white product was filtered, washed and freeze dried. The corresponding synthesis scheme is presented in Figure 1.

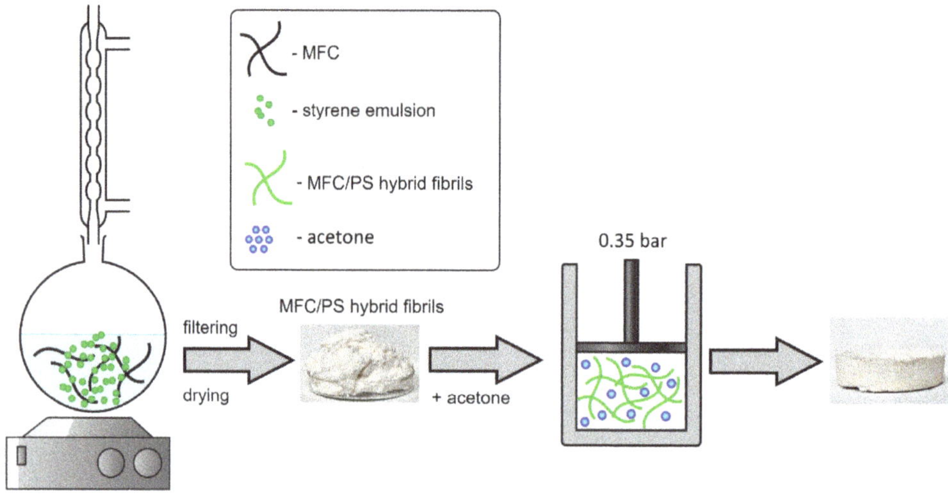

Figure 1. The scheme of synthesis of MFC/PS hybrid fibrils and composite porous matrix. The obtained MFC/PS hybrid fibrils during the first stage were compacted in a piston-cylinder with the addition of a small amount of acetone.

2.3. Preparation of Composite Materials

The porous composite matrices were prepared from 10 g of well-ground hybrid fibrils. MFC/PS was soaked with 15 mL of acetone (118 wt.% in relation to MFC), loaded in a piston-cylinder (diameter of 100 mm) and pressed up to 0.35 bar for 1 min using a universal testing machine TRM-O in manual mode (see Figure 1). No blowing agent was used. The retrieved samples were dried at 55 °C until a constant mass was reached.

Two composite materials were produced by impregnating a porous matrix with AWH materials presented by BMIMCl and CC. A piece of porous matrix (10 g) was placed in a 3M CC solution containing 10 g of CC for 24 h until the solution was completely absorbed. Subsequently, the resulting composite was dried at 60 °C until a constant weight was attained. The composite sample containing BMIMCl was obtained by deposition of 10 g of molten BMIMCl onto the composite matrix for 5 min. Then, the obtained composite sample was dried for 1 h in a forced air-drying cabinet (Binder, Camarillo, CA, USA) at a humidity level of 25%. For convenience, the samples of MFC/PS matrix modified with CC and BMIMCl will be denoted as MFC/PS/CC and MFC/PS/BMIMCl, correspondingly.

2.4. Methods

2.4.1. Electron Microscopy

The morphology of the hybrid fibrous material as well as the porous matrix and composite materials were investigated using scanning electron microscopy (SEM) using a JIB 4501 multi-beam microscope (JEOL Ltd., Tokyo, Japan). The samples were fixed on the SEM stubs using double-sided carbon tape. The surface of the samples was coated with

20 nm Au layer using a Q150R ES Plus ion sputtering system (Quorum Technologies Ltd., Laughton, UK) to avoid undesired surface charging of the samples and the appearance of corresponding artifacts in SEM micrographs. SEM micrographs were obtained in the SE mode at an accelerating voltage of 5 kV in the magnification range of ×40–×1500.

2.4.2. Fourier-Transform Infrared Spectroscopy (FTIR)

The FTIR measurements of the composite samples were performed using a Nicolet iS 10 FTIR spectrometer with germanium ATR crystal (Thermo Fisher Scientific, Waltham, MA, USA). All spectra were acquired in absorption mode in the 4000–600 cm^{-1} range. The spectral resolution was 8 cm^{-1}, the acquisition time was 14 s. The OMNIC Thermo Scientific software (version 7.3) was employed to record and process the acquired data.

2.4.3. The Porosity Study

The density, porosity, and water absorption coefficient values were evaluated for the pristine porous composite matrix and the matrices were modified with AWH agents using the liquid displacement method [37]. Distilled water was used as the displacement liquid. The sample was immersed in water for 5 min until it was saturated at a temperature of 26 °C and a humidity of 25%. The pore volume and porosity were measured three times, and the average value of the obtained indicators was taken as the test result.

The total pore volume (V_p) and porosity (μ) were calculated using Formulas (1) and (2):

$$V_p = (M_{wet} - M_{dry})/\rho(H_2O), \tag{1}$$

$$\mu = (V_p/V) \times 100\%, \tag{2}$$

where M_{wet} is the mass of the wet sample (g), M_{dry} is the mass of the sample before immersion (g), V is the volume of the sample before immersion (cm^3), and $\rho(H_2O)$ is the density of distilled water (g/cm^3) at the study temperature.

The water absorption coefficient (E) was calculated using Formula (3):

$$E = (M_{wet,24} - M_{dry})/M_{dry}, \tag{3}$$

where $M_{wet,24}$ the mass of the wet sample after 24 h of composite soaking in distilled water (g).

2.4.4. Study of Mechanical Characteristics

The tensile mechanical strengths of the pristine porous matrix and the obtained composite materials were studied using a TPM-O testing machine (Tochline, Moscow, Russia) at a speed of 50 mm/min, a temperature of 26 °C, and a humidity of 25%. The samples of 175 mm in length and 5 mm in thickness were prepared and tested.

The shape stability of the porous matrix is of utmost importance during multiple sorption/desorption cycles. To investigate it, the dimensions and weights of the three samples of dry pristine porous matrix (diameter = 100 mm, height = 100 mm, mass = 38.5 g) were noted. Then, the matrix samples were soaked in water for 24 h at 33 °C and dried in forced air-drying cabinet (Binder, USA) at a humidity level of 25% at three different temperatures: 27 °C, 55 °C, and 100 °C. The load of 16 kg was applied for 1 h to each sample to test porous matrix stability after the soaking/desorption cycle. When the load was removed, the corresponding parameters (dimensions and weight) of each sample were noted again.

2.4.5. Thermogravimetric Analysis (TGA)

The thermal stabilities of the pristine MFC/PS porous matrix and the matrices impregnated with BMIMCl and CC were studied using TGA using a STA 449 F5 Jupiter simultaneous thermal analyzer (NETZSCH Instruments, Selb, Germany). The obtained data were compared with the corresponding TGA data acquired for the starting materials (MFC and PS). The experiment was carried out in a nitrogen atmosphere with a heating rate of 10 °C/min in 32–600 °C temperature range.

2.4.6. Study of Water Absorption/Desorption

The study of moisture sorption from the air was performed for the pristine composite porous matrix and matrices impregnated with AWH agents under a transparent cover over a container with water (250 mL) at room temperature. Figure 2 presents the scheme of the experiment. The mass ratio of the CC/BMIMCl:porous matrix was 50:50. All tested samples had a cylindrical shape (diameter = 100 mm) and mass of 20 g; however, the height of matrices impregnated with AWH agents was 26 mm, while the height of the pristine porous composite matrix was 52 mm. The experiment was conducted under conditions of low (φ = 28%) and high (φ = 63%) ambient humidity. The weight of sorbed water was registered every 10 min during the first 150 min of the experiment and in its end after 24 h.

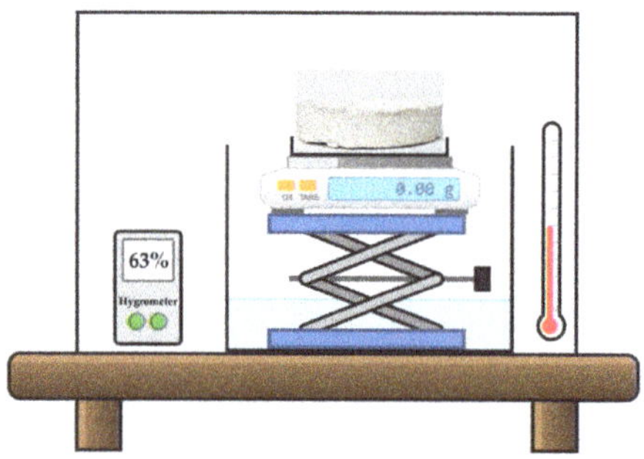

Figure 2. The scheme of water absorption experiment. The sample of pristine (or modified) MFC/PS matrix was placed under a transparent cover over a container of water at constant humidity (φ = 28%, 63%) and temperature. The weight of the sample was measured every 10 min.

The water desorption rates of the pristine composite porous matrix and matrix modified with AWH agents were determined in a forced air-drying cabinet (Binder, USA) at a humidity level of 8% and temperature of 55 °C. The experiment was terminated when the samples reached a constant mass.

3. Results and Discussion

3.1. Structure Investigation

The morphology and structure of the obtained hybrid fibrils, pristine composite porous matrix and the matrices impregnated with CC and BMIMCl were assayed using SEM (Figure 3). According to Figure 3a,b, PS was deposited as a uniform layer on the surface of cellulose fibrils. These results are in a good agreement with our previous observations [37].

Investigation of composite cross-sections revealed the different types of porous matrix impregnation by CC and BMIMCl (Figure 3c–e). CC forms a glaze-like layer on the porous matrix surface without anchoring within its pores. In contrast, BMIMCl exhibits uniform distribution throughout matrix voids, which accounts for the absence of agent-loss tendency during the multi-cycle operation. Such a distribution pattern of BMIMCl contributes to the material's stability and retention of sorption properties over a repeated cycles of use.

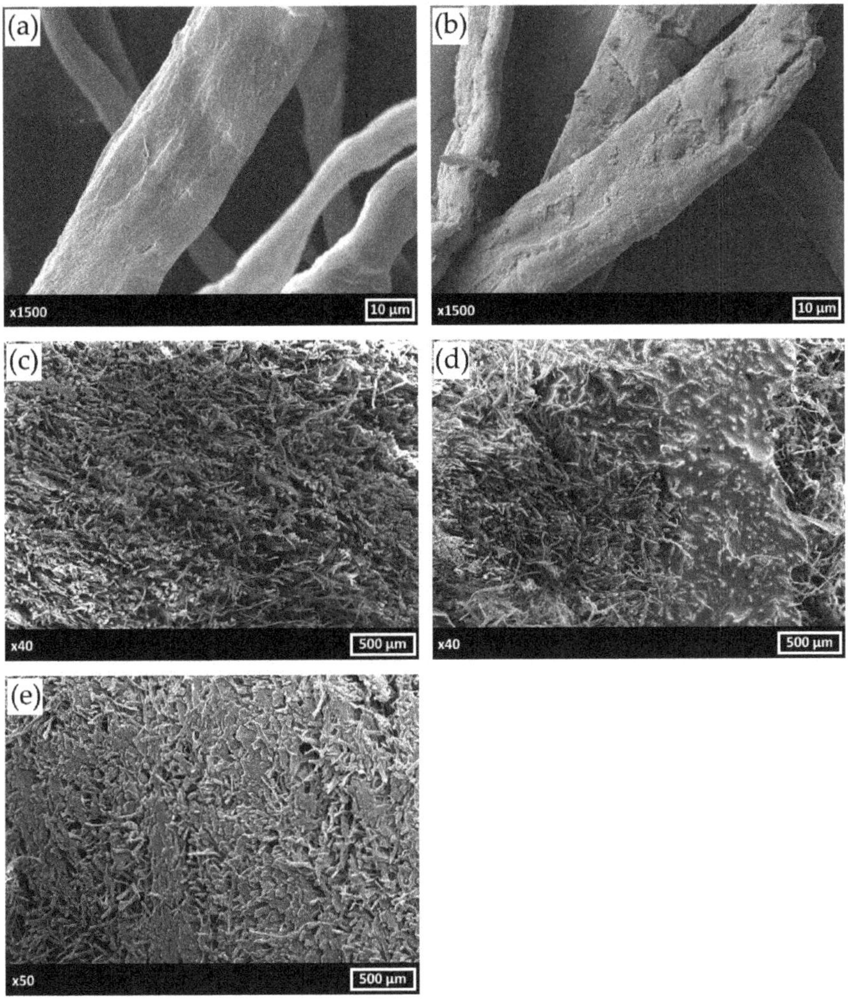

Figure 3. SEM micrographs of bare MFC (**a**), MFC/PS hybrid fibrils (**b**), MFC/PS pristine porous composite matrix (**c**), MFC/PS/CC (**d**), MFC/PS/BMIMCl (**e**).

In accordance with our previous findings, no chemical interaction was observed between the deposited PS and cellulose microfibrils [37] (see Figure 4 and Table 1). The bands at 3025 cm^{-1}, 2917 cm^{-1}, 1492 cm^{-1}, 1452 cm^{-1}, and 697 cm^{-1} were attributed to PS [39], while bands in 3600–3200 cm^{-1} and 1400–900 cm^{-1} wavenumber ranges were attributed to glucose OH groups and glucose ring oscillations, correspondingly [37]. The impregnation of the MFC/PS matrix with CC did not bring any dramatic changes to the FTIR spectrum except from rising intensities of the bands at 3600–3300 cm^{-1} and at 1630 cm^{-1} referring to -OH stretching vibrations and H-O-H bending vibrations of crystal water in CC [40]. The new bands at 1573 cm^{-1} and 1166 cm^{-1} were observed in the MFC/PS/BMIMCl FTIR spectrum. According to the literature data, the observed bands were attributed to the imidazolidinium framework vibrations and C-N stretching vibrations, respectively [41]. As shown in Figure 4, no change in MFC and PS bands has been observed in FTIR spectra of MFC/PS/CC and MFC/PS/BMIMCl which supports assumption about the absence of any chemical interaction.

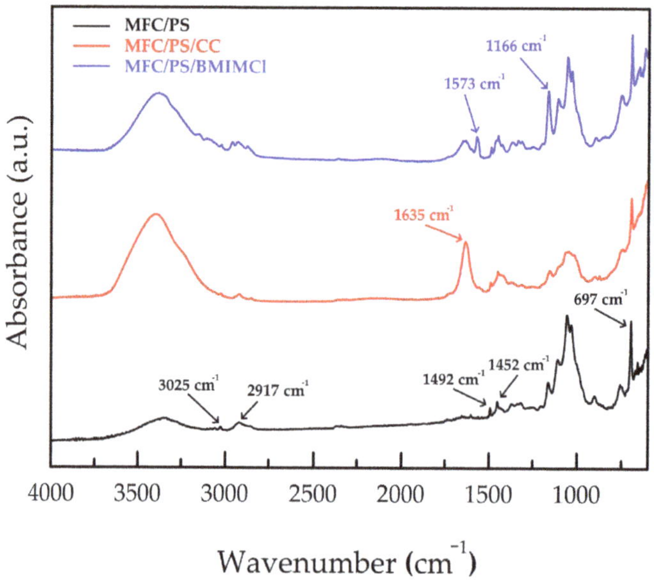

Figure 4. FTIR spectra of MFC/PS, MFC/PS/CC, and MFC/PS/BMIMCl. The wavenumbers of PS and BMIMCl characteristic bands are underwritten.

Table 1. The description of IR characteristic bands of cellulose, PS, CC and BMIMCl.

	Wavenumber, cm^{-1}	Description	Ref.
MFC	1400–900 3700–3200	Glucose ring oscillations Glucose OH stretching oscillations	[37]
PS	697 756 1700–1400 3100–2800	C-H out-of-plane bending oscillations Aromatic ring oscillations Aromatic C-H/CH_2 stretching oscillations	[39]
CC	1635	H-O-H bending oscillations (of crystal water)	[40]
BMIMCl	1166 1573	C-N stretching oscillations Imidazolidinium framework oscillations	[41]

As previously demonstrated, PS amount significantly affects the density (ρ), pore volume (V_p), porosity (μ), and water absorption coefficient (E) of the porous matrix [37]. Matrix impregnation with AWH materials can significantly change these characteristics. As shown in Table 2, the initial matrix is characterized by high porosity and low density. Nevertheless, one should note that these parameters can be considerably influenced by molding conditions. The presence of 70 wt.% hydrophilic MFC in the matrix increased the water absorption coefficient, which provides rapid diffusion of water into the volume of the material. Meanwhile, despite rather low PS content, the composite matrix has excellent shape stability after prolonged exposure to water, similar to the samples described earlier [37]. Such a high shape stability allows the development of new systems for moisture capture from air on the basis of the hybrid materials.

Modification of the matrix by AWH materials significantly affected the porosity and density values of the obtained composites. As shown in SEM images, CC glazes the pores of the matrix, resulting in a threefold decrease in total pore volume and porosity compared to the pristine MFC/PS. In contrast, BMIMCl addition led to porosity and pore volume increase. This phenomenon can be explained by the ability of BMIMCl to interact with cellulose fibrils, making matrix pore space more malleable and capacious. The density value of MFC/PS/BMIMCl composite is greater than the same values of the MFC/PS matrix and MFC/PS/CC by 8.8 and 1.8 times, respectively, which also demonstrates the better permeation of BMIMCl inside the porous matrix.

Table 2. The data on density (ρ), porosity (μ), total pore volume (V_p), and water absorption coefficient (E) of pristine MFC/PS matrix and matrices impregnated with CC and BMIMCl.

Sample	ρ, g/cm^3	V_p, cm^3	μ, %	E
MFC/PS	0.05	60.0 ± 1.3	29.4 ± 0.6	2.80
MFC/PS/CC	0.25	20.0 ± 2.3	9.8 ± 1.1	2.60
MFC/PS/BMIMCl	0.44	93.5 ± 4.2	45.8 ± 2.1	3.80

3.2. Thermal Stability Studies

The TGA data were used to quantify the composition of the obtained composite porous matrix as performed before [37]. According to the obtained data, the content of MFC in the obtained matrix does not exceed 70 wt.%, which is in good agreement with the calculated value (see Figure 5).

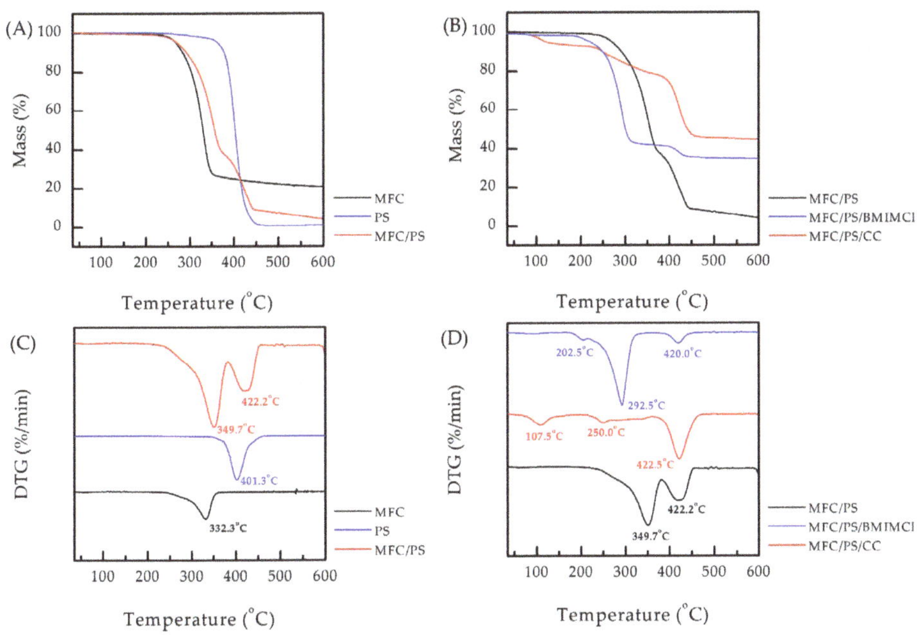

Figure 5. TGA and dTGA curves of (**A**,**C**) MFC, PS, MFC/PS, (**B**,**D**) MFC/PS, MFC/PS/CC, and MFC/PS/BMIMCl. The corresponding dTGA peaks are underwritten.

The thermal stability of the obtained composite porous matrix is of great importance since the desorption procedure can include temperature rising to 90 °C. According to Figure 5A,C, the thermal decomposition of the matrix occurs as a result of successive processes of dehydration of bound moisture at 80–110 °C, followed by destruction of the cellulose component at 250–350 °C and PS decomposition at 350–450 °C. Interestingly, the both dTGA peaks of the components of the MFC/PS porous matrix are shifted towards higher temperatures compared to the data obtained for the bare materials (by 17.4 °C for MFC and by 20.9 °C for PS). The same phenomenon has been observed earlier [38] and can be explained by the different kinetics of the material thermal decomposition, which depends on its amount in the sample.

The TGA and corresponding dTGA curves of the MFC/PS matrix and matrices impregnated with AWH agents are presented in Figure 5B,C. In disagreement with the literature data [42], MFC/PS/CC loses water contained in CC in 50–200 °C. The decomposition of PS in MFC/PS/CC undergoes the same temperature as in the pristine matrix, whereas the decomposition of MFC is not pronounced. The dTGA curve of MFC/PS/BMIMCl contains three peaks referring to water evaporation (202.5 °C), cellulose thermal degradation (292.5 °C), and PS thermal degradation (420.0 °C). The cellulose decomposition at lower temperatures can be explained by the partial disruption of the cellulose crystalline regions by BMIMCl [43]. It was thus demonstrated that both impregnated matrices exhibited adequate thermal stability up to 200 °C, which is the most significant temperature range for potential future implementation.

3.3. *Study of MFC/PS Matrix Mechanical Strength and Cyclic Stability*

The results of the mechanical strength properties investigation of the pristine composite matrix and modified matrix are presented in Table 3 and Figure 6.

Table 3. Tensile strength (σ_{max}), tearing strength (σ_p), maximum tensile load (P_{max}), and breaking load ($\mathbf{P_p}$) of pristine MFC/PS matrix and matrices impregnated with CC and BMIMCl.

Sample	σ_{max}, MPa	σ_p, MPa	P_{max}, N	P_p, N
MFC/PS	2.4 ± 0.4	1.7 ± 0.6	82.0 ± 1.7	55.2 ± 12.5
MFC/PS/CC	2.6 ± 0.2	1.8 ± 0.4	83.6 ± 1.4	53.5 ± 2.9
MFC/PS/BMIMCl	2.0 ± 0.1	1.5 ± 0.7	79.1 ± 1.1	39.8 ± 5.4

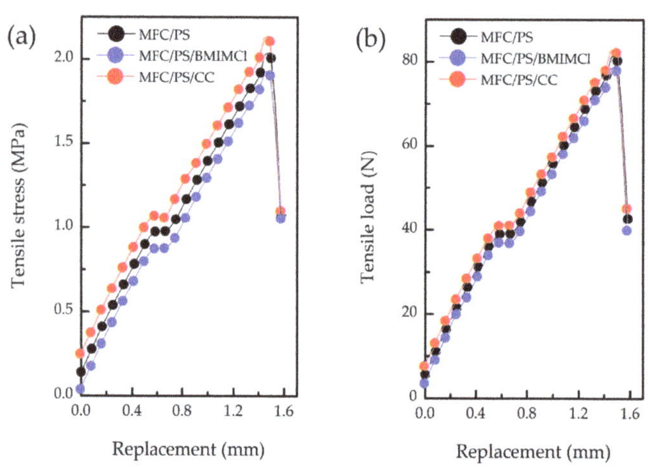

Figure 6. Compressive strength performance of pristine and modified MFC/PS matrices: (**a**) tensile strength, (**b**) tensile load.

Despite the fact that PS content in the pristine composite porous matrix is 30 wt.%, the mechanical properties of the matrix significantly differ from those of pure PS. For instance, the tensile strength (σ_{max}) of pure PS is 10 times higher than the corresponding value of the obtained composite porous matrix, which is first of all linked with its high porosity.

According to Table 3, incorporation of CC in the porous composite matrix led to a slight enhancement of the strength characteristics of composite material compared to the pristine matrix. It can be explained by CC deposition on the surface of the matrix and its vitrification. Unlike CC, impregnation of the porous matrix by BMIMCl resulted in the tensile strength decline. This phenomenon can be attributed to the chemical interaction between BMIMCl and cellulose hydroxyl groups, explaining the better distribution of BMIMCl over the matrix volume. Meanwhile, the chemical binding of BMIMCl with hydroxyl groups can lead to cellulose fibrils softening [44] and, hence, partial disintegration of MFC in a composite matrix.

The integrity of the composite porous matrix is of paramount importance with regard to the stability of the final composite, particularly during its cyclic operation. It was found that the dimensions of the pristine matrix did not change after the load of 16 kg was applied to the matrix sample, after it was subjected to the cycle of water soaking/desorption.

According to Figure 7a, the maximum rate of water absorption is reached within the initial 30 min when the pristine porous composite matrix was immersed in water for 24 h. Then, the absorption rate gradually decreased, reaching a plateau after 3 h of observation. Furthermore, the composite matrix displayed outstanding shape stability even when subjected to prolonged exposure to water, thereby facilitating the potential use of this material in AWH systems for extended operational periods.

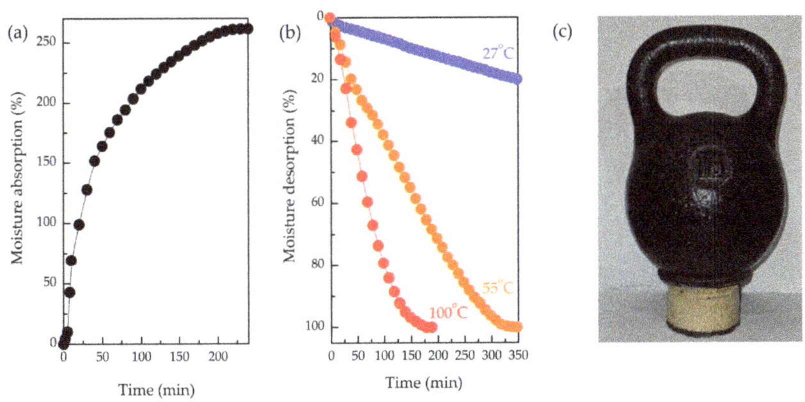

Figure 7. (**a**) Sorption of MFC/PS porous matrix immersed in water; (**b**) desorption at different temperatures; (**c**) demonstration of MFC/PS porous matrix shape stability after 10 sorption/desorption cycles under load of 16 kg.

It is well known that various types of AWH agents require different regeneration temperatures that could reach rather high values. For instance, MOF-303 employed in [35] for vapor harvesting was regenerated at 55–60 °C. Thus, the stability of the MFC/PS matrix at different regenerating temperatures is of great practical importance. In this regard, the desorption of the MFC/PS matrix was carried out at 27 °C, 55 °C, and 100 °C.

As follows from Figure 7b, the MFC/PS matrix ability to desorb accumulated moisture strongly depends on the temperature. Thus, desorption occurs more rapidly at 100 °C, achieving complete desorption within 3 h, whereas desorption at 55 °C takes 6 h. The rate of moisture desorption at room temperature was extremely low (4–5 g/h). The initial mass of the MFC/PS matrix was restored only after 48 h. It should be noted that none of the applied temperature regimes had no effect on the composite matrix stability.

3.4. Moisture Absorption and Desorption Under the Influence of AWH Agents

The summary of moisture absorption tests under low and high humidity levels carried out for the pristine MFC/PS matrix, MFC/PS/BMIMCl, and MFC/PS/CC is presented in Table 4.

Table 4. Moisture sorption by composites at different air humidities.

Absorption Conditions	Weight of Absorbed Moisture per Sample Weight, %			
	Time	**MFC/PS**	**MFC/PS/BMIMCl**	**MFC/PS/CC**
T = 25 °C φ = 28%	150 min 24 h	1.6 ± 0.3 3.9 ± 0.5	3.4 ± 0.1 12.1 ± 0.2	1.4 ± 0.2 4.6 ± 0.4
T = 25 °C φ = 63%	150 min 24 h	1.9 ± 0.2 5.9 ± 0.4	5.3 ± 0.3 16.7 ± 0.5	11.0 ± 0.3 16.9 ± 0.5
Desorption Conditions	**Sample Weight Reduction per Absorbed Water Weight, %**			
	Time	**MFC/PS**	**MFC/PS/BMIMCl**	**MFC/PS/CC**
T = 55 °C φ = 8%	150 min 24 h	99.2 ± 0.1 99.8 ± 0.3	74.7 ± 2.3 94.8 ± 2.4	152.6 ± 3.8 184.8 ± 4.3

The MFC/PS matrix exhibited a modest absorption capacity under both low and high humidities, whereas the incorporation of CC and BMIMCl led to a substantial increase in atmospheric water capture.

In conditions of low humidity, the rate of moisture absorption of MFC/PS/BMIMCl is higher than that of MFC/PS/CC (Figure 8a). Interestingly, the weight of the adsorbed moisture by MFC/PS/CC is close to the same value of the pristine MFC/PS matrix even after 24 h.

In conditions of high humidity (Figure 8b), the sample with CC exhibited rapid moisture uptake within 150 min, while MFC/PS/BMIMCl absorbed only half of that water weight. During the next hours of observation, CC leakage was detected. However, after 24 h of the experiment, the absorbed water weights by both composites were found to be comparable (see Table 3).

The observed differences in the absorption curves of MFC/PS/BMIMCl and MFC/PS/CC can be explained by the various ways the AWH component is distributed within the porous matrix structure. CC (Figure 3d) demonstrates a tendency to glaze the surface of cellulose fibers without anchoring within the pores of the MFC/PS matrix. Since CC is situated closer to the surface of the composite, the rate of water vapor absorption is significantly greater than that of MFC/PS/BMIMCl. However, the water retained by CC does not penetrate inside the porous matrix, inducing salt liquefaction and subsequent leakage. Conversely, water vapor absorption by MFC/PS/BMIMCl is slower due to more uniform distribution of BMIMCl in the matrix volume (Figure 3e) owing to interaction with cellulose hydroxyl groups. Therefore, water absorbed by the MFC/PS/BMIMCl sample is homogeneously distributed throughout the composite volume and, thus, is better retained.

As shown in Table 4 and Figure 8c, the desorption behavior of the MFC/PS matrix and two composites within the first 150 min of the experiment was completely different. Thus, for instance, the sample containing CC lost 153% of its weight per absorbed water weight. Such an unusual mass reduction in the MFC/PS/CC sample can be explained by the loss/leakage of the part of the AWH agent during absorption, which is in agreement with the visual observations. MFC/PS/BMIMCl released only 75% of the captured water, which is less than the value of the pristine MFC/PS matrix. It should be underlined that even after 24 h of drying at 55 °C and a humidity of 8%, MFC/PS/BMIMCl still kept 5% of the captured water. In fact, the reason for this phenomenon is that BMIMCl forms both a strong electrostatic interaction and a hydrogen bonding with the adsorbed water molecules.

The MFC/PS/BMIMCl sample demonstrated outstanding moisture absorption ability compared to the pristine MFC/PS matrix. Unlike CC, no leakage of AWH agent was observed during the moisture absorption/desorption cycle.

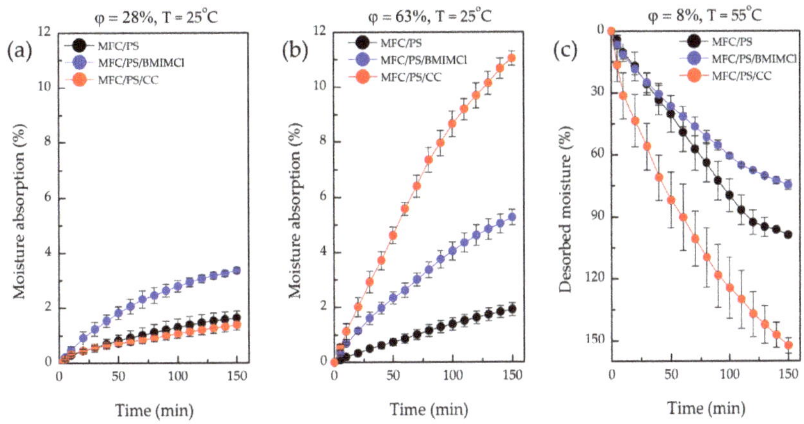

Figure 8. Moisture absorption from the air by pristine MFC/PS, MFC/PS/BMIMCl, and MFC/PS/CC at (**a**) low (28%) and (**b**) high (63%) humidity and 25 °C; (**c**) desorption curves of pristine MFC/PS, MFC/PS/BMIMCl, MFC/PS/CC at 55 °C and humidity of 8%.

A significant challenge of the modern AWH systems is posed by the fact that the desorption stage necessitates a considerable amount of energy input, given that the entire sorbent solution must be heated [26]. The heating of sorbents is typically accomplished through the use of electricity. However, some researchers have suggested the use of solar energy, taking into account its abundance and net zero carbon emissions [35,45]. It should be noted that obtained MFC/PS/BMIMCl composite can effectively desorb 95% of accumulated water at a relatively low temperature (55 °C), which can be reached using solar radiation. The possibility of the use of an alternative energy source as well as a high content of biodegradable cellulose in the obtained composite material makes it an attractive candidate for the modern AWH technologies complying with strict requirements for sustainable development of the economy.

The combination of the hydrophilic (MFC) and hydrophobic (PS) nature of the components of the obtained composite porous matrix improves both the water absorption and release functions. The non-polar PS chains repel the polar water molecules, while the polar MFC chains attract the water molecules [46]. The composite matrix, based on MFC and PS, overcomes the key limitations of existing AWH technologies, offering a sustainable, efficient, and cost-effective method for capturing atmospheric water, thus contributing to the fight against water scarcity.

4. Conclusions

The elaboration and wide application of environmentally friendly polymeric materials can reduce widespread 'plasticization' and pave the way for the development of 'green technologies' in this area of industrial production. Hybrid composite based on microfibrillar cellulose and polystyrene is an effective carrier for moisture capture systems from atmospheric air thanks to its high mechanical strength, porosity, and hydrophilicity. The moisture capture process is provided by using conventional hygroscopic materials (calcium chloride and ionic liquids) as the active components. The distribution of ionic liquid in the porous matrix structure significantly affects the porosity and mechanical properties of the composite material due to the better permeability of BMIMCl and partial dissolution of cellulose fibrils. The water absorbed by the sample with BMIMCl is homogeneously distributed throughout the volume of the composite, ensuring better retention in the pores. The composite material undergoes intensive desorption of water within less than 6 h at temperatures above 55 °C, allowing multiple cyclic uses for water absorption at night and extraction under sunlight during the day without additional energy inputs.

One of the key factors determining the performance characteristics of atmospheric water harvesting (AWH) technology is the absorption efficiency of composite material, as well as its ability to release accumulated moisture from the porous matrix. A promising approach to addressing these challenges involves doping a cellulose matrix with polystyrene and conductive materials, such as carbon felt, carbon nanotubes, melanin, or polyaniline. These materials are characterized by dark surfaces, which, theoretically, could enable the described material to enhance desorption efficiency by harnessing solar radiation.

The described composite matrix can be applied in portable water harvesting devices:

- Lightweight and energy-efficient systems for remote areas with limited access to water infrastructure, utilizing the composite's passive absorption and solar-driven desorption properties;
- Building-Integrated Systems: Incorporating the composite into roofing or facade materials to harvest atmospheric water for domestic or irrigation use, particularly in arid or semi-arid regions;
- Emergency and Disaster Relief: Deployable units for clean water collection in disaster-stricken or drought-affected areas, where conventional water sources are unavailable;
- Agricultural use: Installation in greenhouses or open fields to collect water vapor for irrigation, reducing dependency on traditional water sources;
- Advanced Industrial Systems: Application in cooling systems or industrial processes requiring efficient humidity control, with potential for energy savings and water recovery.

Despite outstanding performance of the atmospheric water absorption by the obtained composites, the issues related to calcium chloride retention must be addressed. One of the possible resolutions is a doping of the composite matrix with additional sorbent or moisture-retaining substance (e.g., polyacrylamide).

Author Contributions: Conceptualization, K.C., R.F. and V.V.; methodology, D.R., V.M. and V.V.; validation, V.V., R.F. and V.M.; formal analysis, Y.G. and V.M.; investigation, D.R., M.G., N.R., W.L. and K.C.; writing—original draft preparation, D.R., W.L. and N.R.; writing—review and editing, K.C., Y.G., R.F. and V.V.; visualization, W.L. and V.M.; supervision, R.F., V.V. and E.I.; project administration, E.I. and Y.G.; funding acquisition, V.V. and E.I. All authors have read and agreed to the published version of the manuscript.

Funding: This work was financially supported by the Ministry of Education and Science of Russia (State assignment; Project FSZE-2023-0005).

Institutional Review Board Statement: Not applicable.

Data Availability Statement: Data are available from the corresponding author by email upon reasonable request.

Acknowledgments: The authors would like to thank Egor Smirnov for the help in TGA and Olga Antipova for the help in SEM investigations.

Conflicts of Interest: The authors declare no conflicts of interest.

References

1. Zeng, W.; You, T.; Wu, W. Passive Atmospheric Water Harvesting: Materials, Devices, and Perspectives. *Nano Energy* **2024**, *125*, 109572. [CrossRef]

2. Kandeal, A.W.; Joseph, A.; Elsharkawy, M.; Elkadeem, M.R.; Hamada, M.A.; Khalil, A.; Eid Moustapha, M.; Sharshir, S.W. Research Progress on Recent Technologies of Water Harvesting from Atmospheric Air: A Detailed Review. *Sustain. Energy Technol. Assess.* **2022**, *52*, 102000. [CrossRef]
3. Ahmad, M.; Nighojkar, A.; Plappally, A. A Review of the Methods of Harvesting Atmospheric Moisture. *Environ. Sci. Pollut. Res.* **2023**, *31*, 10395–10416. [CrossRef]
4. Wang, M.; Liu, E.; Jin, T.; Zafar, S.; Mei, X.; Fauconnier, M.-L.; De Clerck, C. Towards a Better Understanding of Atmospheric Water Harvesting (AWH) Technology. *Water Res.* **2024**, *250*, 121052. [CrossRef]
5. Nikkhah, H.; Azmi, W.M.B.W.; Nikkhah, A.; Najafi, A.M.; Babaei, M.M.; Fen, C.S.; Nouri, A.; Mohammad, A.W.; Lun, A.W.; Yong, N.L.; et al. A Comprehensive Review on Atmospheric Water Harvesting Technologies: From Thermodynamic Concepts to Mechanism and Process Development. *J. Water Process Eng.* **2023**, *53*, 103728. [CrossRef]
6. Zhang, S.; Fu, J.; Xing, G.; Zhu, W.; Ben, T. Porous Materials for Atmospheric Water Harvesting. *ChemistryOpen* **2023**, *12*, e202300046. [CrossRef] [PubMed]
7. Lu, W.; Ong, W.L.; Ho, G.W. Advances in Harvesting Water and Energy from Ubiquitous Atmospheric Moisture. *J. Mater. Chem. A* **2023**, *11*, 12456–12481. [CrossRef]
8. Zhou, X.; Lu, H.; Zhao, F.; Yu, G. Atmospheric Water Harvesting: A Review of Material and Structural Designs. *ACS Mater. Lett.* **2020**, *2*, 671–684. [CrossRef]
9. Pan, T.; Yang, K.; Han, Y. Recent Progress of Atmospheric Water Harvesting Using Metal-Organic Frameworks. *Chem. Res. Chin. Univ.* **2020**, *36*, 33–40. [CrossRef]
10. Vikrant, K.; Kumar, V.; Kim, K.-H.; Kukkar, D. Metal–Organic Frameworks (MOFs): Potential and Challenges for Capture and Abatement of Ammonia. *J. Mater. Chem. A* **2017**, *5*, 22877–22896. [CrossRef]
11. Gutiérrez, M.; Zhang, Y.; Tan, J.-C. Confinement of Luminescent Guests in Metal–Organic Frameworks: Understanding Pathways from Synthesis and Multimodal Characterization to Potential Applications of LG@MOF Systems. *Chem. Rev.* **2022**, *122*, 10438–10483. [CrossRef]
12. Li, T.; Yu, H.; Mi, J.; Li, C.; Meng, H.; Jin, J. Highly Hydrophilic Acrylate Copolymer Supported MIL-160 for Air Water Harvesting. *Chem. Phys. Lett.* **2023**, *816*, 140391. [CrossRef]
13. Feng, A.; Akther, N.; Duan, X.; Peng, S.; Onggowarsito, C.; Mao, S.; Fu, Q.; Kolev, S.D. Recent Development of Atmospheric Water Harvesting Materials: A Review. *ACS Mater. Au* **2022**, *2*, 576–595. [CrossRef]
14. Chen, Y.; Mu, T. Revisiting Greenness of Ionic Liquids and Deep Eutectic Solvents. *Green Chem. Eng.* **2021**, *2*, 174–186. [CrossRef]
15. Lu, H.; Shi, W.; Guo, Y.; Guan, W.; Lei, C.; Yu, G. Materials Engineering for Atmospheric Water Harvesting: Progress and Perspectives. *Adv. Mater.* **2022**, *34*, 2110079. [CrossRef] [PubMed]
16. Gong, F.; Li, H.; Zhou, Q.; Wang, M.; Wang, W.; Lv, Y.; Xiao, R.; Papavassiliou, D.V. Agricultural Waste-Derived Moisture-Absorber for All-Weather Atmospheric Water Collection and Electricity Generation. *Nano Energy* **2020**, *74*, 104922. [CrossRef]
17. Farghal, H.H.; Nebsen, M.; El-Sayed, M.M.H. Exploitation of Expired Cellulose Biopolymers as Hydrochars for Capturing Emerging Contaminants from Water. *RSC Adv.* **2023**, *13*, 19757–19769. [CrossRef] [PubMed]
18. Dong, J.; Chen, X.; Li, Y.; Luan, M.; Yang, X.; Chen, H.; Koosha, M.; Zhai, Y.; Fakhrullin, R.F. Hydrophilic Chitosan: Modification Pathways and Biomedical Applications. *Russ. Chem. Rev.* **2024**, *93*, RCR5120. [CrossRef]
19. Hong, F.; Qiu, P.; Wang, Y.; Ren, P.; Liu, J.; Zhao, J.; Gou, D. Chitosan-Based Hydrogels: From Preparation to Applications, a Review. *Food Chem. X* **2024**, *21*, 101095. [CrossRef] [PubMed]
20. Yu, F.; Chen, Z.; Guo, Z.; Irshad, M.S.; Yu, L.; Qian, J.; Mei, T.; Wang, X. Molybdenum Carbide/Carbon-Based Chitosan Hydrogel as an Effective Solar Water Evaporation Accelerator. *ACS Sustain. Chem. Eng.* **2020**, *8*, 7139–7149. [CrossRef]
21. Zainal, S.H.; Mohd, N.H.; Suhaili, N.; Anuar, F.H.; Lazim, A.M.; Othaman, R. Preparation of Cellulose-Based Hydrogel: A Review. *J. Mater. Res. Technol.* **2021**, *10*, 935–952. [CrossRef]
22. Omidian, H.; Akhzarmehr, A.; Chowdhury, S.D. Advancements in Cellulose-Based Superabsorbent Hydrogels: Sustainable Solutions across Industries. *Gels* **2024**, *10*, 174. [CrossRef]
23. Wei, P.; Chen, W.; Song, Q.; Wu, Y.; Xu, Y. Superabsorbent Hydrogels Enhanced by Quaternized Tunicate Cellulose Nanocrystals with Adjustable Strength and Swelling Ratio. *Cellulose* **2021**, *28*, 3723–3732. [CrossRef]
24. Kallenberger, P.A.; Fröba, M. Water Harvesting from Air with a Hygroscopic Salt in a Hydrogel–Derived Matrix. *Commun. Chem.* **2018**, *1*, 28. [CrossRef]
25. Li, S.; Hernandez, S.; Salazar, N. Biopolymer-Based Hydrogels for Harvesting Water from Humid Air: A Review. *Sustainability* **2023**, *15*, 848. [CrossRef]
26. Bai, Q.; Zhou, W.; Cui, W.; Qi, Z. Research Progress on Hygroscopic Agents for Atmospheric Water Harvesting Systems. *Materials* **2024**, *17*, 722. [CrossRef] [PubMed]
27. Bai, Z.; Wang, P.; Xu, J.; Wang, R.; Li, T. Progress and Perspectives of Sorption-Based Atmospheric Water Harvesting for Sustainable Water Generation: Materials, Devices, and Systems. *Sci. Bull.* **2024**, *69*, 671–687. [CrossRef] [PubMed]
28. Ejeian, M.; Wang, R.Z. Adsorption-Based Atmospheric Water Harvesting. *Joule* **2021**, *5*, 1678–1703. [CrossRef]

29. Nguyen, L.T.; Bai, Z.; Zhu, J.; Gao, C.; Liu, X.; Wagaye, B.T.; Li, J.; Zhang, B.; Guo, J. Three-Dimensional Multilayer Vertical Filament Meshes for Enhancing Efficiency in Fog Water Harvesting. *ACS Omega* **2021**, *6*, 3910–3920. [CrossRef]
30. William, G.E.; Mohamed, M.H.; Fatouh, M. Desiccant System for Water Production from Humid Air Using Solar Energy. *Energy* **2015**, *90*, 1707–1720. [CrossRef]
31. Safoui, R.; Belaaribi, R.; Achahour, O.; Elfanaoui, A.; Ihlal, A.; Mouaky, A.; Amagour, M.E.H.; Abou Oualid, H.; Awad, M.M. Atmospheric Water Harvesting Using a Desiccant-Based Solar Still: Experimental Investigation and Economic Analysis. *Eng. Res. Express* **2024**, *6*, 045559. [CrossRef]
32. Li, J.; Yao, Z.; Liu, X.; Yang, C.; Liu, P.; Liu, P.; Wang, W.; Guo, H.; Huang, G.; Jin, X.; et al. Modified Metal-Organic Framework-Based Polymer Materials for Atmospheric Water Harvesting and Liquid Water Storage. *Surf. Interfaces* **2025**, *56*, 105651. [CrossRef]
33. Wang, X.; Yang, D.; Zhang, M.; Hu, Q.; Gao, K.; Zhou, J.; Yu, Z.-Z. Super-Hygroscopic Calcium Chloride/Graphene Oxide/Poly(N-Isopropylacrylamide) Gels for Spontaneous Harvesting of Atmospheric Water and Solar-Driven Water Release. *ACS Appl. Mater. Interfaces* **2022**, *14*, 33881–33891. [CrossRef]
34. Yang, X.; Chen, Z.; Xiang, C.; Shan, H.; Wang, R. Enhanced Continuous Atmospheric Water Harvesting with Scalable Hygroscopic Gel Driven by Natural Sunlight and Wind. *Nat. Commun.* **2024**, *15*, 7678. [CrossRef] [PubMed]
35. Kim, H.; Rao, S.R.; Kapustin, E.A.; Zhao, L.; Yang, S.; Yaghi, O.M.; Wang, E.N. Adsorption-Based Atmospheric Water Harvesting Device for Arid Climates. *Nat. Commun.* **2018**, *9*, 1191. [CrossRef]
36. Kim, H.; Yang, S.; Rao, S.R.; Narayanan, S.; Kapustin, E.A.; Furukawa, H.; Umans, A.S.; Yaghi, O.M.; Wang, E.N. Water Harvesting from Air with Metal-Organic Frameworks Powered by Natural Sunlight. *Science* **2017**, *356*, 430–434. [CrossRef]
37. Cherednichenko, K.; Bardina, K.; Vishnevich, A.; Gablina, M.; Gataulina, A.; Nikolaev, Y.; Gushchin, P.; Ivanov, E.; Kopitsyn, D.; Vinokurov, V. A Facile One-Step Synthesis of Polystyrene/Cellulose (PS@MFC) Biocomposites for the Preparation of Hybrid Water-Absorbing Sponge Materials. *Polymers* **2023**, *15*, 4328. [CrossRef] [PubMed]
38. Cherednichenko, K.A.; Sayfutdinova, A.R.; Kraynov, A.; Anikushin, B.; Ignatiev, V.; Rubtsova, M.I.; Konstantinova, S.A.; Shchukin, D.G.; Vinokurov, V.A. A Rapid Synthesis of Nanofibrillar Cellulose/Polystyrene Composite via Ultrasonic Treatment. *Ultrason. Sonochem.* **2022**, *90*, 106180. [CrossRef] [PubMed]
39. Fang, J.; Xuan, Y.; Li, Q. Preparation of Polystyrene Spheres in Different Particle Sizes and Assembly of the PS Colloidal Crystals. *Sci. China Technol. Sci.* **2010**, *53*, 3088–3093. [CrossRef]
40. Cao, S.; Luo, X.; Han, X.; Lu, X.; Zou, C. Development of a New Modified $CaCl_2{\cdot}6H_2O$ Composite Phase Change Material. *Energies* **2022**, *15*, 824. [CrossRef]
41. Chen, K.; Xu, W.; Ding, Y.; Xue, P.; Sheng, P.; Qiao, H.; Wang, S.; Yu, Y. Mechanical and Thermal Properties of All-Wood Biocomposites through Controllable Dissolution of Cellulose with Ionic Liquid. *Polymers* **2020**, *12*, 361. [CrossRef]
42. Li, Y.; Liu, Q.; Liu, Y.; Wang, D.; Song, W.; Chen, Y.; Liu, J. Calcium Chloride Hexahydrate/Nano-SiO_2 Composites as Form-Stable Phase Change Materials for Building Energy Conversation: The Influence of Pore Size of Nano-SiO_2. *Energy Build.* **2020**, *208*, 109672. [CrossRef]
43. Kadokawa, J.; Murakami, M.; Kaneko, Y. A Facile Preparation of Gel Materials from a Solution of Cellulose in Ionic Liquid. *Carbohydr. Res.* **2008**, *343*, 769–772. [CrossRef]
44. Aghmih, K.; Bouftou, A.; El Bouchti, M.; Boukhriss, A.; Gmouh, S.; Majid, S. Synthesis and Application of Functionalized Ionic Liquids-Based Imidazolium as Solvent for Cotton Fibre Cellulose Dissolution. *Cellulose* **2023**, *30*, 1467–1481. [CrossRef]
45. LaPotin, A.; Zhong, Y.; Zhang, L.; Zhao, L.; Leroy, A.; Kim, H.; Rao, S.R.; Wang, E.N. Dual-Stage Atmospheric Water Harvesting Device for Scalable Solar-Driven Water Production. *Joule* **2021**, *5*, 166–182. [CrossRef]
46. Nasef, M. Preparation and Applications of Ion Exchange Membranes by Radiation-Induced Graft Copolymerization of Polar Monomers onto Non-Polar Films. *Progress Polym. Sci.* **2004**, *29*, 499–561. [CrossRef]

polymers

MDPI

Article

Electrochemical Switching of Laser-Induced Graphene/Polymer Composites for Tunable Electronics

Maxim Fatkullin [1], Ilia Petrov [1], Elizaveta Dogadina [1], Dmitry Kogolev [1], Alexandr Vorobiev [1], Pavel Postnikov [1], Jin-Ju Chen [2], Rafael Furlan de Oliveira [3], Olfa Kanoun [4,*], Raul D. Rodriguez [1,*] and Evgeniya Sheremet [1]

1 Research School of Chemical and Biomedical Technologies, Tomsk Polytechnic University, Lenin Ave. 30, 634050 Tomsk, Russia; maksim@tpu.ru (M.F.); ilia.ser.petrov@gmail.com (I.P.); elizavetadogadina@gmail.com (E.D.); kogolev@tpu.ru (D.K.); aov4@tpu.ru (A.V.); postnikov@tpu.ru (P.P.); esheremet@tpu.ru (E.S.)
2 School of Materials and Energy, University of Electronic Science and Technology of China, Chengdu 610054, China; jinjuchen@uestc.edu.cn
3 Brazilian Nanotechnology National Laboratory (LNNano), Brazilian Center for Research in Energy and Materials (CNPEM), Campinas 13083-970, Brazil; rafael.furlan@lnnano.cnpem.br
4 Measurement and Sensor Technology, Faculty of Electrical Engineering and Information Technology, Technische Universität Chemnitz, 09126 Chemnitz, Germany
* Correspondence: olfa.kanoun@etit.tu-chemnitz.de (O.K.); raul@tpu.ru (R.D.R.)

Abstract: Laser reduction of graphene oxide (GO) is a promising approach for achieving flexible, robust, and electrically conductive graphene/polymer composites. Resulting composite materials show significant technological potential for energy storage, sensing, and bioelectronics. However, in the case of insulating polymers, the properties of electrodes show severely limited performance. To overcome these challenges, we report on a post-processing redox treatment that allows the tuning of the electrochemical properties of laser-induced rGO/polymer composite electrodes. We show that the polymer substrate plays a crucial role in the electrochemical modulation of the composites' properties, such as the electrode impedance, charge transfer resistance, and areal capacitance. The mechanism behind the reversible control of electrochemical properties of the rGO/polymer composites is the cleavage of polymer chains in the vicinity of rGO flakes during redox cycling, which exposes rGO active sites to interact with the electrolyte. Sequential redox cycling improves composite performance, allowing the development of devices such as electrolyte-gated transistors, which are widely used in chemical sensing applications. Our strategy enables the engineering of the electrochemical properties of rGO/polymer composites by post-treatment with dynamic switching, opening up new possibilities for flexible electronics and electrochemical applications having tunable properties.

Keywords: laser processing; graphene polymer composites; electrochemistry; liquid-gated transistor; tunable properties

1. Introduction

Graphene and its derivatives have attracted significant attention from researchers during the last two decades [1,2]. Among these derivatives, graphene oxide (GO) is particularly promising in transferring graphene technologies from the lab to real-world applications, primarily due to the possibility of tailoring its physicochemical properties and hydrophilic nature. This hydrophilicity makes GO compatible with different solution-based deposition technologies, such as inkjet printing, spray coating, and screen printing, which are ideal for large-area and batch fabrication [3]. GO can be processed using different routes: chemical, thermal, and photoreduction, which result in reduced graphene oxide (rGO) with partially recovered graphene-like properties [4]. Photoreduction by laser processing offers the benefit of creating well-defined and ad hoc patterning that allows the production of free-form rGO structures with tunable properties.

The unique properties of rGO [5,6], such as tunable electrical conductivity, wettability, and chemical stability, make it attractive as a filler in polymer composites for a wide range of applications, including energy storage, electrocatalysis, bioelectronics, and sensors [7–12]. In these applications, the electrochemical properties of rGO play a crucial role. While most research works considering GO and rGO electrochemistry are focused on the applications, several works have also looked at the fundamental electrochemical behavior of GO and rGO [13,14]. However, rGO-based composites have received less attention regardless of their practical relevance, especially for composites obtained by laser processing.

In this work, we investigate the electrochemical properties of laser-induced rGO/polymer composites using polyethylene terephthalate and polyvinylidene fluoride as substrates for GO deposition [15]. We report an unexpected electrochemical behavior observed under potentials exceeding the water electrolysis voltage window. This behavior results in a substrate-predefined "switching" of electrochemical properties such as electrical impedance, charge transfer resistance (R_{ct}), and areal capacitance after a few "activation" electrochemical treatment cycles associated with changes in polymer structure. This switching capability, arising from the electrolyte interaction with freshly exposed rGO, provides a new degree of freedom for tailoring the electrochemical properties of rGO/polymer composites beyond what is achievable by sole laser processing since laser processing, in particular for composite formation has only a small window of parameters to achieve the composite while preserving the structural integrity of both components. In other words, the ability to control properties is limited by the composite formation mechanism [15]. Moreover, after the "activation" cycles, the rGO/polymer composite properties can be reversibly changed upon consecutive electrochemical treatment. This electrochemical treatment allows the development of an electrolyte-gated transistor that otherwise did not work when using non-treated electrodes. The mechanism behind this switching behavior is related to electrochemical polymer erosion, which exposes rGO to the interaction with the electrolyte.

2. Materials and Methods

2.1. Materials

Graphene oxide water dispersion with a concentration of 4 mg mL^{-1} (monolayer content >95%, measured in 0.05 wt%; particle size: D90 5–7 μm, D50 2–4 μm, D10 1–2 μm; Graphenea, San Sebastián, Spain); phosphate-buffered saline (PBS) with pH of 6.86 (Uralkhiminvest, Ufa, Russia); PET sheets with a thickness of 0.65 mm were purchased in local store; PVDF was prepared using fused deposition modeling 3D printing.

2.2. Laser-Induced Composite Fabrication

Both rGO/PET and rGO/PVDF were fabricated following the same protocol. First, GO was drop-casted on a polymer substrate at 90 μL cm^{-2} and left for drying on a heating plate under 40 °C in air. The dry films with a 30 μm GO thickness (Figure S1) were laser processed using a pulsed diode laser with a wavelength of 436 nm operating at 1.6 kHz frequency (pulse energy ~0.17 J, pulse width ~260 μs). After laser processing, the composite structures were sonicated in distilled water for 5 min to remove GO residues and poorly integrated rGO.

2.3. Electrochemical Treatment and Characterization

All the electrochemical experiments were conducted in the three-electrode cell with Pt wire, Ag/AgCl (saturated KCl), and rGO/polymer composite as the counter, reference, and working electrodes. PBS (pH 6.86) was used as an electrolyte throughout all the experiments. Data were recorded using Corrtest CS2350M (Corrtest, Wuhan, China) bipotentiostat-bigalvanostat with the impedance measuring module. Electrochemical impedance spectroscopy (EIS) measurements were carried out with a sinusoidal waveform with a 0 V DC component and a 50 mV AC amplitude. This higher-than-typical amplitude (10–20 mV) was necessary to ensure a sufficient signal-to-noise ratio, as the electrodes, especially after electrochemical treatment, showed resistances of several tens of kOhm). The EIS measurements were performed within the 100 kHz–0.25 Hz frequency range. R_{ct} was determined as an intercept of a semicircle range with the Z′ axis in the Nyquist plot. Cyclic voltammograms (CVs) were measured with a scan rate of 100 mV s^{-1}. Potential windows are specified in the manuscript text. Areal capacitance was calculated from the CVs using the following equation:

$$C = \frac{\int idV}{2 \cdot \Delta E \cdot A \cdot W} \quad (1)$$

where $\int idV$ is the integral over CV, ΔE is a potential window, A is the geometrical area of the electrode, and W is a scan rate.

2.4. Electrolyte-Gated Transistor (EGT) Demonstration

The EGT performance was demonstrated by employing a Zurich MLFI Lock-in Amplifier (Zurich Instruments, Zurich, Switzerland). The EGT channel dimensions were approximately 2×2 mm^2. Drain and source contacts were made with copper tape and conductive silver paste and encapsulated with polydimethylsiloxane (PDMS). The applied drain-source voltage (V_{DS}) was −0.5 V, and the gate-source voltage (V_G) sweep rate was 25 mV s^{-1}. Transfer characteristics were recorded in the V_G range spanning from −0.6 to +0.6 V. Gate voltage equals +0.5 V was used for sensing KCl with different concentrations. This value provides the maximum sensitivity and evades the electrolysis and reduction processes. Pt wire was used as a gate electrode.

EGT operation was performed following the steps: First, a 20 μL deionized (DI) water drop was deposited on the EGT channel, and the I_D current was recorded for 1 min. The liquid was then removed with absorbing paper, and a 20 μL drop of 50 mM KCl was added in the same way. The procedure was repeated for 100 mM and 200 mM KCl solutions.

2.5. Scanning Electron Microscopy

The scanning electron microscopy (SEM) was performed using the Tescan Mira 3LMU (Tescan, Brno, Czech Republic) and COXEM EM-30+ (COXEM, Daejeon, Republic of Korea). All the SEM images shown in the manuscript were acquired by detecting the secondary electrons. The top-view SEM images were acquired with $10^3\times$ magnification, while cross-section images for rGO/PET and rGO/PVDF were acquired with 200× and 190× magnification, respectively.

2.6. X-Ray Photoelectron Spectroscopy

The X-ray photoelectron spectroscopy (XPS) was performed using a Thermo Fisher Scientific XPS NEXSA (Thermo Fisher Scientific, Waltham, MA, USA) spectrometer with a monochromatic Al K Alpha X-ray source at 1486.6 eV. The survey study was carried out employing a pass energy of 200 eV and an energy resolution of 1 eV. For the high-resolution spectra, the pass energy was 50 eV, and the energy resolution was 0.1 eV with a spot diameter of 400 μm.

2.7. Raman Spectroscopy

Raman spectra were recorded using a confocal Raman microscope (NTEGRA Spectra, NT-MDT, Zelenograd, Russia) with 532 nm (for rGO/PVDF) and 633 nm (for rGO/PET) diode excitation lasers. The Mitutoyo objective with 20× magnification was used to focus a laser beam on the sample surface and record spectra with 1 s exposure time and 10 times accumulation.

2.8. Fourier-Transform Infrared Spectroscopy

FTIR was measured on the IRAffinity-1S (Shimadzu, Kyoto, Japan) spectrometer in the total internal reflection mode.

3. Results

3.1. Electrode Fabrication and Characterization

rGO/PET composite electrodes were fabricated by laser processing drop-casted GO films on a PET substrate (Figure 1a). The fabrication protocol and the mechanism behind the composite formation are presented in detail elsewhere [9,15]. Briefly, GO acts as a photothermal transducer during laser irradiation, partially converting light into thermal energy. This energy reduces GO into rGO and partially becomes transferred to the polymer substrate underneath, reducing its viscosity and resulting in rGO intermixing with the liquified polymer due to the Benard–Marangoni convection. This process generates a conductive carbon network encapsulated by the polymer surface.

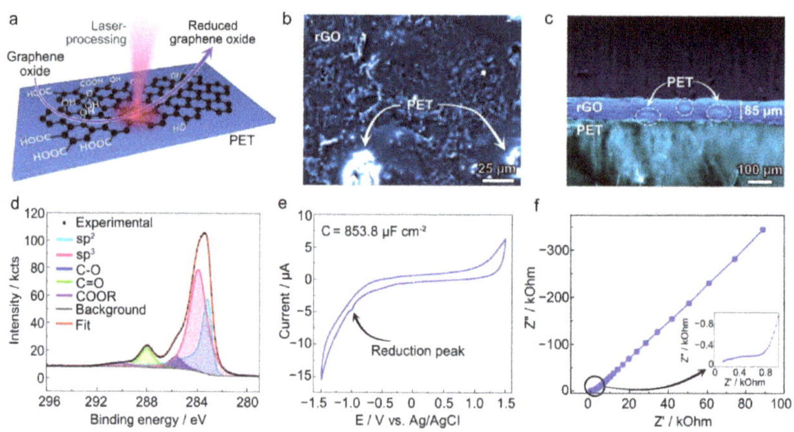

Figure 1. (**a**) Schematic illustration of the electrode fabrication procedure. (**b**) SEM top-view image of rGO/PET composite electrode. (**c**) Cross-sectional SEM image of rGO/PET composite electrode. (**d**) C1s high-resolution XPS spectrum from rGO/PET composite electrode. (**e**,**f**) CV and Nyquist plots, respectively, measured using the rGO/PET composite electrode. Inset represents the zoom into the high-frequency region where semicircle charge transfer resistance is observed.

SEM imaging reveals two different regions after laser processing, in agreement with our previous work [9]: (i) a rough, porous region with low static charging attributed to conductive rGO and (ii) a brighter region with strong static charging attributed to PET islands (Figure 1b). This surface structure has been confirmed by optical microscopy, which also shows the encapsulated black carbon material visible through the transparent PET layer (Figure S2). Cross-sectional SEM imaging also reveals an rGO/PET composite film with a thickness of ca. 80 μm lying on the PET substrate, with some PET islands visible as bright spots within the composite layer (Figure 1c).

The survey XPS spectrum of the rGO/PET electrode (Figure S3) shows the characteristic C1s and O1s bands, with a small amount of sulfur—a residual byproduct from the GO synthesis. High-resolution C1s spectra (Figure 1d) were deconvoluted into peaks related to sp^2 carbon (283.2 eV), sp^3 carbon (283.9 eV), C-O (285.7 eV), C=O (288.0 eV), and -COOR (290.3 eV). Below, we discuss the relative content of these elements and functional groups.

The electrochemical properties of the electrodes have been characterized by cyclic voltammetry and electrochemical impedance spectroscopy. CV (Figure 1e) exhibits nearly pure capacitive behavior with a small reduction peak at approximately −1 V (vs. Ag/AgCl), likely due to the reduction of aldehyde moieties [13] and signs of water electrolysis. Areal capacitance (C_A) has been set as the figure of merit to compare the induced changes. The as-prepared electrode exhibited a C_A of 1.15 ± 0.74 mF cm^{-2}, slightly lower than the state of the art for rGO [16], possibly due to the partial coverage of rGO active sites by dielectric PET islands. The R_{ct} and impedance at 0.25 Hz ($|Z|_{0.25}$) extracted from the Nyquist plot have been used as figures of merit for EIS. R_{ct} calculated from the Nyquist plot (Figure 1f) was several kΩ, followed by a linear region characteristic of diffusion through a semi-infinite diffusion layer, resulting in high impedance values at low frequencies (Figure 1f).

3.2. *Unexpected Findings on the Electrochemical Behavior of rGO/PET Electrodes*

To explore the electrochemical properties of the rGO/PET electrodes, we carried out reduction and oxidation cycles on the sample using a three-electrode cell. We applied high constant voltages of different polarities (−4 V for reduction and +3 V for oxidation) for 200 s each (Figure 2a) and measured CVs and the impedance after each treatment step. We performed eight full treatment cycles, each including one reduction and one oxidation step. The first two cycles, referred to as "activation" cycles, exhibited distinctive behavior in CVs, current versus time curves (i–t), and Nyquist plots (Figure 2b). Subsequent cycles are denoted as "switching" due to the changes observed in $|Z|$ at 0.25 Hz, R_{ct}, and areal capacitance (Figure 2c–e). The values obtained from these experiments are summarized in Table S1.

3.2.1. Activation Cycles

During the first redox cycles, we observed a significant drop in both $|Z|_{0.25}$ and R_{ct} from kΩ (initially in the range of several kΩ to dozens of kΩ) to several hundred Ω upon applying both positive and negative potentials (Figure 2b). At the second redox cycle, both $|Z|_{0.25}$ and R_{ct} start to increase slightly. Interestingly, the sample areal capacitance depends on the polarity of the applied potential, rising by a factor of ~6 within the first two reduction cycles and further decreasing by one order of magnitude after the oxidation step. Importantly, these changes occurred only if the reduction step was performed first. Otherwise, starting with the oxidation step resulted in electrode degradation, as shown by the CV in Figure S4, limiting the possibility of obtaining meaningful data.

First red: CVs from the initial samples exhibit mostly capacitive behavior, with a weak reduction peak found at ca. −1 V, a signature of the electrochemical reduction of residual carbonyl and carboxyl groups remaining after laser modification. Applying the reduction potential (−4 V) results in a typical hydrogen evolution reaction i–t curve, with the current decaying over time (Figure S5, 1st red). After the first reduction step, the electrodes still exhibit pure capacitive behavior, but with no evident peaks in the CVs and with an increase in capacitance contribution (Figure S6, 1st red; Figure 2e).

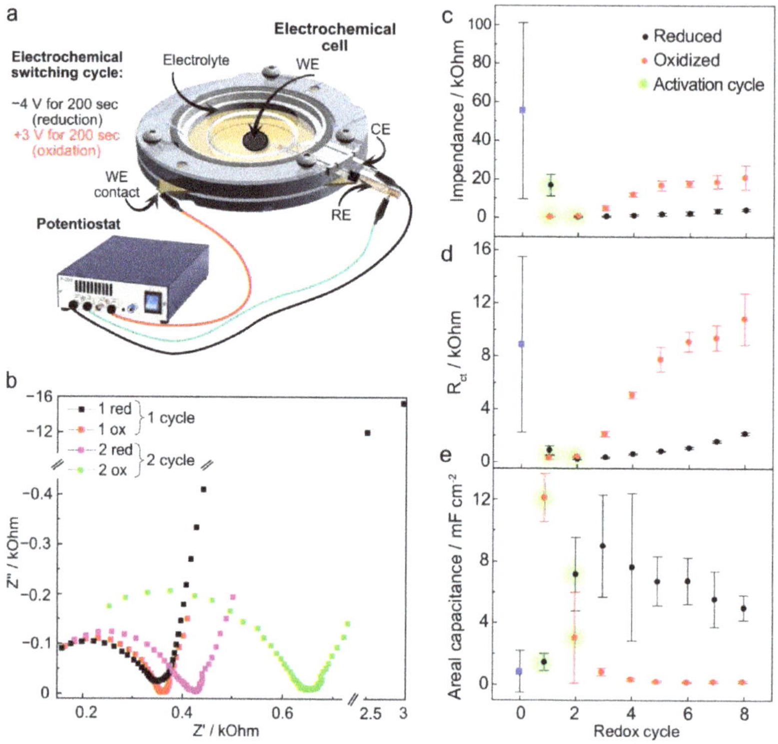

Figure 2. (**a**) Schematic illustration of the experimental setup representing the electrochemical cell and a short description of a sequential electrochemical treatment. (**b**) Nyquist plots from rGO/PET samples after the first two redox cycles, indicating the observed changes in impedance. (**c–e**) rGO/PET sample figures of merit were obtained from CVs and impedance plots through all eight investigated redox cycles: impedance value at 0.25 Hz, charge transfer resistance, and areal capacitance, respectively. The values at the "0" point represent the values for the initial samples, which are also marked with blue squares. The error bars show the STD of the three electrodes.

First ox: The oxidation step with +3 V shows signs of a concomitant oxygen evolution reaction, as indicated by the slight current increase during the first 50 s followed by a decay up to 200 s (Figure S5, 1st ox). The CV recorded after the first oxidation step shows a further increase in the capacitive contribution and the appearance of symmetrical oxidation and reduction peaks at ±0.5 V.

Second red: The i–t curve during the second reduction exhibited behavior similar to that of the 1st red, which is related to the water-splitting process. Even though the underlying processes are similar, they occur differently since the 2nd red i–t curve shows a steep current decay, contrary to the 1st red, associated with an enhanced diffusion of water molecules toward the electroactive rGO network. However, CV recorded after the second reduction reveals the same oxidation and reduction peak profile at the same voltages as observed after the first oxidation step.

Second ox: The second oxidation process differs from previous modifications and shows two distinguishable current peaks in the i–t curves during the first 100 s, indicative of electrochemical reactions. These reactions shift the equilibrium cathodic potential toward zero, as evidenced by the intercept of the backward and forward scan curves in the CV at ~−0.8 V, as well as dynamic changes in CVs from cycle to cycle (Figure S7). To ensure that CV measurements did not significantly affect the changes introduced by the oxidation treatment, we narrowed down the potential range after oxidation from ±1.5 V to ±0.5 V. This range was selected because most available literature, including ours, indicate no redox processes within this range [17]. The CV measuring range for reduction cycles was not changed since keeping the same range did not affect the figures of merit. After narrowing the potential range, we observed a strong reduction peak on the i–t curve during the fourth reduction step, signaling a strong ongoing electrochemical reduction, most likely coming from processes in rGO (Figure S5). This peak confirms the necessity of narrowing the CV potential window after oxidative treatment.

We performed a set of in-depth characterization to understand the structural changes during the "activation" period. Since we did not observe reaction signatures in the i–t curves, we suspected morphological changes, such as electrochemical roughening, could explain the electrode behavior. SEM images show that electrochemical treatment significantly decreases the number of PET fragments, as indicated by the contrast between treated and untreated regions of the sample, while the initial sample had PET islands homogeneously distributed all over the surface (Figure S8). This result is a sign of electrochemical destruction of the PET. We calculated the ratio between sp^2 and sp^3 hybridized carbon atoms from XPS C1s spectra and compared it to the untreated sample with a sp^2/sp^3 ratio of 0.88 ± 0.71.

XPS spectra from the sample after the first reduction step show the typical rGO peaks. The C1s high-resolution region (Figure S9) was deconvoluted into bands corresponding to sp^2 and sp^3 hybridized carbon (283.4 and 284.3 eV, respectively), C-O (285.7 eV), C=O (288.3 eV), -COOR (290 eV), and the π-π* satellite (292 eV). The calculated sp^2/sp^3 ratio is 1.67 ± 0. 77, almost double that of the initial sample (Figure 1d). This increase correlated with an increase in the C/O ratio from 4.42 ± 0.77 (initial) to 5.09 ± 0.31 (1st red). The survey spectrum also revealed the presence of sodium in the sample, likely originating from sodium cations appearing on the surface of PET due to cleavage of ester bond and formation of -COONa groups (PBS electrolyte contains NaCl). The following first oxidation step shows ion trapping from the electrolyte, evidenced by the appearance of phosphorus (0.66 ± 0.08 atomic %) and chlorine (1.00 ± 0.02 atomic %) in the survey spectra. Additionally, the sodium content increased slightly due to entrapment of sodium phosphate and chloride. The most remarkable result from this oxidation step is that the sp^2/sp^3 carbon ratio remained unchanged, while the C/O ratio drastically decreased to 3.38 ± 0.07. This is surprising because an increase in oxygen content is typically related to the change in carbon hybridization from sp^2 to sp^3.

The second reduction and oxidation cycles did not change the sample elemental composition but induced changes in their states. The second reduction significantly increases the sp^2/sp^3 ratio to 5.85 ± 0.71 (by a factor of 3.3) while only slightly affecting the C/O ratio (increased from 3.38 ± 0.07 to 3.54 ± 0.20). This demonstrates that the chemical reactions occurred within the electrode's structure. Other elements' content from the electrolyte (Cl^-, $PO4^-$, Na^+) changed according to the polarity of the applied potential, as summarized in Table S2. The second oxidation decreased the sp^2/sp^3 ratio to 1.45 ± 0.02, while the C/O (3.57 ± 0.05) remained unchanged.

These results are surprising because changes in the sp^2/sp^3 atomic ratio in these systems are usually explained by the formation or cleavage of chemical bonds (reflected by a decrease or increase in sp^2/sp^3, respectively) between carbon and oxygen-containing groups. This demonstrates that the underlying mechanism differs from the typical reduction and oxidation of GO reported in the literature.

3.2.2. Switching Cycles

Further electrochemical treatment cycles result in reversible "switching" of the electrochemical properties. Applying the reduction potential led to $|Z|_{0.25}$ values of ~3 kΩ. In contrast, oxidation potentials caused a constant increase in $|Z|_{0.25}$ up to the 5th cycle, after which the impedance stabilized around 20 kΩ for all following oxidation treatments (Figure 2c). A similar trend is observed for R_{ct}, with the reduced state stabilizing around 1 kΩ and the oxidized state stabilizing at the 5th cycle with values close to 10 kΩ (Figure 2d). For the areal capacitance, the oxidized state values were 0.5 mF cm^{-2}. In contrast, the reduced state exhibited a one-order-of-magnitude increase (Figure 2e).

Analysis of the impedance curves revealed that, after the first two activation cycles, applying the reduction potential leads to the appearance of a second semicircle in the 3rd red Nyquist plot (Figure S10a) with a maximum (~15°) in the phase angle plot located around 1 kHz. The phase angle at frequencies below 1 Hz does not exceed 15° (Figure S10b). The second semicircle (representing an imperfect double-layer capacitor) at the mentioned position with a low phase angle value could be a signature of rGO, as previously observed for photoreduced GO [14]. After oxidation treatment, the second semicircle shifts from 1 kHz toward lower frequencies, with the phase angle rising up to 30°, indicating oxidation of the reduced graphene oxide (Figure S10b).

To determine the mechanism behind this "switching" of electrochemical properties, we characterized the samples up to the 5th oxidation since the key values for all states stabilized at this cycle. We first recorded Raman spectra (Figure S11) showing that the exposed regions (not covered with PET) were characterized by D and G peaks, defect-activated band around 1350 cm^{-1}, and sp^2 carbon band at 1590 cm^{-1}, respectively. The intensity ratio between these bands (I_D/I_G) has often been used to estimate the defect concentration in graphenes. For GO and rGO, this ratio is related to the degree of reduction since oxygen-containing groups are structural defects [18].

The initial rGO/PET electrode shows an uneven distribution of Raman spectra at different locations, with some regions showing high-quality rGO with I_D/I_G much lower than 1, while others had higher I_D/I_G ratios due to highly defective rGO. After the first reduction, the I_D/I_G was calculated to be 0.77, which dropped to 0.57 after the first oxidation, following the trend observed in the electrochemical properties where impedance decreased with both the first reduction and oxidation. Surprisingly, the second reduction resulted in a significant increase in I_D/I_G to 1.82, while further electrochemical treatment resulted in only minor fluctuations in I_D/I_G within the error margin. It is important to note that I_D/I_G can be insensitive to changes in chemical composition for highly defective rGO structures [19,20], which might be the case here since electrochemical treatment could result in a high number of defects. Therefore, we needed more than the Raman spectra to fully elucidate the underlying switching mechanism.

Considering the relatively high potentials and the nature of the PET substrate, we hypothesized that the unusual electrochemical behavior accompanied by changes in the sp^2/sp^3 ratio but not the C/O ratio could be attributed to PET's electrochemical activity; the particular mechanism is discussed below.

We replaced PET with the electrochemically inert polymer polyvinylidene fluoride (PVDF) to elucidate the individual contributions of rGO and PET to this unusual electrochemical behavior. This comparison was possible due to the similar mechanism behind the laser-induced rGO composite formation with PET and PVDF [15].

3.3. Electrochemical Treatment of rGO/PVDF Electrodes

To determine whether the observed electrochemical switching originates from rGO alone or involves some kind of interaction between the rGO surface and specific functional groups in polymer chains, we fabricated rGO/PVDF electrodes using the same laser-induced reduction method, as well as sequential electrochemical reduction and oxidation. Before analyzing the electrochemical properties, we investigated the morphology and chemical composition of the rGO/PVDF composite.

SEM analysis shows a distinct morphology for rGO/PVDF compared to rGO/PET. The rGO/PVDF composite is represented by rGO flakes without noticeable polymer presence on the electrode surface (Figure 3b). Cross-sectional SEM images (Figure 3c) reveal two clearly separated regions: a PVDF layer and a highly porous rGO layer with a thickness of ca. 81 μm on top. This implies that the electrochemical behavior of rGO/PVDF (Figure 3d,e) is dominated by the exposed rGO since there are no visible signs of rGO encapsulation by PVDF, and PVDF is an electrochemically inert material.

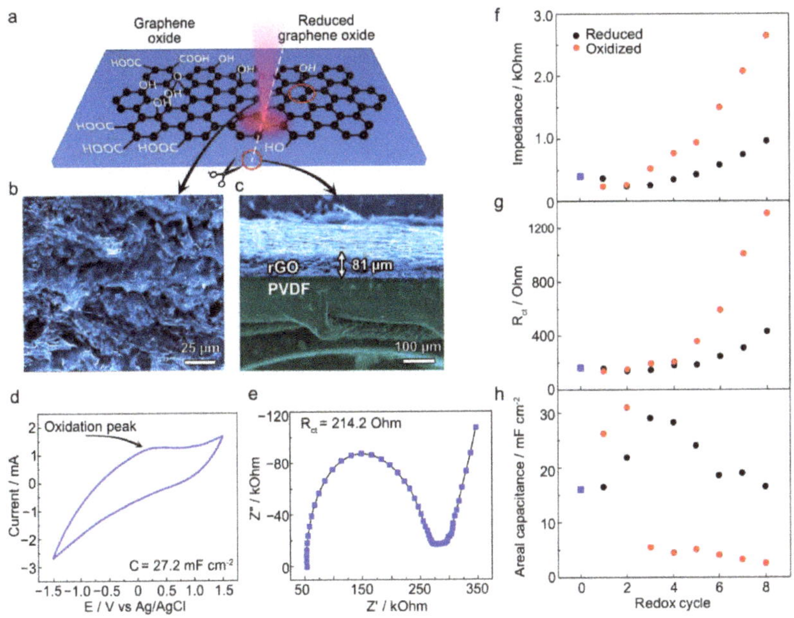

Figure 3. (**a**) Schematic illustration of the rGO/PVDF preparation via laser treatment with the sample (**b**) top-view and (**c**) cross-section obtained by SEM. (**d**,**e**) CV and Nyquist plot for the initial untreated sample, respectively. (**f**–**h**) Plots of calculated sample figures of merit were obtained from CVs and impedance plots through all the investigated 8 redox cycles: impedance values at 0.25 Hz, charge transfer resistance, and areal capacitance, respectively. The values at the "0" point represent the values for the initial samples, which are also marked with blue squares.

We performed an XPS analysis to further investigate the presence of polymer on the electrode's surface. In contrast to rGO/PET, fluorine in PVDF allows for the differentiation between rGO and PVDF in XPS data easily. The survey XPS spectra reveal F1s, C1s, and O1s peaks at 687.5, 285.1, and 532.9 eV (Figure S12). High-resolution C1s spectra were deconvoluted into seven peaks attributed to sp^2 and sp^3 hybridized carbon, C-O, C=O, -COOR, CF_2, and CF_3 at 284.3, 285.2, 286.9, 287.5, 288.5, 290.4, and 292.8 eV (Figure S12). The presence of a peak corresponding to fluorine in the XPS spectra could be explained through the sorption of low molecular weight PVDF destruction products formed as a result of intense heating of the PVDF substrate during the laser reduction of GO [21].

These differences in surface structure resulted in a significantly lower impedance (ca. 300 Ω) of rGO/PVDF compared to the rGO/PET composite. This is attributed to a smaller Warburg impedance related to a higher ion diffusion on the exposed rGO flakes in the PVDF composite (Figure 3d). The R_{ct} is also approximately one order of magnitude lower. Furthermore, the areal capacitance exhibits values one order of magnitude higher than that of rGO/PET, consistent with the greater exposure of the electroactive rGO network to the electrolyte.

The first two redox cycles did not significantly change the electrochemical parameters set as the figures of merit (Figure 3f–h). However, further treatment shows a similar trend to rGO/PET, although much less pronounced. For instance, impedance varied in rGO/PVDF from ~1 kΩ to ~3 kΩ, compared to switching from ~3 kΩ to ~20 kΩ for rGO/PET. Contrary to rGO/PET showing stable reduced and oxidized states after five cycles, the parameters extracted from EIS for rGO/PVDF diverged with each cycle without reaching stable values.

The initial CV of rGO/PVDF shows a prominent oxidation peak around 0–0.2 V (Figure 3d). The phase angle plot exhibits a well-pronounced extremum (-32°) in the high-frequency region related to charge transfer resistance and a rise in phase up to -23° at low frequencies, indicating residual oxygen functionalities in the graphene structure remaining after laser processing [14]. The first reduction did not cause significant changes in the CV or impedance plots. However, the first oxidation increased the areal capacitance and changed the CV shape to a predominantly capacitive behavior with slight reduction and oxidation peaks between $\pm$0.5–1 V (Figure S13). The impedance decreased by 100 Ω with the phase at a low frequency not exceeding -10°, possibly due to the electrochemical roughening of the electrode's surface, which increased the exposure of highly reduced conductive active sites. The second redox cycle does not cause any changes in the impedance plots or CVs.

We narrowed the potential window from the third cycle for measuring CV as performed for rGO/PET. We observed clear signs of rGO reduction and oxidation, with low-frequency phase values reaching -30° after oxidation and -12° after reduction. All CVs, i–t curves, and impedance data are presented in Figures S13–S15.

SEM images from samples subjected to several redox cycles revealed increased surface roughness (Figure S16), which might be responsible for the increased capacitance. Although visually treated and untreated regions might not be well distinguished, the extracted surface profiles confirm the significant difference in roughness introduced by electrochemical treatment. We performed XPS analyses following the same procedure as for rGO/PET to investigate the chemical changes. We calculated the C/O ratio as the figure of merit (Table S3), surprisingly showing that the initial sample (ca. 15.2 $\pm$ 0.5) had a C/O ratio several times higher than that for rGO/PET and most reports in the literature for rGO obtained using similar laser wavelengths [22]. This demonstrates that the C/O is not a reliable indicator of rGO structure when laser-reduced on a PET substrate, as the oxygen of PET itself can significantly influence it.

In rGO/PVDF, the first reduction slightly decreased the C/O ratio to 9.0 $\pm$ 0.1 (Table S3), possibly due to removing poorly integrated but highly reduced rGO flakes from the electrode surface. The first oxidation further decreases the C/O ratio to 5.7 $\pm$ 0.3. Further electrochemical treatment resulted in the expected behavior: reduction potentials increase the C/O ratio, while oxidation decreases it, consistent with the electrochemical reduction and oxidation of rGO absent of underlying PVDF substrate.

3.4. Mechanism Behind Electrochemical Properties Switching

We propose a mechanism for switching electrochemical properties based on the observed differences between rGO/PET and rGO/PVDF. The rGO/PVDF served as a control system to isolate the contribution of rGO. Over eight redox cycles, rGO/PVDF could be reversibly reduced and oxidized, consistent with previous results on GO obtained using the modified Hummers method [23,24]. This was reflected in the impedance plots, particularly in the phase angle behavior and the CVs, where weak, symmetrical reduction, and oxidation peaks replaced the initial oxidation peak. Although these redox reactions resulted in "switching", the effect was much less pronounced than in rGO/PET.

We hypothesize that the switching mechanism is related to PET hydrolytic decomposition. Cathodic decomposition of PET occurs through a two-electron reaction with peaks at −2.5 and −3 V [25]. The switching effect appeared only at potentials exceeding ±2.5 V, further supporting PET's participation in the electrochemical reactions since rGO undergoes redox reactions at lower potentials. Thus, the depolymerization begins during the first redox cycle (activation), and there is no switching effect. As the destruction of thin polymer layers progresses, more conductive rGO is exposed to the electrolyte, increasing the areal capacitance and decreasing impedance, mainly by reducing the diffusional component. Further cycles started showing the switching behavior due to rGO redox reactions of the freshly exposed flakes. However, we do not exclude the possibility of a cascade of various processes involving PET and rGO, such as hydrolysis, inter-material diffusion, and reactions of oxygen-containing groups.

The difference in behavior between rGO/PET and rGO/PVDF could be explained by the presence of polymer on the composite surface. In rGO/PET, erosion exposes a limited amount of the conductive rGO network, significantly contributing to the electrode/electrolyte interface. In contrast, the rGO/PVDF surface is dominated by rGO, where oxidation is a more statistically distributed process, affecting only some regions on the surface and leading to less pronounced changes at the electrode/electrolyte interface. This, along with the electrochemical inertness of PVDF and limited electrolyte diffusion through the rGO layer, contributes to the hindered switching behavior observed in the rGO/PVDF composite.

3.5. Application of Laser-Processed rGO/Polymer Composites in Electrolyte-Gated Transistors

rGO is widely used in chemical sensors and biosensors, where its electrochemical properties define the device's performance. However, the use of rGO/polymer composite is limited since the polymer matrix can significantly hinder the electrode/electrolyte interactions, preventing exposure to active sites. We exploit the electrochemical "activation" phenomenon discovered here to expose rGO in an rGO/PET composite for application as an EGT [26]. EGTs are formidable electronic transducers widely employed as chemical sensors, biosensors, and neuromorphic devices for bioelectronics [27]. They bear very interesting characteristics such as low voltage operation (<1 V), elevated sensitivity to surface perturbations, and signal amplification ability [28]. The EGT working principle relies on the modulation of the charge carriers at the transistor channel via the formation of electrical double layers at the device interfaces (field-effect) or ion intake (electrochemical mode), which are controlled by the gate voltage (V_G) [28–31].

To demonstrate the impact of electrochemical treatment on rGO/polymer composites, we compared the performance of initial rGO/PET and electrochemically-treated (1st red, 1st ox, 3rd red, and 3rd ox cycles) electrodes as a channel material in EGTs. Our strategy enabled the fabrication of millimeter-scale EGTs with rGO channels and adjustable electronic properties. The best performance was obtained for the 1st red sample, which we discuss further below.

The EGT schematic representation is shown in Figure 4a. We first recorded the device transfer characteristics, i.e., the drain current (I_D) vs. V_G in deionized water (DI) and 50 mM KCl solution. The initial rGO/PET sample showed no current modulation under ±0.6 V V_G bias voltage either in DI or KCl (Figure S17a). This result is attributed to the polymer shielding the rGO surface, suppressing the formation of an electrical double-layer or ion intake within the sample. In contrast, the 1st red sample showed significant I_D modulation, especially at positive V_G. For example, at a positive bias of V_G = 0.5 V, a 40 μA difference in I_D is observed between operation in DI water and in 50 mM KCl solution (Figure 4b). This current modulation shows that PET erosion during the 1st red step activates the surface, allowing the interaction between electrolyte ions and the rGO channel.

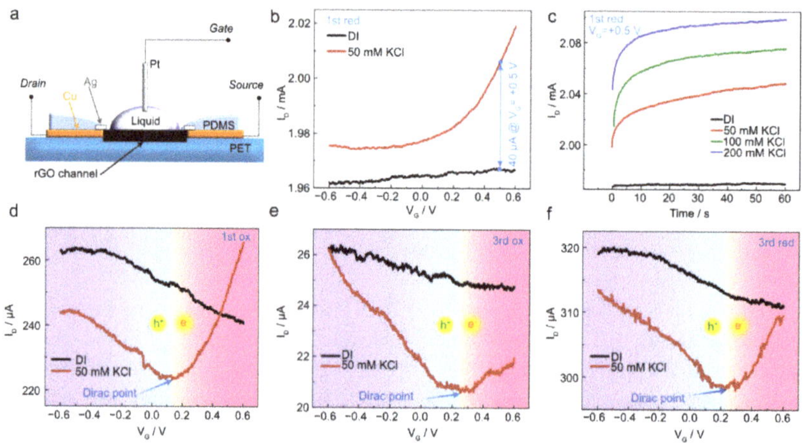

Figure 4. (**a**) Schematic illustration of the setup for transistor characteristics measurements. (**b**) Transfer characteristics in DI water and 50 mM KCl liquid gate for 1st red sample at V_{DS} = −0.5 V. (**c**) Device response at V_G = +0.5 V and V_{DS} = −0.5 V corresponds to different KCl molarity for 1st red samples. (**d–f**) Transfer characteristics in DI water and 50 mM KCl for 1st ox, 3rd ox, and 3rd red samples, respectively.

To further demonstrate the robustness of EGTs produced from the electrochemically activated composite, we measured I_D over time at different KCl concentrations. The initial rGO/PET sample showed no sensitivity to the electrolyte composition (Figure S16b). The 1st red sample showed a reasonably stable response over time at different KCl concentrations (50, 100, and 200 mM) as depicted in Figure 4c. Increasing the electrolyte concentration increases the electrical double-layer capacitance and, therefore, I_D. A stable current response is a key characteristic in sensing, especially for applications targeting real-time monitoring of chemical species. Transfer characteristics of the 1st ox, 3rd red, and 3rd ox samples are presented in Figure 4d–f. We observe a clear charge neutrality point near +0.2 V and the typical rGO ambipolar electrical conduction here, illustrated by the p- and n-type branches of the transfer curve. From the device transfer characteristics, we calculated the hole (μ_h) and electron mobility (μ_e) in EGTs based on rGO/PET composites (Note S1). We found μ_h to amount to 0, $5.9 \cdot 10^{-3}$, $5.6 \cdot 10^{-3}$, and $1.6 \cdot 10^{-2}$ cm^2 V^{-1} s^{-1}, and μ_e was found equal to 0.15, $1.6 \cdot 10^{-2}$, $0.9 \cdot 10^{-2}$, and $0.8 \cdot 10^{-2}$ cm^2 V^{-1} s^{-1} for the 1st red, 1st ox, 3rd red, and 3rd ox, respectively. Details on the calculation of the charge carrier mobilities are given in the Supporting Information. The μ_e for the 1st red sample showed state-of-the-art values for rGO [32], although p-type conduction could not be obtained, possibly due to hole trapping at the interface. Our method demonstrates the potential of electrochemically-activated laser-reduced GO/polymer composites to be used in the fabrication of EGT with potential for sensing applications.

4. Conclusions

In this work, we investigated the electrochemical behavior of laser-induced rGO/PET composite electrodes under harsh electrochemical conditions, using potentials exceeding the water window. We found that this treatment gives rise to unexpected changes in the composite electrode's structure, enabling the tuning of electrochemical properties, including impedance, charge transfer resistance, and areal capacitance. This effect originates from the unique composite structure, with a high surface coverage of PET, due to the Marangoni effect during laser processing, which initially hinders the availability of rGO active sites. The electrochemical treatment involves two processes. First is an activation period where PET erosion and depolymerization expose rGO to the electrolyte. Second, a switching period in which freshly exposed rGO undergoes reversible redox reactions, as deduced from the behavior of rGO/PVDF composite electrodes, where rGO dominates the electroactive surface. The difference in switching efficiency between rGO/PET and rGO/PVDF is attributed to the electrochemical activity of PET and its role in interrupting the conductive rGO network, enhancing the impact of electrochemical treatment on the overall electrode properties.

This electrochemical treatment method, coupled with laser-induced GO reduction and free-form fabrication, provides a new approach to engineering the electrochemical properties of rGO/polymer composites. We illustrated the ability to control ion transfer through PET erosion, opening possibilities for applications such as electrolyte-gated transistors for ion sensing.

Supplementary Materials: The following supporting information can be downloaded at https://www.mdpi.com/article/10.3390/polym17020192/s1, Figure S1. Cross-section optical microscopy image of the GO/PET (sample before laser treatment). Figure S2: Optical microscopy image of pristine rGO/PET electrode; Figure S3. XPS survey spectrum from the rGO/PET pristine sample; Table S1: Calculated values from CVs and Impedance plots after electrochemical treatment; Figure S4: Typical CV after the first reduction in a following sequence 1st ox-1st red (treatment was started with the oxidation step); Figure S5: i–t curves from the rGO/PET electrode during each of the first 5 redox cycles; Figure S6: CVs from rGO/PET electrode after each of the first 5 redox cycles; Figure S7: CV after the third oxidation cycle without changing the potential window; Figure S8: SEM images of the initial sample and sample after the 2nd oxidation. The red mask was created as a grain analysis to illustrate PET islands. Dashed squares show regions from which the profiles were extracted, while the white circle represents the electrochemically treated region; Figure S9: Survey and C1s XPS high-resolution spectra from rGO/PET electrode after two redox cycles. The tables show values of relative atomic content averaged from 3 different points per sample; Table S2: Atomic percent of elements from XPS Survey spectra for rGO/PET samples after first two redox cycles; Figure S10: (a) Nyquist plots and (b) phase angle plots from the rGO/PET electrode after each of the first 5 redox cycles; Figure S11: Raman spectra from the rGO/PET electrodes after each of 5 electrochemically treatment cycles; Table S3: C/O ratio extracted from XPS survey spectra from different treatment cycles of rGO/PVDF; Figure S12: Survey and C1s XPS high-resolution spectra from rGO/PVDF electrode after several redox cycles; Figure S13: CVs from rGO/PVDF electrode after each of the first 5 redox cycles; Figure S14: i–t curves from rGO/PVDF electrode after each of the first 5 redox cycles; Figure S15: (a) Nyquist plots and (b) phase angle plots from the rGO/PVDF electrode after each of the first 5 redox cycles; Figure S16: SEM of rGO/PET electrode after several redox cycles; Figure S17: (a) Transfer characteristic of pristine rGO/PET in DI and 50 mM KCl. (b) Respective current-time at V_G = +0.5 V and V_{DS} = −0.5 V for varied KCl concentrations; Note S1: Charge carriers' mobility for all the samples was calculated according to the Formula (S1), $\mu = \frac{\Delta I_D}{\Delta I_G} \cdot \frac{L}{W \cdot V_{DS} \cdot C}$ (S1), where μ—charge carriers mobility, $\frac{\Delta I_D}{\Delta I_G}$—slope of the linear region of transfer characteristic, L, W—the length and the width of a channel, V_{DS}—source-drain voltage, and C—specific capacitance.

Author Contributions: M.F.: Conceptualization, Methodology, Formal analysis, Investigation, Data Curation, Writing—Original Draft, Writing—Review and Editing. I.P.: Conceptualization, Methodology, Formal analysis, Investigation, Data Curation, Writing—Original Draft, Writing—Review and Editing, Visualization. E.D.: Formal analysis, Investigation, Data Curation, Writing—Review and Editing, Visualization. D.K.: Investigation, Data Curation, Writing—Review and Editing. A.V.: Methodology, Investigation. P.P.: Conceptualization, Formal analysis, Writing—Review and Editing. J.-J.C.: Formal analysis, Writing—Review and Editing. R.F.d.O.: Formal analysis, Writing—Review and Editing. O.K.: Formal analysis, Writing—Review and Editing. R.D.R.: Formal analysis, Writing—Review and Editing, Visualization. E.S.: Conceptualization, Methodology, Formal analysis, Writing—Review and Editing, Supervision, Resources, Project administration. All authors have read and agreed to the published version of the manuscript.

Funding: This research was funded by the Russian Science Foundation grant No. 22-12-20027 (https://rscf.ru/project/22-12-20027/, accessed on 1 November 2024) and funding from the Tomsk region administration. Electronic components based on laser integration for the biocompatible/biodegradable flexible electronic devices.

Institutional Review Board Statement: This study did not require ethical approval.

Data Availability Statement: The original contributions presented in the study are included in the article/Supplementary Material, further inquiries can be directed to the corresponding author.

Acknowledgments: The authors are grateful to the center of TPU's "Physical and chemical methods of analysis" for XPS measurements. Jin-Ju Chen thanks the Chengdu Science and Technology Program (grant No. 2023-GH02-00021-HZ). R.F.O. acknowledges support from FAPESP/Brazil (2021/06238-5) and INCT/INEO/Brazil.

Conflicts of Interest: The authors declare no conflict of interest.

References

1. Dasari, B.L.; Nouri, J.M.; Brabazon, D.; Naher, S. Graphene and Derivatives—Synthesis Techniques, Properties and Their Energy Applications. *Energy* **2017**, *140*, 766–778. [CrossRef]
2. Seekaew, Y.; Arayawut, O.; Timsorn, K.; Wongchoosuk, C. Synthesis, Characterization, and Applications of Graphene and Derivatives. In *Carbon-Based Nanofillers and Their Rubber Nanocomposites*; Elsevier: Amsterdam, The Netherlands, 2019; pp. 259–283, ISBN 9780128132487.
3. Dreyer, D.R.; Park, S.; Bielawski, C.W.; Ruoff, R.S. The Chemistry of Graphene Oxide. *Chem. Soc. Rev.* **2010**, *39*, 228–240. [CrossRef] [PubMed]
4. Pei, S.; Cheng, H.-M. The Reduction of Graphene Oxide. *Carbon* **2012**, *50*, 3210–3228. [CrossRef]
5. Ahmed, A.; Singh, A.; Young, S.-J.; Gupta, V.; Singh, M.; Arya, S. Synthesis Techniques and Advances in Sensing Applications of Reduced Graphene Oxide (rGO) Composites: A Review. *Compos. Part A Appl. Sci. Manuf.* **2023**, *165*, 107373. [CrossRef]
6. Razaq, A.; Bibi, F.; Zheng, X.; Papadakis, R.; Jafri, S.H.M.; Li, H. Review on Graphene-, Graphene Oxide-, Reduced Graphene Oxide-Based Flexible Composites: From Fabrication to Applications. *Materials* **2022**, *15*, 1012. [CrossRef]
7. Hooch Antink, W.; Choi, Y.; Seong, K.-D.; Kim, J.M.; Piao, Y. Recent Progress in Porous Graphene and Reduced Graphene Oxide-Based Nanomaterials for Electrochemical Energy Storage Devices. *Adv. Mater. Interfaces* **2018**, *5*, 1701212. [CrossRef]
8. Radich, E.J.; Kamat, P.V. Origin of Reduced Graphene Oxide Enhancements in Electrochemical Energy Storage. *ACS Catal.* **2012**, *2*, 807–816. [CrossRef]
9. Fatkullin, M.; Menzelintsev, V.; Lipovka, A.; Dogadina, E.; Plotnikov, E.; Brazovskiy, K.; Li, S.; Ma, L.; Cheng, C.; Porokhova, E.; et al. Smart Graphene Textiles for Biopotential Monitoring: Laser-Tailored Electrochemical Property Enhancement. *ACS Sens.* **2024**, *9*, 1809–1819. [CrossRef]
10. Fatkullin, M.; Dogadina, E.; Bril', I.; Ivanov, A.; Matkovic, A.; Rodriguez, R.D.; Sheremet, E. Nanomaterials/polymer-Integrated Flexible Sensors: A Full-Laser-Processing Approach for Real-Time Analyte Monitoring. *IEEE Sens. J.* **2024**, *24*, 13816–13822. [CrossRef]

11. Mosa, I.M.; Pattammattel, A.; Kadimisetty, K.; Pande, P.; El-Kady, M.F.; Bishop, G.W.; Novak, M.; Kaner, R.B.; Basu, A.K.; Kumar, C.V.; et al. Ultrathin Graphene-Protein Supercapacitors for Miniaturized Bioelectronics. *Adv. Energy Mater.* **2017**, *7*, 1700358. [CrossRef]
12. Silva, V.A.O.P.; Fernandes, W.S., Jr.; Rocha, D.P.; Stefano, J.S.; Munoz, R.A.A.; Bonacin, J.A.; Janegitz, B.C. 3D-Printed Reduced Graphene Oxide/polylactic Acid Electrodes: A New Prototyped Platform for Sensing and Biosensing Applications. *Biosens. Bioelectron.* **2020**, *170*, 112684. [CrossRef] [PubMed]
13. Eng, A.Y.S.; Ambrosi, A.; Chua, C.K.; Saněk, F.; Sofer, Z.; Pumera, M. Unusual Inherent Electrochemistry of Graphene Oxides Prepared Using Permanganate Oxidants. *Chemistry* **2013**, *19*, 12673–12683. [CrossRef] [PubMed]
14. Casero, E.; Parra-Alfambra, A.M.; Petit-Domínguez, M.D.; Pariente, F.; Lorenzo, E.; Alonso, C. Differentiation between Graphene Oxide and Reduced Graphene by Electrochemical Impedance Spectroscopy (EIS). *Electrochem. Commun.* **2012**, *20*, 63–66. [CrossRef]
15. Abyzova, E.; Petrov, I.; Bril', I.; Cheshev, D.; Ivanov, A.; Khomenko, M.; Averkiev, A.; Fatkullin, M.; Kogolev, D.; Bolbasov, E.; et al. Universal Approach to Integrating Reduced Graphene Oxide into Polymer Electronics. *Polymers* **2023**, *15*, 4622. [CrossRef]
16. Fu, X.-Y.; Cai, Q.; Ma, J.-N.; Zhu, L.; Han, D.-D.; Zhang, Y.-L. Free-Standing and Flexible Graphene Supercapacitors of High Areal Capacitance Fabricated by Laser Holography Reduction of Graphene Oxide. *Appl. Phys. Lett.* **2021**, *118*, 071601. [CrossRef]
17. Zhou, A. 'an; Bai, J.; Hong, W.; Bai, H. Electrochemically Reduced Graphene Oxide: Preparation, Composites, and Applications. *Carbon* **2022**, *191*, 301–332. [CrossRef]
18. Feicht, P.; Eigler, S. Defects in Graphene Oxide as Structural Motifs. *ChemNanoMat* **2018**, *4*, 244–252. [CrossRef]
19. Arul, R.; Oosterbeek, R.N.; Robertson, J.; Xu, G.; Jin, J.; Simpson, M.C. The Mechanism of Direct Laser Writing of Graphene Features into Graphene Oxide Films Involves Photoreduction and Thermally Assisted Structural Rearrangement. *Carbon* **2016**, *99*, 423–431. [CrossRef]
20. Ma, B.; Rodriguez, R.D.; Ruban, A.; Pavlov, S.; Sheremet, E. The Correlation between Electrical Conductivity and Second-Order Raman Modes of Laser-Reduced Graphene Oxide. *Phys. Chem. Chem. Phys.* **2019**, *21*, 10125–10134. [CrossRef]
21. Lovinger, A.; Reed, D. Inhomogeneous Thermal Degradation of Poly(vinylidene Fluoride) Crystallized from the Melt. *Macromolecules* **1980**, *13*, 989–994. [CrossRef]
22. Murastov, G.V.; Lipovka, A.A.; Fatkullin, M.I.; Rodriguez, R.D.; Sheremet, E.S. Laser Reduction of Graphene Oxide: Local Control of Material Properties. *Phys. Usp.* **2023**, *66*, 1105–1133. [CrossRef]
23. Chen, J.; Yao, B.; Li, C.; Shi, G. An Improved Hummers Method for Eco-Friendly Synthesis of Graphene Oxide. *Carbon* **2013**, *64*, 225–229. [CrossRef]
24. Hummers, W.S., Jr.; Offeman, R.E. Preparation of Graphitic Oxide. *J. Am. Chem. Soc.* **1958**, *80*, 1339. [CrossRef]
25. Pud, A.A. Surface Electrochemical Reactions and the Subsequent Degradation of Solid-Phase Poly(ethylene Terephthalate) at a Cathode. *Polym. Degrad. Stab.* **1990**, *30*, 181–193. [CrossRef]
26. Vasilijević, S.; Boukraa, R.; Battaglini, N.; Piro, B. Graphene-Based Materials and Their Applications in Electrolyte-Gated Transistors for Sensing. *Synth. Met.* **2023**, *295*, 117355. [CrossRef]
27. Fan, Q.; Wang, L.; Xu, D.; Duo, Y.; Gao, J.; Zhang, L.; Wang, X.; Chen, X.; Li, J.; Zhang, H. Solution-Gated Transistors of Two-Dimensional Materials for Chemical and Biological Sensors: Status and Challenges. *Nanoscale* **2020**, *12*, 11364–11394. [CrossRef]
28. Furlan de Oliveira, R.; Montes-García, V.; Livio, P.A.; González-García, M.B.; Fanjul-Bolado, P.; Casalini, S.; Samorì, P. Selective Ion Sensing in Artificial Sweat Using Low-Cost Reduced Graphene Oxide Liquid-Gated Plastic Transistors. *Small* **2022**, *18*, e2201861. [CrossRef]
29. Aspermair, P.; Mishyn, V.; Bintinger, J.; Happy, H.; Bagga, K.; Subramanian, P.; Knoll, W.; Boukherroub, R.; Szunerits, S. Reduced Graphene Oxide-Based Field Effect Transistors for the Detection of E7 Protein of Human Papillomavirus in Saliva. *Anal. Bioanal. Chem.* **2021**, *413*, 779–787. [CrossRef] [PubMed]
30. Reiner-Rozman, C.; Kotlowski, C.; Knoll, W. Electronic Biosensing with Functionalized rGO FETs. *Biosensors* **2016**, *6*, 17. [CrossRef]
31. He, Q.; Wu, S.; Gao, S.; Cao, X.; Yin, Z.; Li, H.; Chen, P.; Zhang, H. Transparent, Flexible, All-Reduced Graphene Oxide Thin Film Transistors. *ACS Nano* **2011**, *5*, 5038–5044. [CrossRef]
32. Furlan de Oliveira, R.; Livio, P.A.; Montes-García, V.; Ippolito, S.; Eredia, M.; Fanjul-Bolado, P.; González García, M.B.; Casalini, S.; Samorì, P. Liquid-gated Transistors Based on Reduced Graphene Oxide for Flexible and Wearable Electronics. *Adv. Funct. Mater.* **2019**, *29*, 1905375. [CrossRef]

polymers

Article

The Staining Susceptibility and Surface Roughness of Teeth Restored by Microabrasion and Resin Infiltration: An In Vitro Study

Treetossatep Inna, Nantawan Krajangta and Thanasak Rakmanee *

Department of Restorative Dentistry, Faculty of Dentistry, Thammasat University, 99 M. 18, Paholyothin Road, Klong Luang, Pathum Thani 12120, Thailand; treetossatepi@gmail.com (T.I.); knantawa@gmail.com (N.K.)
* Correspondence: rthanasa@tu.ac.th; Tel.: +66-29869205

Abstract: This study assessed the susceptibility to staining and surface roughness of white-spot lesions (WSLs) treated with resin infiltration (RIT) and microabrasion (MA) under simulated aging through thermocycling in red wine. Seventy-eight extracted human premolars with artificial WSLs were divided into three groups: untreated WSLs (control), RIT-treated (ICON®, DMG), and MA-treated (Opalustre®, Ultradent). Each group was further split: one subgroup immersed in artificial saliva and the other thermocycled in red wine. The color change (ΔE) and surface roughness (R_a) were measured before and after staining using a spectrophotometer and a non-contact profilometer. Thermocycling in red wine increased color change (ΔE) across all groups, with the highest values observed for MA (43.94 ± 3.57), followed by RIT (31.40 ± 4.89). Surface roughness (R_a) was highest in untreated WSLs (0.61 ± 0.18 μm) and lowest in RIT (0.15 ± 0.03 μm). While RIT and MA similarly improved WSL appearance, RIT exhibited superior smoothness. These findings suggest that RIT provides a more durable surface with reduced roughness, although staining susceptibility remains comparable to MA. Moderate positive correlation was found between ΔE and Ra, indicating that roughness is one of the factors influencing color changes.

Keywords: resin infiltration; microabrasion; staining; surface roughness; thermocycling; spectrophotometer

1. Introduction

Dental caries are a prevalent oral disease caused by the metabolic activity of bacteria, which results in the demineralization of tooth structure through acid production. The earliest clinical manifestation of this process is the appearance of white-spot lesions (WSLs), characterized by white opaque areas on the enamel surface due to subsurface mineral loss [1]. WSLs not only indicate the potential for progression to cavitated caries but also pose significant esthetic concerns, especially when located in visible areas during smiling or speaking. The esthetic impact of WSLs arises from the contrast in refractive indices (RIs) between healthy enamel (RI = 1.62) and the microporosities in the lesion, filled with water (RI = 1.33) or air (RI = 1.0). This disparity causes light scattering, giving WSLs their characteristic chalky appearance [2].

Management of WSLs has evolved from traditional restorative approaches to minimally invasive techniques that focus on prevention and early intervention. When WSLs are in a reversible stage, treatment aims to balance remineralization and demineralization. Promoting remineralization involves guiding patients' oral hygiene care, regulating sugar consumption, and using agents such as fluoride or casein phosphopeptide—amorphous calcium phosphate (CPP-ACP) [1]. These agents can slow lesion progression by enhancing remineralization, but they often fail to improve esthetic outcomes immediately [2], as remineralization tends to be superficial [3], leaving the subsurface porosities and whitish appearance intact [4].

Resin infiltration (RIT) (ICON®) is a technique that uses low-viscosity resin to penetrate the porous enamel caries [5–7]. Once infiltrated, the resin's RI (~1.46) closely matches that of sound enamel, improving the lesion's appearance [2]. Microabrasion (MA) is another minimally invasive treatment involving mechanical abrasion and acid erosion to create a shiny, lustrous enamel surface [8]. However, excessive abrasion can result in enamel loss and tooth sensitivity. While MA has been considered easy to perform, time efficient, and cost effective [8], RIT is regarded as more conservative in terms of enamel preservation [8,9].

A recent study on the treatment of white-spot lesions (WSLs) using RIT and MA have shown that both techniques effectively enhance the esthetic appearance of WSLs, with RIT being more effective in managing WSLs [10,11]. Both treatments can maintain color stability within a short periods of 3 months [10], and patients are satisfied with both treatment results after 6 months [11]. Additionally, current treatments incorporate microabrasion combined with resin infiltration to manage WSLs, yielding good outcomes [12].

The oral environment is subject to thermal stress and pH fluctuations from food and beverage consumption, which can impact the surface properties of resin materials, leading to long-term discoloration [13]. Over time, direct dental restorations, particularly resin composites, may exhibit surface wear and filler–matrix debonding, resulting in void formation [14]. In laboratory settings, the aging of dental material is typically evaluated through factors such as increased water sorption, surface roughness, and color changes, along with reduced strength and hardness [13,15]. Thermocycling is widely used to simulate the physical aging of dental materials by repeatedly subjecting them to alternating hot and cold water baths, rather than relying solely on intraoral thermal variations [16]. This present study has been further refined through the use of thermocycling stain models, which provide a more comprehensive simulation of the oral environment. Such models account for the thermal fluctuations, time-dependent interactions, and exposure to staining agents commonly found in beverages [17]. Despite these advancements, no existing study has investigated the dynamic staining susceptibility of RIT and MA when exposed to thermocycling conditions using staining solutions.

Staining susceptibility is a critical factor affecting the longevity of esthetic outcomes. Among the various factors influencing discoloration [18], surface roughness (R_a) plays a notable role, as it can increase the material's susceptibility to staining, particularly from external sources like food and beverages [19]. Several studies investigated the staining susceptibility of RIT and MA, finding that both treatments may undergo color changes when exposed to staining solutions [20–24]. Most studies on the staining susceptibility of RIT [21–26] and MA [20,23] use static models, where samples are immersed in staining solutions to measure color changes. However, these static models may not replicate the dynamic and multifactorial conditions of the oral environment, including thermal stress, acidic pH, and fluid interactions.

This present study addresses these gaps by evaluating the dynamic staining susceptibility and surface roughness of RIT and MA under thermocycling in red wine, a highly staining-prone agent. This approach integrates thermal stress to simulate real-life oral conditions, enhancing the generalizability and clinical relevance of the findings. By adopting this systematic and comprehensive methodology, this research aims to provide a more nuanced understanding of the esthetic longevity of RIT and MA, ultimately guiding clinicians in making evidence-based decisions for managing WSLs.

2. Materials and Methods

This study aimed to compare the staining susceptibility and surface roughness of teeth treated with RIT and MA under thermal stress induced by thermocycling in red wine. The null hypotheses are as follows: (1) no statistical difference in staining susceptibility between RIT- and MA-treated teeth after thermocycling in red wine; (2) no statistical difference in surface roughness between resin-infiltrated and micro-abraded teeth after thermocycling in red wine; and (3) no statistical correlation between staining susceptibility and surface roughness.

This in vitro study utilized extracted human permanent premolar teeth, obtained for orthodontic purposes and selected based on their color within the A3–A4 range using a spectrophotometer (VITA Easyshade® V from VITA Zahnfabrik H. Rauter GmbH & Co. KG. Bad Säckingen, Germany). Ethical approval was obtained from The Human Research Ethics Committee of Thammasat University (Science) under project number 66DE140.

2.1. Experimental Design

The study design is summarized in Figure 1. The baseline color and surface roughness (R_a) of all sound enamel samples were recorded. To create artificial WSLs, the samples were immersed in a demineralization solution. A total of 78 WSLs were created and equally divided into three groups: untreated WSLs, resin infiltration (RIT), and microabrasion (MA). The samples were kept at room temperature, dried, and shielded from direct sunlight. Each group was further subdivided into two subgroups: one immersed in artificial saliva as a control, and the other subjected to thermocycling in red wine. The color and R_a were measured at each experimental stage, as described below.

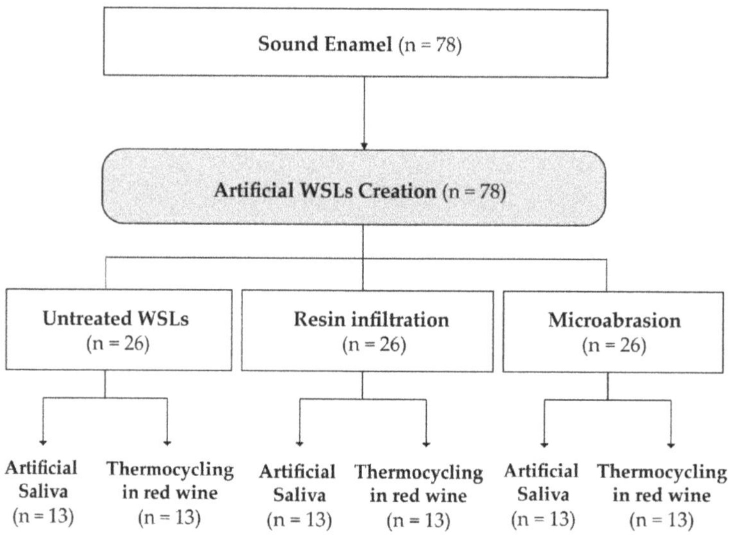

Figure 1. Flow diagram of the experimental procedure.

2.2. Specimen Preparation

A sample size of 78 extracted permanent premolars were calculated using GPower version 3.1.9.7, with a significance level of 0.05 and a power of 0.95 (standard deviation of means from a pilot study: 5.63, unpublished data). The teeth were cleaned with an ultrasonic cleaner and a non-fluoride pumice paste and then stored in 0.1% thymol at 4 °C to disinfect and maintain their structural integrity. Each tooth was sectioned at the cementoenamel junction to remove the root, preserving the crown.

The crowns were mounted in acrylic blocks (Figure 2) with the buccal surface exposed (~5×5 mm^2), protected by black nail varnish [27]. The specimens were immersed in a demineralization solution, as described by Yuan et al. (2014), comprising 6% hydroxylethyl cellulose and 0.1 M lactic acid, at pH 4.5, and at 37 °C for one week to create WSLs [28]. Following demineralization, specimens were rinsed with distilled deionized water for 30 s, and WSL formation was visually inspected and confirmed by a spectrophotometer (VITA Easyshade® V).

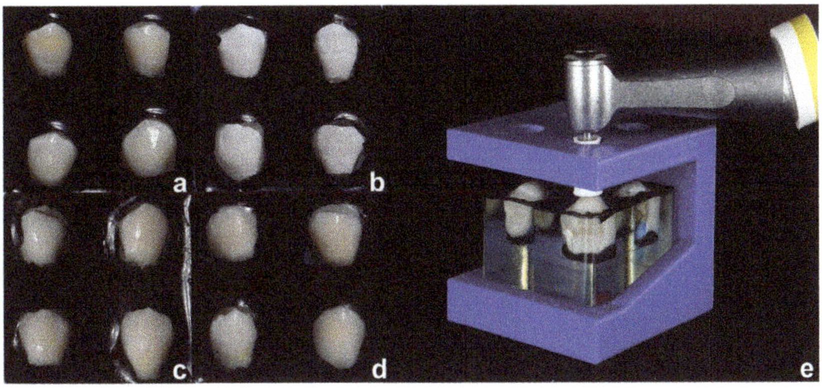

Figure 2. The specimens in each treatment: (**a**) sound enamel, (**b**) artificial white-spot lesion (**c**) after resin infiltration, (**d**) after microabrasion, and (**e**) 3D printing template designed for microabrasion.

2.2.1. Resin Infiltration (RI) Group

The WSLs in this group were treated with RIT (ICON® infiltrant; DMG, Hamburg, Germany) following the manufacturer's instructions. The lesion surface was etched with 15% hydrochloric acid (HCL) (ICON® etch) for 2 min, rinsed with a water spray for 30 s, and dried. Then, 99% ethanol (ICON® dry) was applied for 30 s. A low-viscosity resin (ICON® infiltrant) was applied for 3 min and light cured for 40 s (SmartLite® Focus®, Dentsply Sirona, Bangkok, Thailand). A second resin layer was applied for 1 min and light cured. The samples were polished using Al_2O_3 disks (Sof-Lex Pop on, 3M-ESPE, St. Paul, MN, USA) from fine to superfine for 30 s [29].

2.2.2. Microabrasion (MA) Group

The WSLs in this group were treated with MA (Opalustre®; Ultradent, South Jordan, UT, USA), containing 6.6% HCL and silicon carbide particles (20–160 µm). The slurry was applied using OpalCups® attached to a low-speed handpiece (NSK-EC®; Nakanishi Inc., Tokyo, Japan), with light pressure for 1 min, with three repeated applications (Figure 2d) [20]. After each application, the enamel surface was rinsed with a water spray for 10 s. A 3D-printed template ensured perpendicular application of OpalCups® (Figure 2e). The surface was polished with Al_2O_3 disks, as described above [29] and treated with 2% sodium fluoride gel (Pro-F®, Prominent, Bangkok, Thailand) for 4 min [20], followed by a 30 s rinse.

2.3. Thermocycling in Staining Solution

Each experimental group was divided into control (non-exposed, immersed in artificial saliva) and staining subgroups subjected to thermocycling in red wine [17]. Control samples were kept in artificial saliva (methylcellulose, glycerin, paraben, and purified water) at room temperature. The staining subgroup underwent thermocycling between red wine baths (Nadin Bin777, Cabernet Sauvignon, South Australia, pH = 3.53) at 5 °C and 55 °C, for 30 s each. A pH meter was used to ensure the solution's pH remained stable. Thermocycling (5000 cycles) simulated six months of aging (HWB332R, TC301, King Mongkut's Institute of Technology, Bangkok, Thailand). Color and R_a were then measured after thermocycling.

2.4. Measurement of Staining Susceptibility

Color changes were analyzed using the CIEDE2000 system (ΔE_{00}), based on the formula by Sharma et al. (2005) [30]. Measurements were performed using a spectrophotometer (VITA Easyshade® V), with each sample air-blown for 30 s prior to measurement. All color measurements were conducted under standardized conditions according to ISO/TR 28642-2016 [31], using a viewing box with a D65 illuminant (representing noon daylight) and a controlled illuminance of 1000 lux measured by a lux meter. A 45°/0° optical geometric observation was used for assessment [31,32], with sample positioning standardized by a 3D-printed model to ensure accuracy and reproducibility (Figure 3).

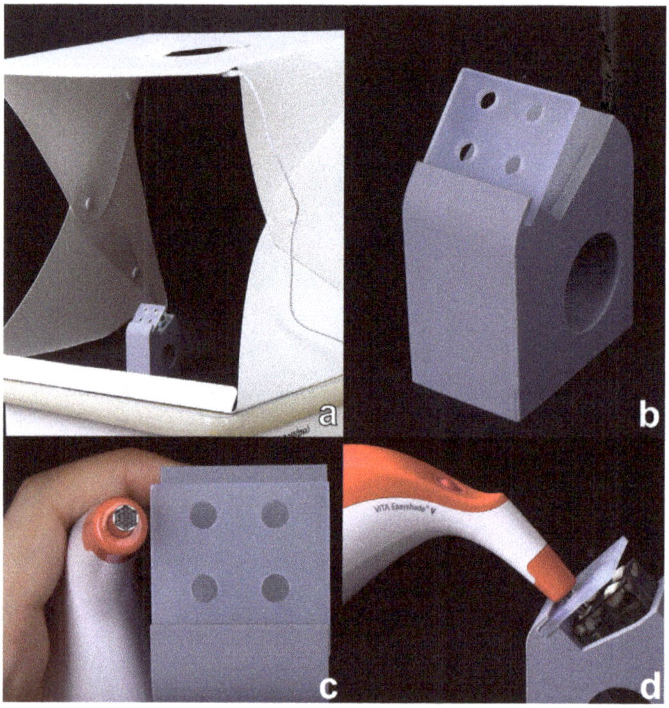

Figure 3. A 3D-printed resin model and color measurement protocols. (**a**) Viewing box with a D65 illuminant. (**b**) A 3D-printed model. (**c**) The template was designed to be the same size as the tip of the VITA Easyshade® V. (**d**) The 45°/0° standard measurement was controlled by the 3D-printed model.

2.5. Surface Roughness Measurement

R_a was assessed using a non-contact profilometer (InfiniteFocus G5, Alicona, Raaba/Graz, Austria), at the mid-buccal surface of the enamel. Measurements were taken at the same position as the baseline.

2.6. Statistical Analysis

Data were analyzed using the SPSS 26.0 (IBM, Chicago, IL, USA) software. Normality was assessed using the Kolmogorov–Smirnov test. Median and median absolute deviation (MAD) described nonparametric data. Wilcoxon signed-rank and Friedman tests were employed for intra-group comparisons, while Kruskal–Wallis's tests were evaluated for inter-group differences in staining susceptibility and R_a. Post hoc Bonferroni correction identified significant differences between paired groups. Spearman correlation analysis examined the relationship between staining susceptibility and R_a. A significant level of 0.05 was used for all tests.

3. Results and Discussion

3.1. Color Values (L*, a*, b*)

Descriptive statistics for the color parameters (L*, a*, and b*) are presented in Table 1. Significant differences were observed in all color parameters after the creation of artificial WSLs compared to sound enamel ($p < 0.05$). The L* value, indicating brightness, significantly decreased after WSL formation, reflecting a loss of brightness. The a* value (red–green) increased significantly, while the b* value (yellow–blue) slightly decreased, signaling a shift in color.

Table 1. Descriptive data of median and ± median absolute deviation (MAD) of color values (L*, a*, b*).

Value	WSLs Creation		Treatment after WSLs					
			Resin Infiltration Group			Microabrasion Group		
	Sound Enamel (N = 78)	WSLs (N = 78)	Sound Enamel (N = 26)	WSLs (N = 26)	After Treatment (N = 26)	Sound Enamel (N = 26)	WSLs (N = 26)	After Treatment (N = 26)
L	85.40 ± 3.50	58.005 ± 5.25 *	84.30 ± 2.85 [a]	58.2 ± 4.65 [b]	86.30 ± 2.5 [a]	86.30 ± 3.35 [a]	55.15 ± 5.4 [b]	91.75 ± 2.90 [a]
a	3.3 ± 1.30	7.1 ± 1.2 *	3.75 ± 1.3 [a]	7.6 ± 1.2 [b]	2.90 ± 1.15 [a]	3.25 ± 0.8 [a]	7.7 ± 0.95 [b]	2.45 ± 1.05 [a]
b	39.75 ± 2.45	36.95 ± 1.85 *	41.20 ± 2.1 [a]	37.7 ± 1.7 [b]	37.60 ± 2.6 [b]	38.8 ± 2.15 [a]	36.3 ± 1.25 [b]	39.85 ± 2.50 [a]

* Significant differences between sound enamel and after WSL creation, using the Wilcoxon signed-ranked test ($p < 0.05$). [a, b] Different lower letters in the same row indicate significant differences among steps in each treatment, using the Friedman test ($p < 0.05$).

Post-treatment with RIT and MA, the color parameters approached those of sound enamel. In the RIT group, L* increased significantly, restoring it closer to the original enamel value. The a* and b* values also shifted towards the baseline, indicating improved color consistency. In the MA group, L* values significantly increased compared to WSL, with similar improvements in a* and b* values. However, the b* value in the RIT remained lower than in sound enamel, reflecting some residual differences in the yellow–blue spectrum.

Post-treatments with RIT and MA, all color parameters returned close to sound enamel, except for b* value in RIT, which might have been due to the slight difference in the refractive index between resin infiltration (1.46) and sound enamel (1.62), which may explain the b* value's persistence [2]. These results suggest that both procedures effectively treat WSLs.

3.2. Surface Roughness (R_a)

Changes in R_a throughout the experiment are summarized in Table 2. Following WSL formation, R_a increased in all groups compared to baseline, with significant differences observed in groups 1, 2, 4, and 5 ($p < 0.05$). Both RIT and MA treatments reduced surface roughness compared to untreated WSLs. Significant differences were noted in groups 3, 5, and 6 ($p < 0.05$). All groups exhibited a decrease in R_a after immersion in artificial saliva (control), with a significant difference observed in groups 2 and 6.

Table 2. Descriptive data of median and ± median absolute deviation (MAD) of surface roughness (R_a, μm).

Timeline	Group					
	Artificial Saliva (Control) (N = 39)			Thermocycling in Red Wine (Staining) (N = 39)		
	1 Untreated WSLs (N = 13)	2 RI (N = 13)	3 MA (N = 13)	4 Untreated WSLs (N = 13)	5 RI (N = 13)	6 MA (N = 13)
Baseline	0.22 ± 0.05 [a]	0.20 ± 0.07 [ac]	0.18 ± 0.04 [ac]	0.15 ± 0.05 [a]	0.18 ± 0.05 [a]	0.18 ± 0.05 [ac]
Artificial WSLs	0.35 ± 0.09 [b]	0.35 ± 0.06 [b]	0.33 ± 0.11 [bc]	0.30 ± 0.07 [b]	0.29 ± 0.05 [b]	0.51 ± 0.12 [ad]
Treatments		0.23 ± 0.05 [ab]	0.14 ± 0.05 [a]		0.19 ± 0.08 [a]	0.09 ± 0.01 [c]
Immersion	0.30 ± 0.09 [ab]	0.08 ± 0.02 [c]	0.09 ± 0.02 [a]	0.61 ± 0.18 [b]	0.15 ± 0.03 [a]	0.55 ± 0.09 [bd]

[a, b, c, d] Different lower letters in the same column indicate significant differences among the group for R_a, using the Friedman test ($p < 0.05$).

Following thermocycling in red wine, the untreated WSL and MA groups experienced substantial increases in R_a. Conversely, R_a in the RIT group modestly decreased, although this change was not statistically significant compared to baseline.

Surface roughness (R_a) is a key parameter for evaluating surface texture and can change over time. In this study, R_a was measured using the InfiniteFocus G5 optical profilometer, which avoids surface contact, preventing damage and measurement errors associated with stylus tip devices [33]. The polishing protocol used significantly influenced R_a values; both RIT and MA followed the same polishing protocol. Previous research shows that Al_2O_3 disks (Sof-Lex Pop on) effectively reduced R_a in resin-infiltrated enamel compared to silicon carbide tips and brushes [29]. While polishing after RIT may not reduce bacterial adhesion, it remains crucial for minimizing the risk of staining [29].

After WSLs were created, the R_a of the samples increased compared to the baseline (Table 2). The lack of significant differences in R_a between groups 3 and 6 may be due to the small sample size, limiting the detection of statistical significance. Larger sample sizes in future studies may help address this limitation. Consistent with previous research, we observed that the R_a of WSLs was higher than that of sound enamel [34–36]. Both RIT and MA reduced surface roughness, similar to findings by Yazkan and Ermis (2018), who reported lower R_a with RIT than MA [36]. However, in our study, R_a for RIT was slightly higher than MA, likely due to variations in the MA protocol. While Yazkan and Ermis used 30 s Opalustre® polishing repeated in two cycles with 2–4 μm diamond paste, our study employed a more intensive protocol—1 min polishing, repeated in three cycles with 24 μm (fine) to 8 μm (superfine) Al_2O_3 disks [20,37]. Despite this, both treatments produced surface roughness values similar to sound enamel.

After thermocycling in a staining model, the R_a of the RIT group trended to increase compared to the control group, although this difference was not statistically significant. Thermal stress from thermocycling may impact the surface integrity of resin-infiltrated enamel lesions, potentially creating microcracks and increasing roughness [24]. This study also showed that R_a in the RIT group was lower than in the MA group, although the difference was not significant, leading to the acceptance of the second null hypothesis. No prior studies have specifically investigated the surface roughness of RIT and MA after thermocycling in red wine staining. In addition to thermal stress, the erosive effect of red wine likely plays a significant role in altering the enamel surface after MA.

3.3. Color Changes (ΔE)

Color changes (ΔE) are presented in Table 3. In the control group (artificial saliva), untreated WSLs exhibited a moderate color change (median ΔE of 2.77). RIT and MA treatments showed lower ΔE values (1.38 and 1.95, respectively) than untreated WSLs, indicating better color stability.

Table 3. Median and ± median absolute deviation (MAD) data of the color change (ΔE) between treatments after red wine thermocycling and surface roughness (R_a) after red wine thermocycling.

	Group	ΔE	R_a (μm)
Control (N = 39)	1. untreated WSLs (N = 13)	2.77 ± 0.58^{b}	0.30 ± 0.09^{bc}
	2. resin infiltration (N = 13)	1.38 ± 0.31^{b}	0.08 ± 0.02^{a}
	3. microabrasion (N = 13)	1.95 ± 0.63^{b}	0.09 ± 0.02^{a}
Staining (N = 39)	4. untreated WSLs (N = 13)	N/A	0.61 ± 0.18^{c}
	5. resin infiltration (N = 13)	31.40 ± 4.89^{a}	0.15 ± 0.03^{ab}
	6. microabrasion (N = 13)	43.94 ± 3.57^{a}	0.55 ± 0.09^{bc}

a, b, c Different lower letters in the same column indicate significant differences among the group for ΔE and R_a ($p < 0.05$). Abbreviations: N/A: not applicable.

In the staining group exposed to red wine thermocycling, color changes were more pronounced. The untreated WSL group did not provide a measurable ΔE due to extreme discoloration. The RIT group had a median ΔE of 31.40, while the MA group exhibited a higher ΔE of 43.94, indicating significant discoloration in both groups. RIT demonstrated slightly better resistance to red wine staining than MA.

The results showed no significant difference in color change between treatments post-thermocycling, leading to the acceptance of the first null hypothesis. The thermocycling in the staining solution simulated the combined effects of thermal stress, acidic pH, and fluid dynamics from beverage consumption [17]. Previous studies confirmed that resin-based materials are prone to discoloration under these factors [14,38]. Red wine, known for its high staining potential due to tannins, alcohol, and a low pH, surpasses beverages like coffee and tea [39]. The tannins in Cabernet sauvignon grapes penetrate resin materials, contributing to significant staining due to the combined effects of alcohol and a low pH [39,40].

In this study, color change was analyzed using the CIEDE2000 system, which more accurately reflects human color perception than the CIELAB formula [41]. The VITA Easyshade® spectrophotometer used in this study provides accurate and reproducible measurements [42]. However, its limited wavelength range (400–700 nm) likely explains why the purplish discoloration (which occurs in the 385–425 nm range) of untreated WSLs was not detected after staining [43,44]. Despite this, both RIT and MA treatments showed marked improvement in color, with RIT displaying slightly better color stability than MA. Silva et al. (2018) found similar results, indicating no significant difference in color stability between RIT and MA after 7 days of coffee exposure [45]. This supports our finding that both treatments experienced color instability after immersion in red wine. A short-term study also reported that MA was more prone to discoloration than RIT, particularly after 24 h immersion in tea and citric acid [23]. The shorter MA protocol may not fully remove porous enamel, leading to greater discoloration. By contrast, a more extensive MA protocol, while effective, showed that RIT still demonstrated better overall color stability, consistent with Lee et al. (2016) [46]. Color stability may be more affected by dietary staining agents than by the treatment type [47].

The discoloration in RIT-treated teeth arises from both intrinsic and extrinsic factors [24]. Intrinsically, RIT materials contain triethylene glycol dimethacrylate (TEGDMA), a low-viscosity, hydrophilic component, which increases water absorption, allowing for pigment penetration and resulting in color changes [7]. Thermal stress from thermocycling can also cause microcracks and micro-fissures, facilitating pigment infiltration [24]. Extrinsically, pigments from food and beverages adhere to the surface but can be mitigated through polishing, which helps reduce surface staining [24].

MA removes enamel porosities through mechanical abrasion and acid erosion and has been widely used to treat WSLs [8]. For example, Opalustre® has been found to be more effective than Prema® in reducing fluorosis-related WSLs [48]. This present study used a post-treatment fluoride polishing protocol based on Bahadir et al. (2022) [20], which improved stain resistance. Despite this, both RIT and MA treatments showed extreme color changes after dynamic staining, although patient satisfaction with both treatments remains high [11].

The CIEDE2000 color change threshold (ΔE_{00}) for 50:50% perceptibility is 0.8, and the acceptability threshold is 1.8 [27]. In this study, all groups exceeded the acceptability threshold, except for the RIT group immersed in artificial saliva. Despite the control groups being in artificial saliva, both untreated WSLs and MA exhibited unacceptable color differences. This may be due to differences in measurement protocols. After immersion, the specimens were air-dried for only 30 s, potentially leaving them slightly moist, which could have affected the color measurements [49]. Thus, these variables should be considered for better control in further experiments.

3.4. Correlation Between Surface Roughness (R_a) and Color Change (ΔE)

A significant correlation was found between R_a and ΔE (Spearman rank correlation coefficient, $r_s = 0.577$, $p < 0.00$), indicating that changes in surface roughness are moderately positively related to the extent of color change. The findings of this study align with those of previous studies [50,51]. Chowdhury et al. (2021) reported that there was a positive correlation between surface roughness and color changes in nanohybrid composite resin when immersed in artificial saliva, tea, coffee, and Coca-Cola [50]. Additionally, their research demonstrated that both color stability and surface roughness are time dependent, with increased immersion time resulting in greater roughness and more pronounced color changes [50,51]. In this present study, the Spearman correlation test showed linear relationship between staining susceptibility (ΔE) and R_a for either RIT or MA, leading to the rejection of the third null hypothesis.

While surface roughness contributes to external staining of dental restorations, it is not the only factor. Other variables, such as water sorption, resin matrix composition, the degree of conversion, polymerization depth, staining agents, and the pH of staining solutions, also affect the discoloration of restorative materials [52].

This study offers valuable insights into the staining susceptibility and surface roughness of RIT and MA, especially for frequent red wine consumers like wine tasters. However, as an in vitro study, it has limitations. The red wine thermocycling model simulates thermal stress, acidic pH, and fluid dynamics [17] but may reflect an extreme discoloration scenario. With 5000 thermocycles simulating six months of aging, no significant difference between both treatments was observed. Future studies with varying thermocycle numbers may reveal differences.

While both RIT and MA effectively restored WSL color to resemble sound enamel, this study did not assess enamel loss, which is critical for determining the best minimally invasive treatment. Additionally, extrinsic staining can be addressed by various methods [53], and further research is also needed to investigate the effectiveness of different polishing or cleaning protocols for reducing stains after RIT and MA.

4. Conclusions

Within the limitations of this study, based on the methodology used in thermocycling in the staining solution model, the following can be concluded:

1. After treatment of WSLs, the color parameters (L*, a*, b* values) of both RIT and MA were restored to levels comparable to sound enamel, except for the b* value in the RIT group. This exception may be attributed to a slight difference in the refractive index between resin infiltration (1.46) and sound enamel (1.62), resulting in the persistence of the b* value. These findings suggest that both RIT and MA are effective treatments for WSLs.
2. After thermocycling in red wine, both treatments exhibited extreme discoloration, with median ΔE values of 31.40 ± 4.89 for RIT and 43.94 ± 3.57 for MA. Although the surface roughness of RIT (0.15 ± 0.03 µm) was lower than that of MA (0.55 ± 0.09 µm), the difference was not statistically significant. These results indicate that RIT and MA have similar staining susceptibility, with RIT demonstrating slightly lower surface roughness than MA.
3. A moderate positive correlation was observed in this thermocycling model, indicating that surface roughness influences color changes in this study, with an r_s value of 0.577. Although surface roughness is a significant determinant in the external staining of dental restorations, it is crucial to recognize that it is not the sole factor influencing this phenomenon. Other contributing variables, as previously discussed, interact synergistically to influence the extent of staining.

Author Contributions: Conceptualization, T.I. and T.R.; methodology, T.I. and T.R.; validation, T.I., N.K. and T.R.; formal analysis, T.I., N.K. and T.R.; investigation, T.I. and T.R.; data curation, T.I. and T.R.; writing—original draft preparation, T.I., N.K. and T.R.; writing—review and editing, T.I. and T.R. All authors have read and agreed to the published version of the manuscript.

Funding: Funding was supported by the Thammasat University Research Unit of Restorative and Esthetic Dentistry and a Post-graduate research fund from the Department of Restorative Dentistry, Thammasat University.

Institutional Review Board Statement: Not applicable.

Data Availability Statement: The original contributions presented in the study are included in the article, further inquiries can be directed to the corresponding author.

Acknowledgments: The authors would like to thank Kamon Budsaba for statistical advice.

Conflicts of Interest: The authors declare no conflicts of interest.

References

1. Sadikoglu, I.S. White Spot Lesions: Recent Detection and Treatment Methods. *Cyprus J. Med. Sci.* **2020**, *5*, 260–266. [CrossRef]
2. Kim, S.; Kim, E.Y.; Jeong, T.S.; Kim, J.W. The evaluation of resin infiltration for masking labial enamel white spot lesions. *Int. J. Paediatr. Dent.* **2011**, *21*, 241–248. [CrossRef] [PubMed]
3. Oliveira, G.M.; Ritter, A.V.; Heymann, H.O.; Swift, E., Jr.; Donovan, T.; Brock, G.; Wright, T. Remineralization effect of CPP-ACP and fluoride for white spot lesions in vitro. *J. Dent.* **2014**, *42*, 1592–1602. [CrossRef] [PubMed]
4. Puleio, F.; Fiorillo, L.; Gorassini, F.; Iandolo, A.; Meto, A.; D'Amico, C.; Cervino, G.; Pinizzotto, M.; Bruno, G.; Portelli, M.; et al. Systematic Review on White Spot Lesions Treatments. *Eur. J. Dent.* **2022**, *16*, 41–48. [CrossRef]
5. Klaisiri, A.; Janchum, S.; Wongsomtakoon, K.; Sirimanathon, P.; Krajangta, N. Microleakage of resin infiltration in artificial white-spot lesions. *J. Oral Sci.* **2020**, *62*, 427–429. [CrossRef] [PubMed]
6. Klaisiri, A.; Vongsang, J.; Leelaudom, T.; Krajangta, N. Methylene Blue Penetration of Resin Infiltration and Resin Sealant in Artificial White-Spot Lesions. *Eur. J. Dent.* **2023**, *17*, 828–833. [CrossRef] [PubMed]
7. Paris, S.; Meyer-Lueckel, H.; Cölfen, H.; Kielbassa, A.M. Penetration coefficients of commercially available and experimental composites intended to infiltrate enamel carious lesions. *Dent. Mater.* **2007**, *23*, 742–748. [CrossRef]
8. Pini, N.I.; Sundfeld-Neto, D.; Aguiar, F.H.; Sundfeld, R.H.; Martins, L.R.; Lovadino, J.R.; Lima, D.A. Enamel microabrasion: An overview of clinical and scientific considerations. *World J. Clin. Cases* **2015**, *3*, 34–41. [CrossRef] [PubMed]
9. Shan, D.; He, Y.; Gao, M.; Liu, H.; Zhu, Y.; Liao, L.; Hadaegh, F.; Long, H.; Lai, W. A comparison of resin infiltration and microabrasion for postorthodontic white spot lesion. *Am. J. Orthod. Dentofac. Orthop.* **2021**, *160*, 516–522. [CrossRef] [PubMed]
10. Asthana, G.; Patel, K.; Parmar, R. Efficacy of microabrasion and resin infiltration techniques for masking of fluorotic white spot lesions: A randomized clinical study. *J. Conserv. Dent. Endod.* **2023**, *26*, 677–681. [CrossRef]
11. ElSayed, H.S.; Mahran, A.; Kadry, W.; Fahim, F. Resin Infiltration Versus Acid Micro-Abrasion in the Treatment of White Spot Lesions in Fixed Orthodontic Patients. *Future Dent. J.* **2022**, *7*, 90–94. [CrossRef]
12. Alrebdi, A.B.; Alyahya, Y. Microabrasion plus resin infiltration in masking white spot lesions. *Eur. Rev. Med. Pharmacol. Sci.* **2022**, *26*, 456–461. [CrossRef]
13. Szczesio-Wlodarczyk, A.; Sokolowski, J.; Kleczewska, J.; Bociong, K. Ageing of Dental Composites Based on Methacrylate Resins-A Critical Review of the Causes and Method of Assessment. *Polymers* **2020**, *12*, 882. [CrossRef] [PubMed]
14. Minami, H.; Hori, S.; Kurashige, H.; Murahara, S.; Muraguchi, K.; Minesaki, Y.; Tanaka, T. Effects of thermal cycling on surface texture of restorative composite materials. *Dent. Mater. J.* **2007**, *26*, 316–322. [CrossRef]
15. Xian, G.; Bai, Y.; Qi, X.; Wang, J.; Tian, J.; Xiao, H. Hygrothermal aging on the mechanical property and degradation mechanism of carbon fiber reinforced epoxy composites modified by nylon 6. *J. Mater. Res. Technol.* **2024**, *33*, 6297–6306. [CrossRef]
16. Eliasson, S.T.; Dahl, J.E. Effect of thermal cycling on temperature changes and bond strength in different test specimens. *Biomater. Investig. Dent.* **2020**, *7*, 16–24. [CrossRef]
17. Ren, Y.F.; Feng, L.; Serban, D.; Malmstrom, H.S. Effects of common beverage colorants on color stability of dental composite resins: The utility of a thermocycling stain challenge model in vitro. *J. Dent.* **2012**, *40* (Suppl. S1), e48–e56. [CrossRef] [PubMed]
18. Menon, A.; Ganapathy, D.M.; Mallikarjuna, A.V. Factors that influence the colour stability of composite resins. *Drug Invent. Today* **2019**, *11*, 744–749.
19. Sirin Karaarslan, E.; Bulbul, M.; Yildiz, E.; Secilmis, A.; Sari, F.; Usumez, A. Effects of different polishing methods on color stability of resin composites after accelerated aging. *Dent. Mater. J.* **2013**, *32*, 58–67. [CrossRef] [PubMed]
20. Bahadir, H.S.; Haberal, M.; Çelik, Ç. Effect of microabrasion on the staining susceptibility of enamel: An in vitro study. *J. Dent. Res. Dent. Clin. Dent. Prospect.* **2022**, *16*, 95–100. [CrossRef] [PubMed]
21. Borges, A.; Caneppele, T.; Luz, M.; Pucci, C.; Torres, C. Color stability of resin used for caries infiltration after exposure to different staining solutions. *Oper. Dent.* **2014**, *39*, 433–440. [CrossRef] [PubMed]
22. Liu, G.; Yang, L.; Gao, L.; Ma, Y.; Wu, X.; Wang, X.; Hao, Z. Durability of infiltrated resin application on white spot lesions after different challenges: An ex vivo study. *J. Prosthet. Dent.* **2024**, *131*, 500–507. [CrossRef] [PubMed]
23. Yetkiner, E.; Wegehaupt, F.; Wiegand, A.; Attin, R.; Attin, T. Colour improvement and stability of white spot lesions following infiltration, micro-abrasion, or fluoride treatments in vitro. *Eur. J. Orthod.* **2014**, *36*, 595–602. [CrossRef] [PubMed]
24. Zhao, X.; Ren, Y.F. Surface Properties and Color Stability of Resin-Infiltrated Enamel Lesions. *Oper. Dent.* **2016**, *41*, 617–626. [CrossRef]

25. Alqahtani, S.; Abusaq, A.; Alghamdi, M.; Shokair, N.; Albounni, R. Colour stability of resin infiltrated white spot lesion after exposure to stain-causing drinks. *Saudi J. Biol. Sci.* **2022**, *29*, 1079–1084. [CrossRef] [PubMed]
26. Emami Arjomand, M.; Hoorizad Ganjkar, M.; Ghamari, R. Effect of Tea on Color Stability of Enamel Lesions Treated with Resin Infiltrant. *Front. Dent.* **2021**, *18*, 4. [CrossRef] [PubMed]
27. Paravina, R.D.; Ghinea, R.; Herrera, L.J.; Bona, A.D.; Igiel, C.; Linninger, M.; Sakai, M.; Takahashi, H.; Tashkandi, E.; Perez Mdel, M. Color difference thresholds in dentistry. *J. Esthet. Restor. Dent.* **2015**, *27* (Suppl. S1), S1–S9. [CrossRef]
28. Yuan, H.; Li, J.; Chen, L.; Cheng, L.; Cannon, R.D.; Mei, L. Esthetic comparison of white-spot lesion treatment modalities using spectrometry and fluorescence. *Angle Orthod.* **2014**, *84*, 343–349. [CrossRef]
29. Esteves-Oliveira, M.; Passos, V.F.; Russi, T.; Fernandes, A.R.R.; Terto, C.N.N.; Mendonça, J.S.; Campus, G.; Wierichs, R.J.; Meyer-Lueckel, H.; Lima, J.P.M. Randomized in situ evaluation of surface polishing protocols on the caries-protective effect of resin Infiltrant. *Sci. Rep.* **2022**, *12*, 20648. [CrossRef] [PubMed]
30. Sharma, G.; Wu, W.; Dalal, E. The CIEDE2000 color-difference formula: Implementation notes, supplementary test data, and mathematical observations. *Color. Res. Appl.* **2005**, *30*, 21–30. [CrossRef]
31. *ISO/TR 28642:2016*; Technical Report(E): Dentistry—Guidance on Colour Measurements. International Organization for Standardization: Geneva, Switzerland, 2016.
32. Krajangta, N.; Klaisiri, A.; Leeraponglit, S.; Intralawan, N.; Tiansuwan, P. Color alteration of CAD/CAM 3D-printed, milled resin-ceramic hybrid material compared to enamel. *Dent. Mater. J.* **2024**, *43*, 386–393. [CrossRef] [PubMed]
33. Lukianowicz, C.; Karpinski, T. Optical system for measurement of surface form and roughness. *Meas. Sci. Rev.* **2001**, *1*, 151–154.
34. Chabuk, M.M.; Al-Shamma, A.M. Surface roughness and microhardness of enamel white spot lesions treated with different treatment methods. *Heliyon* **2023**, *9*, e18283. [CrossRef] [PubMed]
35. Sayed, H.; Abo Elezz, A.; Fahmy, O. Effect of Three Treatment Modalities for White Spot Enamel Lesions on Enamel Surface Roughness and Micro-Hardness. *Dent. Sci. Updates* **2022**, *3*, 127–137. [CrossRef]
36. Yazkan, B.; Ermis, R.B. Effect of resin infiltration and microabrasion on the microhardness, surface roughness and morphology of incipient carious lesions. *Acta Odontol. Scand.* **2018**, *76*, 473–481. [CrossRef]
37. Lins, F.C.; Ferreira, R.C.; Silveira, R.R.; Pereira, C.N.; Moreira, A.N.; Magalhães, C.S. Surface Roughness, Microhardness, and Microleakage of a Silorane-Based Composite Resin after Immediate or Delayed Finishing/Polishing. *Int. J. Dent.* **2016**, *2016*, 8346782. [CrossRef] [PubMed]
38. Valinoti, A.C.; Neves, B.G.; da Silva, E.M.; Maia, L.C. Surface degradation of composite resins by acidic medicines and pH-cycling. *J. Appl. Oral Sci.* **2008**, *16*, 257–265. [CrossRef] [PubMed]
39. Saccucci, M.; Corridore, D.; Di Carlo, G.; Bonucci, E.; Cicciù, M.; Vozza, I. Assessment of Enamel Color Stability of Resins Infiltration Treatment in Human Teeth: A Systematic Review. *Int. J. Environ. Res. Public Health* **2022**, *19*, 11269. [CrossRef]
40. Watrelot, A.A. Tannin Content in Vitis Species Red Wines Quantified Using Three Analytical Methods. *Molecules* **2021**, *26*, 4923. [CrossRef]
41. Gómez-Polo, C.; Portillo Muñoz, M.; Lorenzo Luengo, M.C.; Vicente, P.; Galindo, P.; Martín Casado, A.M. Comparison of the CIELab and CIEDE2000 color difference formulas. *J. Prosthet. Dent.* **2016**, *115*, 65–70. [CrossRef] [PubMed]
42. Horn, D.J.; Bulan-Brady, J.; Hicks, M.L. Sphere spectrophotometer versus human evaluation of tooth shade. *J. Endod.* **1998**, *24*, 786–790. [CrossRef] [PubMed]
43. Chandrasekhar, V.; Reddy, L.P.; Prakash, T.J.; Rao, G.A.; Pradeep, M. Spectrophotometric and colorimetric evaluation of staining of the light cured composite after exposure with different intensities of light curing units. *J. Conserv. Dent.* **2011**, *14*, 391–394. [CrossRef] [PubMed]
44. Tarrant, A.W.S. Chapter 28. In *Optical Measurements*; Boyes, W., Ed.; Butterworth-Heinemann: Boston, MA, USA, 2010.
45. Silva, L.O.; Signori, C.; Peixoto, A.C.; Cenci, M.S.; Faria, E.S.A.L. Color restoration and stability in two treatments for white spot lesions. *Int. J. Esthet. Dent.* **2018**, *13*, 394–403. [PubMed]
46. Lee, J.; Chen, J.W.; Omar, S.; Kwon, S.R.; Meharry, M. Evaluation of Stain Penetration by Beverages in Demineralized Enamel Treated With Resin Infiltration. *Oper. Dent.* **2016**, *41*, 93–102. [CrossRef] [PubMed]
47. Sherwood, A. Effect of Different Food Coloring Additives on the Color Stability of Microabraded, at-Home Bleached and Resin Infiltrated Tooth. *Curr. J. Appl. Sci. Technol.* **2021**, *40*, 66. [CrossRef]
48. Loguercio, A.D.; Correia, L.D.; Zago, C.; Tagliari, D.; Neumann, E.; Gomes, O.M.; Barbieri, D.B.; Reis, A. Clinical effectiveness of two microabrasion materials for the removal of enamel fluorosis stains. *Oper. Dent.* **2007**, *32*, 531–538. [CrossRef]
49. Choi, M.-R. Comparison of Shade Changes According to Dry/Wet Condition of Tooth using Intra-Oral Colorimeter. *J. Dent. Rehabil. Appl. Sci.* **2010**, *26*, 265–271.
50. Chowdhury, D.; Mazumdar, P.; Desai, P.; Datta, P. Comparative evaluation of surface roughness and color stability of nanohybrid composite resin after periodic exposure to tea, coffee, and Coca-cola—An in vitro profilometric and image analysis study. *J. Conserv. Dent.* **2020**, *23*, 395–401. [CrossRef]
51. Peker, O.; Bolgul, B. Evaluation of surface roughness and color changes of restorative materials used with different polishing procedures in pediatric dentistry. *J. Clin. Pediatr. Dent.* **2023**, *47*, 72–79. [CrossRef] [PubMed]

52. Yildiz, E.; Sirin Karaarslan, E.; Simsek, M.; Ozsevik, A.S.; Usumez, A. Color stability and surface roughness of polished anterior restorative materials. *Dent. Mater. J.* **2015**, *34*, 629–639. [CrossRef] [PubMed]
53. Hongsathavij, R.; Kuphasuk, Y.; Rattanasuwan, K. Clinical comparison of the stain removal efficacy of two air polishing powders. *Eur. J. Dent.* **2017**, *11*, 370–375. [CrossRef] [PubMed]

polymers

MDPI

Article

Laser-Induced Silver Nanowires/Polymer Composites for Flexible Electronics and Electromagnetic Compatibility Application

Il'ya Bril' [1,*], **Anton Voronin** [1,2,3], **Yuri Fadeev** [1], **Alexander Pavlikov** [3], **Ilya Govorun** [4,5], **Ivan Podshivalov** [4], **Bogdan Parshin** [2], **Mstislav Makeev** [2], **Pavel Mikhalev** [2], **Kseniya Afanasova** [1,6], **Mikhail Simunin** [1,2,3] and **Stanislav Khartov** [1]

1 Federal Research Center, Krasnoyarsk Scientific Center, Siberian Branch, Russian Academy of Sciences (FRC KSC SB RAS), 660036 Krasnoyarsk, Russia; a.voronin1988@mail.ru (A.V.); daf.hf@list.ru (Y.F.); shabanova.ksenia@mail.ru (K.A.); michanel@mail.ru (M.S.); stas_f1@list.ru (S.K.)
2 Department of Radio-Electronic Systems and Devices, Bauman Moscow State Technical University, 105005 Moscow, Russia; parshbgal@bmstu.ru (B.P.); m.makeev@bmstu.ru (M.M.); pamikhalev@bmstu.ru (P.M.)
3 Siberian Federal University, 660041 Krasnoyarsk, Russia; hahanka@yandex.ru
4 Kirensky Institute of Physics, Federal Research Center KSC SB RAS, 660036 Krasnoyarsk, Russia; govorun-ilya@mail.ru (I.G.); podshivalov.ivan@gmail.com (I.P.)
5 Reshetnev Siberian State University of Science and Technology, 660037 Krasnoyarsk, Russia
6 Institute of Space Technologies, FRC KSC SB RAS, 660037 Krasnoyarsk, Russia
* Correspondence: ellaijiah@gmail.com

Abstract: Nowadays, the Internet of Things (IOT), electronics, and neural interfaces are becoming an integral part of our life. These technologies place unprecedentedly high demands on materials in terms of their mechanical and electrical properties. There are several strategies for forming conductive layers in such composites, e.g., volume blending to achieve a percolation threshold, inkjet printing, lithography, and laser processing. The latter is a low-cost, environmentally friendly, scalable way to produce composites. In our work, we synthesized AgNW and characterized them using Ultraviolet-visible spectroscopy (UV-vis), Transmission electron microscopy (TEM), and Selective area electron diffraction (SAED). We found that our AgNW absorbed in the UV-vis range of 345 to 410 nm. This is due to the plasmon resonance phenomenon of AgNW. Then, we applied the dispersion of AgNW on the surface of the polymer substrate, dried them and we got the films of AgNW.. We irradiated these films with a 432 nm laser. As a result of the treatment, we observed two processes. The first one was the sintering and partial melting of nanowires under the influence of laser radiation, as a consequence of which, the sheet resistance dropped more than twice. The second was the melting of the polymer at the interface and the subsequent integration of AgNW into the substrate. This allowed us to improve the adhesion from 0–1 B to 5 B, and to obtain a composite capable of bending, with radius of 0.5 mm. We also evaluated the shielding efficiency of the obtained composites. The shielding efficiency for 500–600 nm thick porous film samples were 40 dB, and for 3.1–4.1 μm porous films the shielding efficiency was about 85–90 dB in a frequency range of 0.01–40 GHz. The data obtained by us are the basis for producing flexible electronic components based on AgNW/PET composite for various applications using laser processing methods.

Keywords: flexible electronics; silver nanowires; sheet resistance; shielding efficiency

1. Introduction

Flexible electronics have become a common part of our life. Radio Frequency IDentification (RFID) tags, neural interfaces, and wearable sensors have all been included in our lives thanks to the impressive rate of development in this industry. In addition to targeted applications as electronic components, flexible electronic materials can be used as electromagnetic compatibility components. With the development of the flexible electronics industry, the demand for flexible shielding materials has also increased. Flexible electronics, just like traditional ones, consists of three main types of components: 1—substrate; 2—conductive paths; 3—the basic components of which the electronic circuit consists [1].

Currently, polymer materials are used as substrates. There are several reasons for this: polymers are inexpensive to produce and there is a possibility to recycle them, and creating conductive paths is one of the most important tasks in the manufacture of flexible electronics. We can formulate the requirements for this element: 1—it must have the lowest possible resistance; 2—it must be durable and retain its properties under various mechanical influences; 3—it must retain its electrical properties under various bending conditions.

The most advanced technologies make conductive paths and patterns. There are various types of lithography and printing with electrically conductive inks [2]. These inks are dispersions of conductive agents in liquid. Types of inks can be categorized by the type of conductive agents. These can be: 1—carbon materials, such as carbon nanotubes [3,4], carbon black [5], and graphene [6]; 2—modern nanomaterials such as MXenes [7] and other materials [8]; 3—metal materials, such as metal nanoparticles [9] and metal nanowires [10]. Nanoscale conductive agents are preferred, as they allow an increased resolution of printing [11]. The point is that inkjet printing allows the creation of conductive paths with good homogeneity but does not contribute in any way to the adhesion of conductive paths to the substrate. As a result, such paths may not withstand mechanical stress without additional protection. Additionally, such methods are neither inexpensive nor environmentally friendly and they are technically challenging because they require the use of preformed masks [12]. In recent years, laser-based approaches to creating flexible electronics have become increasingly popular as cost-effective, environmentally friendly, and low-cost alternatives to traditional methods [13–15]. Some examples of the amazing results of this approach can be found in the results of Liu and Lee in the formation of silver templates on polyimide substrate [16], the sintering of silver nanoparticles on polyethylene terephthalate (PET) substrate by Kim et al. [17], and the laser-formed composite of aluminum nanoparticles and PET shown in the work of Rodriguez et al. [12]. These and other works related to laser processing of nanomaterials/polymers inspired us to expand our approach to creating electrically conductive nanocomposites using the silver nanowires (AgNW). We expect silver nanowires to act as a photothermal converter due to the effect of localized surface plasmon resonance in the UV-blue wavelength [18], which will make it possible to create a conductive composite of AgNW and any thermoplastic polymer using laser processing. Such composites must have phenomenal wear resistance, and their flexibility will be determined by the mechanical properties of the substrate. To test this hypothesis, we synthesized low aspect ratio nanowires, deposited films on thermoplastic polymers, laser treated them, and assessed their optical, electrical, and mechanical properties. The results obtained will be a significant addition to the rapidly developing topic of laser-induced conductive composites.

2. Materials and Methods

2.1. Synthesis of AgNW

AgNW were synthesized by the following method. First, 1.3 g polyvinylpyrrolidone (PVP) (100 kDa) (ITW Reagents Panreac Castellar del Vallès, Barcelona, Spain) was dissolved in 40 mL of ethylene glycol (EG) (EKOS, Moscow, Russia) into three-necked flask. The flask was slowly heated to 160 °C with vigorous magnetic stirring. When the temperature had stabilized at 160 °C for 1 h, an ethylene glycol solution of $FeCl_3$ was added. Then, 10 mL of 1.47 M ethylene glycol solution of $AgNO_3$ was added to the flask. Then the flask was sealed until the solution became glistening, indicating the formation of AgNW [19]. After the synthesis, we replaced the dispersant with AgNW. For this, the resulting dispersion in ethylene glycol was centrifuged at 10,000 rpm for 15 min using a laboratory centrifuge OPN 16 (Labtex, Moscow, Russia). After centrifugation, the supernatant was drained, and the same amount of isopropyl alcohol was added. This operation was repeated 5 times until the complete removal of ethylene glycol, NO^{3-}, Cl^-, Fe^{3+} ions, and PVP residues.

2.2. Samples Preparation

We used polyethylene terephthalate (PET) (Hi-Fi Industrial Film Ltd., Stevenage, Hertfordshire, UK), polyethylene naphthalate (PEN) (Toray industries, Tokyo, Japan), and polyvinylidene fluoride (PVDF) (3 M, Saint-Paul, MN, USA) as substrates for the application of films by drip casting. We applied different volumes of base dispersion (~20 gm/mL concentration): 12.5, 25, 37.5, 50, 62.5, and 75 μL/cm^2.

2.3. Microscopy

Optical microscopy photographs were taken on an Altami 104 microscope (Altami, Saint Petersburg, Russia). The morphology and geometric characteristics of individual AgNW were studied by transmission electron microscopy (TEM) on a HT 7700 (Hitachi, Tokyo, Japan) at an accelerating voltage of 40–300 kV. The morphology of AgNW films was studied by scanning electron microscopy (SEM) on a SU3500 microscope (Hitachi, Japan) at an accelerating voltage of 20 kV. EDX images and spectra were obtained on a SU3500 (Hitachi, Japan) microscope equipped with an energy-dispersive X-ray spectrometer XFlash 430 (Brucker, MA, USA).

2.4. XRD

X-ray diffractograms of the samples were taken using a X'Pert Pro MPD diffractometer (PANalytical, Almelo, The Netherlands) with a high-speed PIXcel detector in the angular range 30–90° 2Θ, with a step of 0.013°. The Ag lattice parameters were determined and refined using the full Rietveld approach by the difference derivative minimization (DDM) method.

2.5. Optical Properties

The optical density of AgNW was measured in the range of 280–800 nm using a UV-3600 spectrophotometer (Shimadzu, Kyoto, Japan). For this purpose, the base dispersion of AgNW with a concentration of ~20 mg/mL was diluted 100 times. Then, 1 mL of the obtained diluted dispersion was poured into a quartz cuvette. The spectrum was taken with preliminary acquisition of the baseline of pure dispersant (ethanol).

2.6. Electrical Properties

The sheet resistance of AgNW films was measured by the four-probe method using a JG ST2258 four-point probe station (Suzhou Jingge Electronics Co., Suzhou, China) and a JG ST2558-F01 four-probe head (Suzhou Jingge Electronics Co., Suzhou, China).

2.7. Shielding Efficiency

The S_{21} and S_{11} coefficients of the materials were measured by the waveguide method. The investigated sample was placed in the break of a waveguide transmission line.

This investigation used a wide range band (0.01–7 GHz) and a special air coaxial cell with a diameter of 16.00/6.95 mm (type II, 50 Ω, GOST RV 51914-2002). The measurements were carried out in the range of 10 MHz to 7 GHz; this frequency range includes the L (1–2 GHz), S (2–4 GHz) and C (4–8 GHz) bands. The ability to measure at low frequencies with a relatively simple and convenient measurement technique provides good quality results. The measurements were carried out on a Keysight FieldFox N9916A vector network analyzer (Keysight Technologies, Santa-Rosa, CA, USA).

K band (18–26.5 GHz) and a Ka band (26.5–40 GHz) were used. The waveguide-coaxial transitions had a rectangular cross section with dimensions of 4.3 mm × 10.65 mm for K band and 3.55 mm × 7.1 mm for Ka band. Measurements were performed on an R&S ZVA 40 vector circuit analyzer (GmbH & Co. KG, Großheringen, Germany).

2.8. Mechanical Properties

The adhesion strength of AgNW/PET composites was evaluated using a tape test according to ASTM D 3359, in geometry B (notch lattice period was 2 mm). The effect of the reusable tape test on resistance AgNW/PET composites was also studied. In all studies, 3 M tape was used, and the test sample was firmly fixed on the table. The mechanical properties of AgNW/PET composites were studied in comparative experiments on a laboratory bench in single load and cyclic modes [20]. For the single bend mode, we took several bending radii: 10 mm, 5 mm, 2 mm, 1 mm, and 0.5 mm. The bending templates were fabricated using 3D printing. Cyclic bending allowed us to investigate fatigue accumulation in AgNW/PET composites. In our experiments, the number of bending cycles with a radius of 0.5 mm was 1000.

3. Results

3.1. AgNW Characterization

We used TEM, selective area electron diffraction (SAED) and UV-Vis spectrophotometry techniques to study the structure and morphology of AgNW. Figure 1a shows the TEM image of the resulting AgNW.

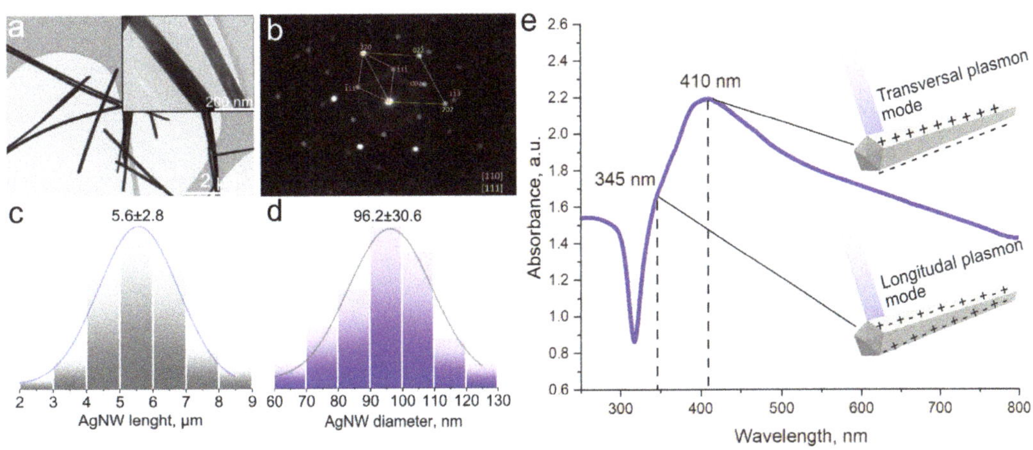

Figure 1. AgNW characterization. (**a**)—TEM Image, (**b**)—SAED, (**c**)—AgNW diameter, (**d**)—AgNW length, (**e**)—UV-vis absorption spectra.

Knowing the average AgNW length (5.6 ± 2.8 µm) and average AgNW (96 ± 30.6 nm) (Figure 1c,d) diameter, we can calculate the average aspect ratio–58.3. Using electron diffraction on the selected region, we see the presence of reflections 111 and 002 which show that the AgNW have more than one silver single crystal, i.e., on the twin structure. The fivefold twinning indicates that we have nanowires with a non-symmetrical pentagon in the cross section. (Figure 2b).

We see a peak and a satellite on the UV-vis absorption spectrum (Figure 1e). They are related to the phenomenon of surface plasmon propagation in nanowires [20,21]. Twinization is a typical growth behavior of nanowires, often found in the literature [22,23]. The complexity of the shape of the plasmon absorption peak increases with the decreasing symmetry of the pentagon in the cross section. We see a peak and a satellite on the UV-vis absorption spectrum. The peak maximum is at a wavelength of 410 nm. This is different from most results found in the literature [24,25]. According to Todd's work, the absorption peak moves towards the blue–green wavelength region with increasing AgNW diameters, [26] which is what we see in our case. A surface plasmon can heat AgNW to 1000 °C, which is higher than the melting point of silver [27]. This, as well as the high degree of absorption in the blue wavelength region, inspired us to attempt laser-induced sintering of nanowires and their simultaneous integration into a thermoplastic polymer substrate with a blue microsecond laser.

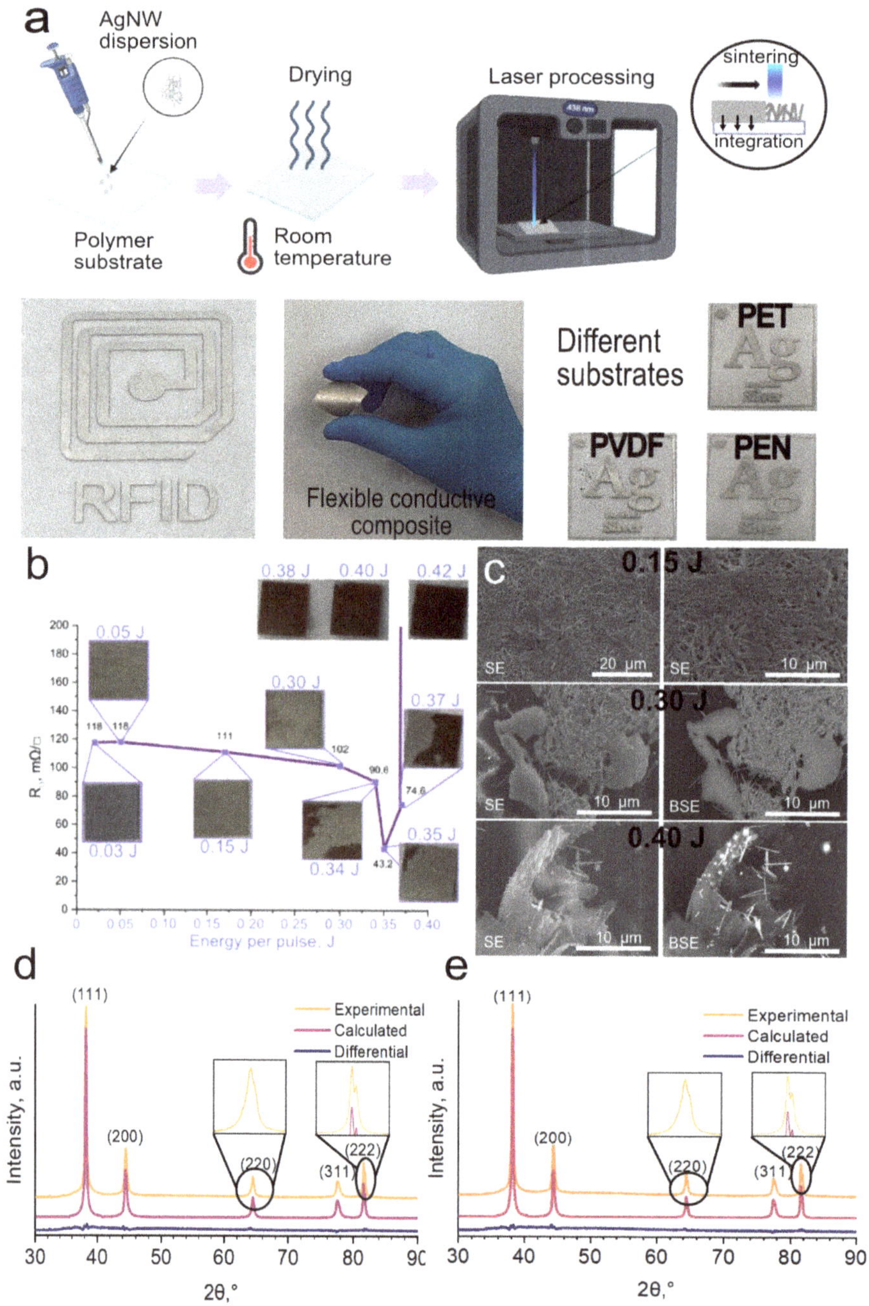

Figure 2. Samples preparation and characterization. (**a**)—Samples obtaining scheme, (**b**)—effect of power density on electrical parameters and morphology, (**c**)—SEM images at different power densities, (**d**)—XRD of original film, (**e**)—XRD of laser-processed film.

3.2. AgNW/Polymer Flexible Composites Obtaining and Characterization

We used a dispersion of AgNW in ethanol, with a concentration of ~20 μL/cm^2, to deposit AgNW on a polymer substrate using a drop-casting method. We chose polyethylene terephthalate (PET), polyvinylidene fluoride (PVDF), and polyethylene naphthalate (PEN) as substrates. After AgNW deposition, our films were dried at room temperature for 24 h. To make conductive free-form patterns on substrates we used a diode laser with 438 nm wavelength. The process scheme can be seen in Figure 2a. This process is possible for any thermoplastics. We have fabricated conductive patterns on commercial PEN and PVDF (Figure 2a). We chose PET as a substrate for further research as it is one of the most used polymers.

The quality of the composite can be varied by modifying two main parameters: the thickness of the AgNW coating and the amount of energy transferred per laser pulse.

It is possible to consider the influence of these two mechanisms on the target parameters of the composite separately. The selection of suitable laser processing parameters for the formation of laser-induced composites is a critical step. By provoking surface plasmon excitation, silver nanowires can reach high temperatures, which can provoke various phenomena occurring with the polymer substrate. The first is melting the nanowires and sintering the contacts. This affects the mechanical and electrical properties of the composite and is described in detail in the sections on electrical and mechanical properties. The second process is the melting of the polymer at the polymer/AgNW interface and the integration of the nanowires into the polymer. The third process is the pyrolysis of the polymer substrate, occurring due to excess energy received from laser radiation.

In Figure 2b, we can see the effect of laser irradiation energy on the morphology and electrical properties of the AgNW/PET composite. With increasing energy, we can see a trend of decreasing resistance. This is due to the sintering of nanowires and, as a consequence, the increase of the contact spot between nanowires (Figure 2c). For example, with increasing energy from 0.03 J/pulse to 0.35 J/pulse, we see a drop in resistance from 118 to 43 mΩ/sq However, further increase of energy leads to pyrolysis of the substrate and intensive melting of nanowires. For example, at an energy of 0.40 J/pulse we obtain a fully pyrolyzed region with pronounced carbon structures, inside which are integrated AgNW globules formed after melting (Figure 2c). Such phenomenon of Ag nanoscale objects is described by Rodriguez et al. [12].

An important parameter for flexible electronics manufacturing is reproducibility. Figure 2b shows that after the energy transition to 0.30 J/pulse, the reproducibility even on small areas becomes not satisfactory. This is due to the density heterogeneity of the films. Highly concentrated areas absorb laser radiation and heat up more than low concentrated areas, causing local pyrolysis of PET.

Taking this into account, it is reasonable to determine the production of subsequent samples at a power insufficient to cause the undesirable effects of the imperfections of the drop-casting method but that allows the minimum resistance to be obtained, which is 0.30 J/pulse.

For a more detailed investigation of the processes occurring during laser sintering and composite formation, we compared the XRD spectra of the original and laser-treated films. In Figure 2d,e we see peaks attributed to metallic Ag. Five distinct diffraction peaks at 2θ = 38.13, 44.4, 64.4, 77.47, and 81.5° were indexed to the (111), (200), (220), (311), and (222) reflections of metallic Ag. Ag in our case has a tetragonal distorted lattice and parameters: a = b=4.088(1), c = 4.095(2) Å. This indicates the preservation of the crystalline structure of AgNW after exposure to 0.30 J/pulse (0.19 J/cm^2), which is in agreement with the literature on laser welding and sintering of AgNW [28,29].

3.3. Laser Sintering AgNW/ PET Composite Morphology and Electrical Properties

To study the property of our obtained AgNW films, we fabricated 12 samples with six different thicknesses on glass substrate and PET, respectively.

The thickness is an important parameter. Changing the thickness allows the indirect changing of the amount of energy delivered per unit area per impulse [30,31]. Recent literature has reported the influence of the thickness of metallic nanoparticle and nanowire films. Reducing the thickness of the layer has a positive effect in most cases, as it allows more energy to be delivered per unit volume, resulting in a more uniform sintered film, reduced porosity, and improved electrical and mechanical parameters. [31,32].

The reduction of porosity and free volume in the AgNW layer can influence the thickness of the AgNW layer obtained after laser treatment. In order to study this, we deposited six different thicknesses of AgNW films on a glass substrate, after which we split the sample into two parts and looked at the cross-section in SEM, as well as taking the EDX spectra. Glass was chosen as the substrate because it is much easier to split it into two parts than any thermoplastic polymer substrate.

Figure 3a shows the concentration dependence of the layer thickness of irradiated nanowires. Taking into account that small concentrations of the basic dispersion give low homogeneity of the coating, we took several points and calculated the average thickness. As we can see, the dependence can be described by a straight-line equation, but low concentrations (12.5 and 25 $\mu L/cm^2$) do not fit the trend due to the high inhomogeneity of the layer.

In Figure 3b we see the EDX of the AgNW film, with a concentration of 62.5 $\mu L/cm^2$. At the top we see the scan area of the sample in the BSE picture, at the bottom the intensity of the element peaks of each mapping point. We see the expected dominance of silver intensity in the region related to the irradiated AgNW film, in the other regions we see the presence of spectra of silicon, oxygen, and sodium, which are components of the glass substrate.

In Figure 3c we can see the EDX spectra. They clearly show the X-ray series of the silver spectrum corresponding to the laser-irradiated AgNW film. In addition to them, we can also see a series of spectra of the glass substrate components (Si, Na, O, etc.).

The most important characteristic for materials used in flexible electronics applications is the electrical resistance of the material. In our case, the conductive layer is a thin integrated AgNW layer.

Usually, such samples are characterized by the sheet resistance, which is the resistance of a rectangular surface with a certain thickness [33,34].

A schematic representation of the four-probe sheet resistance measurement method is shown in Figure 4a.

Figure 4b shows the dependence of sheet resistance of AgNW films before and after laser treatment with energy of 0.30 J per pulse (0.19 J/cm^2). As we said earlier, laser treatment leads to the fusion of AgNW at their contact points and, as a consequence, to an increase in the contact spot, which leads to a drop in resistance after laser treatment. This effect is particularly noticeable in films with minimal AgNW content per unit area. From surface concentrations above 50 $\mu L/cm^2$ we almost observe a plateau in resistance.

We see that the resistance drops with increasing thickness for both treated and untreated films, which follows from Ohm's law. It is also possible to achieve low sheet resistance using this approach. However, this strategy is not suitable for flexible electronics applications because thick conductive coatings will affect the flexibility of the products. From this point of view, it is worthwhile to compare conductive coatings and films in terms of surface resistivity and coating thickness.

We also calculated the resistivity of the composite, which in the case of an ideal homogeneous film should remain constant and be equal to the resistivity of crystalline silver. In Figure 4b, we see that in our case, the resistivity close to the resistivity of crystalline silver is demonstrated by samples with a thickness of 4.1 $\pm$ 0.2 μm (75 $\mu L/cm^2$) and 3.1 $\pm$ 0.2 μm (62.5 $\mu L/cm^2$). This is due to the inhomogeneity of our AgNW conductive layer. As the thickness of the conductive layer decreases, the resistivity increases. This is due to the increase in AgNW conductive layer inhomogeneity at small thicknesses.

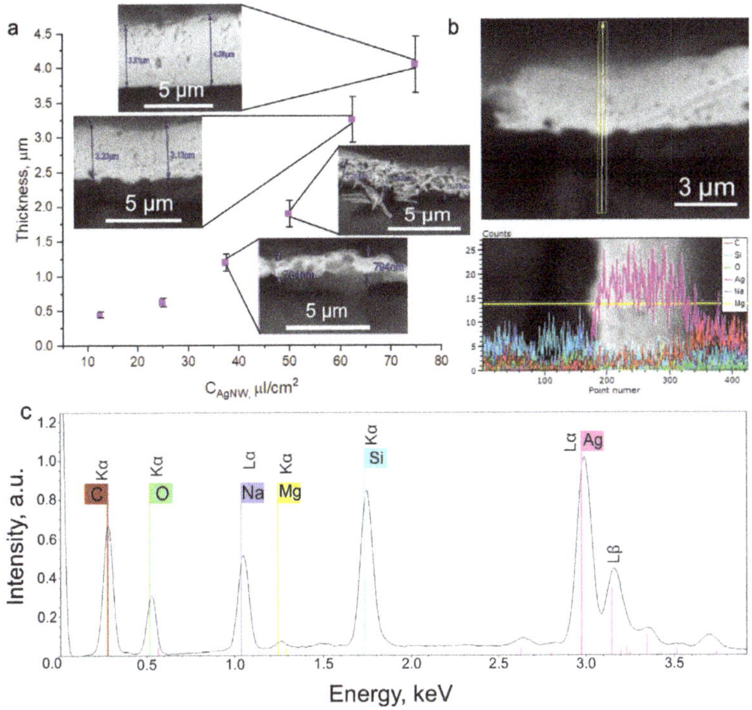

Figure 3. Thickness effect study. (**a**)—Dependence of thickness from concentration of AgNW basic dispersion, (**b**)—SEM image of x-section of 62.5 µL/cm2 BSE (top image), (**c**)—EDX spectra.

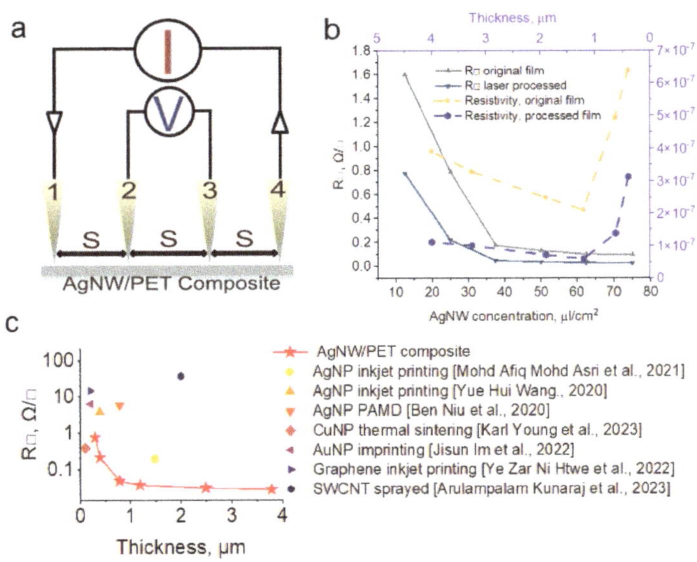

Figure 4. Electrical properties. (**a**)—Four-point probe measurements illustration, (**b**)—sheet resistance/thickness dependence and resistivity of AgNW/PET composite, (**c**)—sheet resistance/thickness comparison [35–42].

In Figure 4c we can see the comparison of conductive coatings and films with our work [35–42]. Note that for the whole range of thicknesses, our results are superior to most of the results mentioned in the literature and better than all the results obtained with nanoscale silver. Only the results of Park et al. with imprinted Au NP are superior to our AgNW/PET composites. Laser allows for much better results than inkjet and lithography, and the technology is more cost-effective than lithography, more environmentally friendly, and easily scalable [43].

3.4. Electromagnetic Shielding Performance of AgNW/PET Composite

The purpose of EMI shielding is to encapsulate the area to be shielded with a shielding sheath of some conductive or magnetic material to provide EM insulation. The first purpose is to limit the leakage of EM energy outside the area and its effect on external equipment. The second purpose is to prevent EM energy from outside the area from entering the area and affecting internal equipment [44]. Quantitatively, the shielding efficiency is evaluated by the shielding efficiency SE. This is defined in decibels (dB) according to the formula:

$$SE = 20 \lg E_1/E_2 \quad (1)$$

where E_1 is the electromagnetic wave (EMW) amplitude at an arbitrary point of the shielding space without the screen and E_2 is the EMW amplitude at the same point with the screen. For a material, the SE can be interpreted as the inverse of the trans-mittance S_{21} (dB):

$$SE = -10\lg(S_{21}) \quad (2)$$

Figure 5 shows the dependence of the transmittance coefficient (S_{21}) in all investigated frequency ranges (0.01–40 GHz). The SE increases with increasing surface density of AgNW deposited from the base dispersion. The shielding efficiency increases from 40 dB to 90 dB when the surface density increases from 12.5 $\mu L/cm^2$ to 75 $\mu L/cm^2$ respectively. It can be seen that samples with concentrations of 12.5 and 25 mL/cm^2 tend to decrease in shielding efficiency with increasing frequency. This is due to the fact that laser-treated thin films are structures containing dielectric and conducting regions [44]. The influence of inhomogeneity increases with increasing frequency, which creates a specific slope of the curves for these samples. However, in the context of shielding values, this weak slope does not have a significant effect.

At surface densities of 37.5–75 $\mu L/cm^2$, we see a trend of increasing shielding efficiency with increasing frequency. Thus, in films with a surface density of 37.5 $\mu L/cm^2$, the shielding efficiency shows a weak increase from 70 to 74 dB and 75 $\mu L/cm^2$ film shows an increase in shielding efficiency from 77 dB to 90 dB. The increase in shielding efficiency with increasing frequency for high continuity films is attributed to the skin effect.

The interaction between incident electromagnetic waves and the surface/interface of an electromagnetic shield can be divided into reflection (R), absorption (A), multiple reflection, and transmission (T). The coefficients are calculated using parameters S_{11} and S_{21}. The formulas are described below:

$$R = P_r/P_i = 10^{0.1S11} \cdot 100\%, \quad (3)$$

$$T = P_t/P_i = 10^{0.1S21} \cdot 100\%, \quad (4)$$

$$A = 100\% - T - R, \quad (5)$$

where P_i, P_r, P_t are the incident, reflected, and transmitted wave powers. The scattering matrix parameters S_{11} and S_{21} (the scattering matrix parameter, which is the transmission coefficient) should be taken in dB.

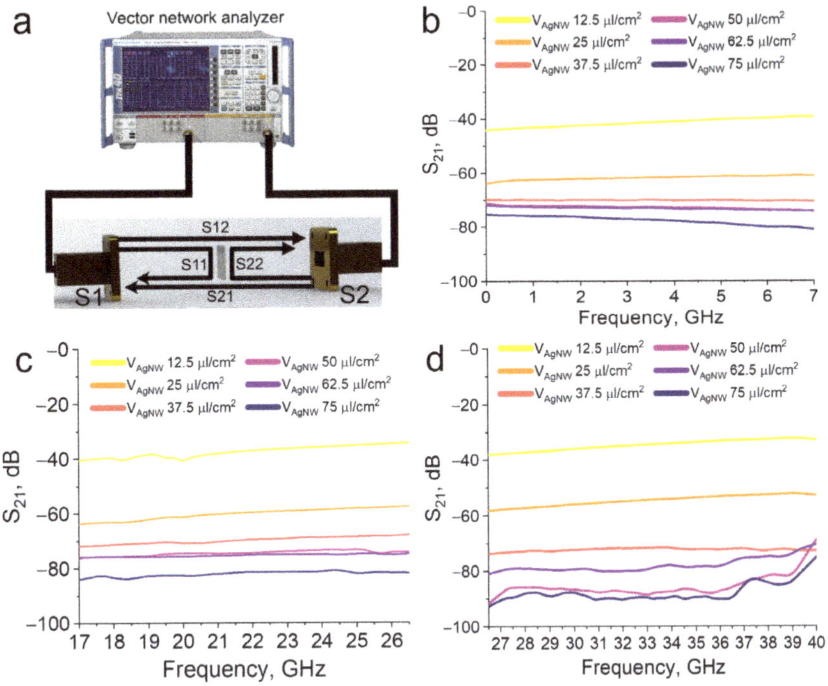

Figure 5. Electromagnetic compatibility. (**a**)—Waveguide method illustration, (**b**)—S_{21} parameter at 0.01–7 GHz range, (**c**)—S_{21} parameter at 17–26.5 GHz range, (**d**)—S_{21} parameter at 26.5–40 GHz range.

It is possible to determine the energy balance for these three components that make up the interaction of the electromagnetic wave with the screen. Using the Formulas (3)–(5), we calculated these components and made the reflection, absorption, transmission (RAT) diagram (Figure 6). The results of the calculations are shown in Figure 6a–c. Since there is a weak tendency for the shielding efficiency to vary up or down, we averaged the RAT values in the following ranges: 0.01–7 GHz, 17–26.5 GHz, and 26.5–40 GHz and listed the corresponding results as columns. We can conclude that the main mechanism of shielding is reflection. The scheme of the interaction of radio waves with AgNW/PET composite is shown in Figure 6d. The increase of transmittance components with increasing frequency is related to the morphology of AgNW/PET composite. It is a porous structure which can be described as a network. Examples of mesh structures with similar transmittance component growth have already been mentioned in the literature [45].

Modern science places extremely stringent demands on materials. In particular, shielding materials must not only have a high shielding capacity, but also be flexible and, more importantly, thin. In view of such requirements, we have compared those mentioned in the literature with ours, correlating shielding effectiveness with thickness.

In Figure 7 you can see the graph. It can be seen that our AgNW/PET composite has a shielding efficiency superior to the coatings mentioned in the analysis. The graphene nanosheet coatings obtained by Panda et al. and Yuan et al. surpass our results, but only at thicknesses many times greater than ours [45–60].

In general, similar results can be achieved using ordinary thick materials. However, their high density and high weight are not suitable for aerospace, telecommunication applications, and materials for wearable and flexible electronics.

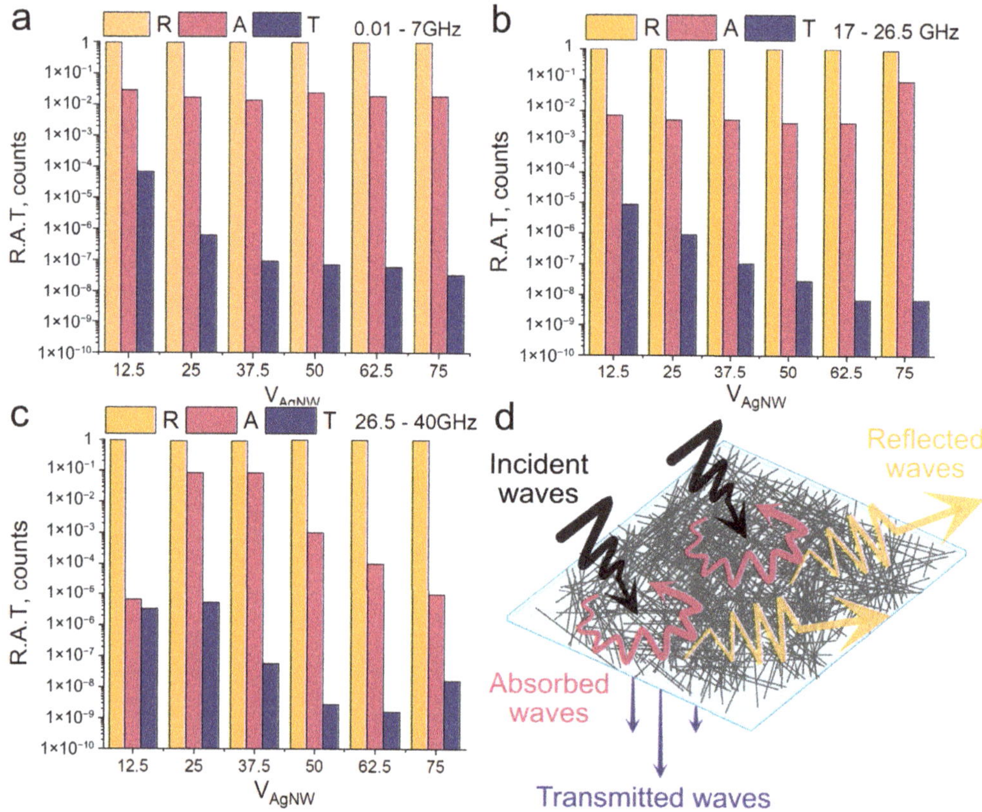

Figure 6. Reflection, absorption, transmission (RAT) diagrams. (**a**)—RAT at 0.01–7 GHz, (**b**)—RAT at 17–26.5 GHz, (**c**)—RAT at 26.5–40 GHz, (**d**)—AgNW/PET composite EMI shielding illustration.

To evaluate the shielding efficiency, taking into account the thickness of the screen, the parameter of specific shielding efficiency (SSE) is used, which is calculated by the following formula [44]:

$$SSE_t = SE_T/\rho \times t \quad (6)$$

where ρ—density in g/cm^3 and t—thickness in mm.

In Figure 7b, you can see a graph comparing SSE for different manuscripts. In the comparison of SSE_t values, we used only metallic nano-objects of different shapes (nanowires, nanoplates, nanoparticles, etc.). As we can see, our coating shows one of the best results for thicknesses below 5 microns.

3.5. Mechanical Properties AgNW/PET Composite

Flexible electronics involves the use of materials under a load that bends or curls the product. We have conducted various mechanical stability tests to evaluate the suitability of our composites for flexible electronics applications. The results of these tests are shown in Figure 8.

Figure 8a–c shows optical photographs of AgNW films before laser treatment and after laser treatment. The images of maximum and minimum concentrations (12.5 μL/cm^2 and 75 μL/cm^2) for laser treated films are also shown. Adhesion was tested according to ASTM D3359. The adhesion of the films without laser treatment is at 0–1B, which means a peel percentage of 35–65%. This level of adhesion is not suitable for use in flexible electronics applications. The treated films show adhesion level 5B for concentrations of 12.5–50 μL/cm^2 and 4B for 62.5 and 75 μL/cm^2. The decrease in adhesion level is due to the fact that the large thickness (about 4 μm) prevents the sintering of AgNW located in different planes. The high adhesion of AgNW to the substrate indirectly confirms the formation of a composite at the PET/AgNW interface.

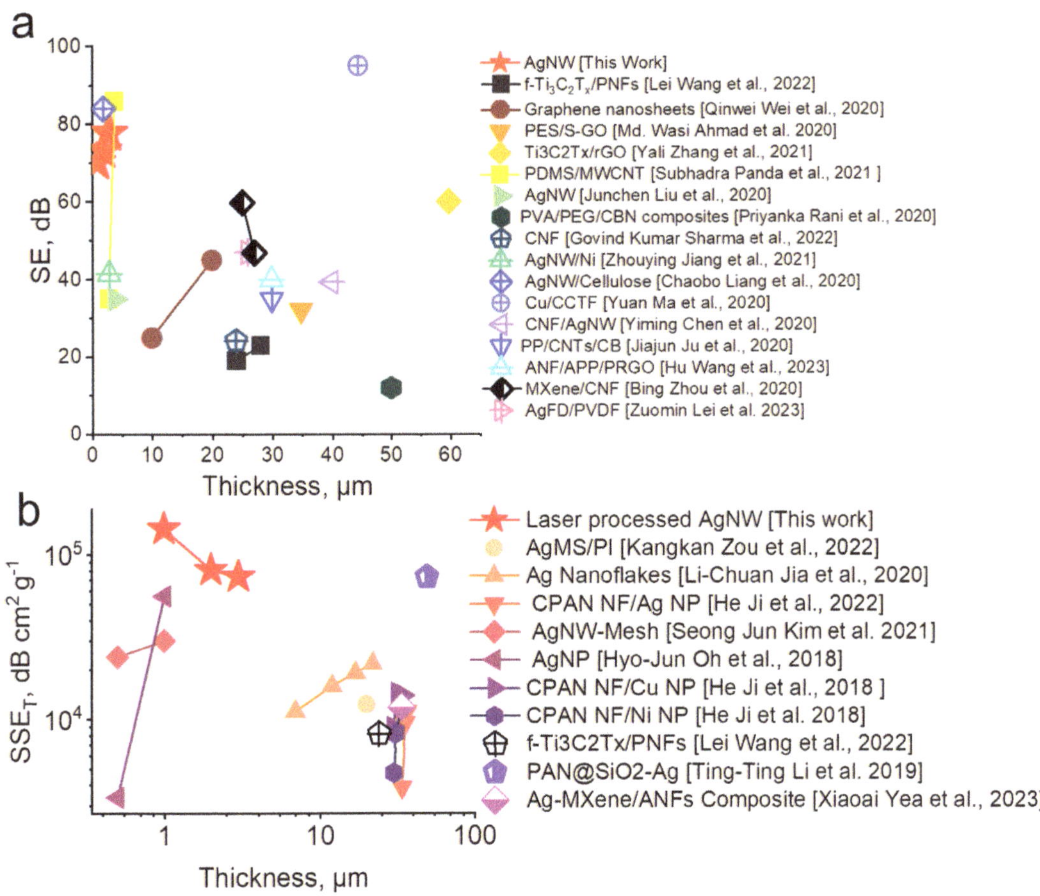

Figure 7. AgNW/PET composite comparisons with alternative materials. (**a**)—SE comparison [45–61], (**b**)—SSE$_T$ comparison [61–68].

We tested AgNW/PET composites for resistance to tape tear-off. We fabricated a conductive pattern, in the form of paths with different widths (500 μm, 1 mm, 3 mm, and 5 mm), at the ends of which were contact pads for measuring equipment (Figure 8d). The results of resistance changing after each detachment are shown in the histogram in Figure 8d. Each of the conductors retained electrical conductivity after five cycles of the taping test. The resistance of the 500 μm track changed the most. Removal of the same amount of AgNW from paths of different widths left different amounts of surviving contacts, which is associated with the increase in resistance.

High adhesion of the conductive layer to the substrate is one of the main requirements for flexible electronics. However, it is not the only one. The material must be able to retain its properties after bends of various radii. To test the durability of the AgNW/PET composite, we performed bending tests with radii of 0.5 mm, 1 mm, 2 mm, 5 mm, and

10 mm. Figure 8e shows the graph of resistance variation as a function of bending radius. The resistance remained at the same level regardless of the bending radius. The sintered contacts between single AgNW after laser treatment retained high tensile strength.

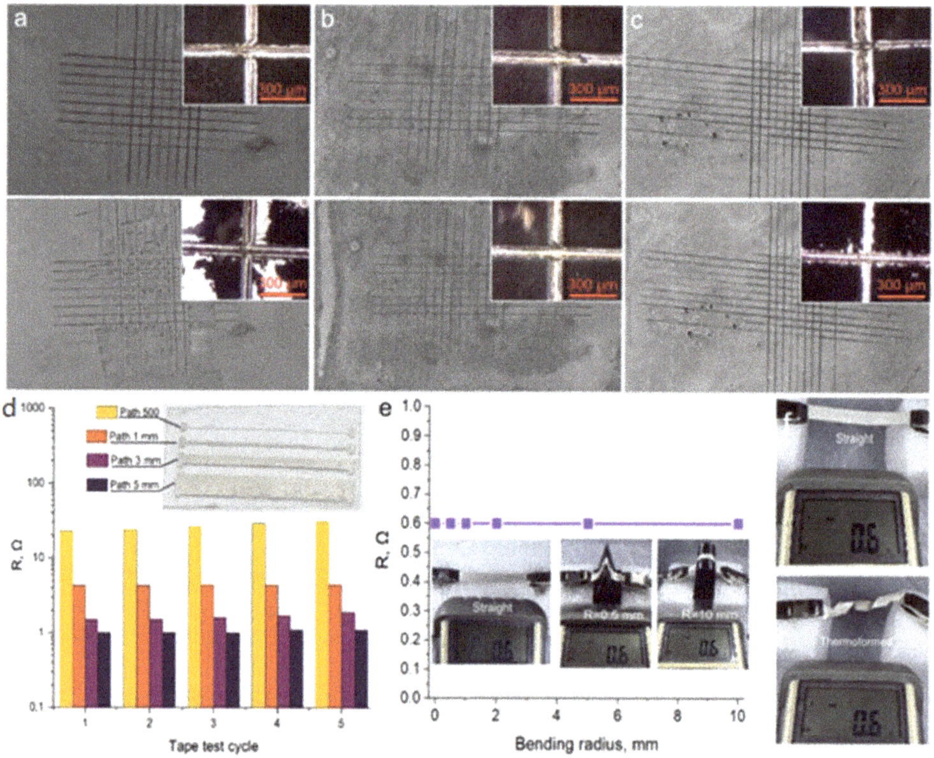

Figure 8. Mechanical properties. (**a**)—Optical image of untreated film before and after the ASTM D3359 test; (**b**)—optical image of processed film with C_{AgNW} = 12.5 µL/cm^2 before and after the ASTM D3359 test; (**c**)—optical image of processed film with C_{AgNW} = 75 µL/cm^2 before and after the ASTMD D3359 test; (**d**)—change in resistance after a tape test; (**e**)—changing of resistance at different bending radius; (**f**)—thermoforming demonstration.

Flexible electronics involves the constant bending and unbending of devices, which requires materials to be resistant to fatigue damage. To test the durability of AgNW/PET composites, we tested them for 1000 cycles of bending-extension with a radius of 0.5 mm. We measured the resistance of the AgNW/PET composite after every 100 bends (Figure S3).

AgNW/PET composite retains the thermoplastic substrate's thermoforming ability, without loss of electrical properties. This expands the scope of its potential applications and makes it possible to use it in various technological processes [69]. Demonstration of the thermoforming process together with detailed characterization of mechanical properties provides a basis for the integration of such composites into industry.

4. Conclusions

In this work, we demonstrated the laser sintering of AgNW with their subsequent integration into the substrate and formation of AgNW/PET composites. In the UV-Vis spectroscopy, we found absorption peaks at 345 and 410 nm, respectively. These two peaks are related to the plasmon surface resonance phenomenon. Considering this, we made AgNW films on a polymer substrate and treated them with a laser with a wavelength of

432 nm. After treatment, we characterized our films by SEM, EDX, and XRD and measured the sheet resistance. From the results, we found that the AgNW were heated by the laser exposure, which led to their sintering or melting, and the melting of the surface of the polymer substrate, followed by the integration of wires and the formation of a composite. The obtained AgNW/PET composite has a low sheet resistance of about 30 mΩ/sq. We also conducted shielding efficiency measurements. Our composite is capable of shielding in wide frequency ranges at the level of 89 dB with thickness about 3.1 ± 0.2 μm. In order to fully evaluate the applicability of the obtained material in flexible electronics applications, we conducted comprehensive mechanical tests, including an ASTM D3359 test, tape test, bending resistance test, cyclic bending test, and thermoforming. As a result, the AgNW/PET composite demonstrated high adhesive strength and bending resistance, including cyclic loads, while maintaining the ability to thermoform. The results obtained are an important addition to the field of manufacturing thermoplastic polymer composites for flexible electronics applications.

Supplementary Materials: The following supporting information can be downloaded at: https://www.mdpi.com/article/10.3390/polym16223174/s1, Figure S1: Photo before and after washing off excess AgNW after laser treatment; Figure S2: Search for modes for laser processing. The figure shows that we can only work in a narrow range of laser speeds and power. 100% speed corresponds to 100 mm per minute; Figure S3: Cyclic load for 1000 cycles. Measurements were taken every 100 bending-extension cycles; Video S1: Sample washing process.

Author Contributions: I.B.—Conceptualization, methodology, validation, investigation, data curation, writing—original draft preparation, writing—review and editing, visualization; A.V.—conceptualization, investigation, data curation, writing—original draft preparation, writing—review and editing, supervision; Y.F.—data curation, investigation, formal analysis, hardware; A.P.—validation, investigation, data curation, writing—original draft preparation, writing—review and editing, visualization; I.G.—methodology, data curation, formal analysis; I.P.—data curation, formal analysis; B.P.—data curation, writing—reviewing and editing, formal analysis; M.M.—data curation, writing—reviewing and editing; P.M.—supervision, project administration, funding acquisition; K.A.—investigation; M.S.—investigation; S.K.—supervision, project administration, and funding acquisition. All authors have read and agreed to the published version of the manuscript.

Funding: The work on synthesis, purification, and characterization of AgNW and shielding measurments was performed with support of the state assignment of the Ministry of Science and Higher Education of the Russian Federation (topic No. FSFN-2024-0016). Studies on the preparation of film structures by laser sintering were carried out with support of the state assignment of the Ministry of Science and Higher Education of the Russian Federation (topic No. FWES-2024-0026).

Institutional Review Board Statement: Not applicable.

Data Availability Statement: Data is contained within the article or Supplementary Materials.

Acknowledgments: SEM and TEM were carried out on the equipment of the Krasnoyarsk Regional Centre for Collective Use of FRC KSC SB RAS. We thank L.A. Soloviev for help in XRD analysis.

Conflicts of Interest: The authors declare no conflicts of interest.

References

1. Mercante, L.A.; Andre, R.S.; Facure, M.H.M.; Correa, D.S.; Mattoso, L.H.C. Recent Progress in Conductive Electrospun Materials for Flexible Electronics: Energy, Sensing, and Electromagnetic Shielding Applications. *Chem. Eng. J.* **2023**, *465*, 142847. [CrossRef]
2. Secor, E.B.; Hersam, M.C. Emerging Carbon and Post-Carbon Nanomaterial Inks for Printed Electronics. *Appl. Phys.* **2016**, *353*, 6298. [CrossRef] [PubMed]
3. Tortorich, R.P.; Choi, J.-W. Inkjet Printing of Carbon Nanotubes. *Nanomaterials* **2013**, *3*, 453–468. [CrossRef] [PubMed]
4. Kordás, K.; Mustonen, T.; Tóth, G.; Jantunen, H.; Lajunen, M.; Soldano, C.; Talapatra, S.; Kar, S.; Vajtai, R.; Ajayan, P.M. Inkjet Printing of Electrically Conductive Patterns of Carbon Nanotubes. *Small* **2006**, *2*, 1021–1025. [CrossRef]
5. Phillips, C.; Al-Ahmadi, A.; Potts, S.-J.; Claypole, T.; Deganello, D. The Effect of Graphite and Carbon Black Ratios on Conductive Ink Performance. *J. Mater. Sci.* **2017**, *52*, 9520–9530. [CrossRef]
6. Tran, T.S.; Dutta, N.K.; Choudhury, N.R. Graphene Inks for Printed Flexible Electronics: Graphene Dispersions, Ink Formulations, Printing Techniques and Applications. *Adv. Colloid Interface Sci.* **2018**, *261*, 41–61. [CrossRef]

7. Jiang, X.; Li, W.; Hai, T.; Yue, R.; Chen, Z.; Lao, C.; Ge, Y.; Xie, G.; Wen, Q.; Zhang, H. Inkjet-Printed MXene Micro-Scale Devices for Integrated Broadband Ultrafast Photonics. *Npj 2D Mater. Appl.* **2019**, *3*, 1–9. [CrossRef]
8. de Moraes, A.C.M.; Hyun, W.J.; Seo, J.-W.T.; Downing, J.R.; Lim, J.-M.; Hersam, M.C. Ion-Conductive, Viscosity-Tunable Hexagonal Boron Nitride Nanosheet Inks. *Adv. Funct. Mater.* **2019**, *29*, 1902245. [CrossRef]
9. Ko, S.H.; Chung, J.; Hotz, N.; Nam, K.H.; Grigoropoulos, C.P. Metal Nanoparticle Direct Inkjet Printing for Low-Temperature 3D Micro Metal Structure Fabrication. *J. Micromech. Microeng.* **2010**, *20*, 125010. [CrossRef]
10. Finn, D.J.; Lotya, M.; Coleman, J.N. Inkjet Printing of Silver Nanowire Networks. *ACS Appl. Mater. Interfaces* **2015**, *7*, 9254–9561. [CrossRef]
11. Dimitriou, E.; Michailidis, N. Printable Conductive Inks Used for the Fabrication of Electronics: An Overview. *Nanotechnology* **2021**, *32*, 33735843. [CrossRef] [PubMed]
12. Rodriguez, R.D.; Shchadenko, S.; Murastov, G.; Lipovka, A.; Fatkullin, M.; Petrov, I.; Tran, T.-H.; Khalelov, A.; Saqib, M.; Villa, N.E.; et al. Ultra-Robust Flexible Electronics by Laser-Driven Polymer-Nanomaterials Integration. *Adv. Funct. Mater.* **2021**, *31*, 2008818. [CrossRef]
13. Lin, L.; Peng, H.; Liu, Z. Synthesis Challenges for Graphene Industry. *Nat. Mater.* **2019**, *18*, 520–524. [CrossRef] [PubMed]
14. Trusovas, R.; Račiukaitis, G.; Niaura, G.; Barkauskas, J.; Valušis, G.; Pauliukaite, R. Recent Advances in Laser Utilization in the Chemical Modification of Graphene Oxide and Its Applications. *Adv. Opt. Mater.* **2016**, *4*, 37–65. [CrossRef]
15. Senyuk, B.; Behabtu, N.; Martinez, A.; Lee, T.; Tsentalovich, D.E.; Ceriotti, G.; Tour, J.M.; Pasquali, M.; Smalyukh, I.I. Three-Dimensional Patterning of Solid Microstructures through Laser Reduction of Colloidal Graphene Oxide in Liquid-Crystalline Dispersions. *Nat. Commun.* **2015**, *6*, 1–7. [CrossRef]
16. Liu, Y.-K.; Lee, M.-T. Laser Direct Synthesis and Patterning of Silver Nano/Microstructures on a Polymer Substrate. *ACS Appl. Mater. Interfaces* **2014**, *6*, 14576–14582. [CrossRef]
17. Kim, K.K.; Ha, I.; Won, P.; Seo, D.-G.; Cho, K.-J.; Ko, S.H. Transparent Wearable Three-Dimensional Touch by Self-Generated Multiscale Structure. *Nat. Commun.* **2019**, *10*, 1–8. [CrossRef]
18. Liu, X.; Wu, B.; Zhang, Q.; Yip, J.N.; Yu, G.; Xiong, Q.; Mathews, N.; Sum, T.C. Elucidating the Localized Plasmonic Enhancement Effects from a Single Ag Nanowire in Organic Solar Cells. *ACS Nano* **2014**, *8*, 10101–10110. [CrossRef]
19. Xiao, N.; Chen, Y.; Weng, W.; Chi, X.; Chen, H.; Tang, D.; Zhong, S. Mechanism Understanding for Size Regulation of Silver Nanowires Mediated by Halogen Ions. *Nanomaterials* **2022**, *12*, 2681. [CrossRef]
20. Ding, S.; Tian, Y.; Jiang, Z.; Wang, C. Joining of Silver Nanowires by Femtosecond Laser Irradiation Method. *Mater. Trans.* **2015**, *56*, 981–983. [CrossRef]
21. Luu, Q.N.; Doorn, J.M.; Berry, M.T.; Jiang, C.; Lin, C.; May, S.P. Preparation and Optical Properties of Silver Nanowires and Silver-Nanowire Thin Films. *J. Colloid Interface Sci.* **2011**, *356*, 151–158. [CrossRef] [PubMed]
22. Gao, Y.; Jiang, P.; Liu, D.F.; Yuan, H.J.; Yan, X.Q.; Zhou, Z.P.; Wang, J.X.; Song, L.; Liu, L.F.; Zhou, W.Y.; et al. Synthesis, Characterization and Self-Assembly of Silver Nanowires. *Chem. Phys. Lett.* **2003**, *380*, 146–149. [CrossRef]
23. Kottmann, J.P.; Martin, O.J.F.; Smith, D.R.; Schultz, S. Plasmon Resonances of Silver Nanowires with a Nonregular Cross Section. *Phys. Rev. B Condens. Matter* **2001**, *64*, 235402. [CrossRef]
24. Liu, H.-S.; Pan, B.-C.; Liou, G.-S. Highly Transparent AgNW/PDMS Stretchable Electrodes for Elastomeric Electrochromic Devices. *Nanoscale* **2017**, *9*, 2633–2639. [CrossRef] [PubMed]
25. Kostova, I.; Nachkova, S.; Tonchev, D.; Kasap, S. Synthesis, UV-VIS Spectroscopy and Conductivity of Silver Nanowires. In *Nanoscience Advances in CBRN Agents Detection, Information and Energy Security*; NATO science for peace and security series; Springer: Dordrecht, The Netherlands, 2015; pp. 175–182. ISBN 9789401796965.
26. Todd, C.S.; Chen, X. Chemometric Evaluation of Ultraviolet–Visible (UV–Vis) Spectra for Characterization of Silver Nanowire Diameter and Yield. *Appl. Spectrosc.* **2019**, *74*, 204–209. [CrossRef]
27. Zhao, Y.-Y.; Ren, X.-L.; Zheng, M.-L.; Jin, F.; Liu, J.; Dong, X.-Z.; Zhao, Z.-S.; Duan, X.-M. Plasmon-Enhanced Nanosoldering of Silver Nanoparticles for High-Conductive Nanowires Electrodes. *OEA* **2021**, *4*, 200101–200109. [CrossRef]
28. Saw, M.J.; Ghosh, B.; Nguyen, M.T.; Jirasattayaporn, K.; Kheawhom, S.; Shirahata, N.; Yonezawa, T. High Aspect Ratio and Post-Processing Free Silver Nanowires as Top Electrodes for Inverted-Structured Photodiodes. *ACS Omega* **2019**, *4*, 13303–13308. [CrossRef]
29. Ding, Y.; Cui, Y.; Liu, X.; Liu, G.; Shan, F. Welded Silver Nanowire Networks as High-Performance Transparent Conductive Electrodes: Welding Techniques and Device Applications. *Appl. Mater. Today* **2020**, *20*, 100634. [CrossRef]
30. Xiong, W.; Hao, L.; Li, Y.; Tang, D.; Cui, Q.; Feng, Z.; Yan, C. Effect of Selective Laser Melting Parameters on Morphology, Microstructure, Densification and Mechanical Properties of Supersaturated Silver Alloy. *Mater. Des.* **2019**, *170*, 107697. [CrossRef]
31. de Souza, A.F.; Al-Rubaie, K.S.; Marques, S.; Zluhan, B.; Santos, E.C. Effect of Laser Speed, Layer Thickness, and Part Position on the Mechanical Properties of Maraging 300 Parts Manufactured by Selective Laser Melting. *Mater. Sci. Eng. A* **2019**, *767*, 138425. [CrossRef]
32. Smichi, A. Direct Laser Sintering of Metal Powders: Mechanism, Kinetics and Microstructural Features. *Mater. Sci. Eng. A* **2006**, *428*, 148–158. [CrossRef]
33. Chandra, H.; Allen, S.W.; Oberloier, S.W.; Bihari, N.; Gwamuri, J.; Pearce, J.M. Open-Source Automated Mapping Four-Point Probe. *Materials* **2017**, *10*, 110. [CrossRef] [PubMed]

34. Naftaly, M.; Das, S.; Gallop, J.; Pan, K.; Alkhalil, F.; Kariyapperuma, D.; Constant, S.; Ramsdale, C.; Hao, L. Sheet Resistance Measurements of Conductive Thin Films: A Comparison of Techniques. *Electronics* **2021**, *10*, 960. [CrossRef]
35. Mohd Asri, M.A.; Ramli, N.A.; Nordin, A.N. Electrical Performance and Reliability Assessment of Silver Inkjet Printed Circuits on Flexible Substrates. *J. Mater. Sci. Mater. Electron.* **2021**, *32*, 16024–16037. [CrossRef]
36. Wang, Y.H.; Du, D.X.; Xie, H.; Zhang, X.B.; Lin, K.W.; Wang, K.; Fu, E. Printability and Electrical Conductivity of Silver Nanoparticle-Based Conductive Inks for Inkjet Printing. *J. Mater. Sci. Mater. Electron.* **2020**, *32*, 496–508. [CrossRef]
37. Niu, B.; Hua, T.; Xu, B. Robust Deposition of Silver Nanoparticles on Paper Assisted by Polydopamine for Green and Flexible Electrodes. *ACS Sustain. Chem. Eng.* **2020**, *8*, 12842–12851. [CrossRef]
38. Young, K.; Chowdhury, R.; Jang, S. Copper Nanoparticle Conductive Patterns Fabricated by Thermal Sintering Using Carboxylic Acid Vapors and Their Application for Radio-Frequency Identification Antennas. *Appl. Phys. A* **2023**, *129*, 1–15. [CrossRef]
39. Im, J.; Trindade, G.F.; Quach, T.T.; Sohaib, A.; Wang, F.; Austin, J.; Turyanska, L.; Roberts, C.J.; Wildman, R.; Hague, R.; et al. Functionalized Gold Nanoparticles with a Cohesion Enhancer for Robust Flexible Electrodes. *ACS Appl. Nano Mater.* **2022**, *5*, 6708–6716. [CrossRef]
40. Htwe, Y.Z.N.; Mariatti, M. Printed Graphene and Hybrid Conductive Inks for Flexible, Stretchable, and Wearable Electronics: Progress, Opportunities, and Challenges. *J. Sci. Adv. Mater. Devices* **2022**, *7*, 100435. [CrossRef]
41. Kunaraj, A.; Chelvanathan, P.; Bakar, A.A.A.; Yahya, I. Single-Walled Carbon Nanotube (SWCNT) Thin Films via Automatic Spray Coating and Nitric Acid Vapor Treatment. *J. Eng. Res.* **2023**. Available online: https://www.sciencedirect.com/science/article/pii/S2307187723001621 (accessed on 5 November 2024). [CrossRef]
42. Li, Y.; Hong, M. Parallel Laser Micro/Nano-Processing for Functional Device Fabrication. *Laser Photonics Rev.* **2020**, *14*, 1900062. [CrossRef]
43. Tan, D.; Jiang, C.; Li, Q.; Bi, S.; Wang, X.; Song, J. Development and Current Situation of Flexible and Transparent EM Shielding Materials. *J. Mater. Sci. Mater. Electron.* **2021**, *32*, 25603–25630. [CrossRef]
44. Voronin, A.S.; Fadeev, Y.V.; Ivanchenko, F.S.; Dobrosmyslov , S.S.; Makeev, M.O.; Mikhalev , P.A.; Osipkov, A.S.; Damaratsky, I.A.; Ryzhenko, D.S.; Yurkov, G.Y.; et al. Original Concept of Cracked Template with Controlled Peeling of the Cells Perimeter for High Performance Transparent EMI Shielding Films. *Surf. Interfaces* **2023**, *38*, 102793. [CrossRef]
45. Wei, Q.; Pei, S.; Qian, X.; Liu, H.; Liu, Z.; Zhang, W.; Zhou, T.; Zhang, Z.; Zhang, X.; Cheng, H.-M.; et al. Superhigh Electromagnetic Interference Shielding of Ultrathin Aligned Pristine Graphene Nanosheets Film. *Adv. Mater.* **2020**, *32*, 1907411. [CrossRef] [PubMed]
46. Zeng, Z.; Jin, H.; Chen, M.; Li, W.; Zhou, L.; Zhang, Z. Lightweight and Anisotropic Porous MWCNT/WPU Composites for Ultrahigh Performance Electromagnetic Interference Shielding. *Adv. Funct. Mater.* **2016**, *26*, 303–310. [CrossRef]
47. Sharma, G.K.; James, N.R. Carbon Black Incorporated Carbon Nanofiber Based Polydimethylsiloxane Composite for Electromagnetic Interference Shielding. *Carbon Trends* **2022**, *8*, 100177. [CrossRef]
48. Kumar, A.; Singh, A.P.; Kumari, S.; Srivastava, A.K.; Bathula, S.; Dhawan, S.K.; Dutta, P.K.; Dhar, A. EM Shielding Effectiveness of Pd-CNT-Cu Nanocomposite Buckypaper. *J. Mater. Chem. A* **2015**, *3*, 13986–13993. [CrossRef]
49. Liang, C.; Ruan, K.; Zhang, Y.; Gu, J. Multifunctional Flexible Electromagnetic Interference Shielding Silver Nanowires/Cellulose Films with Excellent Thermal Management and Joule Heating Performances. *ACS Appl. Mater. Interfaces* **2020**, *12*, 18023–18031. [CrossRef]
50. Ma, Y.; Lv, C.; Tong, Z.; Zhao, C.F.; Li, Y.S.; Hu, Y.Y.; Yin, Y.H.; Liu, X.B.; Wu, Z.P. Single-Layer Copper Particles Integrated with a Carbon Nanotube Film for Flexible Electromagnetic Interference Shielding. *J. Mater. Chem. C* **2020**, *8*, 9945–9953. [CrossRef]
51. Chen, Y.; Pang, L.; Li, Y.; Luo, H.; Duan, G.; Mei, C.; Xu, W.; Zhou, W.; Liu, K.; Jiang, S. Ultra-Thin and Highly Flexible Cellulose Nanofiber/silver Nanowire Conductive Paper for Effective Electromagnetic Interference Shielding. *Compos. Part A Appl. Sci. Manuf.* **2020**, *135*, 105960. [CrossRef]
52. Ju, J.; Kuang, T.; Ke, X.; Zeng, M.; Chen, Z.; Zhang, S.; Peng, X. Lightweight Multifunctional Polypropylene/carbon Nanotubes/carbon Black Nanocomposite Foams with Segregated Structure, Ultralow Percolation Threshold and Enhanced Electromagnetic Interference Shielding Performance. *Compos. Sci. Technol.* **2020**, *193*, 108116. [CrossRef]
53. Liu, Y.; Zhang, K.; Mo, Y.; Zhu, L.; Yu, B.; Chen, F.; Fu, Q. Hydrated Aramid Nanofiber Network Enhanced Flexible Expanded Graphite Films towards High EMI Shielding and Thermal Properties. *Compos. Sci. Technol.* **2018**, *168*, 28–37. [CrossRef]
54. Zhou, B.; Zhang, Z.; Li, Y.; Han, G.; Feng, Y.; Wang, B.; Zhang, D.; Ma, J.; Liu, C. Flexible, Robust, and Multifunctional Electromagnetic Interference Shielding Film with Alternating Cellulose Nanofiber and MXene Layers. *ACS Appl. Mater. Interfaces* **2020**, *12*, 4895–4905. [CrossRef]
55. Lei, Z.; Wei, J.; Lin, Z.Q.; Zhao, T.; Hu, Y.G.; Zhu, P.L.; Rong, S. Ultra-Thin and Highly Flexible Silver Fractal Dendrite/polyvinylidene Fluoride Composite Film for Efficient Electromagnetic Interference Shielding. *Chem. Eng. J.* **2023**, *452*, 139129. [CrossRef]
56. Wang, L.; Ma, Z.; Zhang, Y.; Qiu, H.; Ruan, K.; Gu, J. Mechanically Strong and Folding-Endurance Ti_3C_2Tx MXene/PBO Nanofiber Films for Efficient Electromagnetic Interference Shielding and Thermal Management. *Carbon Energy* **2022**, *4*, 200–210. [CrossRef]
57. Ahmad, M.W.; Dey, B.; Sammar, A.A.; Choudhury, A. Sulfonic-Functionalized Graphene Oxide Reinforced Polyethersulfone Nanocomposites with Enhanced Dielectric Permittivity and EMI Shielding Effectiveness. *J. Macromol.Sci. Part A* **2020**, *57*, 778–790. [CrossRef]

58. Zhang, Y.; Ruan, K.P.; Shi, X.T.; Qiu, H.; Pan, Y.; Yan, Y.; Gu, J.W. Ti3C2Tx/rGO Porous Composite Films with Superior Electromagnetic Interference Shielding Performances. *Carbon* **2021**, *175*, 271–280. [CrossRef]
59. Panda, S.; Acharya, B. PDMS/MWCNT Nanocomposites as Capacitive Pressure Sensor and Electromagnetic Interference Shielding Materials. *J. Mater.Sci. Mater. Electron.* **2021**, *32*, 16215–16229. [CrossRef]
60. Liu, J.; Lin, S.; Huang, K.; Jia, C.; Wang, Q.; Li, Z.; Song, J.; Liu, Z.; Wang, H.; Lei, M.; et al. A Large-Area AgNW-Modified Textile with High-Performance Electromagnetic Interference Shielding. *NPJ Flex. Electron.* **2020**, *4*, 1–7. [CrossRef]
61. Jiang, Z.; Zhao, S.; Chen, L.; Liu, Y.-H. Freestanding "Core-Shell" AgNWs/metallic Hybrid Mesh Electrodes for a Highly Efficient Transparent Electromagnetic Interference Shielding Film. *Opt Express* **2021**, *29*, 18760–18768. [CrossRef]
62. Zou, K.; Sun, H.; Li, X.; Yi, S.; Li, J.; Zhou, Z.; Wang, H.; Yan, D.-X. Extreme Environment-Bearable Polyimide Film with a Three-Dimensional Ag Microfiber Conductive Network for Ultrahigh Electromagnetic Interference Shielding. *Sci. China Mater.* **2022**, *66*, 1578–1586. [CrossRef]
63. Jia, L.C.; Zhou, C.-G.; Sun, W.-J.; Xu, L. Water-Based Conductive Ink for Highly Efficient Electromagnetic Interference Shielding Coating. *Chem. Eng. J.* **2020**, *384*, 123368. [CrossRef]
64. Ji, H.; Zhao, R.; Zhang, N.; Jin, C.; Lu, X.; Wang, C. Lightweight and Flexible Electrospun Polymer Nanofiber/metal Nanoparticle Hybrid Membrane for High-Performance Electromagnetic Interference Shielding. *NPG Asia Mater.* **2018**, *10*, 749–760. [CrossRef]
65. Kim, S.J.; Yoon, H.G.; Kim, S.W. Extremely Robust and Reliable Transparent Silver Nanowire-Mesh Electrode with Multifunctional Optoelectronic Performance through Selective Laser Nanowelding for Flexible Smart Devices. *Adv. Eng. Mater.* **2021**, *23*, 2001310. [CrossRef]
66. Oh, H.-J.; Dao, V.-D.; Choi, H.-S. Electromagnetic Shielding Effectiveness of a Thin Silver Layer Deposited onto PET Film via Atmospheric Pressure Plasma Reduction. *Appl. Surf. Sci.* **2018**, *435*, 7–15. [CrossRef]
67. Li, T.-T.; Wang, Y.; Peng, H.-K.; Zhang, X.; Shiu, B.-C.; Lin, J.-H.; Lou, C.-W. Lightweight, Flexible and Superhydrophobic Composite Nanofiber Films Inspired by Nacre for Highly Electromagnetic Interference Shielding. Compos. *Part A Appl. Sci. Manuf.* **2020**, *128*, 105685. [CrossRef]
68. Ye, X.; Zhang, X.; Zhou, X.; Wang, G. Asymmetric and Flexible Ag-MXene/ANFs Composite Papers for Electromagnetic Shielding and Thermal Management. *Nanomaterials* **2023**, *13*, 2608. [CrossRef]
69. Abyzova, E.; Petrov, I.; Bril', I.; Cheshev, D.; Ivanov, A.; Khomenko, M.; Averkiev, A.; Fatkullin, M.; Kogolev, D.; Bolbasov, E.; et al. Universal Approach to Integrating Reduced Graphene Oxide into Polymer Electronics. *Polymers* **2023**, *15*, 4622. [CrossRef]

polymers

MDPI

Article

Study of Polysulfone-Impregnated Hydroxyapatite for Ultrafiltration in Whey Protein Separation

Tutik Sriani [1], Muslim Mahardika [2], Budi Arifvianto [2], Farazila Yusof [3,4], Yudan Whulanza [5], Gunawan Setia Prihandana [6,*] and Ario Sunar Baskoro [5,*]

1 Department of Research and Development, PT. Global Meditek Utama-IITOYA, Sardonoharjo, Ngaglik, Sleman, Yogyakarta 55581, Indonesia; tsriani@gmail.com
2 Department of Mechanical and Industrial Engineering, Faculty of Engineering, Universitas Gadjah Mada, Jalan Grafika No. 2, Yogyakarta 55281, Indonesia; muslim_mahardika@ugm.ac.id (M.M.); budi.arif@ugm.ac.id (B.A.)
3 Centre of Advanced Manufacturing & Material Processing (AMMP Centre), Department of Mechanical Engineering, Faculty of Engineering, Universiti Malaya, Kuala Lumpur 50603, Malaysia; farazila@um.edu.my
4 Centre for Foundation Studies in Science, University Malaya, Kuala Lumpur 50603, Malaysia
5 Department of Mechanical Engineering, Faculty of Engineering, Universitas Indonesia, Kampus UI, Depok 16425, Indonesia; yudan@eng.ui.ac.id
6 Department of Industrial Engineering, Faculty of Advanced Technology and Multidiscipline, Universitas Airlangga, Jl. Dr. Ir. H. Soekarno, Surabaya 60115, Indonesia
* Correspondence: gunawan.prihandana@ftmm.unair.ac.id (G.S.P.); ario@eng.ui.ac.id (A.S.B.); Tel.: +62-881-0360-00830 (G.S.P.); +62-21-78888430 (A.S.B.)

Abstract: Polysulfone (Psf) ultrafiltration flat-sheet membranes were modified with hydroxyapatite (HA) powder during preparation using the wet-phase inversion method. HA was incorporated to enhance the protein separation capabilities. The asymmetric Psf membranes were synthesized using NMP as the solvent. Through Scanning Electron Microscopy (SEM) analysis, it was revealed that HA was distributed across the membrane. Incorporating HA led to higher flux, the improved rejection of protein, and enhanced surface hydrophilicity. The permeability flux increased with HA concentration, peaking at 0.3 wt.%, resulting in a 38% improvement to 65 LMH/bar. Whey protein separation was evaluated using the model proteins BSA and lysozyme, representing α-Lactalbumin. The results of protein rejection for the blend membranes indicated that the rejection rates for BSA and lysozyme increased to 97.2% and 73%, respectively. Both the native and blend membranes showed similar BSA rejection rates; however, the blend membranes demonstrated better performance in lysozyme separation, indicating superior selectivity compared to native membranes. The modified membranes exhibited improved hydrophilicity, with water contact angles decreasing from 66° to 53°, alongside improved antifouling properties, indicated by a lower flux decline ratio value. This simple and economical modification method enhances permeability without sacrificing separation efficiency, hence facilitating the scalability of membrane production in the whey protein separation industry.

Keywords: polysulfone; hydroxyapatite; ultrafiltration; whey; clean water; protein separation

1. Introduction

Regular bovine milk comprises at least 3.5% fat, 32 g/L of proteins and an approximately 78% fraction of caseins, while the remaining consists of whey proteins [1]. The dairy business produces a substantial amount of whey, predominantly seen as waste, primarily originating from cheese production [2]. Isolating and refining whey proteins enhances the value of what would otherwise be considered waste, hence increasing the sustainability and profitability of the dairy sector [3]. Whey is regarded as a complete protein, encompassing all essential amino acids. Whey protein consists of multiple distinct proteins, each with unique characteristics and advantages. β-Lactoglobulin constitutes the predominant whey protein, comprising approximately 50–55% of the total protein content, essential for muscle recovery and growth owing to its high concentration of branched-chain amino acids [4]. α-Lactalbumin comprises around 20–25% of whey protein, consists of 123 amino acids, and contains tryptophan, which aids in serotonin generation and may enhance relaxation [5]. Glycomacropeptide (GMP) constitutes around 10–15% of whey protein, with 64 amino acids, which function to trigger hormones that enhance satiety and facilitate digestion [6]. Immunoglobulins, bovine serum albumin (BSA), and lactoferrin comprise around 10%, 5% to 10%, and less than 1% of whey protein, respectively [7]. Whey also comprises several proteins found at relatively low concentrations that exhibit significant biological action [8]. Protein powders are considered a quick and high-quality protein source for enhancing post-exercise recovery [9]. Given that each protein type offers distinct advantages, it is essential to isolate each protein to optimize its benefits.

Membrane systems are extensively used in the dairy industry for efficiently regulating the lactose, fat, and protein content of various products due to their highly efficient separation and purification methods [10]. Ultrafiltration (UF) employs polymeric or ceramic membranes that completely retain whey proteins, effectively eliminating superfluous water and lactose [11]. Polysulfone (Psf) is distinguished as a preferred polymer for ultrafiltration membrane fabrication owing to its remarkable physicochemical characteristics [12]. Psf membranes exhibit resistance to numerous contaminants, endure a wide temperature range while preserving their integrity, possess favorable electrical properties, and are easy to fabricate [13]. Nonetheless, despite their beneficial characteristics, Psf membranes are susceptible to fouling, which leads to a reduction in water flux as it advances [14]. The primary issue encountered in the separation of whey proteins is membrane fouling. Researchers frequently incorporate chemicals during membrane manufacture to produce membranes with enhanced characteristics, such as exceptional water permeability, effective rejection, and anti-fouling surfaces. Various additives/fillers have been examined to enhance the anti-fouling characteristics of UF membranes, including polydopamine nanofillers [15], metal oxide nanoparticles [16], carbon-dot-grafted silica [17], graphene oxide [18], hydroxyapatite nanotubes [19], and others.

Hydroxyapatite (HA) has garnered significant attention in contemporary material chemistry owing to its exceptional biocompatibility, eco-friendliness, cost-effectiveness, and abundance of hydroxyl groups. Ouda et al. [15] incorporated hyaluronic acid (HA) and polydopamine (PDA) into polylactic acid membranes, discovering that the enhancement in flux and membrane hydrophilicity resulted from the addition of HA powder and the application of a PDA coating. Instead of using PDA, Anwar and Arthanareeswaran [20] applied silver as an addition to the polyphenylsulfone membrane, coating hydroxyapatite (HA) to treat Palm Oil Mill Effluent (POME). The incorporation of 2% silver-coated HA demonstrated a rejection efficiency of 89.74% and possessed exceptional separation characteristics. The incorporation of a filler comprising hyaluronic acid and boron nitride into a polyethersulfone/polyvinylpyrrolidone membrane resulted in a two-fold increase in flow, while achieving a 90% rejection rate of bovine serum albumin solution [21]. Mu et al. [19] examined the effect of hydroxyapatite nanotubes on the hydrophilicity of PES/PVP membranes. The HA nanotubes were coated with PDA, subsequently followed by taurine grafting. This intricate change enabled the engineered membranes to achieve enhanced permeability, a 97% rejection of BSA, and improved antifouling characteristics.

This research examines the impact of adding hydroxyapatite (HA) to a Polysulfone (Psf) dope solution to produce ultrafiltration membranes. The purchased HA was utilized without alteration to prevent additional production expenses. The objective was to improve the membrane selectivity characteristics while preserving elevated permeability at the ultrafiltration level. The membrane was made using the non-solvent-induced phase separation (NIPS) technique. The influence of hydroxyapatite on the membranes' surface morphology, membrane pore characteristics, permeate flux, and hydrophilicity membrane was analyzed. Scanning Electron Microscopy (SEM) was used to visually study the membrane surface morphology. To evaluate the membranes' efficacy in protein separation, the fabricated Psf membranes were subjected to aqueous protein solutions of different molecular weights, and the protein retention percentages of the membranes were measured.

2. Materials and Methods

2.1. Materials

The polymer used in the study was Polysulfone (Psf), with a molecular weight of 78,000 to 84,000 g/mol (P.T. Solvay Chemical, Cilegon, Indonesia). The solvent, NMethyl-2-Pyrrolidone (NMP), was purchased from Merck KGaA, Darmstadt, Germany. Hydroxyapatite powder (HA) with a molecular weight of 502.31 g/mol and a purity of at least 90% was purchased from Sigma Aldrich, St. Louis, MO, USA. Additionally, Bovine serum albumin (BSA) and lysozyme were acquired from HiMedia Ltd., Mumbai, India. Pure water was utilized throughout the whole fabrication process. All chemicals, unless stated otherwise, were of analytical grade with over 98% purity and were used as received.

2.2. Membrane Fabrication

The ultrafiltration membranes were fabricated using the non-solvent-induced phase separation (NIPS) technique, a recognized method used to produce asymmetric membranes [22]. Initially, a solution comprising Psf as the base polymer and NMP as the solvent was continuously stirred at 70 °C until a homogeneous solution was obtained. The dope solution was maintained at a consistent 20 wt.% concentration of Polysulfone. Thereafter, modest amounts of HA particles (Sigma Aldrich) were added into the solution as a membrane additive. The Psf solution was subsequently combined with 0.1 wt.% to 0.5 wt.% of HA particles, heated, and agitated at 70 °C for 3 h to produce homogeneous dope solutions. Following this procedure, the dope solutions were then cooled to room temperature naturally for the membrane fabrication procedure. Subsequent to cooling, the dope solution was deposited onto a glass plate and evenly distributed with a membrane applicator (M43 6BU, Elcometer Ltd., Manchester, UK) to attain a consistent wet thickness of 200 μm. The glass plate containing the solution was carefully placed in a pure water container to commence the gelatinization process. Subsequently, the membrane was submerged in distilled water for 24 h prior to testing. The membranes were designated according to their HA content: HA-0.0 (native polysulfone), HA-0.1, HA-0.2, HA-0.3, HA-0.4, and HA-0.5.

2.3. Membrane Characterization

This work assessed membrane effectiveness using pure water as the testing fluid. The test for each membrane lasted 30 min to reach steady flux in a stirred dead-end cell (HP4750, Sterlitech Corp., Kent, WA, USA). The effective membrane area used for filtration was 13.4 cm^2, as depicted in Figure 1. Nitrogen gas with a fixed pressure of 2 bars was introduced to the pure water chamber in order to ensure that consistent pressure was applied throughout the test.

During the test, the weight of the permeate water that traversed the examined membrane was logged using a Weighing Environment Logger (AD-1687, A&D, RoHS, Tokey, Japan). The volumetric flux (J_v) and permeability flux (L_p) were computed utilizing the following formulas [23]:

$$J_v = \frac{Q}{A \times \Delta t} \quad (1)$$

$$L_p = \frac{J_v}{\Delta P} \quad (2)$$

where Q represents the volume of the penetrated water collected throughout the collecting period, Δt denotes the collection period, A signifies the membrane, and ΔP indicates the pressure differential.

Subsequent to the permeate flux experiment, a membrane selectivity assessment was executed utilizing a 1 g/L protein solution under the same conditions. To assess the protein separation, Bovine Serum Albumin (BSA, Mw = 66.5 kDa) and lysozyme (Mw = 14.3 kDa), which serves as a representative of α-Lactalbumin, were utilized. The proteins were formulated in a phosphate-buffered solution at a pH of 7.2. The protein separation experiment was conducted in a dead-end cell filtration system, maintaining a constant pressure of 2 bar. The protein concentration in the permeated solutions was measured using an N4S UV-Visible Spectrophotometer (Ningbo Hinotek Instrument Co., Ltd., Ningbo, China), in which the wavelength was set to 280 nm. Solute rejection (SR) was evaluated using the following formula [24]:

$$\%SR = \left[1 - \frac{C_p}{C_f}\right] \times 100 \quad (3)$$

where C_p and C_f denote the concentrations of protein in the permeating and feed solutions, respectively. The data presented in this study reflect the mean values from three samples per membrane.

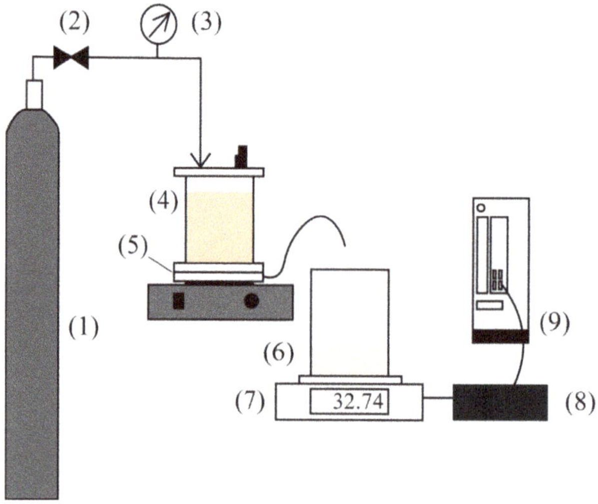

Figure 1. The schematic diagram of the dead-end cell filtration. (1) Nitrogen gas cylinder; (2) Pressure regulator; (3) Pressure gauge; (4) Dead-end cell; (5) Membrane; (6) Permeated water beaker; (7) Electronic weight balance; (8) Weighing environment logger; (9) Computer.

Five iterations of BSA tests were conducted to evaluate the antifouling efficacy of the membranes. Each cycle comprises a 30 min water flux followed by a 10 min BSA flux. The flux decline ratio (FDR) and flux recovery ratio were determined by [25]:

$$FDR = \frac{J_v - J_{vt}}{J_v} \times 100\% \quad (4)$$

$$FRR = \frac{J_{v\prime}}{J_v} \times 100\% \quad (5)$$

where J_{vt} is the permeate flux at the end of the fifth cycle, and $J_{v\prime}$ is the permeate flux of the second cycle. A lower FDR value signifies the superior adhesion resistance of foulants or the enhanced antifouling performance of the membrane. A higher FRR value denotes improved efficiency in membrane cleaning.

The hydrophilicity of the membrane was assessed by determining the angle between a water droplet and the membrane surface. Then, 5 µL of distilled water was dropped carefully onto the top of the membrane surface using a micropipette. A digital microscope (Dinolite Edge 3.0 AM73915MZTL) subsequently recorded an image of the contact angle of the water droplet. The angle was subsequently gauged using AutoCAD software (v2022).

To guarantee precision, the contact angles were assessed four times at random spots of the membrane surface, and the average value was determined. The morphologies of the membrane surfaces and their cross-sectional structures were examined utilizing scanning electron microscopy (SEM Phenom ProX, ThermoFischer, Waltham, MA, USA). The cross-sectional micrographs were obtained by cracking wet membrane samples subsequent to their rapid freezing in liquid nitrogen.

2.4. Pore Characterization

To evaluate the porosity of the membrane surface, the dry membranes were initially trimmed to a predetermined size, then soaked in glycerol for 24 h. The glycerol was further extracted to weigh the wet membrane (W_w). The moist membrane was thereafter dried in a desiccator and allowed to desiccate for one day to acquire the weight of the dry membrane (W_d). The porosity of the membrane surface was examined to evaluate the effect of hydroxyapatite incorporation on the membrane pore size. This investigation was performed utilizing a gravimetric approach, which involves the following formula [26]:

$$\varepsilon(\%) = \frac{W_w - W_d}{\rho H_2O \times A \times L} \times 100 \quad (6)$$

where A represents the effective filtration area of the membrane, L denotes the thickness of the membrane, and ρH_2O signifies the density of the water (0.998 g/cm^3). The average pore size (r_m) of the prepared membranes can be determined using the Guerout–Elford–Ferry Equation, including the porosity data and permeate flux [27]:

$$r_m = \sqrt{\frac{(2.9 - 1.75\varepsilon)8\mu_{H_2O} \times L \times Q_{H_2O}}{\varepsilon \times A \times \Delta P}} \quad (7)$$

where ε denotes the porosity of the membrane, μ_{H_2O} represents the dynamic viscosity of water at an ambient temperature, L indicates the membrane thickness, Q_{H_2O} is the volume of water traversing through the membrane per unit time, A refers to the active filtration area of the membrane, and ΔP is the transmembrane pressure.

3. Results and Discussion

3.1. Membrane Morphology

Five membrane samples were fabricated with differing amounts of hydroxyapatite. Among this collection of fabricated membranes, one was designated as HA-0.0, which functioned as a reference without the inclusion of HA. The remaining membranes were designated as HA-0.1, HA-0.2, HA-0.3, HA-0.4, and HA-0.5, reflecting the HA weight content in the Psf dope solution. The whole samples were fabricated with a wet thickness of 200 μm. The surface morphologies of the membranes were analyzed using a SEM Phenom ProX microscope.

Cross-sectional scanning electron microscopy (SEM) examination is a prevalent method for visually assessing membrane morphology. Figure 2 presents the cross-sectional micrographs of the native Psf membrane and the Psf/HA blend membranes, with magnifications varying from 1000× to 20,000× throughout the first to fourth rows. The fifth row displays the micrographs of the membrane surface. At 1000× magnification, all membranes have a thin surface layer atop a vertically extending, finger-like intermediate structure. The lower section of the membrane exhibits a sparse, spongiform architecture characterized by discernible wall pores. In the membrane gelatinization phase of the NIPS technique, the interaction between NMP as the solvent and pure water as the non-solvent induces solidification due to the thermodynamic instability of the cast solution [28]. The cast solution quickly interfaces with water, resulting in immediate vitrification and permitting water ingress, which generates macrovoids inside the membrane structure. Hackett et al. [29] explored how the composition of the coagulation bath impacts the efficacy of the Psf membrane. They found that utilizing solely pure water during membrane submersion retains a significant amount of solvent in the central region due to NMP–water incompatibility. Kang et al. [30] discovered that blending hydrophilic agents into the membrane dope solution, such as silica or HA, improves the interchange rate of solvent and non-solvent throughout the solidification phase, thus enlarging the macrovoids at deeper depths, which mimic the form of pre-inflated balloons. The lower section of the membrane exhibits a distinct coagulation rate when water permeates from below; this is attributable to the glass support, and leads to the formation of spongy structures with bigger pores akin to those on the upper surface.

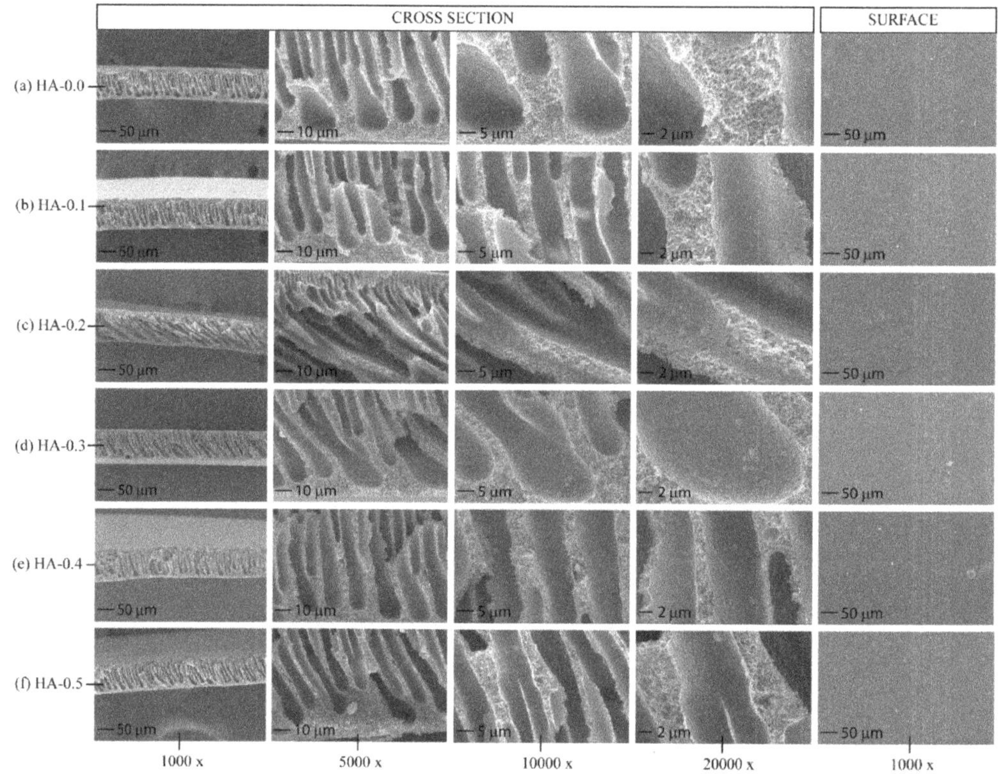

Figure 2. Morphologies of membrane samples. Cross-sectional images at both low and high magnification are displayed in the first to fourth rows, while the fifth row features top surface micrographs: (**a**) HA-0.0; (**b**) HA-0.1; (**c**) HA-0.2; (**d**) HA-0.3; (**e**) HA-0.4; (**f**) HA-0.5.

The SEM cross-sectional micrographs demonstrate the existence of hydroxyapatite (HA) on the surface of finger-like structures and within wall pores, exhibiting a random distribution from the top region to the bottom region of the membrane. HA is also present on the surface and affixed irregularly, resulting in surface protuberances. The SEM cross-sectional views reveal no detectable alterations in the dimensions of the membrane wall pores or macrovoids attributable to HA. Nonetheless, the blend membranes possess HA globules that are randomly affixed to the macrovoid surfaces. The variation in globule diameters indicates the aggregation of HA particles during gelatinization. The heightened viscosity of the Psf blend membranes, resulting from the enhanced HA concentration, influences the morphology of the Psf/HA blend membranes. Certain particles infiltrate the wall pores, becoming trapped within the membrane. A greater quantity of HA particles is clearly detected at the membrane's base, especially in HA-0.5, which is attributable to the gravitational settling and agglomeration of HA particles during the casting procedure [31,32].

The zeolite-nanoparticle-impregnated Psf membrane exhibited particle agglomeration at an elevated loading level [33]. Yurekli further indicated that the membrane might achieve a saturation point at which the maximal loading of zeolite nanoparticles was reached.

3.2. Water Contact Angle Analysis

A primary disadvantage of Psf membranes is the accumulation of build-up on their surface, leading to pore obstruction [31]; thus, hydrophilicity is crucial for membrane efficacy. The measurement of the water contact angle is frequently applied to evaluate the hydrophilicity and wettability of membrane surfaces. This method assesses the angle formed between the water droplet and a planar membrane surface, signifying hydrophilicity when the angle is between 0° to 90°, or hydrophobicity when the angle is larger than 90°. A decrease in the contact angle with water indicates an enhancement in the membrane's hydrophilic characteristics [34]. If the contact angle is below 90°, the liquid adheres to the surface, and at 0 degrees, the liquid uniformly disperses throughout the surface.

Figure 3 illustrates the measurement of water contact angles for different concentrations of HA. The native Psf membrane examined in this study exhibits a contact angle of 66°, aligning with previous results indicating that native Psf membranes possess water contact angles ranging from 60° to 70° [29,33–35]. Figure 3 indicates that hydrophilicity enhances with an elevated HA content. The Psf/HA-0.5 membrane had the highest hydrophilicity, with a contact angle of 54°, whereas the contact angles of Psf/HA-0.1 to 0.3 were comparable to that of the native membrane. This tendency indicates that elevated HA concentrations increase hydrophilicity owing to the hydroxyl groups on the particle surfaces that facilitate hydrogen bonding to the sulfonic groups of Psf [33,36]. During the phase separation process, the hydrophilic agent will draw in molecules of water, creating more spaces for water infiltration [37]. SEM micrographs of the composite membrane surface revealed a random distribution of HA particles. Due to its hydrophilic nature, HA tends to attract and engage with water rather than repel it. Astala and Stott [38] explained that HA surfaces are distinguished by undercoordinated oxygen atoms in surface phosphate groups. A key process for water adsorption entails the interaction between O atoms and H_2O molecules through hydrogen, with Ca-rich surfaces exhibiting the highest reactivity. Most studies propose that the enhancement in membrane hydrophilicity is directly linked to the existence of hydrophilic nanofillers located near or spread on the surface of the membrane [39].

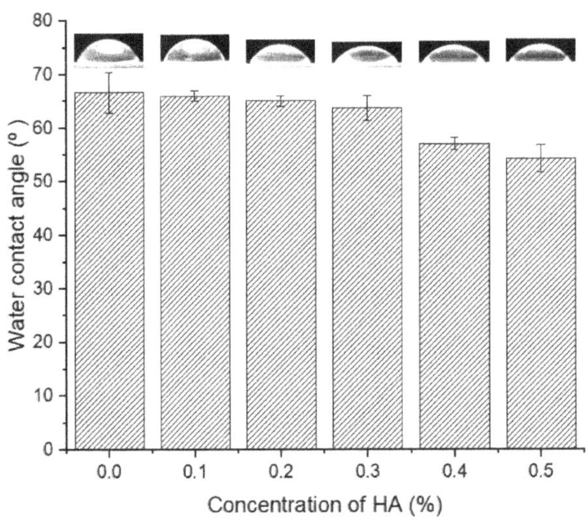

Figure 3. Water contact angle of Psf membranes with varied hydroxyapatite concentrations.

3.3. Permeate Flux Test Experiments

Figure 4 presents the permeate flux of Psf/HA blend membranes conducted in a dead-end filtration cell at a pressure of 2 bar. The permeate flux was observed for 30 min, repeated four times, and the average was determined. The permeate flux value utilized for computation was documented following 30 min of filtering. The native Psf membrane demonstrated a permeate flux of 45.5 LMH/bar. The permeate flux augmented with HA concentration, attaining a peak flux of 63 LMH/bar at 0.3 wt.%, then declining to a value comparable to that of the native Psf membrane. The trend in the permeate flux demonstrated that flux rises with the HA content up to 0.3 wt.% and thereafter declines, nearing the value of the native membrane. This reduction is probably attributable to the existence of HA. At reduced concentrations, HA particles generate open voids by infiltrating wall pores, hence augmenting the flux. At concentrations beyond 0.3 wt.%, the surplus HA particles obstruct the pores, hence diminishing water movement. Eslami et al. [31] attribute the diminished flux values to a surge in viscosity, which subsequently reduces the solvent/non-solvent exchange rate, resulting in lower flux values.

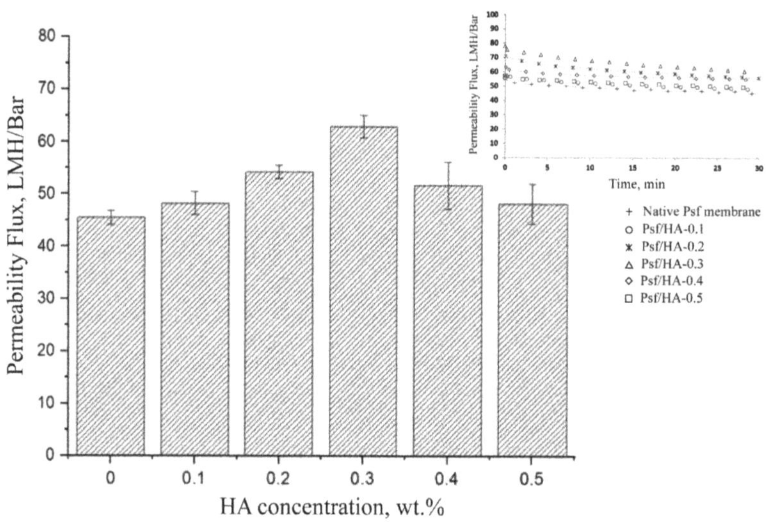

Figure 4. Permeability flux of Psf membranes with varied HA concentrations.

3.4. Protein Separation Performance Analysis

Ultrafiltration is the predominant industrial technique used for isolating whey proteins. The extent of protein adherence on the surface of the membrane profoundly influences the assessment of anti-fouling capacity. The reduced adsorption of protein results in enhanced membrane resistance to fouling. Proteins often exhibit stronger adsorption on hydrophobic surfaces and are less easily desorbed compared to hydrophilic surfaces. Cayot and Lorient [40] assert that α-Lactalbumin is the smallest predominant whey protein, possessing a molecular weight of 14 kDa and comprising around 20–25% of whey protein. The predominant whey protein is Immunoglobulins, possessing a molecular weight of 150–1000 kDa and comprising approximately 10% of whey protein.

All membranes were initially compressed at a 2 bar pressure with pure water for 30 min prior to testing. Figure 5 illustrates the protein separation of BSA, a large protein with a molecular weight of 66.5 kDa, and lysozyme, a tiny protein (14.3 kDa), which signifies α-Lactalbumin. Upon BSA rejection, the rejection rate of all membranes exceeds 85%. The native Psf membrane exhibits a BSA rejection rate of 89.6%, which increases to 97.2% with the incorporation of HA at a concentration of 0.2 wt% in Psf/HA blend membranes. The rejection diminishes beyond this juncture. Lysozyme rejection demonstrates considerable enhancement with a reduced protein size. The native Psf membrane exhibits a rejection rate of 36%, while the blend membranes demonstrate enhanced rejection capabilities. The peak rejection is 0.2 wt.% at 73% (double-fold value of the native Psf membrane), and the rejection diminishes to 54% at HA 0.5 wt.%.

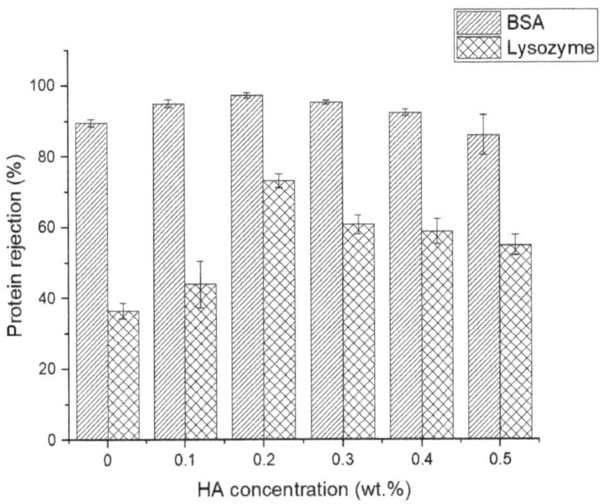

Figure 5. Protein rejection of Psf membranes with varied HA concentrations.

Protein flow was observed for 20 min for both BSA and lysozyme, each at a concentration of 1000 mg/L (Figure 6). The BSA flux exhibited comparable patterns and values for both the native and blend membranes. The fluxes of the Psf blend membranes were initially marginally different to those of the native membrane but progressively diminished to align with the flux of the native membrane. This reduction is presumably because of BSA molecules occupying the membrane's interstices, progressively obstructing them over time. At the conclusion of the experiment, all membranes had comparable flux levels.

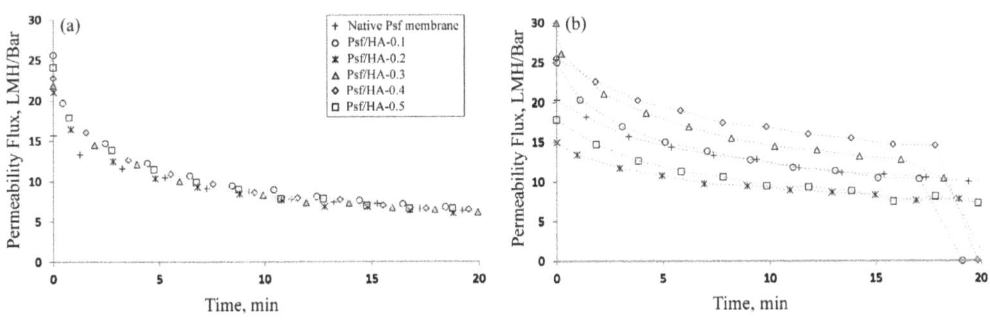

Figure 6. Protein flux of blend membranes: (**a**) BSA flux; (**b**) Lysozyme flux.

The flux of lysozyme across the native membrane exceeded that of BSA, presumably because lysozyme has a smaller molecular size than BSA. All membranes exhibited a more gradual reduction in lysozyme flux relative to BSA flux, which is ascribed to the smaller molecular size of lysozyme. The native membrane exhibited a lysozyme rejection rate of merely 36.5%, which increased to 73.2% upon elevating the HA concentration to 0.2 wt.%. The enhancement in lysozyme rejection is attributed to protein adsorption by HA across the membrane. Nevertheless, as the HA concentration was elevated, lysozyme rejection diminished by as much as 26% at 0.5 wt.% HA. This decrease is probably attributed to inadequate space for the liquid to flow, as the majority of open holes were occupied by HA particles. Both 0.2 and 0.5 wt.% HA exhibited reduced fluxes, maybe attributable to distinct factors. At 0.2 wt.%, the reduced flux was attributable to the adsorption of lysozyme by HA, leading to optimal rejection. At 0.5 wt.%, the diminished flux was attributed to the open spaces being obstructed by HA powder, impeding liquid flow through the membrane, thereby leading to reduced flux and simultaneous rejection. Salimi et al. [41] further indicated that augmenting the HA content in the membranes improved BSA rejection. The incorporation of a higher HA concentration led to a thicker membrane skin layer, smaller pore size, and lower permeation flux.

3.5. Membrane Pore Analysis

Table 1 outlines the surface porosity and mean pore size of the native Psf and blend membranes. Notably, the porosity of the blend membranes, with the exception of Psf/HA-0.5, is inferior to that of the original Psf membrane. As HA comprises solid particles that remain undissolved, it is anticipated that the membrane will grow increasingly compact with higher HA concentrations. The porosity values ranging from 0.1 wt.% to 0.4 wt.% are comparable; however, the porosity at 0.5 wt.% exhibits a considerable rise, exceeding that of the native Psf membrane. The equation for determining porosity incorporates the wet and dry membrane weights, the effective membrane area, and the membrane thickness. A significant rise in membrane thickness was noted during measurement, possibly attributable to the substantial existence of HA on the surface of the membrane. This presence influences the thickness measurement, resulting in a calculation that suggests the membrane is more porous. Nevertheless, according to the water flux and protein separation measurements, the porosity of Psf/HA-0.5 should indeed be smaller. The size of macropores within the finger-like structures also influenced the membrane porosity value. The visible macropores were quantified using the SEM images for each sample. The macropore sizes are somewhat comparable, with Psf/HA-0.2 exhibiting the smallest macropore size among all samples examined. The flux recovery ratio (FRR) was calculated using Equation (5). A higher FRR value indicates enhanced efficiency in membrane cleaning [25], which in this case was when 0.1 wt.% HA was applied. All blend membranes, with the exception of Psf/HA-0.5, exhibit a better membrane cleaning efficiency relative to the native Psf membrane.

Table 1. Surface porosity, pore size, and FRR of Psf/HA blend membranes.

Membrane Code	Porosity, ε (%)	Mean Pore Size, r_m (Å)	Macropores Size, μm	FRR, (%)
Psf/HA-0.0	68.91	42.41	0.44 ± 0.093	48.72 ± 7.558
Psf/HA-0.1	65.72	40.04	0.48 ± 0.090	79.6 ± 4.845
Psf/HA-0.2	65.28	38.97	0.42 ± 0.074	73.2 ± 4.663
Psf/HA-0.3	66.28	39.91	0.49 ± 0.066	69.05 ± 4.621
Psf/HA-0.4	66.92	41.12	0.49 ± 0.055	68.70 ± 6.603
Psf/HA-0.5	71.01	43.82	0.49 ± 0.020	30.23 ± 6.999

Figure 5 illustrates that the Psf/HA-0.2 membrane demonstrated the highest protein rejection. To evaluate the effect of HA on the membrane lifespan, we performed five cycles of protein filtration, each consisting of a 30 min pure water flux test followed by a 10 min BSA filtration. Upon completion of the test, the flux decline ratio (FDR) could be determined in accordance with Equation (4). A lower FDR value generally indicates that the membrane exhibits more resistance to foulants, signifying enhanced antifouling properties [25]. Figure 7 illustrates that, while the flux trends for both the native Psf membrane and the blend membrane were comparable during the first and second cycles, the third cycle and subsequent cycles exhibited distinct values and trends. The native membrane exhibited a flatter flow pattern, indicating the significant onset of pore blockage. In the blend membranes, the beginning fluxes consistently exceeded the final fluxes over all five cycles. Despite the reduction in the flux value with each cycle, the downward trend in flux suggested that the amount of foulants adhering to the membrane surface was less than that of the native membrane. The incorporation of HA into the Psf membrane enhanced its hydrophilicity, which is advantageous for the whey separation industry.

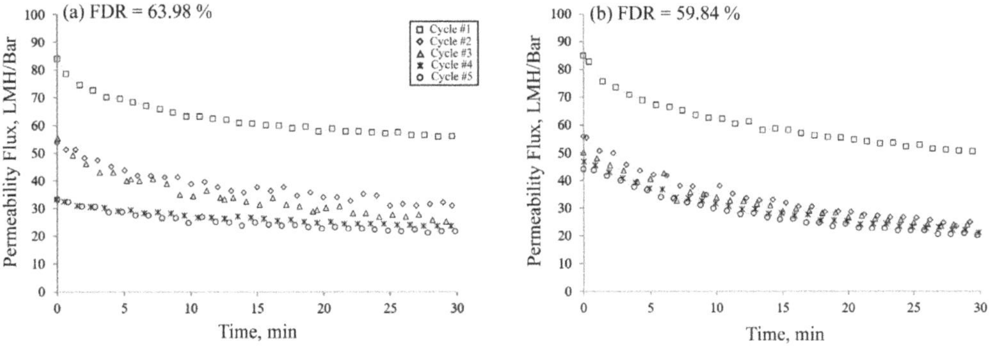

Figure 7. Flux decline ratio (FDR); (**a**) Native Psf membrane; (**b**) Psf/HA-0.2.

Table 2 compares prior studies on membrane performance that utilized HA as the primary modulator. Each column presents the value of the natural membrane in comparison to its modified counterpart. A variety of polymer-based solutions were utilized, including polyethersulfone (PES), polysulfone (PSf), and polyphenylsulfone (PPSU). While most studies employed PVP as a pore former, this experiment did not utilize PVP, hence influencing porosity and flux. Besides employing PVP as a pore former, HA was also modified by coating and grafting techniques. Each approach enhances the flux, pore size, rejection, and hydrophilicity, facilitating whey protein separation.

Table 2. Overview of research on HA blending in polymeric membranes comparing the native membrane and modified membranes.

Polymer	Modifier	Permeability Flux, LMH/Bar	Pore Size, nm	Rejection (%)	ε (%)	WCA (°)	Ref.
PES/PVP	HA nanotubes–PDA coating–taurine grafting	169/439	34.1/43.2	~97 (BSA)	55/73	70/53	[19]
PPSU	Silver-coated HA	39/57	-	89.74 (oil)	-	-	[20]
PES/PVP	HA/boron nitride	286/620	54/66.7	90 (BSA)	-	-	[21]
PES/PVP	HA	67/148.7	81/84	97.3 (BSA)	-	-	[41]
Psf/PVP	HA-modified PDA and PEI	152.8/481.6	5.36/7.41	94.5 (BSA)	52/74	81/60	[42]
Psf	HA	45.5/63	4.24/3.89	97.2 (BSA) 73.2 (Lysozyme)	68/65	66/65	This work

The optimal membrane for this study exhibits a performance equivalent to earlier research on protein rejection and hydrophilicity, despite the hydroxyapatite not undergoing any modifications before usage. Future research should utilize pore-forming agents, such as PVP, to enhance flow. The straightforward membrane fabrication procedure presented in this study should be the focal point for scaling up membrane production in the dairy industry.

4. Conclusions

The performance of polysulfone (Psf) ultrafiltration membranes was investigated after modification with hydroxyapatite (HA). These flat-sheet membranes were produced using the wet-phase inversion method. Scanning Electron Microscopy (SEM) analysis revealed that the HA particles were randomly distributed throughout the membrane. The surface hydrophilicity increased with HA concentration due to hydrogen bonding with HA's hydrophilic properties. The highest water flux was observed at 0.3 wt.% HA, but it decreased at higher concentrations, likely due to pore blockage by HA particles. Both native and modified membranes showed high Bovine Serum Albumin (BSA) rejection, with the highest rejection at 0.2 wt.% HA. A significant improvement was noted in lysozyme separation, with the maximum rejection increasing from 36.5% in the native membrane to 73.2% in the modified membrane. This enhancement is attributed to protein adsorption by HA. Contact angle measurements validate the enhanced hydrophilicity of the Psf membrane with increasing HA concentration. This cost-effective and facile approach to enhancing permeability while maintaining the separation efficiency of the membrane has the potential to scale up industrial application.

Author Contributions: Conceptualization, G.S.P. and A.S.B.; methodology, B.A. and F.Y.; software, T.S.; validation, Y.W., A.S.B. and M.M.; formal analysis, G.S.P.; investigation, G.S.P. and T.S.; resources, G.S.P., Y.W., A.S.B., F.Y. and M.M.; data curation, T.S.; writing—original draft preparation, G.S.P. and T.S.; writing—review and editing, Y.W., A.S.B. and M.M.; visualization, T.S.; supervision, Y.W., G.S.P. and A.S.B.; project administration, M.M.; funding acquisition, G.S.P., Y.W., A.S.B., B.A. and M.M. All authors have read and agreed to the published version of the manuscript.

Funding: This research was funded by the Hibah Riset Kolaborasi Indonesia Scheme: Universitas Airlangga (Grant No. 958/UN3.LPPM/PT.01.03/2023), the Universitas Gadjah Mada (Grant No. 2692/UN1/DITLIT/Dit-Lit/PT.01.03/2023), and the Universitas Indonesia (Grant No. NKB-1078/UN2.RST/HKP.05.00/2023).

Institutional Review Board Statement: Not applicable.

Data Availability Statement: The original contributions presented in the study are included in the article, further inquiries can be directed to the corresponding author.

Conflicts of Interest: Author Tutik Sriani was employed by the company PT. Global Meditek Utama-IITOYA. The remaining authors declare that the research was conducted in the absence of any commercial or financial relationships that could be construed as a potential conflict of interest.

References

1. Martini, M.; Altomonte, I.; Sodi, I.; Vasylieva, Y.; Salari, F. Sterol, tocopherol, and bioactive fatty acid differences between conventional, high-quality, and organic cow milk. *J. Dairy Sci.* **2023**, *106*, 8239–8248. [CrossRef]
2. Lappa, I.; Papadaki, A.; Kachrimanidou, V.; Terpou, A.; Koulougliotis, D.; Eriotou, E.; Kopsahelis, N. Cheese whey processing: Integrated biorefinery concepts and emerging food applications. *Foods* **2019**, *8*, 347. [CrossRef] [PubMed]
3. Barba, F.J. An integrated approach for the valorization of cheese whey. *Foods* **2021**, *10*, 564. [CrossRef] [PubMed]
4. Roth-Walter, F.; Afify, S.M.; Pacios, L.F.; Blokhuis, B.R.; Redegeld, F.; Regner, A.; Petje, L.M.; Fiocchi, A.; Untersmayr, E.; Dvorak, Z.; et al. Cow's milk protein β-lactoglobulin confers resilience against allergy by targeting complexed iron to immune cells. *J. Allergy Clin. Immunol.* **2021**, *147*, 321–334.e4. [CrossRef] [PubMed]
5. Layman, D.K.; Lönnerdal, B.; Fernstrom, J.D. Applications for α-lactalbumin in human nutrition. *Nutr. Rev.* **2018**, *76*, 444–460. [CrossRef] [PubMed] [PubMed Central]
6. Ebrahimi, A.; Andishmand, H.; Huo, C.; Amjadi, S.; Khezri, S.; Hamishehkar, H.; Mahmoudzadeh, M.; Kim, K.H. Glycomaropeptide: A comprehensive understanding of its major biological characteristics and purification methodologies. *Compr. Rev. Food Sci. Food Saf.* **2024**, *23*, e13370. [CrossRef]
7. Zhao, C.; Chen, N.; Ashaolu, T.J. Whey proteins and peptides in health-promoting functions—A review. *Int. Dairy J.* **2022**, *126*, 105269. [CrossRef]
8. de Castro, R.J.S.; Domingues, M.A.F.; Ohara, A.; Okuro, P.K.; dos Santos, J.G.; Brexó, R.P.; Sato, H.H. Whey protein as a key component in food systems: Physicochemical properties, production technologies and applications. *Food Struct.* **2017**, *14*, 17–29. [CrossRef]
9. Hogan, S.A.; O'Loughlin, I.B.; Kelly, P.M. Soft matter characterization of whey protein powder systems. *Int. Dairy J.* **2016**, *52*, 1–9. [CrossRef]

10. Galanakis, C.M.; Chasiotis, S.; Botsaris, G.; Gekas, V. Separation and recovery of proteins and sugars from Halloumi cheese whey. *Food Res. Int.* **2014**, *65*, 477–483. [CrossRef]
11. Ganju, S.; Gogate, P.R. A review on approaches for efficient recovery of whey proteins from dairy industry effluents. *J. Food Eng.* **2017**, *215*, 84–96. [CrossRef]
12. Hoslett, J.; Massara, T.M.; Malamis, S.; Ahmad, D.; Boogaert, I.V.D.; Katsou, E.; Ahmad, B.; Ghazal, H.; Simos, S.; Wrobel, L.; et al. Surface water filtration using granular media and membranes: A review. *Sci. Total Environ.* **2018**, *639*, 1268–1282. [CrossRef] [PubMed]
13. Guillen, G.R.; Pan, Y.; Li, M.; Hoek, E.M.V. Preparation and characterization of membranes formed by nonsolvent induced phase separation: A review. *Ind. Eng. Chem. Res.* **2011**, *50*, 3798–3817. [CrossRef]
14. Yunos, M.Z.; Harun, Z.; Basri, H.; Ismail, A.F. Studies on fouling by natural organic matter (NOM) on polysulfone membranes: Effect of polyethylene glycol (PEG). *Desalination* **2014**, *333*, 36–44. [CrossRef]
15. Ouda, M.; Ibrahim, Y.; Kallem, P.; Govindan, B.; Banat, F.; Hasan, S.W. Highly permeable, environmentally-friendly, antifouling polylactic acid-hydroxyapatite/polydopamine (PLA-Hap/PDA) ultrafiltration membranes. *J. Clean. Prod.* **2022**, *330*, 129871. [CrossRef]
16. Yogarathinam, L.T.; Usman, J.; Abba, S.I.; Lawal, D.; Baig, N.; Aljundi, I.H. Evaluating nano-metal oxide mixed matrix membranes for whey protein separation using hybrid intelligent optimization learning. *Chem. Eng. Res. Des.* **2024**, *205*, 388–400. [CrossRef]
17. Heng, Z.W.; Tan, Y.Y.; Chong, W.C.; Mahmoudi, E.; Mohammad, A.W.; Teoh, H.C.; Sim, L.C.; Koo, C.H. Preparation of a novel polysulfone membrane by incorporated with carbon dots grafted silica from rice husk for dye removal. *J. Water Process Eng.* **2021**, *40*, 101805. [CrossRef]
18. Yasir, A.T.; Benamor, A.; Ba-Abbad, M.; Hawari, A.H. Performance of polysulfone ultrafiltration membranes incorporating graphene oxide nanoparticles functionalized with different generations of poly(amido amine). *J. Water Process Eng.* **2024**, *60*, 105095. [CrossRef]
19. Mu, Y.; Zhu, K.; Luan, J.; Zhang, S.; Zhang, C.; Na, R.; Yang, Y.; Zhang, X.; Wang, G. Fabrication of hybrid ultrafiltration membranes with improved water separation properties by incorporating environmentally friendly taurine modified hydroxyapatite nanotubes. *J. Membr. Sci.* **2019**, *577*, 274–284. [CrossRef]
20. Anwar, F.; Arthanareeswaran, G. Silver nano-particles coated hydroxyapatite nano-composite membrane for the treatment of palm oil mill effluent. *J. Water Process Eng.* **2019**, *31*, 100844. [CrossRef]
21. Kallem, P.; Ibrahim, Y.; Hasan, S.W.; Show, P.L.; Banat, F. Fabrication of novel polyethersulfone (PES) hybrid ultrafiltration membranes with superior permeability and antifouling properties using environmentally friendly sulfonated functionalized polydopamine nanofillers. *Sep. Purif. Technol.* **2021**, *261*, 118311. [CrossRef]
22. Deng, X.; Yang, F.; Ma, J.; Li, Y.; Dang, J.; Ouyang, M. Ternary phase field model and characterization of water/NMP/polysulfone membrane prepared by non-solvent induced phase separation. *Sep. Purif. Technol.* **2024**, *330*, 125307. [CrossRef]
23. Saraswathi, M.S.S.A.; Rana, D.; Alwarappan, S.; Gowrishankar, S.; Kanimozhi, P.; Nagendran, A. Cellulose acetate ultrafiltration membranes customized with bio-inspired polydopamine coating and in situ immobilization of silver nanoparticles. *New J. Chem.* **2019**, *43*, 4216–4225. [CrossRef]
24. Kanagaraj, P.; Nagendran, A.; Rana, D.; Matsuura, T.; Neelakandan, S.; Malarvizhi, K. Effects of Polyvinylpyrrolidone on the permeation and fouling-resistance properties of Polyetherimide ultrafiltration membranes. *Ind. Eng. Chem. Res.* **2015**, *54*, 4832–4838. [CrossRef]
25. Xu, S.; Wang, P.; Sun, Z.; Liu, C.; Lu, D.; Qi, J.; Ma, J. Dual-functionalization of polymeric membranes via cyclodextrine-based host-guest assembly for biofouling control. *J. Membr. Sci.* **2019**, *569*, 124–136. [CrossRef]
26. Manorma; Ferreira, I.; Alves, P.; Gil, M.H.; Gando-Ferreira, L.M. Lignin separation from black liquor by mixed matrix polysulfone nanofiltration membrane filled with multiwalled carbon nanotubes. *Sep. Purif. Technol.* **2021**, *260*, 118231. [CrossRef]
27. Bagheripour, E.; Moghadassi, A.R.; Hosseini, S.M.; Ray, M.B.; Parvizian, F.; Van der Bruggen, B. Highly hydrophilic and antifouling nanofiltration membrane incorporated with water-dispersible composite activated carbon/chitosan nanoparticles. *Chem. Eng. Res. Des.* **2018**, *132*, 812–821. [CrossRef]
28. Hołda, A.K.; Vankelecom, I.F.J. Understanding and guiding the phase inversion process for synthesis of solvent resistant nanofiltration membranes. *J. Appl. Polym. Sci.* **2015**, *132*, 42130. [CrossRef]
29. Hackett, C.; Hale, D.; Bair, B.; Endeboh, G.D.M.; Hao, X.; Qian, X.; Wickramasinghe, S.R.; Thompson, A. Polysulfone ultrafiltration membranes fabricated from green solvents: Significance of coagulation bath composition. *Sep. Purif. Technol.* **2024**, *332*, 125752. [CrossRef]
30. Kang, Y.; Obaid, M.; Jang, J.; Ham, M.H.; Kim, I.S. Novel sulfonated graphene oxide incorporated polysulfone nanocomposute membranes for enhanced-performance in ultrafiltration process. *Chemosphere* **2018**, *207*, 581–589. [CrossRef]
31. Eslami, S.; Norpuzbahari, S.; Vatanpour, V.; Ghadimi, A.; Rostamizadeh, M. Synthesis and characterization of enhanced polysulfone-based mixed matrix membranes containing ZSM-5 zeolite for protein and dye removal. *J. Ind. Eng. Chem.* **2024**, *137*, 455–467. [CrossRef]
32. Wu, H.; Mansouri, J.; Chen, V. Silica nanoparticles as carriers of antifouling ligands for PVDF ultrafiltration membranes. *J. Membr. Sci.* **2013**, *433*, 135–151. [CrossRef]
33. Yurekli, Y. Removal of heavy metals in wastewater by using zeolite nano-particles impregnated polysulfone membranes. *J. Hazard. Mater.* **2016**, *309*, 53–64. [CrossRef] [PubMed]

34. Mulder, J. *Basic Principles of Membrane Technology*, 2nd ed.; Kluwer Academic Publishers: Dordrecht, The Netherlands, 1996; p. 584.
35. Elhamarnah, Y.; Hey, T.; Lipnizki, F.; Qiblawey, H. Investigating the impact of stormwater fouling on polysulfone ultrafiltration membranes modified with deep eutectic solvents. *J. Water Process Eng.* **2023**, *56*, 104362. [CrossRef]
36. Bouiahya, K.; Oulguidoum, A.; Laghzizil, A.; Shalabi, M.; Nunzi, J.M. Hydrophobic chemical surface functionalization of hydroxyapatite nanoparticles for naphthalene removal. *Colloids Surf. A Physicochem. Eng. Asp.* **2020**, *595*, 124706. [CrossRef]
37. Yang, L.; Tang, B.; Wu, P. UF membrane with highly improved flux by hydrophilic network between graphene oxide and brominated poly(2,6-dimethyl-1,4-phenylene oxide). *J. Mater. Chem. A* **2014**, *2*, 18562. [CrossRef]
38. Astala, R.; Stott, M.J. First-principles study of hydroxyapatite surfaces and water adsorption. *Phys. Rev. B* **2008**, *78*, 075427. [CrossRef]
39. Yin, J.; Deng, B.L. Polymer-matrix nanocomposite membranes for water treatment. *J. Membr. Sci.* **2015**, *479*, 256–275. [CrossRef]
40. Cayot, P.; Lorient, D. Structure-function relationships of whey proteins. In *Food Proteins and Their Applications*; Damodaran, S., Paraf, A., Eds.; Marcel Dekker: New York. NY, USA, 1997; pp. 225–256.
41. Salimi, E.; Ghaee, A.; Ismail, A.F. Performance and antifouling enhancement of polyethersulfone hollow fiber membranes incorporated with highly hydrophilic hydroxyapatite nanoparticles. *RSC Adv.* **2016**, *6*, 44480–44488. [CrossRef]
42. Mu, Y.; Feng, H.; Zhang, S.; Zhang, C.; Lu, N.; Luan, J.; Wang, G. Development of highly permeable and antifouling ultrafiltration membranes based on the synergistic effect of carboxylated polysulfone and bio-inspired co-deposition modified hydroxyapatite nanotubes. *J. Colloid. Interface Sci.* **2020**, *572*, 48–61. [CrossRef]

polymers

MDPI

Article

Effects of BET Surface Area and Silica Hydrophobicity on Natural Rubber Latex Foam Using the Dunlop Process

Danvanichkul Assadakorn [1], Gongxu Liu [1], Kuanfa Hao [1], Lichen Bai [1], Fumin Liu [1], Yuan Xu [2], Lei Guo [1,2,*] and Haichao Liu [2,*]

[1] College of Electromechanical Engineering, Qingdao University of Science & Technology, Qingdao 266061, China; assadakorndan@gmail.com (D.A.); 118765487207@163.com (G.L.); jayhao0730@163.com (K.H.); 15684195159@163.com (L.B.); liufumin@qust.edu.cn (F.L.)

[2] National Engineering Research Center of Advanced Tire Equipment and Key Materials, Qingdao University of Science & Technology, Qingdao 266061, China; 18792469648@163.com

* Correspondence: 06weny@163.com (L.G.); liuhaichao@qust.edu.cn (H.L.)

Abstract: To reinforce natural rubber latex foam, fumed silica and precipitated silica are introduced into latex foam prepared using the Dunlop process as fillers. Four types of silica, including Aerosil 200 (hydrophilic fumed silica), Reolosil DM30, Aerosil R972 (hydrophobic fumed silica), and Sipernat 22S (precipitated silica), are investigated. The latex foam with added silica presents better mechanical and physical properties compared with the non-silica foam. The hydrophobic nature of the fumed silica has better dispersion in natural rubber compared to hydrophilic silica. The specific surface area of silica particles (BET) also significantly influences the properties of the latex foam, with larger specific surface areas resulting in better dispersity in the rubber matrix. It was observed that exceeding 2 phr led to difficulties in the foaming process (bulking). Furthermore, higher loading of silica also affected the rubber foam, resulting in an increased shrinkage percentage, hardness, compression set, and crosslink density. The crosslink density increased from 11.0 ± 0.2 mol/cm^3 for non-silica rubber to 11.6 ± 0.6 mol/cm^3 for Reolosil DM30. Reolosil DM30 also had the highest hardness, with a hardness value of 52.0 ± 2.1 IRHD, compared to 45.0 ± 1.3 IRHD for non-silica foam rubber and 48 ± 2.4 IRHD for hydrophilic fumed silica Aerosil 200. Hydrophobic fumed silica also had the highest ability to return to its original shape, with a recovery percentage of 88.0% $\pm$ 3.5% compared to the other fumed silica. Overall, hydrophobic fumed silica had better results than hydrophilic silica in both fumed and precipitated silica.

Keywords: latex foam; silica; Dunlop process; hydrophobicity; BET surface area

1. Introduction

Natural rubber, sourced from rubber trees, is a versatile material known for its elasticity. It finds wide applications in diverse industries, including bedding, thermal insulation, and automotive components [1]. Various types of foam rubber exist, each tailored to meet specific needs. The manufacturing process of foam rubber is relatively straightforward, with fillers playing a pivotal role in shaping the properties of the final product. Common fillers include calcium carbonate, silica, and carbon black [2,3]. These fillers serve as primary reinforcing agents, contributing to the structural integrity and functional characteristics of rubber compounds. Calcium carbonate, silica, and carbon black each exhibit unique properties that influence the mechanical and physical attributes of the rubber, as well as its surface chemistry and coloration [2,4]. The choice of filler depends on the industry and the specific type of rubber. In the tire industry, for example, carbon black is the primary filler of choice due to its exceptional reinforcing properties and compatibility with rubber compounds [5]. The addition of carbon black significantly enhances the mechanical strength, wear resistance, and overall performance of rubber used in tire manufacturing [6,7]. In the production of foam rubber, the utilization of fillers takes a different trajectory. Calcium carbonate stands out as a cost-effective and highly compatible filler. Its widespread use is attributed to its affordability and versatility, making it a preferred choice in various formulations. Similarly, silica plays a crucial role in shaping the characteristics of foam rubber. Beyond its reinforcing capabilities, silica introduces specific surface interactions, altering the material's properties and surface chemistry [8].

Silica is an oxide of silicon with the chemical formula SiO_2, commonly presented as a white powder. Its polar structure results in strong filler–filler interactions and adsorption [8,9]. As non-black fillers, silica performs well as a rubber reinforcing filler. Generally, carbon black reinforcement provides a higher modulus than silica reinforcement, but silica-reinforced rubber excels in several aspects, such as aging resistance, abrasion resistance, and adhesion properties [10,11]. Synthetic amorphous silica can be primarily categorized into two types: precipitated silica and fumed silica. The differentiation between these two variants lies in their distinct manufacturing processes and particle dimensions. Fumed silica, produced through the hydrolysis of silicon tetrachloride in a flame, is inherently hydrophilic due to its method of synthesis. The high surface area and fine particle size of fumed silica contribute to its excellent moisture absorption properties [12]. Hydrophilic fumed silica is particularly advantageous in applications requiring moisture absorption, such as in pharmaceuticals, cosmetics, and certain adhesive formulations [13,14]. Recognizing the demand for hydrophobic characteristics, surface modifications are often employed in the production of hydrophobic fumed silica. The surface treatment involves the introduction of hydrophobic functionalities, enhancing fumed silica's resistance to moisture absorption, expanding the range of applications, improving dispersion, impacting rheological properties, and tailoring surface chemistry [15,16]. By contrast, precipitated silica is typically hydrophilic in its natural state. The production process of precipitated silica involves the acid precipitation of sodium silicate, resulting in a material with a hydrophilic surface [17]. Unlike fumed silica, which can be inherently hydrophilic or hydrophobic depending on its production process and subsequent surface treatments, the particle dimensions of precipitated silica are influenced by the specific conditions during precipitation. This type of synthetic silica has found diverse applications owing to its versatility and cost-effectiveness [17,18].

To the best of our knowledge, fumed and precipitated silica have not been widely used as fillers in the production of natural foam rubber using the Dunlop method. However, some researchers have explored the use of fumed silica and precipitated silica as fillers in the rubber and polymer industry. Bayat and Fasihi (2019) [19] studied the effect of coupling agents in natural rubber silica composite foam. They used different contents of precipitated silica and found that it was well-dispersed in natural rubber. The results showed that as the silica concentration increased, the density also increased, while the cell size decreased with higher silica content. Luo et al. [20] investigated the interaction between fumed silica and epoxidized natural rubber. Their research indicated that fumed silica had a strong interaction with epoxidized natural rubber and effectively contributed to crosslink density. Prasertsri et al. [21] studied the reinforcement of natural rubber using fumed silica and precipitated silica mixed by two-roll mills. The results showed that the samples using fumed silica as a filler had higher hardness, stiffness, and tensile strength compared to those using precipitated silica with the same content. However, the heat build-up in the samples containing fumed silica was significantly higher, indicating that the use of fumed silica as a filler must also consider the application requirements of the final product.

The aim of this present work is to study the reinforcement of natural rubber latex foam using fumed silica and precipitated silica. Latex foams were produced using different types of silica, including fumed silica (both hydrophilic and hydrophobic varieties) and precipitated silica. Zaborski et al. [12] highlighted the importance of considering the surface area of silica particles. Consequently, the effect of specific surface area of silica particles (BET) was also considered. The crosslink density, mechanical properties, as well as surface morphologies of latex foam were investigated to gain further insights into the reinforcing mechanism of silica in latex foam.

2. Experiment

2.1. Material

Table 1 displays the materials used in this work. Natural rubber latex with a high ammonia concentration (60% dry rubber content) was provided by Sri Trang Agro-Industry Public Company Limited ("STA") (Surat Thani, Thailand). Three types of fumed silica and one type of precipitated silica were obtained from Evonik Degussa Co., Ltd. (Essen, Germany). Specifically, these included hydrophilic fumed silica (Aerosil 200, purity > 99.8%, BET surface area 200 $\pm$ 25 m^2/g) and hydrophobic fumed silica (Aerosil R972, purity > 99.8%, BET surface area 110 $\pm$ 20 m^2/g). Hydrophobic fumed silica (Reolosil DM30, purity > 99.8%, BET surface area 230 $\pm$ 20 m^2/g) was purchased from Tokuyama Chemicals (Zhejiang) Co., Ltd. (Jiaxing, China). The hydrophilic precipitated silica (Sipernat 22s, purity > 97%, BET surface area 190 m^2/g) was also purchased from Evonik Degussa Co., Ltd. (Essen, Germany). All silica samples were sourced from the same suppliers to ensure consistency; however, Tokuyama Chemicals did not have the specific silica specifications required for this study. Consequently, three types of silica were obtained from Tokuyama Chemicals and one type from Evonik Degussa Co., Ltd. (Essen, Germany), which met the specifications for pH, tamped density, and other critical parameters. Key characteristics of silica samples are displayed in Table 2. Sulfur was used as a vulcanizing agent and various additives were used, such as zinc oxide (ZnO_2), potassium, zinc-N-diethyldithiocarbamate, zinc 2-mercaptobenzothiazole, Wingstay L, diphenyl guanidine, and sodium silicofluoride. These chemicals were prepared by Qingdao Amita Natural Latex., Co. Ltd. (Qingdao, China).

Table 1. Formulation of latex foam in parts per hundred of rubber (phr).

Materials	Weight (phr)
High Ammonia concentrated natural latex 60% DRC	100
Potassium-oleate solution 10%	1
Sulphur dispersion 50%	2.5
Zinc-N-diethyldithiocarbamate dispersion 50%	1
Zinc 2-mercaptobenzothiazole dispersion 50%	1
Wing stay L dispersion	1
Zinc Oxide dispersion of 50%	4
Diphenyl guanidine dispersion 50%	1
Sodium silicofluoride dispersion 20%	2.5
Filler dispersion fumed or precipitated silica	0.5, 1, 1.5, 2

Table 2. Key characteristics of silica samples used in this study.

Silica Name	Specific Surface Area (m^2/g)	Carbon Content (%)	Tamped Density (g/cm^3)	Ph Value (4% Suspension)	SiO_2 Content (wt.%)	Hydrophilic/ Hydrophobic
Aerosil 972	110 $\pm$ 20	0.6–1.2	Approx. 50	3.6–5.5	$\geq$99.8	hydrophobic
Reolosil Dm30	235 $\pm$ 20	1.7	Approx. 50	4.8	$\geq$99.8	hydrophobic
Aerosil 200	200 $\pm$ 25	N/A *	Approx. 50	3.7–4.7	$\geq$99.8	hydrophilic
Sipernat 22s	190	N/A *	Approx. 90	6.5	$\geq$97	hydrophilic

* N/A indicates that the information is not available in the product specifications.

2.2. Rubber Compounds and Foam Preparation

Figure 1 shows the schematic diagram of the foaming process using the Dunlop method. The NR compound was prepared by using a stirring machine to mix the components as in Table 1 with a mechanical stirring speed of 450 rpm at room temperature of 15 °C. This initial low speed was chosen to ensure homogeneous mixing without risking premature vulcanization.

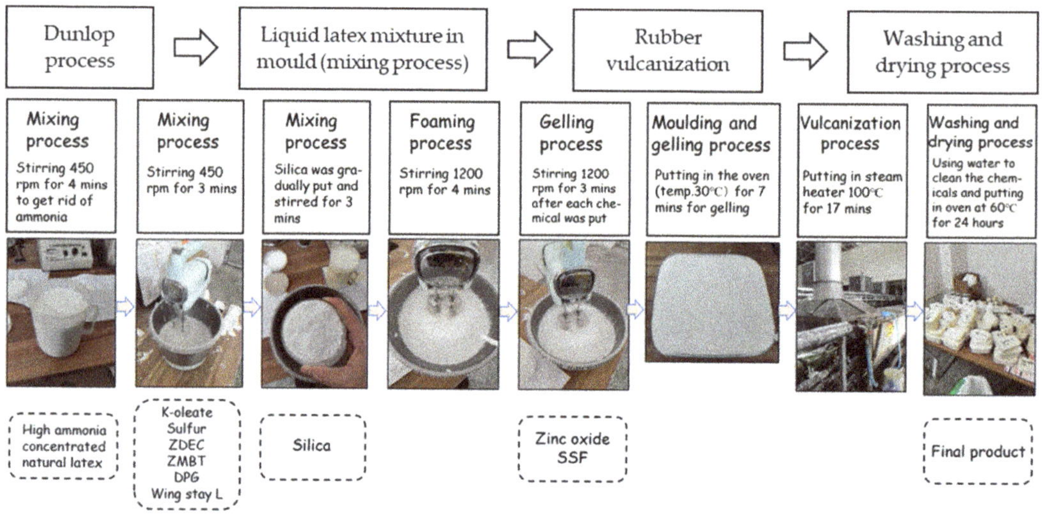

Figure 1. Schematic diagram of the Dunlop process.

Mixing process: Natural rubber was added into the stirring machine, stirred for 4 min to remove the ammonia, which is essential for improving the viscosity and stability of the latex and facilitating better foam expansion. Afterward, the other ingredients consisting of K-oleate, sulfur, zinc Diethyldithiocarbamate, zinc 2-mercaptobenzothiazole, Wingstay L, and diphenyl guanidine were added and stirred for 3 min. These ingredients were included to enhance the crosslinking and stability of the rubber matrix during the vulcanization process.

Silica concentration: Silica was gradually added, while continuously stirring for 3 min. The concentration of each type of silica was controlled by adding four levels of content: 0.5 phr, 1 phr, 1.5 phr, and 2 phr. This range was selected to systematically assess the effects of varying silica concentrations on the foam's mechanical properties. Concentrations exceeding 2.5 phr were avoided based on preliminary experiments conducted with this specific formulation, which indicated that higher silica content resulted in poor mixing and the onset of pre-vulcanization. This behavior compromised the uniformity and performance of the rubber compound. It is important to note that these observations are specific to the formulation used in this study, and other rubber formulations may exhibit different behaviors under similar conditions. Therefore, while the selected silica concentrations are appropriate for this particular system, further research may be necessary to explore the optimal silica levels for different formulations.

Foaming process: After mixing all these ingredients, the stirring speed was increased to 1200 rpm and the mixture was stirred for 4 min during the foaming process. This higher speed facilitates the uniform distribution of air bubbles, crucial for achieving the desired foam structure. The gelling agent, zinc oxide, was added and stirred for 3 min, and then Sodium silicofluoride dispersion was added and stirred for 2 min, respectively. The total mixing time for all samples was set to 19 min.

Temperature control: To mitigate the risk of premature vulcanization and gelation associated with heat generation, the room temperature was maintained at 15 °C throughout the mixing process. This temperature control helps to maintain the stability of the latex and prevent premature curing, which could otherwise adversely affect the foam structure.

Gelling process: After mixing all the materials according to the formulation in Table 1, foam rubber was put in the rectangular glass mold of $13 \times 19 \times 9$ cm^3, and the foam sample was placed in the oven at a temperature of 30 °C for 7 min as a gelling process. This gelling temperature was selected to promote the initial setting of the foam structure without triggering full vulcanization, which is essential for maintaining the desired expansion characteristics

Vulcanization process: Curing of samples was conducted at 100 °C for 17 min in a steam oven. This temperature and time were selected based on preliminary studies and experimental trials that demonstrated optimal crosslinking and mechanical performance under these conditions. These parameters were fine-tuned to ensure effective vulcanization, as the experiments indicated that this duration allowed for the best balance between curing and maintaining the desired material properties.

Washing and drying process: All samples after curing were washed with deionized water and placed in the oven at 60 °C for 24 h for drying.

2.3. *Characterizations*

2.3.1. Foam Density

Foam densities were determined by cutting samples into 30 mm × 30 mm × 30 mm, according to ASTM D3574 [22]. The density was calculated using the mass and volume by the following equation:

$$\rho_f = \frac{m_f}{v_f} \tag{1}$$

where, m_f is the mass of the foam sample and v_f is the volume.

2.3.2. Crosslink Density

The samples were cut to a small size (4 × 2 × 25 mm^3) for a crosslink density test based on the equilibrium swelling method. The crosslink density was calculated by using the Flory–Rehner equation [23,24]:

$$-\ln(1 - V_r) - V_r - \chi v_r^2 = v_s \eta_{swell} \left(v_r^{1/3} - \frac{v_r}{2} \right) \tag{2}$$

where V_r is the volume fraction of rubber in swollen gel, χ is the rubber–solvent interaction parameter, v_s is the molar volume of toluene (106.8 cm^3/mol^{-1}), η_{swell} is the swelling of the compounded rubber (mol/cm^{-3}).

V_r was tested according to ISO 1817 [25]. Samples were weighed and then swollen in toluene for 1 week. Then, the weights of the samples were measured, and the liquid on the surface of the samples was removed using filter paper. The samples were then dried at a temperature of 80 °C for 48 h. The weight after drying was measured and used to calculate the volume fraction of rubber in the swollen gel using the following equation [26,27]:

$$V_r = \frac{\left[w_0 \varnothing \frac{1-\alpha}{\rho_r} \right]}{\left[\left(w_0 \varnothing \frac{1-\alpha}{\rho_r} \right) + \frac{(w_1 - w_2)}{\rho_s} \right]} \tag{3}$$

$$\alpha = \frac{w_1 - w_0}{w_1} \tag{4}$$

$$\varnothing = 1 - \alpha \tag{5}$$

where w_0, w_1 are the weights (g) of samples before and after immersed in toluene, respectively, $\varnothing$ is the mass fraction of natural rubber (amount of natural rubber/total quantity), α is the mass fraction loss during swelling in toluene, and ρ_s and ρ_r are the densities (g/cm^3) of toluene and the rubber composites, respectively.

$$\chi = 0.34 + V_s (\delta_s - \delta_p)^2 / RT \tag{6}$$

where δ_s and δ_p are the solubility parameters of toluene (8.26) and natural rubber, which are 8.26 and 8.91 $cal^{0.5}/cm^{1.5}$, respectively. R is the gas constant (cal/mol-K) and T is the absolute temperature (K) [28,29].

2.3.3. Microstructural Analysis Using Scanning Electron Microscope (SEM)

The foam morphologies of the samples were determined using a scanning electron microscope (SEM). The samples were first surface-coated with gold using a sputter coater to ensure good electrical conductivity between the samples and the aluminum stub, thus preventing any charging effect during the observation. The SEM used was JEOL JSM-7500F (Tokyo, Japan). The acceleration voltage was set at 5 kV, and magnifications of ×50 and ×10,000 were used to observe the pore structure of the latex foam and the distribution of silica filler in the rubber matrix. The average diameter of cells was measured using ImageJ software (version 1.54g) from at least 300 different pores.

2.3.4. Fourier Transform Infrared Spectroscopy Analysis (FTIR)

The chemical functional groups of the foam rubber samples were determined using attenuated total reflection Fourier transform infrared (ATR-FTIR) spectroscopy with a VERTEX 70 FT-IR spectrometer (Waltham, MA, USA). The foam rubber samples, sized 10 mm square with a 10 mm thickness, were placed on a Ge crystal probe and analyzed in the wave number range of 4000–600 cm^{-1}.

2.3.5. Compression Set

In the compression test according to the ASTM D3574 standard, the specimen size was 50 ± 1 mm (2-inch) square with a thickness of 25 ± 1 mm (1 inch). The samples were compressed by using a force transducer to reduce the thickness of the original sample to 50% and placed in an oven at a temperature of 70 °C ± 1 °C for 22 h, then the samples were taken out of the oven and the transducer and stayed in the atmosphere for 30 min. Then, the thickness of samples was measured. Figure 2 shows the schematic diagram for the test of the compression set. Three samples of each type were tested and the average was reported. The thickness measured was calculated by using the following two equations [30,31]:

$$C_d = [(t_0 - t_1)/t_0] \times 100 \quad (7)$$

$$recovery\ percentage = \frac{t_1}{t_0} \times 100 \quad (8)$$

Figure 2. Schematic diagram for testing the compression set.

C_d is the calculated percentage expressing the permanent deformation in relation to the total deformation, accounting for the transducer height (%). t_0 is the original height or thickness of the foam sample before compression, t_1 is the height or thickness of the foam sample after compression.

2.3.6. Hardness Test

A hardness test was conducted using the International Rubber Hardness Degree (IRHD) tester, specifically the Wallace H14 Macro IRHD tester (Dorking, UK), in accordance with ASTM D1415 standards [32]. Specimens were precision-cut to dimensions of 10 mm in thickness, forming squares with sides measuring 20 mm at a controlled room temperature of 23 °C. For each specimen, hardness measurements were taken at five distinct points distributed across the surface. The median value of these measurements, rounded to the nearest IRHD, was calculated to determine the final hardness value. This approach ensures a representative and accurate characterization of the material's hardness based on multiple points on each specimen.

2.3.7. Percentage Shrinkage

The percentage shrinkage was determined in accordance with the standard of ASTM D1055 [33] by measuring the dimensions of the sample. All sides, including width, length, height, and the height of the center, were measured. The percentage shrinkage was calculated by comparing the changes in these dimensions between the sample after vulcanization and the dimensions of the mold. The equation below was used to calculate the percentage shrinkage:

$$\text{Percentage shrinkage} = \left(\frac{X_1 - X_2}{X_1}\right) \times 100 \quad (9)$$

where X_1 = dimension of sides (cm), X_2= dimensions of sides of rubber after vulcanization.

3. Discussion and Results

The experiment was divided into two major parts for analysis. The first part investigated different types of silica, including fumed and precipitated silica, with a focus on both hydrophilic and hydrophobic varieties, as well as particle size. These factors were found to influence foam morphology. The second part studied the impact of silica content on the vulcanization of the latex foam produced using the Dunlop process.

3.1. Microstructural Analysis

The surface morphologies of the latex foams were examined using SEM, as shown in Figure 3. The figure illustrates the structures of the foam samples produced by different types and concentrations of silica. According to Figure 3, it is evident that a higher concentration of silica results in a smaller cell size and a more uniform distribution of the cell structure. At a concentration of 1 phr, the SEM results demonstrate that the BET specific surface area and hydrophobicity/hydrophilicity of the silicas play a decisive role in determining the foam microstructure. Latex foams containing fumed hydrophobic silicas, such as Reolosil DM30 and Aerosil R972, exhibited a mixture of larger and smaller pore sizes; while fumed hydrophilic silica (Aerosil 200) and precipitated silica (Sipernat 22S) produced predominantly medium and smaller pore sizes (Figure 3). Increasing the silica concentration to 2 phr made the distinctions between the silica types less pronounced, as all the silicas were able to effectively promote a finer and more uniform cellular structure in the latex foam [34,35]. After mixing with natural rubber, it increased the viscosity of the rubber mixture, which reduced the bubble size during the foaming process [36]. Furthermore, a higher concentration of silica increased the nucleation sites for bubble formation, leading to the formation of a larger number of smaller bubbles, rather than fewer larger ones [37,38].

The ImageJ program was used to analyze the cell diameters, as shown in Figure 4. Fumed silica, Aerosil R972, has the smallest average pore diameters at both 1 phr and 2 phr concentrations. By contrast, fumed silica, Reolosil DM30, has the largest average pore diameter in both concentrations. These results suggest that particles with a larger BET specific surface area tend to produce larger average pore diameters. However, the hydrophilic fumed silica, Aerosil 200, which has nearly the same BET specific surface area as Reolosil DM30, resulted in smaller average pore diameters. Specifically, the average pore diameter of Aerosil 200 was 38.1 ± 1.3 μm at 1 phr and 29.8 ± 2.3 μm at 2 phr. With regard to Reolosil DM30, the average pore diameters were 40.8 ± 7.9 μm at 1 phr and 34.9 ± 2.0 μm at 2 phr. The precipitated silica, Sipernat 22S, which has a similar BET specific surface area, exhibited slightly different average pore diameters (40.3 ± 1.4 μm at 1 phr and 32.8 ± 1.7 μm at 2 phr) compared to hydrophobic fumed silica, Reolosil DM30.

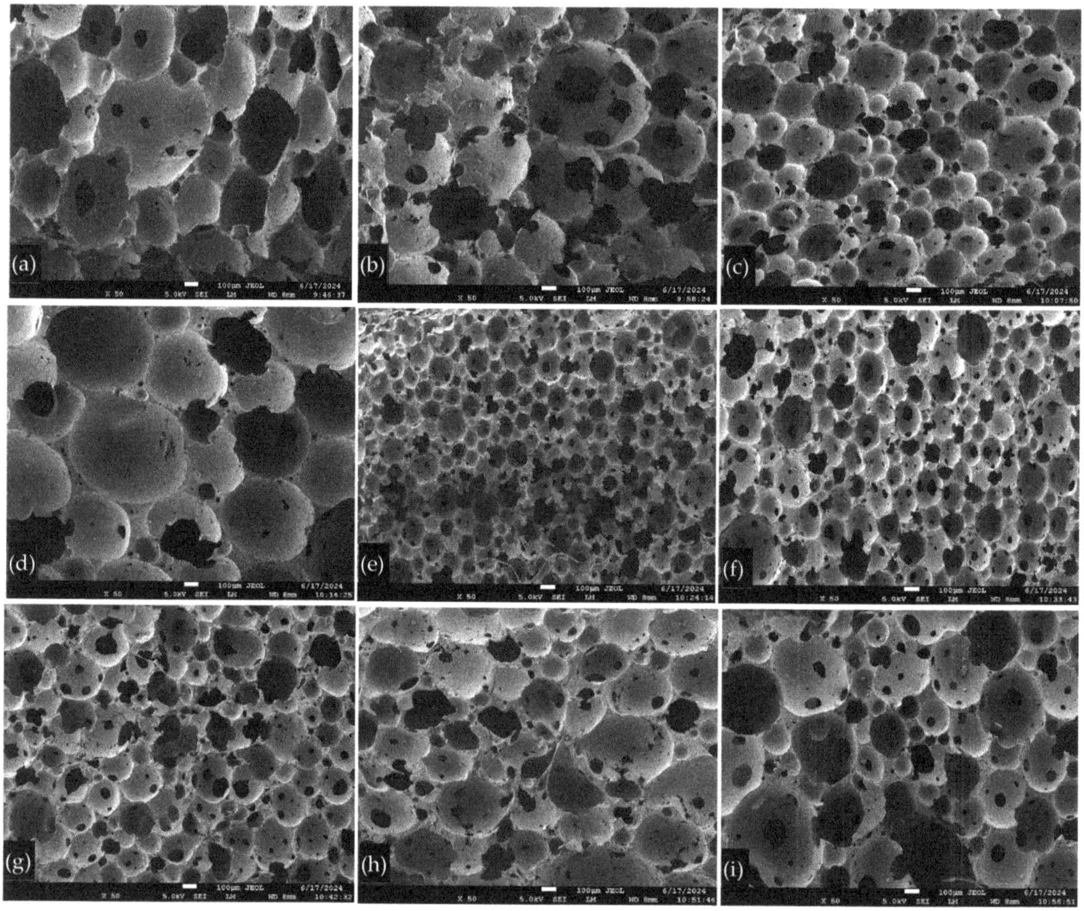

Figure 3. SEM image of natural latex foam (**a**) without silica, (**b**) with DM30 at 1 phr, (**c**) with DM30 at 2 phr, (**d**) with Aerosil R972 at 1 phr, (**e**) with Aerosil R972 at 2 phr, (**f**) with Aerosil 200 at 1 phr, (**g**) with Aerosil 200 at 2 phr, (**h**) with Sip22s at 1 phr, and (**i**) with Sip22s at 2 phr. SEM images of foam rubber at 50× magnification. A 100 μm scale bar is included for reference.

There were slight differences between hydrophobic and hydrophilic types in the average pore diameters of silica. However, the pore structures of the foam rubber were distinct, as can be seen from Figure 3. While increasing the silica concentration resulted in smaller and more uniform cell sizes overall, there was notable variability between different silica types, particularly at 1 phr concentrations. Fumed hydrophobic silicas, such as Reolosil DM30, exhibited a broader range of pore sizes, reflected in larger standard deviations. By contrast, the hydrophilic fumed silica, Aerosil 200, and the precipitated silica, Sipernat 22S, produced more consistently sized, smaller pores.

From the SEM figures at 10,000× magnification, as shown in Figure 5, the hydrophobic fumed silica was observed to disperse more evenly within the natural rubber matrix compared to the hydrophilic fumed silica. The hydrophilic silica showed more pronounced aggregation, suggesting that hydrophobic silica had better compatibility with the natural rubber matrix [38]. Additionally, it can be seen that the BET surface area of the silica particles is a critical factor in their distribution within the rubber matrix. Silica particles with a larger BET surface area tended to have better dispersions [39].

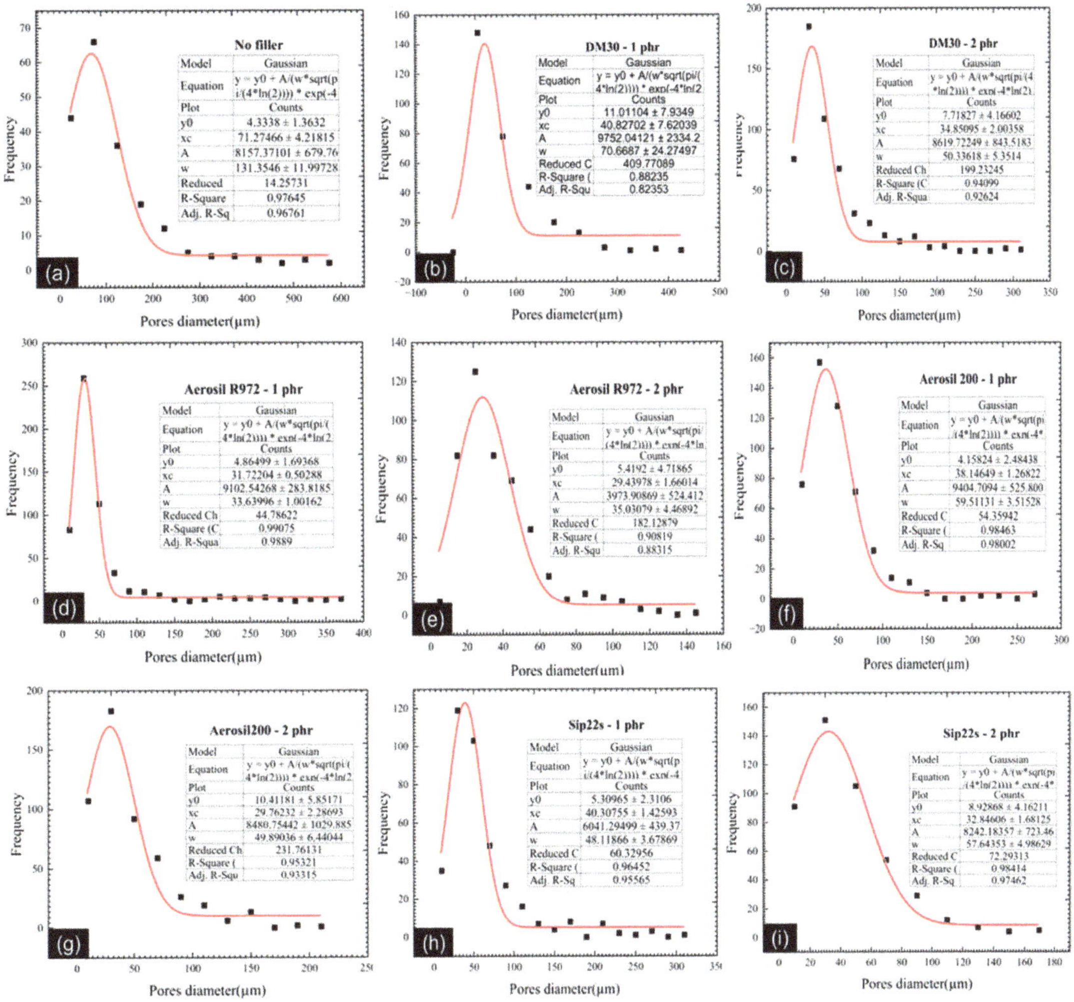

Figure 4. Average pores diameter of silicas (**a**) without silica, (**b**) with DM30 at 1 phr, (**c**) with DM30 at 2 phr, (**d**) with Aerosil R972 at 1 phr, (**e**) with Aerosil R972 at 2 phr, (**f**) with Aerosil 200 at 1 phr, (**g**) with Aerosil 200 at 2 phr, (**h**) with Sip22s at 1 phr, and (**i**) with Sip22s at 2 phr.

Note: SEM images at 50× magnification were captured with a 100 μm scale bar in Figure 3 for accurate interpretation of the pore structures, while SEM images at 10,000× magnification were captured with a 1 μm scale bar in Figure 5 to provide detailed visualization of silica dispersion.

3.2. FTIR

ATR-FTIR spectroscopy was used to analyze the chemical interactions between natural rubber and different types and concentrations of silica (Figure 6). While the FTIR spectra did not reveal the presence of new functional groups across different silica types, variations in the intensity of existing peaks were observed, indicating possible physical interactions between silica and the rubber matrix. A band was observed at 1084 cm^{-1}, corresponding to C–O–Si stretching vibrations. The presence of this band, even though subtle, suggests possible interactions between the silica particles and the natural rubber matrix [40]. These changes in peak intensity, rather than the appearance of new peaks, imply that the interactions are likely physical rather than chemical in nature at the tested silica concentrations.

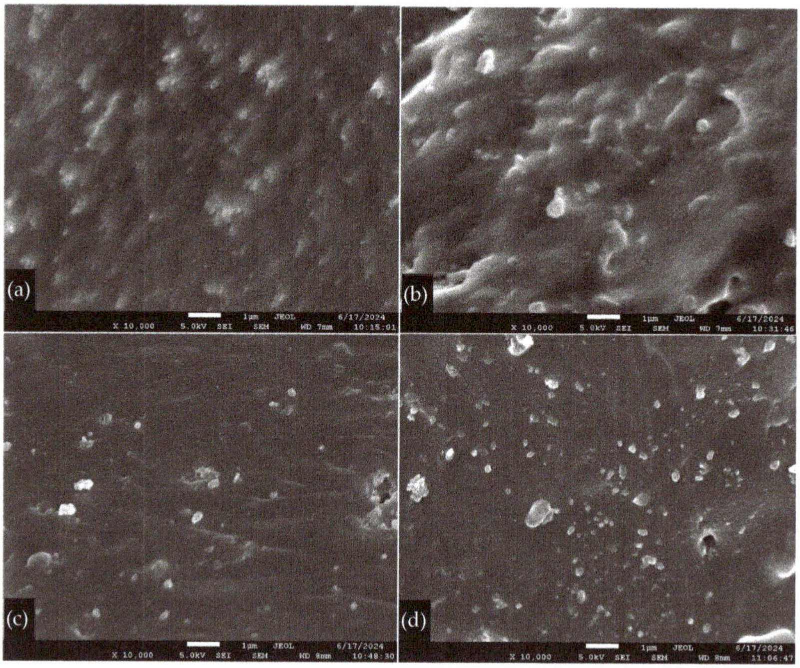

Figure 5. Morphology of the distribution of silicas in natural rubber. (**a**) DM30 at 2 phr, (**b**) Aerosil R972 at 2 phr, (**c**) Aerosil 200 at 2 phr, and (**d**) Sip22s at 2 phr. SEM images of foam rubber at 10,000× magnification. A 1 μm scale bar is included for reference.

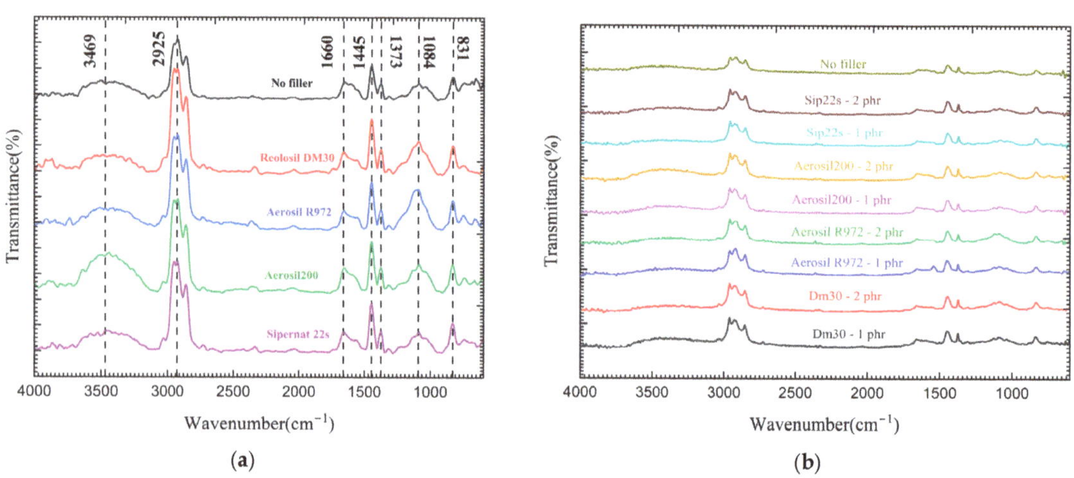

Figure 6. FTIR results of latex foam with different types of silica at 1 phr (**a**) and with different concentrations of silica (**b**).

Additionally, the peak intensity at 3469 cm^{-1}, associated with the vibrations of hydroxyl groups (O-H), was notably higher in samples containing hydrophilic silicas, such as Aerosil 200 and precipitated, Sipernat 22S. This is consistent with their higher surface hydroxyl group density. Conversely, the hydrophobic silicas (Aerosil R972 and Reolosil DM30) exhibited a lower intensity in this region, which aligns with their lower hydroxyl group content. This distinction between hydrophilic and hydrophobic silicas reflects the expected behavior of these materials, where hydrophobic silicas have fewer surface hydroxyl groups due to their treatment to reduce moisture affinity.

The changes in peak intensities support the idea that silica interacts with the natural rubber matrix in a physical manner, affecting the overall dispersion of the silica within the rubber. The better dispersion of hydrophobic silicas, such as Aerosil R972 and Reolosil DM30, likely contributes to the improved mechanical properties observed in these samples. It is also important to note that FTIR's sensitivity may not be sufficient to detect more subtle interactions at the low silica concentrations used in this study. Therefore, while no significant chemical bonding was detected between the silica and natural rubber, The observed variations in peak intensities at specific wavenumbers (1084 cm^{-1} and 3469 cm^{-1}) indicate that physical interactions significantly influence the performance of the silica-filled rubber foam [41].

3.3. Foam Density

Figure 7 shows the densities of different types of silica with different contents. The density of the foam samples increased linearly with the increase in silica content for all types of silica. The highest density was observed for the samples containing 2 phr of Reolosil DM30 hydrophobic fumed silica (0.120 ± 0.005 g/cm^3), followed by those with 2 phr of Aerosil 200 hydrophilic fumed silica (0.119 ± 0.006 g/cm^3), then 2 phr of Aerosil R972 hydrophobic silica (0.113 ± 0.003 g/cm^3), and finally 2 phr of Sipernat 22S precipitated silica (0.110 ± 0.004 g/cm^3). The density of the non-silica foam rubber was 0.100 ± 0.002 g/cm^3. The differences in density between the samples can be attributed to the ability of the silicas to create a finer and more uniform cellular structure, as observed in the SEM images. The higher the silica loading, the greater the reinforcement of the rubber matrix, leading to higher foam densities [42,43].

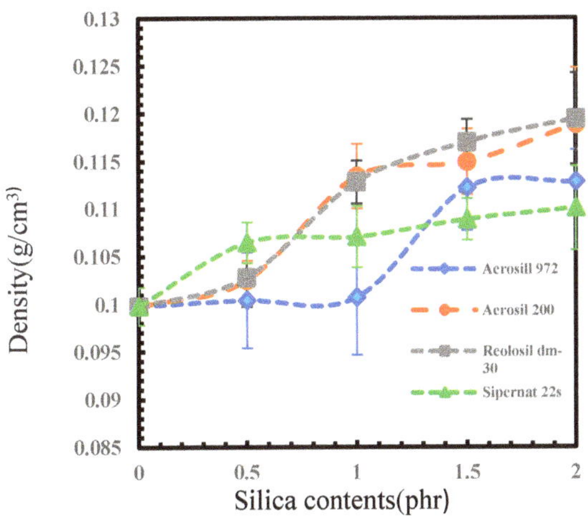

Figure 7. Density of different types of silica.

3.4. Crosslink Density

The crosslink density of unfilled and silica-filled samples was estimated using the swelling method in toluene solvent, as shown in Figure 8. The swelling rate of samples with different silica loadings was nearly uniform, suggesting that the loading of silica had minimal influence on the swelling rate [44]. Despite slight differences in crosslink density, it can be observed that higher silica loadings corresponded to higher crosslink densities. The addition of silica promoted the dispersion of the vulcanization system in latex. During the mixing process of silica and latex, the fusion of rubber molecular chains with silica occurred, enhancing the crosslinking uniformity and efficiency [45].

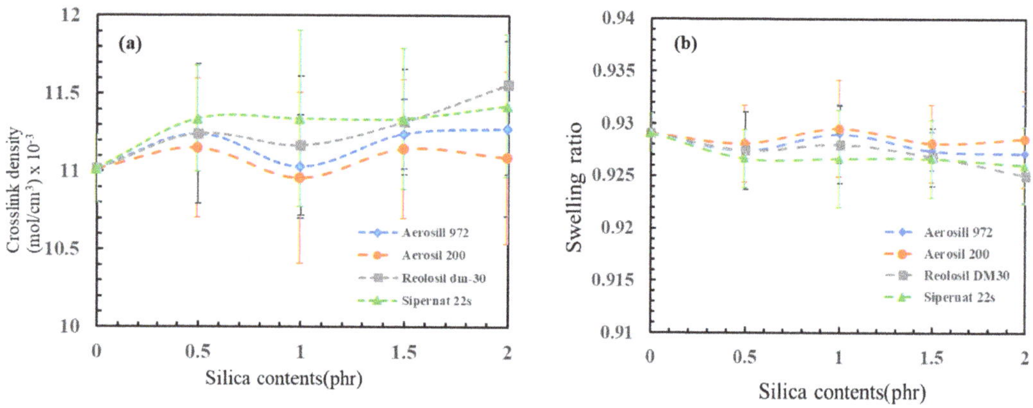

Figure 8. (**a**) Crosslink densities of different types of silica, (**b**) swelling ration of different types of silica.

At a silica concentration of 1 phr, the crosslink density was comparable to that of the non-silica sample. However, with increasing silica concentrations, crosslink density also increased. This is because the rubber cells decreased while the silica cells increased, leaving some silica cells unswollen. The crosslink density of non-silica foam rubber was $(11.0 \pm 0.2) \times 10^{-3}$ mol/cm^3. Among the types of silica, the fumed silica, Reolosil DM30, exhibited the highest crosslink density values $(11.6 \pm 0.6) \times 10^{-3}$ mol/cm^3 due to its high BET surface area and hydrophobic properties.

Hydrophobic silica particles, such as Reolosil DM30, repel water and have a higher affinity for non-polar solvents [46,47]. When incorporated into a rubber matrix, hydrophobic silica is less likely to interact with water molecules and may instead interact with non-polar segments of the rubber chains. This interaction promotes the formation of crosslinks between polymer chains during vulcanization, resulting in a higher crosslink density compared to systems without silica or with hydrophilic silicas [48].

Although the precipitated silica, Sipernat 22S, has a lower specific surface area, its higher bulk density compared to all three types of fumed silica can affect crosslink density [39]. A higher bulk density reduces cell space, resulting in higher silica cell concentrations. This suggests that the properties of silica, including surface chemistry and bulk density, play crucial roles in influencing crosslink density within rubber composites.

3.5. Compression Set

The compression set values, as shown in the above figures, varied with silica loading (0, 0.5, 1, 1.5 and 2 phr) across the four different types of silica. It was observed that the compression set values increased with higher silica loading. The hydrophobic fumed silica, Reolosil DM30, exhibited the highest values, followed by the hydrophobic fumed silica, Aerosil R972, the hydrophilic precipitated silica, Sipernat 22S, and finally the hydrophilic fumed silica, Aerosil 200.

Compression set values serve as an indicator of the elasticity of the latex foam, with lower values indicating higher elasticity and ability to return to its original shape [49]. The compression set value of the non-silica sample was the highest (24.0% ± 0.7%), due to the unevenly distributed pore characteristics within the rubber matrix [50,51]. By contrast, silica-reinforced latex foam exhibited a more even distribution of cells. The increased amount of silica affected the rubber matrix, resulting in foam rubber with smaller and evenly arranged rubber cells. The hydrophobic fumed silica, Reolosil DM30, with its high specific surface area, influenced the performance of the reinforced rubber foam. A larger BET typically results in a higher contact area between rubber cells and silica, meaning stronger interaction between the silica filler and natural rubber, potentially leading to lower compression set properties. This means that the foam rubber retains its shape better after compression [7,12]. Conversely, a lower specific surface area may reduce the propensity for agglomeration, thus enhancing the compression set properties. At 2 phr, the compression set of Reolosil DM30 was 12.0% ± 1.0%, while Aerosil R972, another hydrophobic fumed silica with a lower BET, had a compression set value of 14.0% ± 0.6% (Figure 9).

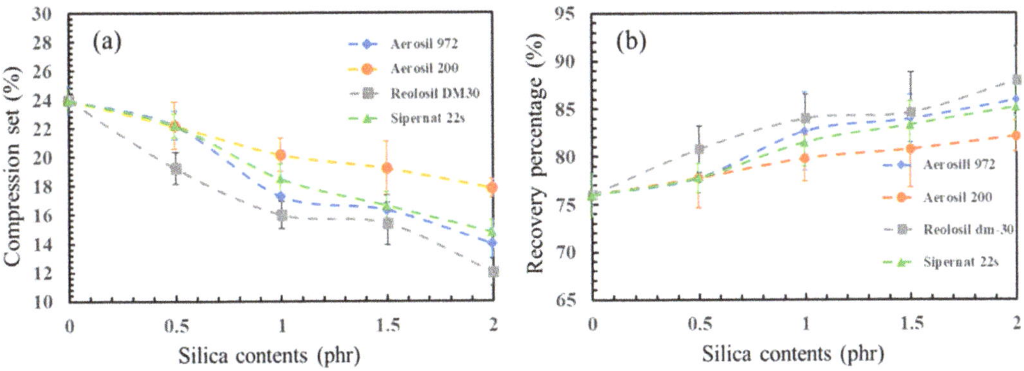

Figure 9. (**a**) Compression set of different types of silica; (**b**) recovery percentage of different types of silica.

Additionally, hydrophilic silica has a higher affinity for water and may retain water molecules within the material. This water absorption can lead to swelling of the rubber matrix, affecting its elastic recovery properties. Consequently, rubber composites containing hydrophilic silica may exhibit higher compression set values due to increased permanent deformation under compression. This is reflected in the results: at 2 phr, Aerosil 200, a fumed silica with hydrophilic nature, had the highest compression set value of 17.9% ± 0.6% because of its poorer and more uneven, porous distribution [48], as shown in Figure 3. Similarly, the compression set of precipitated silica, which is also hydrophilic in nature, had a higher compression set value (14.8% ± 0.8%) than both hydrophobic silicas.

3.6. Hardness

Increasing the loading of silica in rubber composites typically results in increased hardness, as demonstrated in Figure 10. Non-silica foam exhibited a hardness value of 45.0 ± 1.3 IRHD, and by incorporating silica into the rubber, the hardness of the foams increased accordingly. As silica loading increases, more filler particles disperse throughout the rubber matrix. The addition of silica contributes to reducing the size of bubbles or cells within the foam structure. This reduction in cell size is facilitated by the increased surface area available for interaction between the silica particles and the rubber chains [52]. Consequently, stronger interfacial adhesion forms between the silica particles and the rubber matrix. With higher silica loading, the foam structure allows less space between the rubber cells, leading to a denser foam structure. This densification significantly contributes to increased hardness [50]. By minimizing the space between cells, the foam becomes less compressible and more resistant to deformation.

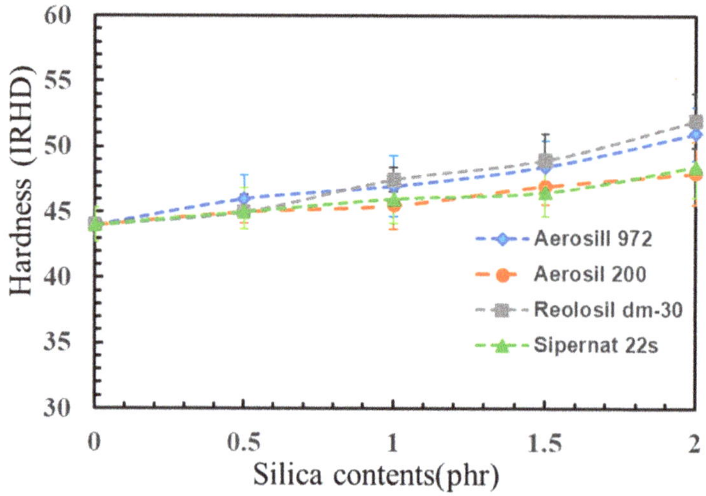

Figure 10. Hardness of different types of silica with different contents.

A comparison of different silica fillers revealed that the hydrophobic silica, particularly Reolosil DM30 and Aerosil R972, exhibited the highest enhancement in rubber properties compared to the hydrophilic silica, Sipernat 22S and Aerosil 200. At a concentration of 2 phr, Aerosil R972 and Reolosil DM30 demonstrated hardness values of 51.0 $\pm$ 2.0 IRHD and 52.0 $\pm$ 2.1 IRHD, respectively, while Sipernat 22S and Aerosil 200 exhibited hardness values of 48.5 $\pm$ 2.4 IRHD and 48.0 $\pm$ 2.4 IRHD, respectively. One significant factor contributing to this observation is the BET surface area of the silica fillers. Reolosil DM30, known for its hydrophobic nature, possesses the highest BET surface area among the tested silica fillers. The higher BET surface area allows for a greater number of active sites to be available for interaction with the rubber matrix, promoting stronger filler–matrix adhesion and ultimately resulting in increased hardness [38,53].

3.7. Percentage Shrinkage

Figure 11 depicts the percentage shrinkage of latex foam for four distinct silica concentrations. It is evident that the various silica concentrations resulted in different percentage shrinkage values. The addition of silica led to a notable reduction in the shrinkage of the center height in comparison to non-silica rubber. However, the height and length of the samples exhibited a gradual increase in percentage shrinkage with higher silica loadings. In the width dimension, there was no clear trend, but adding silica loadings at 0.5 and 1 phr could reduce the percentage shrinkage. An intriguing observation was made regarding the height of the center, where the percentage shrinkage exhibited a decreasing trend as the silica loadings increased. The samples' height and length were smaller and shorter with higher silica loadings, but the center of the samples showed an opposite trend, with the height being greater.

This phenomenon of shrinkage may be attributed to the rubber's relatively low resistance to swelling in solvents and incomplete foaming rates, resulting in varying densities of the rubber [54]. Furthermore, variations in the quantity of the solids and the types of chemicals employed in the formula, including the foaming agent, can give rise to uncertainties in the mechanical properties of the rubber and affect the percentage of shrinkage [55].

In this experiment, the incorporation of silica as a filler resulted in a reduction in the percentage shrinkage of rubber in the compound. The incorporation of silica as a powder is effectively linked with rubber particles, resulting in a thicker and harder rubber matrix. Consequently, as the silica loadings increased, the shrinkage percentage decreased due to the densification and reinforcement effects of silica on the rubber matrix [56].

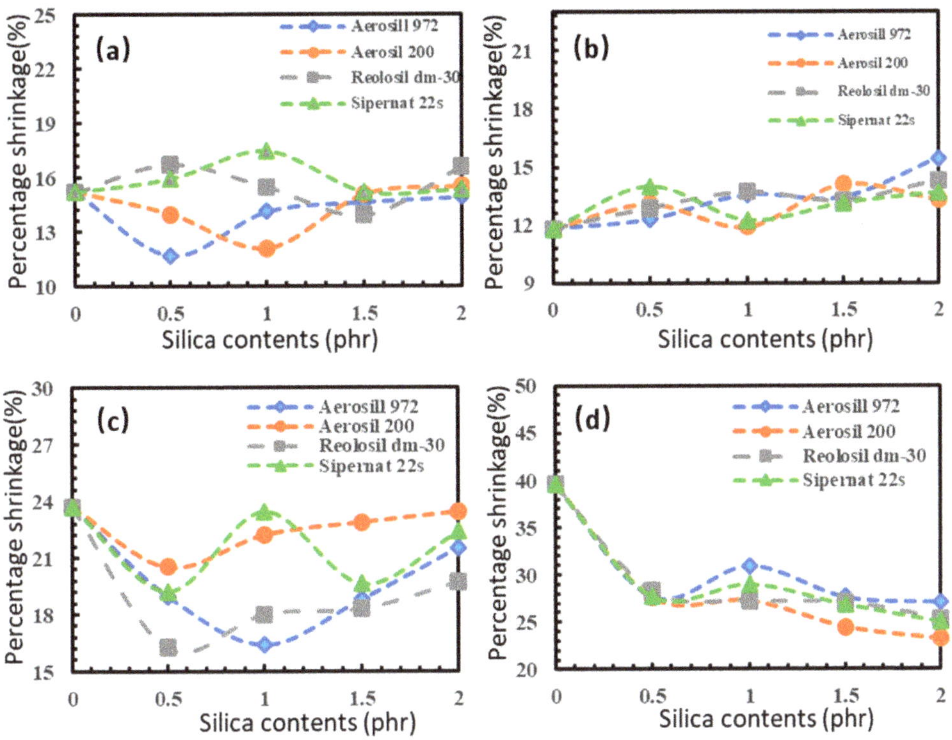

Figure 11. Percentage shrinkage of different types of silica compared with sides of sample: (**a**) width; (**b**) length; (**c**) height; and (**d**) height of center.

4. Conclusions

This study investigated the influence of different types and concentrations of silica as fillers, including hydrophilic and hydrophobic fumed silicas, as well as precipitated silica, on the properties of natural rubber foam produced using the Dunlop process. The physical properties, swelling, hardness, and shrinkage behavior of the natural rubber foam were determined. Four different types of silica were added to the rubber formulation with loadings ranging from 0 to 2 phr. The results indicated that the incorporation of silica fillers—both fumed and precipitated—could be evenly distributed in the rubber matrix. Hydrophobic fumed silicas exhibited better dispersion and enhanced the physical and mechanical properties of the natural rubber foam. In the SEM analysis, it was observed that the hydrophobic silicas had a more homogeneous dispersion and integration with the rubber matrix, leading to improved reinforcement and physical/mechanical performance compared to the hydrophilic silica. Even with a lower specific surface area (BET), the hydrophobic fumed silica, Aerosil R972, was still better dispersed compared to hydrophilic silica with a higher BET. Increasing loadings of silica had a noticeable effect on the porosity of the foam rubber; higher silica loadings resulted in smaller foam pores due to an increase in crosslink density as the silica concentration increased, which led to a more consistent pore structure in the foam rubber matrix. By contrast, a lower specific surface area made the average pore diameter smaller, as observed with the hydrophilic fumed silica, Aerosil 200 (29.4 $\pm$ 1.7 μm). Higher loadings of silica also increased the foam density and hardness. The compression set decreased as silica loading increased, meaning the foam rubber could recover its original shape better when silica filler was added. Hydrophobic fumed silica with a higher specific surface area, Reolosil DM30, provided the best overall reinforcement and improvement in the properties of natural rubber foam. Even with its lower specific surface area, the hydrophobic fumed silica, Aerosil R972, still resulted in greater hardness and recovery percentages compared to hydrophilic silica. However, excessive silica loading resulted in an increase in the shrinkage of the foam, except for the height of the center foam, which showed a decrease in shrinkage. In this experiment, the Dunlop process was used as the foaming method. Loadings of silica higher than 2 phr were not suitable, as excessive fillers impaired the foaming process and prevented some silica from mixing well with the natural rubber. FTIR spectra analysis was used to try to confirm the chemical interaction between silica and natural rubber. However, the results of the FTIR analysis were not clear enough to identify differences in the chemical bonds between hydrophobic and hydrophilic silica and natural rubber, as the silica loading was too low. It only showed a lower peak of the O-H band in hydrophobic silica compared to hydrophilic silica.

Author Contributions: Conceptualization, L.G. and H.L.; funding acquisition, L.G. and H.L.; investigation, D.A., G.L., K.H. and L.B.; methodology, G.L. and F.L.; project administration, Y.X.; writing—original draft, D.A. and Y.X.; writing—review and editing, F.L. and H.L. All authors have read and agreed to the published version of the manuscript.

Funding: This research was funded by the National Natural Science Foundation of China (No. 52373101), Natural Science Foundation of Shandong Province, China (No. ZR202102180625 and ZR2020QE207), and Postdoctoral Research Foundation of China (No. 2023M733754).

Institutional Review Board Statement: Not applicable.

Data Availability Statement: The original contributions presented in the study are included in the article, further inquiries can be directed to the corresponding authors.

Conflicts of Interest: There are no conflicts of interest to declare.

References

1. Sobhy, M.; El-Nashar, D.E.; Maziad, N.A. Cure Characteristics and Physicomechanical Properties of Calcium Carbonate Reinforcement Rubber Composites. *Egypt. J. Solids* **2003**, *26*, 241–257.
2. Cui, C.; Ding, H.; Cao, L.Y.; Chen, D. Preparation of $CaCO_3$-SiO_2 Composite with Core-Shell Structure and Its Application in Silicone Rubber. *Przem. Chem.* **2015**, *17*, 128–133. [CrossRef]
3. Kraus, G. Interactions of Elastomers and Reinforcing Fillers. *Rubber Chem. Technol.* **1965**, *38*, 1070–1114. [CrossRef]
4. Prasertsri, S.; Rattanasom, N. Mechanical and Damping Properties of Silica/Natural Rubber Composites Prepared from Latex System. *Polym. Test.* **2011**, *30*, 515–526. [CrossRef]
5. Joon-Hyung, K. The Foaming Characteristics and Physical Properties of Natural Rubber Foams: Effects of Carbon Black Content and Foaming Pressure. *J. Appl. Polym. Sci.* **2016**, *12*, 795–801.
6. Ghosh, A.K.; Maiti, S.; Adhikari, B.; Ray, G.S.; Mustafi, S. Effect of Modified Carbon Black on the Properties of Natural Rubber Vulcanizate. *J. Appl. Polym. Sci.* **1997**, *66*, 683–693. [CrossRef]
7. Chuayjuljit, S.; Imvittaya, A.; Na-Ranong, N.; Potiyaraj, P. Effects of Particle Size and Amount of Carbon Black and Calcium Carbonate on Curing Characteristics and Dynamic Mechanical Properties of Natural Rubber. *J. Met. Mater. Miner.* **2022**, *12*, 51–57.
8. Callister, W.D. *Materials Science and Engineering: An Introduction*, 7th ed.; John Wiley & Sons, Inc.: New York, NY, USA, 2000; p. 871.
9. Rattanasom, N.; Saowapark, T.; Deeprasertkul, C. Reinforcement of Natural Rubber with Silica/Carbon Black Hybrid Filler. *Polym. Test.* **2007**, *26*, 369–377. [CrossRef]
10. Taikum, O.; Scholz, M. The Last 100 Years of Fumed Silica in Rubber Reinforcement. *Rubber World* **2020**, *242*, 35–44.
11. Huang, L.; Li, Q.; Wang, X.; Zhang, J.; Yang, Z.; Zhou, W. Structural Analyses of the Bound Rubber in Silica-Filled Silicone Rubber Nanocomposites Reveal Mechanisms of Filler-Rubber Interaction. Compos. *Sci. Technol.* **2023**, *233*, 109905.
12. Zaborski, M.; Vidal, A.; Ligner, G.; Balard, H.; Papirer, E.; Burneau, A. Comparative Study of the Surface Hydroxyl Groups of Fumed and Precipitated Silicas. I. Grafting and Chemical Characterization. *Langmuir* **1989**, *5*, 447–451. [CrossRef]
13. Protsak, I.; Tertykh, V.A.; Pakhlov, E.; Deryło-Marczewska, A. Modification of Fumed Silica Surface with Mixtures of Polyorganosiloxanes and Dialkyl Carbonates. *Prog. Org. Coat.* **2017**, *106*, 163–169. [CrossRef]
14. Maciá-Agulló, T.G.; Fernández-García, J.C.; Torró-Palau, A.M.; Barceló, A.C.O.; Martín-Martínez, J.M. Hydrophobic or Hydrophilic Fumed Silica as Filler of Polyurethane Adhesives. *J. Adhes. Sci. Technol.* **1995**, *50*, 265–277. [CrossRef]
15. Hassanajili, S.; Sajedi, M.T. Fumed Silica/Polyurethane Nanocomposites: Effect of Silica Concentration and Its Surface Modification on Rheology and Mechanical Properties. *J. Appl. Polym. Sci.* **2016**, *25*, 697–710. [CrossRef]
16. Gun'ko, V.M.; Zarko, V.I.; Leboda, R.; Chibowski, E.; Grzegorczyk, W. Features of Fumed Silica Coverage with Silanes Having Three or Two Groups Reacting with the Surface. *Colloids Surf.* **2000**, *166*, 187–201. [CrossRef]
17. Ettlinger, M.; Ladwig, T.; Weise, A. Surface Modified Fumed Silicas for Modern Coatings. *Prog. Org. Coat.* **2000**, *40*, 31–34. [CrossRef]

18. Evans, L.R.; Hope, J.C.; Okel, T. Use of Precipitated Silica to Improve Brass-Coated Wire-to-Rubber Adhesion. *Rubber World* **2020**, *214*, 21–26.
19. Bayat, H.; Fasihi, M. Effect of Coupling Agent on the Morphological Characteristics of Natural Rubber/Silica Composite Foams. *ePolymers* **2019**, *19*, 430–436. [CrossRef]
20. Luo, Y.; Wang, Y.Q.; Zhong, J.; He, C.Z.; Li, Y.Z.; Peng, Z. Interaction Between Fumed-Silica and Epoxidized Natural Rubber. *J. Inorg. Organomet. Polym. Mater.* **2011**, *21*, 777–783. [CrossRef]
21. Prasertsri, S.; Rattanasom, N. Fumed and Precipitated Silica Reinforced Natural Rubber Composites Prepared from Latex System: Mechanical and Dynamic Properties. *Polym. Test.* **2012**, *31*, 593–605. [CrossRef]
22. *ASTM D3574*; Standard Test Methods for Flexible Cellular Materials—Slab, Bonded, and Molded Urethane Foams. ASTM International: West Conshohocken, PA, USA, 2020.
23. Hong, Y.; Weng, G.; Liu, Y.; Fu, K.; Chang, A.; Chen, Z. Effect of Silane Coupling Agent on the Fatigue Crack Propagation of Silica-Filled Natural Rubber. *J. Appl. Polym. Sci.* **2015**, *132*, 41980.
24. Lee, J.; Kim, K.; Kim, J.; Lee, H.; Lee, D.; Jeon, H. Influence of the Silanes on the Crosslink Density and Crosslink Structure of Silica-Filled Solution Styrene Butadiene Rubber Compounds. *Mol. Cryst. Liq. Cryst.* **2016**, *24*, 711–727. [CrossRef]
25. *ISO 1817*; Rubber, Vulcanized or Thermoplastic—Determination of the Effect of Liquids. International Organization for Standardization: Geneva, Switzerland, 2015.
26. Liang, T.; Isayev, A.I. Effect of Ultrasonic Extrusion of Star Styrene-Butadiene Rubber on Properties of Carbon Black- and Silica-Filled Compounds and Vulcanizates. *J. Appl. Polym. Sci.* **2019**, *136*, 47451. [CrossRef]
27. Abd-El-Messieh, S.L.; El-Nashar, D.E.; Khafagi, M. Compatibility Investigation of Microwave Irradiated Acrylonitrile Butadiene/Ethylene Propylene Diene Rubber Blends. *Polym. Plast. Technol. Eng.* **2004**, *43*, 135–158. [CrossRef]
28. Bernal-Ortega, P.; Anyszka, R.; Morishita, Y.; di Ronza, R.; Blume, A. Determination of the Crosslink Density of Silica-Filled Styrene Butadiene Rubber Compounds by Different Analytical Methods. *Polym. Bull.* **2024**, *81*, 995–1018. [CrossRef]
29. Tangboriboonrat, P.; Suchiva, K.; Kuhakarn, S. Characterization of Non-Crosslinked Natural Rubber Latex by Phase Transfer Technique. *Polymer* **1994**, *35*, 5144–5145. [CrossRef]
30. Timothy, J.J.; Meschke, G. A Cascade Continuum Micromechanics Model for the Effective Elastic Properties of Porous Materials. *Int. J. Solids Struct.* **2016**, *83*, 1–12. [CrossRef]
31. Bashir, A.S.; Manusamy, Y.; Chew, T.L.; Ismail, H.; Ramasamy, S. Mechanical, Thermal, and Morphological Properties of (Eggshell Powder)-Filled Natural Rubber Latex Foam. *J. Vinyl Addit. Technol.* **2015**, *23*, 3–12. [CrossRef]
32. *ASTM D1415*; Standard Test Method for Rubber Property—International Hardness. ASTM International: West Conshohocken, PA, USA, 2018.
33. *ASTM D1055*; Standard Specification for Rubber Latices—Synthetic. ASTM International: West Conshohocken, PA, USA, 2021.
34. Xiang, B.; Li, J.; Dong, X.; Wang, Y.; Zhou, Y. Microcellular Silicone Rubber Foams: The Influence of Reinforcing Agent on Cellular Morphology and Nucleation. *Polym. Eng. Sci.* **2018**, *59*, 5–14. [CrossRef]
35. Sholeh, M.; Rochmadi, R.; Sulistyo, H.; Budhijanto, B.; Doloksaribu, M. The Effect of Nanostructured Silica Synthesis Temperature on the Characteristics of Silica Filled Natural Rubber Composite. *IOP Conf. Ser. Mater. Sci. Eng.* **2021**, *1143*, 012011. [CrossRef]
36. Kumkrong, N.; Dittanet, P.; Sae-Oui, P.; Loykulnant, S.; Prapainainar, P. Properties of Silica/Natural Rubber Composite Film and Foam: Effects of Silica Content and Sulfur Vulcanization System. *J. Polym. Res.* **2022**, *29*, 302. [CrossRef]
37. Wang, M.; Tu, H.; Murphy, L.J.; Mahmud, K. Carbon—Silica Dual Phase Filler, A New Generation Reinforcing Agent for Rubber: Part VIII. Surface Characterization by IGC. *J. Polym. Sci. Part A Polym. Chem.* **2000**, *73*, 666–677. [CrossRef]
38. Boonkerd, K.; Chuayjuljit, S.; Abdulraman, D.; Jaranrangsup, W. Silica-Rich Filler for the Reinforcement in Natural Rubber. *J. Polym. Sci. Part A Polym. Chem.* **2012**, *85*, 1–13. [CrossRef]
39. Kawaguchi, M. Stability and Rheological Properties of Silica Suspensions in Water-Immiscible Liquids. *Colloids Surf. A Physicochem. Eng. Asp.* **2020**, *278*, 102139. [CrossRef] [PubMed]
40. Kralevich, M.L.; Koenig, L.J. FTIR Analysis of Silica-Filled Natural Rubber. *J. Polym. Sci. Part A Polym. Chem.* **1998**, *71*, 300–309. [CrossRef]
41. Heideman, G.; Datta, R.N.; Noordermeer, J.W.M.; Van Baarle, B. Fundamental aspects of sulfur vulcanization. *Rubber Chem. Technol.* **2005**, *78*, 245–268. [CrossRef]
42. Greenwood, N.N.; Earnshaw, A. *Chemistry of the Elements*, 2nd ed.; Butterworth-Heinemann: Oxford, UK, 1997.
43. Bayat, H.; Fasihi, M. Curing Characteristics and Cellular Morphology of Natural Rubber/Silica Composite Foams. *J. Mater. Sci.* **2019**, *77*, 3171–3184. [CrossRef]
44. Thammathadanukul, V.; O'Haver, J.H.; Harwell, J.H.; Osuwan, S.; Na-Ranong, N.; Waddell, W.H. Comparison of Rubber Reinforcement Using Various Surface-Modified Precipitated Silicas. *Rubber Chem. Technol.* **1996**, *59*, 1741–1750. [CrossRef]
45. Sombatsompop, N. Investigation of Swelling Behavior of NR Vulcanisates. *Polym. Eng. Sci.* **1998**, *37*, 19–39. [CrossRef]
46. Mishra, S.; Mitra, R. Rational Design in Suspension Rheology of Cellular Silica. *Adv. Colloid Interface Sci.* **2010**, *370*, 102–113. [CrossRef]
47. Jhang, C. Preparation and Characterization of Waterborne Polyurethane with Fumed Silica Composites. Master's Thesis, National Taipei University of Technology, Taipei, Taiwan, 2013, pp. 1–80. Available online: https://www.airitilibrary.com/Article/Detail/U0006-2706201311253000 (accessed on 18 October 2024).
48. Mathias, J.; Wannemacher, G. Basic Characteristics and Applications of Aerosil. *J. Colloid Interface Sci.* **1988**, *125*, 61–68. [CrossRef]

49. Suethao, S.; Phongphanphanee, S.; Wong-ekkabut, J.; Smitthipong, W. The Relationship between the Morphology and Elasticity of Natural Rubber Foam Based on the Concentration of the Chemical Blowing Agent. *Materials* **2021**, *13*, 1091. [CrossRef] [PubMed]
50. Wei, C.; Lu, A.; Su-ming, S.; Wei, X.; Zho, X.; Sun, J. Establishment of Constitutive Model of Silicone Rubber Foams Based on Statistical Theory of Rubber Elasticity. *Chin. J. Polym. Sci.* **2018**, *36*, 1077–1083. [CrossRef]
51. Suethao, S.; Ponloa, W.; Phongphanphanee, S.; Wong-ekkabut, J.; Smitthipong, W. Current Challenges in Thermodynamic Aspects of Rubber Foam. *Nature* **2021**, *11*, 6097. [CrossRef] [PubMed]
52. Sattayanurak, S.; Noordermeer, J.W.; Sahakaro, K.; Kaewsakul, W.; Dierkes, W.K.; Blume, A. Silica-Reinforced Natural Rubber: Synergistic Effects by Addition of Small Amounts of Secondary Fillers to Silica-Reinforced Natural Rubber Tire Tread Compounds. *Adv. Mater.* **2019**, *2019*, 5891051. [CrossRef]
53. Liu, D.; Zhao, Y.; Zhang, Y.; Xu, X.; Zhang, H.; Gao, Y. Correlation between Mechanical Properties and Microscopic Structures of an Optimized Silica Fraction in Silicone Rubber. *Compos. Part B* **2018**, *165*, 373–379. [CrossRef]
54. Ariff, M.Z.; Zakaria, Z.; Tay, L.H.; Lee, S.Y. Effect of Foaming Temperature and Rubber Grades on Properties of Natural Rubber Foams. *J. Appl. Polym. Sci.* **2007**, *107*, 2531–2538. [CrossRef]
55. Pajarito, B.B.; Quizon, M.A.E.; Marquez, R.A.L.; Bermio, P.B.C. Modeling the Hardness and Mechanical Properties of Moulded Natural Rubber Pad. Mater. *Sci. Forum* **2017**, *890*, 54–58. [CrossRef]
56. Muromtsev, N.D.; Gertseva, O.; Pichkhidze, Y.S. The Prediction of the Shrinkage of a Profile Based on Ethylene Propylene Rubber. *Rubber Chem. Technol.* **2014**, *41*, 59–62. [CrossRef]

polymers

MDPI

Article

Design and Characterization of Poly(ethylene oxide)-Based Multifunctional Composites with Succinonitrile Fillers for Ambient-Temperature Structural Sodium-Ion Batteries

Vasan Iyer [1,2,*], Jan Petersen [1,2], Sebastian Geier [1,2] and Peter Wierach [1,2,3]

1. Cluster of Excellence SE2A—Sustainable and Energy-Efficient Aviation, Technische Universität Braunschweig, 38108 Braunschweig, Germany; jan.petersen@dlr.de (J.P.); sebastian.geier@dlr.de (S.G.); peter.wierach@dlr.de (P.W.)
2. Department of Multifunctional Materials, German Aerospace Center (DLR), Institute of Lightweight Systems, Lilienthalplatz 7, 38108 Braunschweig, Germany
3. Institute of Polymer Materials and Plastics Engineering, Technische Universität Clausthal, Agricolastrasse 6, 38678 Clausthal-Zellerfeld, Germany

* Correspondence: vasan.iyer@dlr.de; Tel.: +49-5312952984

Abstract: A new approach to developing structural sodium batteries capable of operating in ambient-temperature conditions has been successfully achieved. The developed multifunctional structural electrolyte (SE) using poly(ethylene oxide) (PEO) as a matrix integrated with succinonitrile (SN) plasticizers and glass-fiber (GF) reinforcements identified as GF_PEO-SN-$NaClO_4$ showed a tensile strength of 32.1 MPa and an ionic conductivity of 1.01×10^{-4} S cm^{-1} at room temperature. It displayed a wide electrochemical stability window of 0 to 4.9 V and a high sodium-ion transference number of 0.51 at room temperature. The structural electrode (CF | SE) was fabricated by pressing the structural electrolyte with carbon fibers (CFs), and it showed a tensile strength of 72.3 MPa. The fabricated structural battery half-cell (CF | | SE | | Na) demonstrated good cycling stability and an energy density of 14.2 Wh kg^{-1}, and it retained 80% capacity at the end of the 200th cycle. The cycled electrodes were observed using scanning electron microscopy, which revealed small dendrite formation and dense albeit uniform deposition of the sodium metal, helping to avoid a short-circuit of the cell and providing more cycling stability. The developed multifunctional matrix composites demonstrate promising potential for developing ambient-temperature sodium structural batteries.

Keywords: structural sodium batteries; structural energy storage; multifunctional materials; fiber-reinforced composites; plasticizers; carbon-fiber electrode

1. Introduction

The transportation sector, which is largely reliant on fossil fuels for energy, results in large greenhouse gas emissions that contribute to climate change effects. These effects could be mitigated by transitioning to e-mobility. It is necessary to achieve this transition in the European Union (EU) by 2035, according to the proposed Green Deal [1]. Though the automobile sector is in the phase of transition to electric vehicles from conventional fossil fuel-based vehicles, the aviation sector has much catching up to do. The implementation plan to achieve zero emissions in the aviation sector has been put forward, and the deadline is set for 2050 [2,3].

The current key technology for energy storage in the e-mobility sector is mostly based on Li-ion batteries. As the demand for electric vehicle range increases, it leads to an increase in the battery mass to compensate for the increased energy density requirements. As the vehicle structure and battery remain separated, it contributes to an overall increase in the vehicle mass and reduced energy efficiency. For example, the total battery pack mass for a standard-range (54 kWh version) Tesla Model 3 is 324 kg, or 18.4% of the car mass, whereas, for the long-range model (75 kWh version), the battery mass is increased to 480 kg, or 26.3% of the mass of the car [4]. A possible solution to mitigate such a problem is to integrate the battery within the structure, i.e., to view the battery as a function rather than as a separate component [5]. Such a multifunctional structure with integrated energy storage capability is called a structural battery, and it has the potential to provide massless energy storage, high performance, high overall efficiency, and increased vehicle utility space [6].

The integration of batteries within the structure can be achieved in varied degrees of integration (DOI), with DOI level 0 being the current-day electric vehicle architecture, where the batteries and structure are totally separated. In DOI level I, the batteries are integrated into the empty spaces within the structure, and in DOI level II, the thin film battery is attached to the surface of the structure [7–9]. The complexity of the structural battery design increases with higher DOI levels (III and IV), where true multifunctionality is achieved, in which the battery is now seen as a function and cannot be separated from the structure [10,11]. In higher DOI structural batteries, each component of the battery plays multiple roles and contributes both to structural strength and ion storage and provides a pathway for the movement of ions. The main challenges in achieving higher DOI designs include the optimization of both the mechanical and the electrochemical properties of the battery components. Currently, the reported energy density values of higher DOI structural batteries are comparatively lower than conventional batteries by one or two orders of magnitude, and typical values lie in the range of 1–50 Wh kg^{-1} [12]. The ultimate goal of structural battery research is to achieve both high DOI and energy density to provide massless storage and overall high energy efficiency.

Previous studies have used carbon-fiber (CF) electrodes in developing multifunctional energy storage materials [12–15]. CF is interesting as it possesses both mechanical and electrochemical properties and is also an important component for manufacturing composites for transport applications [16,17]. In the state-of-the-art DOI III structural battery architecture, carbon fibers are used as they contribute to the structural strength as reinforcements, act as electrodes, and store ions. Previous studies mainly focused on structural battery designs based on lithium-ion intercalation into CF and investigated bi-continuous phase-type electrolytes and all solid-state-type electrolytes with glass-fiber separators as the multifunctional matrix material [14–19]. These materials contribute to structural strength as well as facilitate ion transport.

In contrast, there are few structural battery architectures based on sodium-ion intercalation into CF, thus requiring further investigation. This study addresses this gap and investigates structural battery designs based on sodium-ion intercalation using all solid-state composite-type electrolytes as multifunctional matrix materials. The most important consideration for designing sodium structural batteries lies in the way in which the sodium-ion intercalates into carbon fibers. Lithium ion can intercalate in between graphitic layers, edges, and lattice surfaces, whereas sodium ion mainly intercalates onto lattice surfaces and in pores/defect sites of the carbon-fiber microstructure. Hard carbon possesses such characteristic pores/defect sites that favor sodium-ion intercalation [12,17,20]. The investigation of intermediate modulus (IM)-type carbon fibers reveals that they possess a microstructure that resembles hard carbon, and hence, IM-type carbon fiber is used to construct electrodes for sodium-ion intercalation in the present study [17,21].

The choice of polymer matrix plays a crucial role in the design of a multifunctional matrix composite for structural batteries. In the present study, high-molecular-weight poly(ethylene oxide) or PEO is chosen as a matrix for the preparation of solid polymer electrolytes (SPEs). PEO has the unique advantages of good mechanical properties, high safety, and a low glass transition temperature, and it offers good thermal, electrochemical, and interfacial stability due to the high solubility of sodium salts [22,23]. Even though PEO possesses many good characteristics, the room-temperature ionic conductivity of the PEO-based electrolytes with alkali metal salts is significantly low. This is mainly due to the crystalline nature of PEO [22,23]. The addition of inorganic active fillers has been proven to reduce the crystalline nature of PEO and enhance ion transport [24,25].

In a previous study of sodium-ion-based structural batteries involving the fabrication of multifunctional composites using inorganic ceramic NZSP ($Na_3Zr_2Si_2PO_{12}$) nanoparticle fillers, the battery showed good cycling stability and retained 80% capacity at the end of 225 cycles. However, it operated at high temperatures [12]. Therefore, a strategy is needed for designing a sodium structural battery for ambient-temperature operation. Previous studies reported that the plastic crystalline phase of succinonitrile (SN) showed high solubility for various salts due to its high polarity. The investigated electrolytes displayed high ionic conductivities in a wide range of temperatures, with values reaching 3×10^{-3} cm^{-1} at 25 °C, which are two to three orders of magnitude higher than other polymer-based matrices at 25 °C [26]. Furthermore, studies involving PEO-based matrices with SN fillers also reported high ionic conductivities at room temperature [27–29]. These results give the motivation for the present study to fabricate multifunctional composites using SN as plasticizers.

This study reports the first-of-its-kind multifunctional composites fabricated using all solid-state PEO-based electrolytes with succinonitrile (SN) plasticizers for ambient-temperature sodium structural batteries. The composites are investigated for their multi-functional performance. The structural electrolyte was prepared using a two-step approach, where first, the polymer membrane or the composite solid electrolyte was prepared using a solution-casting technique, and then it was reinforced with glass fibers to manufacture the structural electrolyte (GF_PEO-SN-$NaClO_4$). The structural electrolyte is a multifunctional composite matrix material with the tasks of transferring mechanical loads and transporting sodium ions. The glass-fiber reinforcement is done to further improve the mechanical characteristics of the structural electrolyte, and it displayed a tensile strength of 32.1 MPa and ionic conductivity of 1.01×10^{-4} S cm^{-1} at room temperature. The structural electrode component was fabricated by laminating the prepared structural electrolyte with intermediate modulus (IM) type carbon fibers, which showed a tensile strength of 72.3 MPa. The fabricated structural battery half-cell (CF | | GF_PEO-SN–$NaClO_4$ | | Na) provided a typical energy density of 14.2 Wh kg^{-1} at 0.1 C rate, and the cycling performance tested at 0.9 C rate showed 80% capacity retention after 200 cycles.

2. Experimental Section

2.1. Materials and Methods

Chemicals and materials for the synthesis of structural electrolytes include poly (ethylene oxide) (PEO (molecular weight (M_w) of 10^6 g mol^{-1}), Sigma-Aldrich, Steinheim, Germany), anhydrous acetonitrile(ACN with 99% purity, Sigma-Aldrich, Steinheim, Germany), Succinonitrile ($C_2H_4(CN)_2$, Sigma-Aldrich, Steinheim, Germany), sodium perchlorate salt ($NaClO_4$, Sigma-Aldrich, Steinheim, Germany), glass-fiber woven fabric with an aerial weight of 163 g m^{-2} (Aero, R&G Faserverbundwerkstoffe GmbH, Waldenbuch, Germany), spread tow carbon fibers (CF) Tenax IMS65 24k tows with aerial weight of 55 g m^{-2} was acquired from R&G Faserverbundwerkstoffe GmbH, Waldenbuch, Germany for the fabrication of the structural electrode. Sodium (Na) metal sticks for the preparation of sodium foils were purchased from Alfa Aesar, Karlsruhe, Germany. Siliconized (double-sided) papers for preparing structural electrolytes and electrodes using a press were obtained as a gift from Laufenberg GmbH, Krefeld, Germany.

2.2. Material Characterization and Analysis

The SEM (scanning electron microscopy using FEI, Thermo Fisher Scientific Inc., Dreieich, Germany) and energy dispersive X-ray analysis (EDAX) techniques were used for studying the morphology and characterizing the fabricated structural electrolyte and electrode.

2.3. Preparation of Polymer Membranes and Structural Electrolytes

The polymer membranes were prepared as described in the schematic drawing shown in Figure 1a, and Figure S1 shows the complete process flow for the fabrication of structural electrolytes. In the first step, the thin electrolyte membranes were prepared using the solution-casting technique. Initially calculated amounts of PEO and $NaClO_4$ were added to the (ACN) organic solvent in a sealed container and stirred for 4 h. The ratio of ethylene oxide (-CH_2-CH_2-O-) to sodium ion (EO:Na^+) is maintained at 15:1. Then various amounts (30–40% by weight) of succinonitrile (SN) were added to the mixture, and it was sonicated (using a Branson 250 Digital Sonifier, Branson, New Hampshire, USA) for 30 min and further stirred for 12 h until a homogenous mixture is obtained. The solution is then cast into a PTFE evaporation dish and set aside in the fume hood for 10 h until a layer with a uniform surface and thickness is formed. The membranes so obtained are vacuum dried for 24 h at 50 °C to evaporate the ACN solvent completely and to get thin polymer membranes identified as PEO-SN–$NaClO_4$ (Figure S2) with thickness measured to be 180 μm. In the second step, the glass fibers were reinforced and sandwiched between the two thin polymer membranes (Figure 1a (step 2)) utilizing a press (Collins P500S press, COLLIN Lab & Pilot Solutions GmbH, Maitenbeth, Germany) to fabricate the structural electrolyte with a thickness of 360 μm. The pressing was carried out for 25 min at room temperature and 0.2 MPa pressure. The adhesive nature of the thin polymer membrane helps in achieving good integration with glass fibers. The glass-fiber-reinforced structural electrolyte is identified as GF_PEO-SN-$NaClO_4$ (Figure S2).

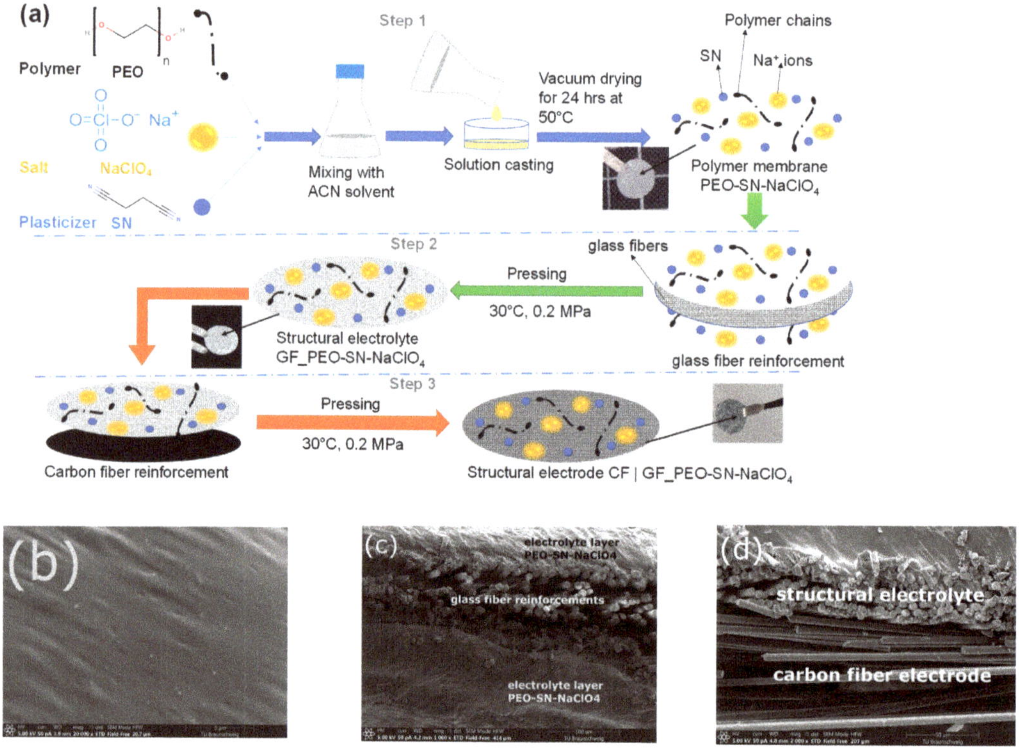

Figure 1. *Cont.*

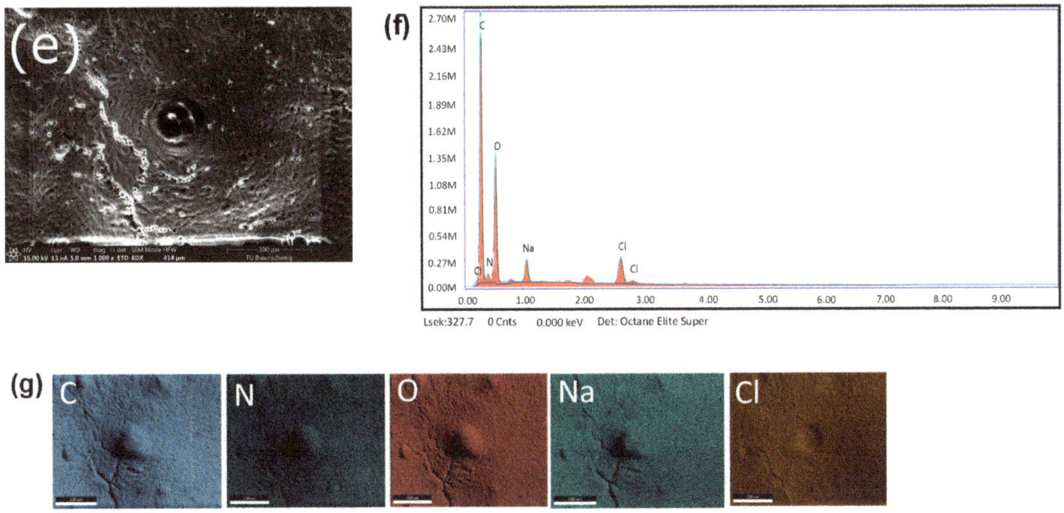

Figure 1. Multifunctional composite preparation and characterization: (**a**) schematic drawing showing the manufacturing process of the structural electrolyte and structural electrode. SEM images: (**b**) Polymer membrane PEO-SN–$NaClO_4$. (**c**) cross-sectional view of the structural electrolyte GF_PEO-SN–$NaClO_4$. (**d**) cross-sectional view of the structural electrolyte CF | | GF_PEO-SN–$NaClO_4$. (**e**) structural electrolyte surface after EDAX. (**f**) EDAX pattern for the structural electrolyte GF_PEO-SN–$NaClO_4$. (**g**) SEM–EDAX element mapping for the structural electrolyte.

2.4. Preparation of Structural Electrode

The structural electrodes were prepared by placing the structural electrolyte over a layer of spread tow carbon fiber in between the siliconized papers to prevent sticking to the press plates. It was then pressed at room temperature with 0.2 MPa pressure using a Collins P500S press for 25 min, as depicted in Figure 1a (step 3). The adhesive nature of the prepared structural electrolytes allowed seamless integration and bonding with the carbon-fiber layer. Figure S3 shows both the top and bottom surfaces of the prepared structural electrode.

2.5. Electrochemical Characterization and Analysis

The sodium-ion conductivities of the prepared electrolytes were evaluated using electrochemical impedance spectroscopy (EIS) using the four-point electrode cell technique. An RHD measuring cell TSC Battery (manufactured by RHD Instruments GmbH & Co. KG, Darmstadt, Germany) with stainless-steel blocking electrodes was used to record the EIS spectra. The electrolytes, with a size of 8 mm diameter, were used to make the EIS recordings in the set frequency ranges from 1 Hz to 1 MHz, and the sinusoidal perturbation was set to 10 mV in the Zahner potentiostat (Zhennium, Zahner-Elektrik GmbH & Co. KG, Kronach, Germany).

The sodium-ion transference numbers were assessed using a symmetrical cell (Na | | electrolyte | | Na) prepared in the 2032-coin cell configuration (Figure S4). The AC–DC (alternating current–direct current) polarization technique was used, where initially, a 10 mV AC polarization voltage was applied to the symmetrical cell, and the EIS spectrum was recorded in the frequency range 1 Hz to 1 MHz. From the obtained Nyquist plot, the charge transfer resistance (R_0) was then calculated. A DC polarization voltage of 10 mV is applied, and the chronoamperogram (which records the evolution of current with respect to the applied DC polarization voltage) is observed until the current reaches a constant steady-state value (I_{SS}). The starting current (I_0) in the current evolution is obtained from the experiment. The EIS spectrum was recorded once again, and the steady-state charge transfer resistance (R_{SS}) was calculated from the Nyquist plot [30].

The electrochemical stability window (ESW) of the fabricated electrolytes was evaluated using linear sweep voltammetry (LSV) using the prepared cells in asymmetric configuration (SS | | Electrolyte | | Na) with stainless-steel (SS) electrode and sodium (Na) foil counter electrode (Figure S4). A positive scan from 2.5 V to 6.5 V and a negative scan from 2.5 V to 0 V was conducted with a scan rate of 1 mV s^{-1} at 30 °C.

2.6. Mechanical Characterization

The mechanical strength of the multifunctional components (structural electrolyte and structural electrode) was assessed by performing tensile tests with a 5 kN load cell (using Zwick-Roell universal testing machine, UTM). A high-quality 3D industrial camera (ZEISS ARAMIS 3D, Carl Zeiss IQS GmbH, Oberkochen, Germany) setup was used to capture precise strain measurements. To pixelate the samples with a stochastic black-and-white pixel distribution, paint in a speckle pattern was used to cover the surface of the sample, which was then exposed to the camera. This helps in more accurate elongation measurements. All samples were prepared according to the ASTM D-638 Type V standard, and the loading rate of the cell was set at 2 mm min^{-1}.

2.7. Structural Battery Half-Cell Assembly and Testing

The structural battery performance was evaluated using the 2032-coin cell configuration (Figure S5). The fabricated structural battery (Na | | GF_PEO-SN-$NaClO_4$ | | CF) consists of carbon fiber as a positive electrode and sodium foil as the counter electrode. For the construction of the cell, a structural electrode with an 8 mm diameter was used, and a disc-shaped sodium (Na) foil measuring 13 mm in diameter was positioned on top of the electrolyte surface. The sodium foil was initially pressed firmly against the stainless-steel (SS) spacer surface. A second SS spacer was placed below the structural electrode (carbon-fiber surface) to serve as a current collector. The coin cell assembly was conducted inside the glove box (in a strict protective environment with humidity of 0 ppm and oxygen of 0 ppm) using the MTI MSK-160E pressure adjustable electric crimper (manufactured by MTI Corporation, Richmond, CA, USA). The cell was inserted into a Gamry dual 2032-coin cell battery holder (Part number: 992-00159) and was charged–discharged galvanostatically using the potentiostat Reference 3000 (Gamry LCC, Warminster, PA, USA). The capacity of the structural battery was assessed based on the weight of the carbon fiber used.

3. Results and Discussion

Figure 1b,c depict the surface and cross-section of the structural electrolyte captured using scanning electron microscopy (SEM). Figure 1b is magnified at 20,000×, while Figure 1c is magnified at 1000×. SEM images were taken with a landing energy of 5000 V. In Figure 1c, the cross-section image shows the glass-fiber reinforcements located between the two electrolyte layers. Figure 1e displays the SEM image of the structural electrolyte post-EDAX analysis (20,000× magnification, 15,000 V Landing energy), while Figure 1f shows the EDAX pattern for the same. The EDAX was performed with the following parameters (kV: 15, Magnification: 876; Angle of view: 35.6; Filter time (μs): 0.9; Resolution: (eV) 126.9). Figure 1g displays the distribution of elements, with both the EDAX pattern and SEM–EDAX mapping confirming the presence of SN plasticizers and sodium perchlorate salt in the structural electrolyte. Figure 1d displays the structural electrode cross-section through SEM imaging at a magnification of 2000, revealing the structural electrolyte with embedded glass fiber above and the carbon-fiber electrode below.

3.1. Electrochemical Characterization of Structural Electrolyte

Some key factors of the multifunctional structural electrolyte to be considered for structural battery application include ionic conductivity, ion transference number, the electrochemical stability window, cycling stability, and mechanical property. The prepared electrolytes are initially evaluated for their sodium-ion conductivities using the electrochemical impedance spectroscopy (EIS) technique, as explained in Section 2.

The ionic conductivities of the electrolytes are evaluated using the bulk resistance (R_b) determined from the Nyquist plots (Figures 2a and S6). The value of R_b is determined by where the Nyquist plot intercepts the real axis and is also evaluated using the EIS equivalent circuit (as shown in Figure S7). The calculation for ionic conductivity is made using Equation (1)

$$\sigma = \frac{t}{AR_b} \quad (1)$$

where t is the thickness of the electrolyte, and A is the geometrical contact area of the electrolyte with the stainless-steel electrodes. The polymer composite electrolytes are prepared using various proportions of PEO and the plasticizer succinonitrile (SN). The electrolyte, composed of 55% PEO and 35% SN by weight, exhibited the highest ionic conductivity of 1.32×10^{-4} S cm^{-1} at room temperature, along with good structural integrity. Hence, this is utilized for the fabrication of structural electrolytes. Though the electrolyte composition with 50% PEO and 40% SN showed maximum ionic conductivity of 3.09×10^{-4} S cm^{-1} at room temperature, the texture is too soft and not suitable for structural electrolyte preparation. Table S1 summarizes the room-temperature ionic conductivities of investigated electrolyte compositions.

Figure 2a shows the Nyquist plots of the prepared structural electrolyte GF_PEO-SN–$NaClO_4$ and polymer membrane (with 55%PEO-35%SN) designated as PN-SN-$NaClO_4$. The structural electrolyte showed an ionic conductivity of 1.01×10^{-4} S cm^{-1} at room temperature, suitable for ambient-temperature battery operation. However, it is lower than PEO-SN-$NaClO_4$ due to the presence of an insulative glass-fiber layer. The amorphous nature of the electrolyte helps in maintaining high ionic conductivity even after glass-fiber reinforcements. Figure 2b shows the Arrhenius profile of the electrolytes, and the comparison of ionic conductivities of various prepared electrolytes is summarized in Figure S8.

The linear sweep voltammetry (LSV) measurements were conducted using asymmetric cells (SS | | GF_PEO-SN–$NaClO_4$ | | Na) to evaluate the electrochemical stability window (ESW) of the structural electrolyte. The positive scan was conducted from 2.5 V to 6.5 V, while the negative scan ranged from 2.5 V to 0 V with a scan rate of 1 mV s^{-1} at room temperature. The LSV curve for the structural electrolyte is shown in Figure 2c. The structural electrolyte showed a wide ESW of 0 to 4.9 V, beyond which the current steadily increases with each applied millivolt. The ESW was discovered to be greater than the value obtained for pure PEO-NaClO4 (Figure S9), which is between 0 and 3.95 V [12]. This indicates that both the addition of the SN plasticizer and the glass-fiber reinforcements have widened the ESW due to their stabilization effects [31].

The transference ion number of the structural electrolyte (GF_PEO-SN-$NaClO_4$) was evaluated using the AC–DC experiments described in Section 2. The chronoamperogram captured data for a 10 mV DC polarization voltage in Figure 2d, indicating values for initial current (I_0) and steady-state current (I_{SS}). The EIS Nyquist plot (Figure S10) gives the initial charge transfer resistance (R_0) and the final steady-state resistance (R_{SS}) before and after applying DC polarization, respectively.

The transference ion number was obtained using Equation (2) [30].

$$t_{Na^+} = \frac{I_{ss}(\Delta V - I_0 R_0)}{I_0(\Delta V - I_{ss} R_{ss})} \quad (2)$$

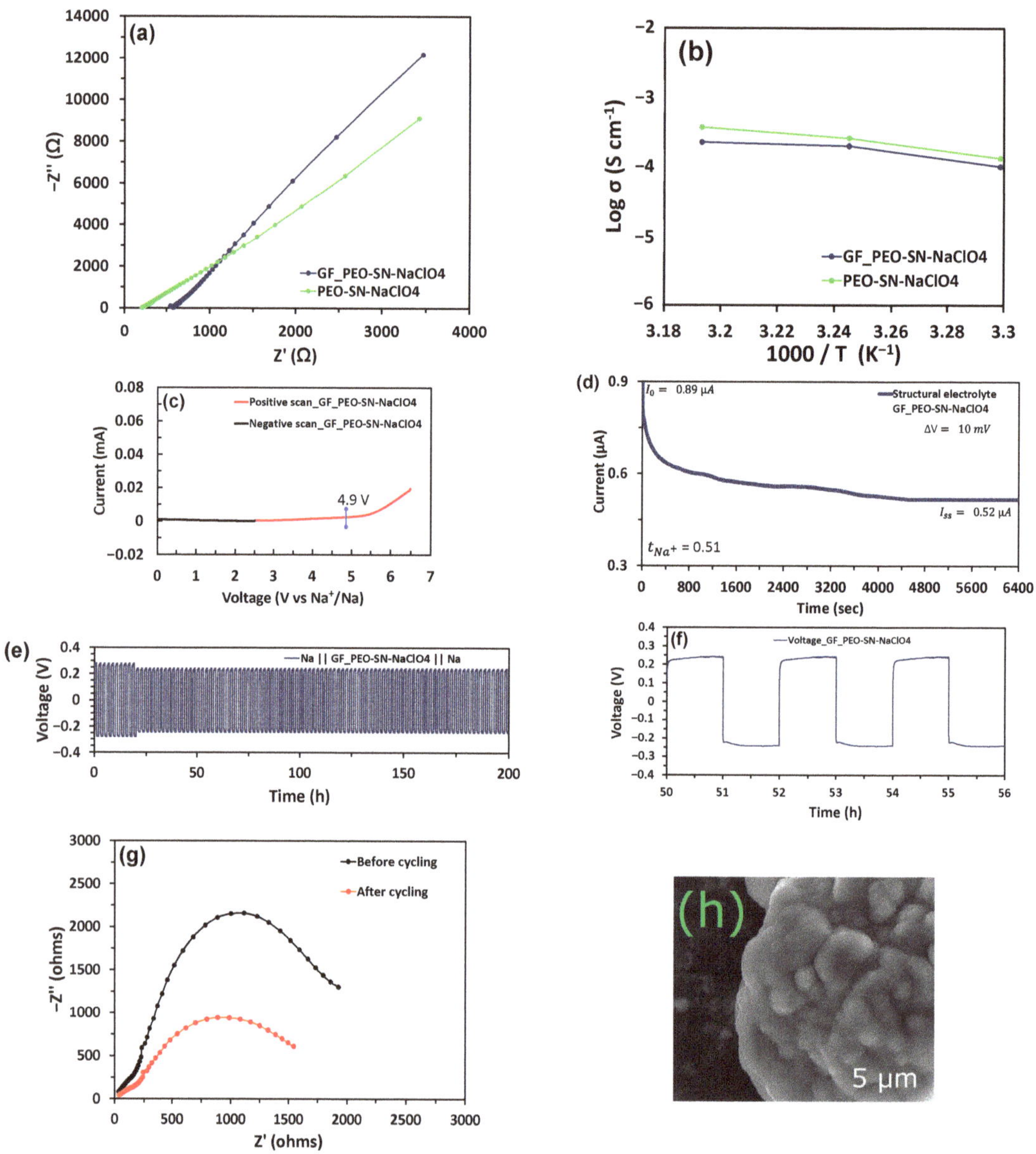

Figure 2. Electrochemical characterization: (**a**) EIS of structural electrolyte (GF_PEO-SN-$NaClO_4$) and PEO-SN-NaClO4 at room temperature. (**b**) Arrhenius plots of electrolytes from 30 °C to 45 °C. (**c**) Linear sweep voltammetry (LSV) curve of Structural electrolyte GF_PEO-SN-$NaClO_4$. (**d**) Chronoamperogram or current response of Na | | GF_PEO-SN-$NaClO_4$ | | Na symmetrical cell to applied DC polarization. (**e**) Galvanostatic cycling of symmetrical cells Na | | GF_PEO-SN-$NaClO_4$ | | Na for a current density of 0.1 mA cm^{-2}. (**f**) Typical voltage profile of the Na | | GF_PEO-SN–$NaClO_4$ | | Na cell from 50 to 56 h. (**g**) EIS of Na | | GF_PEO-SN-$NaClO_4$ | | Na symmetrical cell before and after cycling. (**h**) SEM image of the Na electrode for the Na | | GF_PEO-SN-$NaClO_4$ | | Na cell after cycling at 5 μm range.

The structural electrolyte exhibited a sodium-ion transference number of 0.51 at room temperature, comparable to the reported value of 0.48 in the literature [29]. This high value can be attributed to the addition of succinonitrile plasticizers and their ability to provide sodium-ion pathways and to the addition of glass fibers, which augments sodium-ion movement [19].

To assess the performance and cycling stability of the structural electrolyte, the symmetrical cell configuration (Na | | GF_PEO-SN-$NaClO_4$ | | Na) was galvanostatically charged and discharged using a current density of 0.1 mA cm^{-2}. Figure 2e displays the cycling curves for the symmetrical cell, while Figure 2g shows the EIS Nyquist plots before and after cycling. The cell displayed a high initial charge transfer resistance (R_{CT}) value of 2010 Ω, which is evident in the initial voltage rise in the symmetrical cell (Figure 2e). The localized sodium plating and unstable interface account for this early high R_{CT} [32,33]. After approximately 20 h, the cell voltage drops as the interface stability improves, and a structured voltage profile is displayed (as depicted in Figure 2f). The reduced R_{CT} value of 1490 Ω obtained from the Nyquist plot (Figure 2g) indicates the stability of the interface. In total, the cell operated continuously for 200 h, and the voltage profile between 50 and 56 h is displayed in Figure 2f. Figure 2h, with a magnification of 25,000 and a landing energy of 10,000 V, depicts an SEM image of the Na electrode in the Na | | GF_PEO-SN-$NaClO_4$ | | Na symmetrical cell after 200 cycles, illustrating compact yet uniform sodium depositions with minor dendrites to prevent cell short-circuiting which is usually caused by larger dendrite formations.

3.2. *Mechanical Characterization of Multifunctional Composites*

The tensile strength of the fabricated composites was determined by conducting mechanical tests on the prepared structural electrolyte and electrode samples using a tensile testing machine, as outlined in Section 2. A ZEISS ARAMIS industrial camera is utilized to acquire accurate strain measurements during the test.

Figure S11 displays the strain measurements of the samples from the start point to the maximum load point. The structural electrolyte's tensile strength was assessed using a 5 kN load cell arrangement illustrated in Figure 3a. The stress–strain graph in Figure 3b displays a maximum tensile strength of 32.1 MPa, which is attributed to incorporating glass fibers into the production of the structural electrolyte. The average tensile stress measured in the samples was 29.4 MPa, with a standard deviation of 5.73 MPa, as illustrated in Figure S13. The Young's modulus is determined to be 1.08 GPa. The anticipated result showed that the tensile stress value is less than the 40.9 MPa reported in the prior study on multifunctional polymer composite with NZSP nanoparticle fillers [12]. The main reason for this is the incorporation of succinonitrile plasticizers, which increase the amorphous nature of the matrix, resulting in a decrease in the overall tensile strength of the multifunctional matrix composite.

The tensile test setup with a 5 kN load cell for the structural electrode is shown in Figure 3c, while the stress–strain curve is shown in Figure 3d. The strain measurements of the structural electrode sample are shown in Figure S12. The structural electrode material exhibited a typical tensile strength of 72.3 MPa. The increase is due to the lamination of carbon fiber onto the structural electrolyte. Figure S13 displays a mean tensile stress of 76.6 MPa with a standard deviation of 9.6 MPa. The calculation of Young's modulus gives a value of 2.12 GPa. Once again, the structural electrode's tensile strength is slightly lower at 91.3 MPa than the previous value reported [12]. The primary reason for this is the variation in the types of fillers utilized in these research studies. Figure S14 displays the force–displacement curves for the different structural electrolyte and structural electrode samples that were tested.

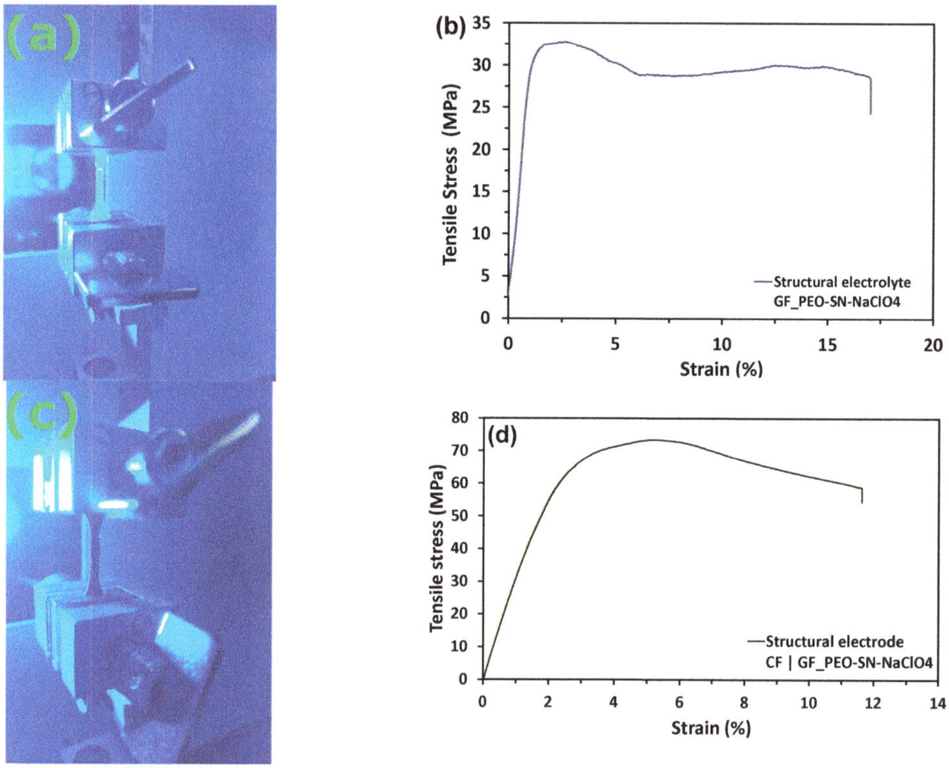

Figure 3. Mechanical characterization: (**a**) Experimental setup for the tensile test of structural electrolyte. (**b**) Typical stress–strain curve of structural electrolyte GF_PEO-SN-$NaClO_4$. (**c**) Experimental setup for the tensile test of structural electrodes. (**d**) Typical stress–strain curve of structural electrode CF| | GF_PEO-SN-$NaClO_4$.

3.3. Structural Battery Fabrication and Performance

The structural battery was fabricated as described in Section 2 in the 2032-coin cell configuration (Figure S3), with spread tow carbon-fiber (CF) cathode and sodium metal anode (CF | | GF_PEO-SN-$NaClO_4$ | | Na). Figure 4a shows the cell's charge-discharge profiles at a 0.1 C rate from 0.2 V to 3 V, with a nominal voltage of 1.9 V during discharge. The aerial weight of the carbon fibers (55 g m^{-2}) was used for the computation of cell capacities. The cell exhibited a high discharge capacity of 14 mAh g^{-1} in the first cycle because of the stable formation of the solid interface layer (SEI), with a Coulombic efficiency of 58% [34]. From successive cycles, the Coulombic efficiency improved significantly in the range of 96–98%. The typical discharge capacity was found to be 7.5 mAh g^{-1}, and the corresponding energy density at 0.1 C rate was estimated to be 14.2 Wh kg^{-1}. The energy density of the current work is comparatively lower than our prior study on a PEO-based multifunctional composite containing active NZSP nanoparticle fillers, which achieved an energy density of 23 Wh kg^{-1}. This is because NZSP is sodium-rich, and the vacant sites in its solid-state structure aid in the diffusion of ions, contributing to high ionic conductivity and higher energy density [12].

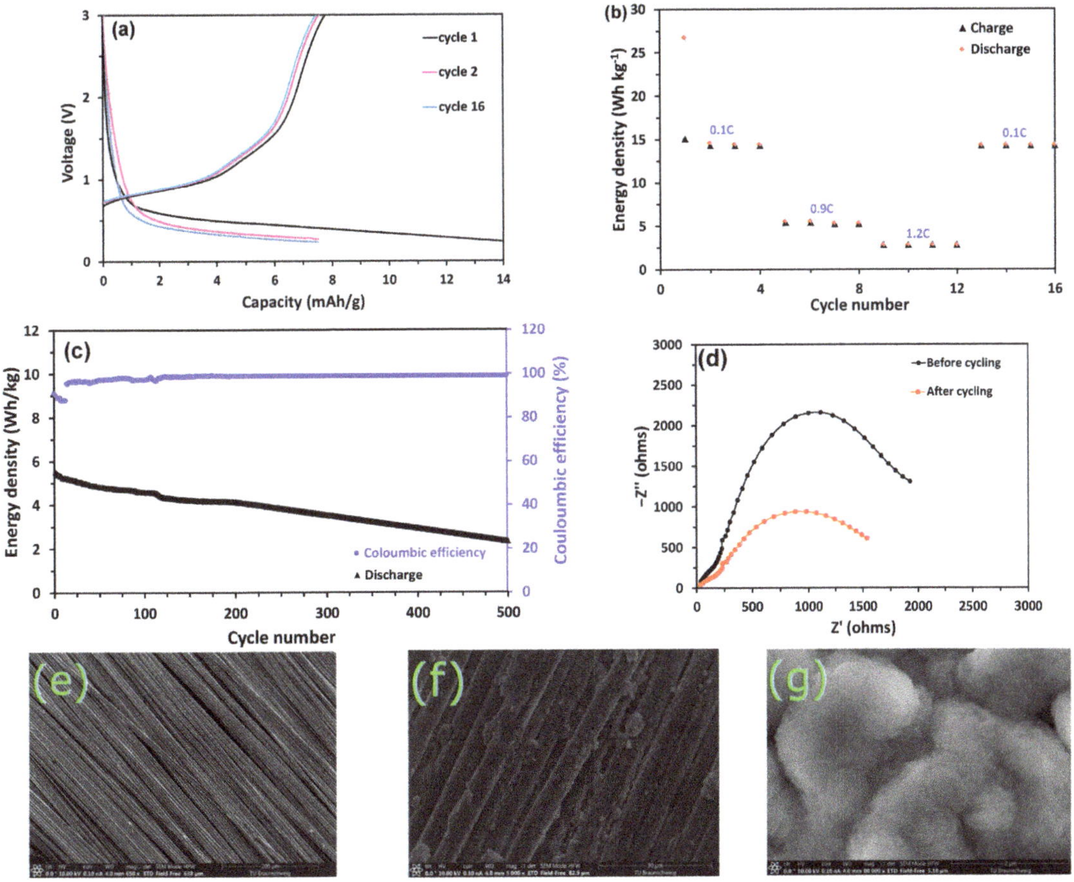

Figure 4. Structural battery performance: (**a**) Charge-discharge profiles of CF | | GF_PEO-SN-$NaClO_4$ | | Na cell for 0.1 C rate. (**b**) Rate capability tests showing specific capacity Vs cycle number of CF | | GF_PEO-SN-$NaClO_4$ | | Na cell at different C rates. (**c**) Stability tests showing charge-discharge energy density and Coulombic efficiency Vs cycle number of CF | | GF_PEO-NZSP–$NaClO_4$ | | Na cell for 0.9C rate. (**d**) EIS of CF | | GF_PEO-SN-$NaClO_4$ | | Na cell before and after cycling. (**e**) SEM image of pure carbon-fiber electrode: (**f**) SEM image of carbon-fiber electrode after cycling. (**g**) SEM image of Na electrode after cycling.

To assess the performance of the cell at different C rates, rate capability tests were conducted. The cell was charged–discharged for 4 cycles at each C rate, and the energy density vs cycle number plot is shown in Figure 4b. For 0.1 C, 0.9 C, and 1.2 C, the cell displayed energy densities of 14.2 Wh kg^{-1}, 5.5 Wh kg^{-1}, and 2.6 Wh kg^{-1}, respectively. The energy density at 0.1 C rate remained fairly constant from the 13th to the 16th cycle.

These values reflect the current trends in DOI III structural battery design, with reported energy densities ranging from 1 to 50 Wh kg^{-1} [12]. Table 1 summarizes and compares the various reported energy densities for DOI III structural batteries with the values reported in this study, while Figure S15 summarizes the energy density versus elastic modulus plot of various DOI III structural battery designs [10,35–39]. The primary difficulty in designing structural batteries is enhancing both their mechanical and electrochemical capabilities. In Figure 4c, the discharge profile for a 0.9 C rate is displayed along with the Coulombic efficiencies measured for each cycle. The cell exhibited good cycling performance and could retain 80% capacity at the end of 200 cycles, and the Coulombic efficiency remained high at 98% consistently after the 12th cycle once the structural electrolyte-carbon-fiber interface was stabilized. The cell capacity degradation plot is shown in Figure S16. The capacity degradation occurs because of the alteration in charge transfer resistance (R_{CT}), as seen in the EIS Nyquist plot in Figure 4d. At the start of the initial cycle, the resistance of the R_{CT} was 980 Ω, and by the conclusion of the 200th cycle, it had risen to 2200 Ω, which can be attributed to the formation of sodium dendrites. Figure 4e (Magnification: 650; Landing Energy: 10,000 V) and Figure 4f (Magnification: 5000; Landing Energy: 10,000 V) compare the SEM images of the carbon-fiber electrodes before and after cycling, respectively, while Figure 4g (Magnification: 80,000; Landing Energy: 10,000 V) shows the Na electrodes after cycling. The formation of dendrites is evident from these images. It is small with dense but uniform Na deposition (as shown in Figure 4g). This helps in avoiding short circuits of the cell and enhances cycling stability. The cycling stability of the structural battery could be further credited to the formation of a stable solid electrolyte interphase (SEI) layer. It is expected that the cell energy density could be further boosted by the treatment of carbon fibers and further stacking of cells, which is essential for high energy density structural battery architectures, along with using less resistive materials instead of glass fibers. Further steps in realizing the full structural battery cell include the usage of suitable carbon fibers coated with sodium-abundant cathode material instead of Na metal as an active material providing a sodium-ion source. In this case, the carbon fibers will act both as electrodes and as current collectors. Such an architecture is expected to provide pathways to design structural batteries capable of operating in ambient-temperature conditions. Since succinonitrile plasticizers are used in the present multifunctional composite system with a melting point of 58 °C, it is necessary to adapt low-temperature curing resin systems for the manufacture of multifunctional structural components using these multifunctional composites [40].

Table 1. Table showing the comparison of this work with reported structural electrolyte type and typical energy density values for DOI III type structural batteries.

References	Electrolyte Type	Ion Type	Reported Values		
			C Rate	Specific Capacity ($Ah\ kg^{-1}$)	Energy Density ($Wh\ kg^{-1}$)
Liu et al. [10]	Gel-polymer-type electrolyte	Li-ion	0.05C	DNA [(a)]	35
Meng et al. [35]	Kevlar reinforced gel type electrolyte	Li-ion	DNA [(a)]	DNA [(a)]	1.4
Thakur and Dong [36]	Solid polymer-type electrolyte coated on individual carbon fibers	Li-ion	0.5C	23.4	7.6
Moyer et al. [37]	Liquid electrolyte impregnated on Celgard separator	Li-ion	0.1C	30	35
Asp et al. [38]	Bi-continuous phase-type electrolyte with GF plain weave separator	Li-ion	0.05C	8.55	23.6
Asp et al. [38]	Bi-continuous phase-type electrolyte with Whatman GF/A separator	Li-ion	0.05C	4.13	11.6
Siraj et al. [39]	Bi-continuous phase-type electrolyte with GF plain weave separator	Li-ion	0.05C	14.7	41.2
Siraj et al. [39]	Bi-continuous phase-type electrolyte with Whatman GF/A separator	Li-ion	0.05C	9.82	25.9

Table 1. *Cont.*

References	Electrolyte Type	Ion Type	Reported Values		
			C Rate	Specific Capacity ($Ah\ kg^{-1}$)	Energy Density ($Wh\ kg^{-1}$)
Iyer et al. [12]	PEO-based all-solid-state composite structural electrolyte with NZSP ($Na_3Zr_2Si_2PO_{12}$) nanoparticle fillers embedded with glass-fiber woven fabric separator	Na-ion	0.1C	10.8	23 *
This work	PEO-based all-solid-state composite structural electrolyte with succinonitrile ($C_2H_4(CN)_2$) plasticizer fillers embedded with glass-fiber woven fabric separator	Na-ion	0.1C	7.5	14.2 *

* The energy density values mentioned here are for structural battery half-cells (testing of structural cathode with sodium metal anode); [(a)] Data not available.

4. Conclusions

This work developed and characterized multifunctional composites suitable for designing ambient-temperature sodium-ion structural batteries. Some of the key findings are summarized below:

- The structural electrolyte (GF_PEO-SN-$NaClO_4$) was fabricated using a poly(ethylene oxide) (PEO)-based matrix with succinonitrile fillers, and it was reinforced with glass fibers. This is important to achieve high tensile strength electrolytes suitable for designing structural batteries.
- The mechano-electrochemical characterization of the structural electrolyte showed a tensile strength of 32.1 MPa and an ionic conductivity of 1.01×10^{-4} S cm^{-1} at room temperature. It displayed a wide electrochemical stability window of 0 to 4.9 V and a high sodium-ion transference number of 0.51. The multifunctional performance shown by the electrolyte shows its capability to both transport ions (which is necessary for energy storage functionality) and to withstand applied loads necessary for structural integrity.
- The structural electrode (CF | | GF_PEO-SN–$NaClO_4$) component was fabricated by pressing the intermediate modulus carbon fibers (possessing hard carbon type microstructures) with the prepared structural electrolyte to boost sodium-ion storage capabilities. It exhibited a considerable tensile strength of 72.3 MPa.
- To investigate sodium-ion insertion capabilities, the structural battery (CF | | GF_PEO-SN–$NaClO_4$ | | Na) was fabricated, and its performance was tested. It provided a typical energy density of 14.2 Wh kg^{-1} at 0.1 C rate, tested for cycling performance, showed good cycling stability, and retained 80% capacity at the end of 200 cycles. The microstructure analysis of the cycled electrodes using scanning electron microscopy revealed small dendrite formation, but the small, uniform Na metal deposition avoided cell short circuits and hence ensured good cycling stability.

The fabricated composites (structural electrolyte and electrode) showed multifunctional characteristics necessary for the design of ambient-temperature sodium-ion structural batteries.

Supplementary Materials: The following supporting information can be downloaded at: https://www.mdpi.com/article/10.3390/polym16192806/s1, Figure S1: Process chain for the preparation of structural electrolyte; Figure S2: Images showing prepared electrolytes samples; Figure S3: Images of prepared structure electrode samples; Figure S4: Coin Cell assembly for asymmetrical and symmetrical cells; Figure S5: Cell assembly for structural battery half-cell; Figure S6: EIS Nyquist plots of solid polymer electrolytes (SPE) with varying amounts of PEO and SN; Figure S7: Electrochemical impedance spectroscopy (EIS) equivalent circuit for the various electrolytes; Figure S8: Ionic conductivities of electrolytes at different temperatures; Figure S9: Linear sweep voltammetry (LSV) curves of PEO-$NaClO_4$ membrane; Figure S10: EIS Nyquist plots of Na | | GF_PEO-SN-$NaClO_4$ | | Na with applied ac perturbation before and after dc polarization; Figure S11: Images from GOM viewer showing the strain measurement on the surface of the structure electrolyte; Figure S12: Images from GOM viewer showing the strain measurement on the surface of the structural electrode; Figure S13: Mean and standard deviation values of tensile strength for the structural electrolyte and structural electrode using five samples; Figure S14: Force–displacement curves of Structural electrolyte (GF_PEO-SN–$NaClO_4$) and Structural electrode (CF | | GF_PEO-SN–$NaClO_4$); Figure S15: Reported cell level energy density and elastic modulus of various structural battery architectures with literature references; Figure S16: Percentage cell capacity retention versus cycle number for the CF | | GF_PEO-SN–$NaClO_4$ | | Na cell; title; Table S1: Table showing the ionic conductivities of various polymer electrolyte compositions at room temperature (25 °C).

Author Contributions: Conceptualization, V.I.; methodology, V.I.; software, V.I.; validation, V.I., J.P., and S.G.; formal analysis, V.I.; investigation, V.I.; resources, P.W. and J.P.; data curation, V.I.; writing—original draft preparation, V.I.; writing—review and editing, S.G., J.P. and P.W.; visualization, V.I.; supervision, P.W.; project administration, P.W.; funding acquisition, P.W. All authors have read and agreed to the published version of the manuscript.

Funding: We would like to acknowledge the funding by the Deutsche Forschungsgemeinschaft (DFG, German Research Foundation) under Germany's Excellence Strategy—EXC 2163/1-Sustainable and Energy-Efficient Aviation—Project-ID 390881007. Furthermore, we acknowledge support from the Open Access Publication Funds of the Technical University Braunschweig. The APC was funded by both Technical University Braunschweig and DLR (German Aerospace Center) separately.

Institutional Review Board Statement: Not applicable.

Data Availability Statement: The original contributions presented in the study are included in the article. Further inquiries can be directed to the corresponding author.

Conflicts of Interest: The authors declare no conflicts of interest.

References

1. European Commission. The European Green Deal. Available online: https://europa.eu/!DG37Qm (accessed on 31 August 2024).
2. European Commission. Clean Sky Benefits. Available online: https://www.clean-aviation.eu/benefits-for-cleaner-greener-healthier-skies (accessed on 31 August 2024).
3. European Commission. Flightpath 2050. Available online: https://data.europa.eu/doi/10.2777/15458 (accessed on 31 August 2024).
4. Tesla Battery Pack Architecture. Available online: https://futurism.com/take-an-in-depth-look-at-the-tesla-model-3s-new-battery-pack-architecture (accessed on 31 August 2024).
5. Wetzel, E.D. Reducing Weight: Multifunctional Composites Integrate Power, Communications, and Structure. *AMPTIAC Q.* **2004**, *8*, 91–95.
6. Adam, T.J.; Liao, G.; Petersen, J.; Geier, S.; Finke, B.; Wierach, P.; Kwade, A.; Wiedemann, M. Multifunctional Composites for Future Energy Storage in Aerospace Structures. *Energies* **2018**, *11*, 335. [CrossRef]
7. Thomas, J.P.; Qidwai, W.R.; Pogue, W.R., III; Pham, G. Multifunctional structure battery composites for marine systems. *J. Compos. Mater.* **2012**, *47*, 5–26. [CrossRef]
8. Pereira, T.; Guo, Z.; Nieh, S.; Arias, J.; Hahn, T. Embedding thin-film lithium energy cells in structural composites. *Compos. Sci. Technol.* **2008**, *68*, 1935–1941. [CrossRef]
9. Petersen, J.; Kube, A.; Geier, S.; Wierach, P. Structure-Integrated Thin-Film Supercapacitor as a Sensor. *Sensors* **2022**, *22*, 6932. [CrossRef]
10. Liu, P.; Sherman, E.; Jacobsen, A. Design and fabrication of multifunctional structural batteries. *J. Power Sources* **2009**, *189*, 646–650. [CrossRef]
11. Leijonmarck, S.; Carlson, T.; Lindbergh, G.; Asp, E.S.; Maples, H.; Bismarck, A. Solid polymer electrolyte coated carbon fibers for structural and novel micro batteries. *Compos. Sci. Technol.* **2013**, *89*, 149–157. [CrossRef]
12. Iyer, V.; Petersen, J.; Geier, S.; Wierach, P. Development and multifunctional characterization of a structural sodium-ion battery using a high-tensile-strength poly(ethylene oxide)-based matrix composite. *ACS Appl. Energy Mater.* **2024**, *7*, 3968–3982. [CrossRef]
13. Geier, S.; Petersen, J.; Iyer, V.; Wierach, P. Challenges of integrating supercapacitors into structures for space qualification. In Proceedings of the 16th European Conference on Spacecraft Structures, Materials and Environmental Testing (ECSSMET 2021), Braunschweig, Germany, 23–25 March 2021.
14. Asp, L.E.; Johansson, M.; Lindbergh, G.; Xu, J.; Zenkert, D. Structural battery composites: A review. *Funct. Compos. Struct.* **2019**, *1*, 042001. [CrossRef]
15. Muñoz, B.K.; Lozano, J.; Sánchez, M.; Ureña, A. Hybrid Solid Polymer Electrolytes Based on Epoxy Resins, Ionic Liquid, and Ceramic Nanoparticles for Structural Applications. *Polymers* **2024**, *16*, 2048. [CrossRef]

16. Jacques, E.; Kjell, M.H.; Zenkert, D.; Lindbergh, G.; Behm, M.; Willgert, M. Impact of electrochemical cycling on the tensile properties of carbon fibres for structural lithium-ion composite batteries. *Compos. Sci. Technol.* **2012**, *72*, 792–798. [CrossRef]
17. Fredi, G.; Jeschke, S.; Boulaoued, A.; Wallenstein, J.; Rashidi, M.; Liu, F.; Harnden, R.; Zenkert, D.; Hagberg, J.; Lindbergh, G.; et al. Graphitic microstructure and performance of carbon fibre Li-ion structural battery electrodes. *Multifunct. Mater.* **2018**, *1*, 015003. [CrossRef]
18. Schneider, L.M.; Ihrner, N.; Zenkert, D.; Johansson, M. Bicontinuous Electrolytes via Thermally Initiated Polymerization for Structural Lithium Ion Batteries. *ACS Appl. Energy Mater.* **2019**, *2*, 4362–4369. [CrossRef]
19. Dong, G.; Mao, Y.; Yang, G.; Li, Y.; Song, S.; Xu, C.; Huang, P.; Hu, N.; Fu, S. High-Strength Poly(ethylene oxide) Composite Electrolyte Reinforced with Glass Fiber and Ceramic Electrolyte Simultaneously for Structural Energy Storage. *ACS Appl. Energy Mater.* **2021**, *4*, 4038–4049. [CrossRef]
20. Stevens, D.A.; Dahn, J.R. The Mechanisms of Lithium and Sodium Insertion in Carbon Materials. *J. Electrochem. Soc.* **2001**, *148*, A803–A811. [CrossRef]
21. Kjell, M.H.; Jacques, E.; Zenkert, D.; Behm, M.; Lindbergh, G. PAN-based carbon fiber electrodes for structural lithium-ion batteries. *J. Electrochem. Soc.* **2011**, *158*, A1455–A1460. [CrossRef]
22. Zhao, Q.; Stalin, S.; Zhao, C.Z.; Archer, L.A. Designing solid-state electrolytes for safe, energy-dense batteries. *Nat. Rev. Mater.* **2020**, *5*, 229–252. [CrossRef]
23. Goikolea, E.; Palomares, V.; Wang, S.; de Larramendi, I.R.; Guo, X.; Wang, G.; Rojo, T. Na-Ion Batteries—Approaching Old and New Challenges. *Adv. Energy Mater.* **2020**, *10*, 2002055. [CrossRef]
24. Hayashi, A.; Noi, K.; Sakuda, A.; Tatsumisago, M. Superionic glass-ceramic electrolytes for room-temperature rechargeable sodium batteries. *Nat. Commun.* **2012**, *3*, 856. [CrossRef]
25. Ma, Q.L.; Guin, M.; Naqash, S.; Tsai, C.L.; Tietz, F.; Guillon, O. Scandium-Substituted $Na_3Zr_2(SiO_4)_2(PO_4)$ Prepared by a Solution-Assisted Solid-State Reaction Method as Sodium-Ion Conductors. *Chem. Mater.* **2016**, *28*, 4821–4828. [CrossRef]
26. Alarco, P.J.; Abu-Lebdeh, Y.; Abouimrane, A.; Armand, M. The plastic-crystalline phase of succinonitrile as a universal matrix for solid-state ionic conductors. *Nat. Mater.* **2004**, *3*, 476–481. [CrossRef] [PubMed]
27. Fan, L.Z.; Maier, J. Composite effects in poly(ethylene oxide)–succinonitrile based all-solid electrolytes. *Electrochem. Commun.* **2006**, *8*, 1753–1756. [CrossRef]
28. Park, C.W.; Ryu, H.S.; Kim, K.W.; Ahn, J.H.; Lee, J.Y.; Ahn, H.J. Discharge properties of all-solid sodium–sulfur battery using poly (ethylene oxide) electrolyte. *J. Power Sources* **2007**, *165*, 450–454. [CrossRef]
29. Yu, X.; Xue, L.; Goodenough, J.B.; Manthiram, A. Ambient-Temperature All-Solid-State Sodium Batteries with a Laminated Composite Electrolyte. *Adv. Funct. Mater.* **2020**, *31*, 2002144. [CrossRef]
30. Evans, J.; Vincent, C.A.; Bruce, P.G. Electrochemical measurement of transference numbers in polymer electrolytes. *Polymer* **1987**, *28*, 2324–2328. [CrossRef]
31. Li, D.; Chen, L.; Wang, T.; Fan, L.Z. 3D Fiber-Network-Reinforced Bi-continuous Composite Solid Electrolyte for Dendrite-free Lithium Metal Batteries. *ACS Appl. Mater. Interfaces* **2018**, *10*, 7069–7078. [CrossRef]
32. Lee, B.; Paek, E.; Mitlin, D.; Lee, S.W. Sodium Metal Anodes: Emerging Solutions to Dendrite Growth. *Chem. Rev.* **2019**, *119*, 5416–5460. [CrossRef]
33. Qiang, Z.; Chen, Y.M.; Xia, Y.F.; Liang, W.F.; Zhu, Y.; Vogt, B.D. Ultra-long cycle life, low-cost room temperature sodium-sulfur batteries enabled by highly doped (N,S) nanoporous carbons. *NanoEnergy* **2017**, *32*, 59–66. [CrossRef]
34. Feng, M.; Wang, S.; Yu, Y.; Feng, Q.; Yang, J.; Zhang, B. Carboxyl functionalized carbon fibers with preserved tensile strength and electrochemical performance used as anodes of structural lithium-ion batteries. *Appl. Surf. Sci.* **2017**, *392*, 27–35. [CrossRef]
35. Meng, C.; Muralidharan, N.; Teblum, E.; Moyer, K.E.; Nessim, G.D.; Pint, C.L. Multifunctional Structural Ultrabattery Composite. *Nano Lett.* **2018**, *18*, 7761–7768. [CrossRef]
36. Thakur, A.; Dong, X. Printing with 3D continuous carbon fiber multifunctional composites via UV-assisted coextrusion deposition. *Manuf. Lett.* **2020**, *24*, 1–5. [CrossRef]
37. Moyer, K.; Meng, C.; Marshall, B.; Assal, O.; Eaves, J.; Perez, D.; Karkkainen, R.; Roberson, L.; Pint, C.L. Carbon fiber reinforced structural lithium-ion battery composite: Multifunctional power integration for CubeSats. *Energy Storage Mater.* **2020**, *24*, 676–681. [CrossRef]
38. Asp, L.E.; Bouton, K.; Carlstedt, D.; Duan, S.; Harnden, R.; Johannisson, W.; Johansen, M.; Johansson, M.K.G.; Lindbergh, G.; Liu, F.; et al. A Structural Battery and its Multifunctional Performance. *Adv. Energy Sustain. Res.* **2021**, *2*, 2000093. [CrossRef]
39. Siraj, M.S.; Tasneem, S.; Carlstedt, D.; Duan, S.; Johansen, M.; Larsson, C.; Xu, J.; Liu, F.; Edgren, F.; Asp, L.E. Advancing Structural Battery Composites: Robust Manufacturing for Enhanced and Consistent Multifunctional Performance. *Adv. Energy Sustain. Res.* **2023**, *4*, 2300109. [CrossRef]
40. Danzi, F.; Salgado, R.M.; Oliveira, J.E.; Arteiro, A.; Camanho, P.P.; Braga, M.H. Structural Batteries: A Review. *Molecules* **2021**, *26*, 2203. [CrossRef]

polymers

MDPI

Article

Multifunctional Aspects of Mechanical and Electromechanical Properties of Composites Based on Silicone Rubber for Piezoelectric Energy Harvesting Systems

Vineet Kumar , Md. Najib Alam , Manesh A. Yewale and Sang-Shin Park *

School of Mechanical Engineering, Yeungnam University, 280 Daehak-Ro, Gyeongsan 38541, Gyeongbuk, Republic of Korea; vineetfri@gmail.com (V.K.); mdnajib.alam3@gmail.com (M.N.A.); maneshphd@gmail.com (M.A.Y.)

* Correspondence: pss@ynu.ac.kr

Abstract: Energy harvesting systems fabricated from rubber composite materials are promising due to their ability to produce green energy with no environmental pollution. Thus, the present work investigated energy harvesting through piezoelectricity using rubber composites. These composites were fabricated by mixing titanium carbide (TiC) and molybdenum disulfide (MoS_2) as reinforcing and electrically conductive fillers into a silicone rubber matrix. Excellent mechanical and electromechanical properties were produced by these composites. For example, the compressive modulus was 1.55 ± 0.08 MPa (control) and increased to 1.95 ± 0.07 MPa (6 phr or per hundred parts of rubber of TiC) and 2.02 ± 0.09 MPa (6 phr of MoS_2). Similarly, the stretchability was 133 ± 7% (control) and increased to 153 ± 9% (6 phr of TiC) and 165 ± 12% (6 phr of MoS_2). The reinforcing efficiency (R.E.) and reinforcing factor (R.F.) were also determined theoretically. These results agree well with those of the mechanical property tests and thus validate the experimental work. Finally, the electromechanical tests showed that at 30% strain, the output voltage was 3.5 mV (6 phr of TiC) and 6.7 mV (6 phr of MoS_2). Overall, the results show that TiC and MoS_2 added to silicone rubber lead to robust and versatile composite materials. These composite materials can be useful in achieving higher energy generation, high stretchability, and optimum stiffness and are in line with existing theoretical models.

Keywords: titanium carbide; molybdenum disulfide; silicone rubber; reinforcing efficiency; electromechanical properties

1. Introduction

Piezoelectric energy harvesting is a novel technique that can convert kinetic energy into an electric charge using piezoelectric materials. Integrating this technology into composite materials leads to multifunctional possibilities, such as creating flexible, durable, stretchable, and lightweight materials [1]. Piezoelectric materials commonly involve ceramics like lead zirconate titanate (PZT) and polymer matrices like polyvinylidene fluoride (PVDF). For rubber composites, dielectric rubbers like silicone rubber are mixed with piezoelectric fillers like PZT [2,3]. This integration allows the composite to maintain the flexibility and durability of the rubber while gaining the ability to harvest energy. There are various applications for piezoelectric rubber composites. They include wearable electronics, structural health monitoring, energy harvesting mats, and automobile applications. Wearable electronics are used in power sensors and portable electronics through human motion [4]. Energy harvesting mats are embedded flooring materials that capture energy from foot traffic. This mechanical energy, obtained through a foot press, is converted from footsteps into electric power [5]. However, there are various limitations to using these rubber composites for energy harvesting. These include poor efficiency, a low coupling coefficient, low power density, scalability, and integration [6]. Thus, seamlessly integrating these composites into existing systems and ensuring their compatibility with other materials and components is crucial before commercialization. Overall, this piezoelectric energy harvesting system with rubber composites provides a promising area for research and development, with potential applications across various industries [7].

The mechanical properties of these piezoelectric rubber composites are crucial for their performance and multifunctional applications. These properties determine the composite material's behavior under mechanical deformation. They also determine the ability of these composites to maintain their piezoelectric features under cyclic mechanical deformation [8]. Many mechanical parameters, such as the modulus, tensile strength, fracture toughness, and elongation at break, are critical for the overall performance of the composite. For example, for piezoelectric rubber composites, achieving a balance between flexibility (low elastic modulus) and sufficient mechanical strength is essential [9,10]. Moreover, the tensile strength and elongation at break are critical for applications that involve stretching or bending, ensuring that the composite can endure mechanical stress without failure. These mechanical properties are influenced by various factors. These factors include the filler loading, the type of rubber matrix, filler dispersion, filler–rubber compatibility, and filler–rubber interfacial interactions [11]. A balance among these factors is required to obtain robust energy harvesting performance. Overall, understanding and optimizing these mechanical properties of piezoelectric rubber composites is crucial for their effective application in various fields. These applications include strain sensors, actuators, and smart textiles [12].

The electromechanical properties of piezoelectric rubber composites are characterized by the interaction between mechanical deformation and an electrical response. This piezoelectric effect can be direct or inverse depending upon the required application [13]. The direct effect converts mechanical deformation into an electrical response, such as in energy harvesting. However, the inverse effect converts electrical energy into mechanical displacement, as in actuators. Moreover, mechanical properties such as stiffness, flexibility, stretchability, and fatigue resistance are very important in influencing the electromechanical properties [14]. Various key factors influence the electromechanical properties. These are the nature of the piezoelectric material, the type of rubber matrix, filler dispersion, filler–rubber interfacial adhesion, and filler–rubber compatibility [15]. For example, the filler–rubber interfacial bonding assists in enhancing the stress transfer from the rubber matrix to the filler particles when subjected to mechanical deformation. In addition, filler–rubber compatibility is very important as it strongly influences the filler dispersion and interfacial bonding in composites [16]. Moreover, the processing technique for the fabrication of composites also critically influences the electromechanical behavior. Overall, the electromechanical properties of piezoelectric rubber composites are critical to their functionality and application potential. These applications include sensors, energy harvesting, actuators, and health monitoring systems [17,18].

Various studies demonstrate the use of rubber composites for energy harvesting applications [19–22]. For example, Liu et al. [19] studied composites based on lead zirconate titanate, graphene, and polydimethylsiloxane as a rubber matrix. The results show that a high dielectric constant of 30 was achieved. These improved electrical properties helped in achieving an output voltage of ~40 mV using a universal testing machine and ~3 V through human motion. In another interesting study by Kumar et al. [20], the composites were prepared by mixing silicone rubber with barium titanate, carbon black, and multiwall carbon nanotubes. The composites fabricated exhibited high tensile strength of 0.2 MPa, hardness of 33, and electrical resistance of 260 Ω. The energy harvesting properties showed an output voltage of ~1.8 V at an electrical poling value of 10 kV/mm. In another study, Deng et al. [21] reinforced silicone rubber with carbon nanotubes and achieved great mechanical stretchability of ~900%. The authors achieved a gauge factor of 2.1 at 130% and finally a power density of 21.7 W cm^{-1}. Lastly, Chung et al. [22] reported on composites that were prepared by mixing poly[styrene-b-isoprene-b-styrene] with carbon black and multiwall carbon nanotubes. The results showed that a robust stretching ability of >1000% was obtained. Moreover, a gauge factor of 69.8 at ~450% strain and durability of 10,000 cycles were reported. In addition, the best output voltage of 0.5 V, electrical resistance of 100 Ω, and durability of >25,000 cycles were reported [22].

With these aspects in mind, the present work involves the fabrication of composites. A comparative study on electromechanical properties like the response time and sensitivity for silicone-rubber-reinforced titanium carbide (TiC) and molybdenum disulfide (MoS_2) is performed for the first time. Moreover, the reinforcing factor, reinforcing efficiency, and their multifunctionalities are reported for silicone rubber with TiC and MoS_2 as they are also not well understood yet. Besides this, the study focuses on harvesting mechanical energy through the piezoelectric principle. The results show a robust improvement in the stretchability of the composites. For example, the stretchability is 125% (control) and increases to 153% (6 phr of TiC) and 165% (6 phr of MoS_2). The electromechanical results show that the output voltage is 3.5 mV (6 phr of TiC) and 6.7 mV (6 phr of MoS_2). Moreover, the theoretical models show that the experimental modulus is in fair agreement with the filled composites up to 6 phr and then deviates. Finally, the biomechanical results show that the output voltage is the highest for thumb pressing. For example, thumb pressing results in an output voltage of ~10 mV (6 phr of TiC) and 5.6 mV (6 phr of MoS_2). Thus, this paper is novel as it describes various aspects that are not understood fully in the literature on TiC- and MoS_2-based silicone rubber composites. Therefore, they are reported in the present work experimentally and validated theoretically.

2. Experimental Section

2.1. Materials and Methods

A transparent room-temperature-vulcanized silicone rubber with the commercial name "KE-441-KT" was used as a rubber matrix. A curing agent with the commercial name "CAT-RM" was used. Both the rubber and curing agents were purchased from the Shin-Etsu Chemical Corporation, Japan. A metal-like ceramic and 3-dimensional (3D) material named "titanium carbide (TiC)" was purchased from Sigma-Aldrich, Saint Louis, MI, USA. Two-dimensional (2D) molybdenum disulfide (MoS_2) with a particle size of ~2 μm and chemical purity of >98% was bought from Sigma-Aldrich, USA. TiC and MoS_2 were used as reinforcing fillers and to improve the electrical conductivity. Finally, an anti-adhesive spray was sprayed on the molds as a mold-releasing agent and was purchased from Nabakem, Pyeongtaek-si, Korea.

2.2. Preparation of Rubber Composites

The preparation of rubber composites is a very important step in influencing the mechanical and electromechanical properties of the composites. This fabrication step was optimized from a previous study and used without further modification [23]. The steps of preparation were as follows.

(1) As detailed in Table 1 below, different amounts of the filler and silicone rubber were taken and mixed for nearly 10 min through the solution mixing technique.

Table 1. Table of formulations.

Sample	Amount of SR (phr)	Amount of TiC (phr)	Amount of MoS_2 (phr)	Amount of Curing Agent (phr)
Control *	100	-	-	2
TiC-based composites	100	2, 4, 6, 8 **	-	2
MoS_2-based composites	100	-	2, 4, 6, 8 **	2

* Here, the control sample refers to the unfilled sample, which did not contain any filler and only a vulcanized silicone rubber matrix was present. ** These values correspond to the composite samples containing 2 phr, 4 phr, 6 phr, and 8 phr of both TiC and MoS_2 as fillers in silicone rubber.

(2) Then, a known amount of curing agent was added and mixed for 1 min. The final composites were placed inside the molds and kept for 24 h for the vulcanization process.

This vulcanization process was performed in a room-temperature vulcanization system at 25 °C.

(3) The final vulcanized composites were taken out of the molds and tested for their properties and application prospects.

2.3. Characterization Techniques

The mechanical property tests were performed under different types of strain, which were the compression and tensile modes. These tests were undertaken under a load cell of 1 kN using a universal testing machine (UTS, Lloyd instruments, Bognor Regis, UK). The static and cyclic compressive tests were performed on cylindrical samples with a diameter of 20 mm and thickness of 10 mm. The compressive tests were performed at a strain rate of 4 mm/min, pre-load of 0.5 N, and up to the maximum compressive strain of 35% for static and 30% for cyclic tests.

Other static tests were performed under uniaxial tensile strain at a pre-load of 0.1 N. The strain rate was 200 mm/min and dumbbell-shaped samples were used with a thickness of 2 mm, gauge length of 20 mm, and width of 4 mm. These mechanical tests were performed following the DIN 53 504 standard [24]. The electromechanical properties were also tested using the UTS machine on cylindrical samples under compressive cyclic deformation. These tests were performed at 10–30% cyclic strain using cylindrical samples. The mechanical and electromechanical long-term stability was tested at 30% compressive cyclic strain using the UTS machine. The output voltage generated through piezoelectricity was recorded using a multimeter (model 34401A, Agilent Technologies Inc., Santa Clara, CA, USA). The assembly of the energy harvesting set-up is provided in Scheme 1. Finally, the biomechanical properties were studied through real-time monitoring using a thumb press, index finger press, and middle finger press. For real-time monitoring, a cylindrical sample was used and the output voltage generated through piezoelectricity was recorded using the multimeter, as detailed above. Finally, the filler dispersion was studied using SEM (model S-4800, Hitachi, Tokyo, Japan). The fractured surface of the composite was sectioned into 0.2-mm-thick samples using a surgical blade. These sections were then mounted on an SEM stub and sputtered with platinum for 2 min before recording the SEM micrographs.

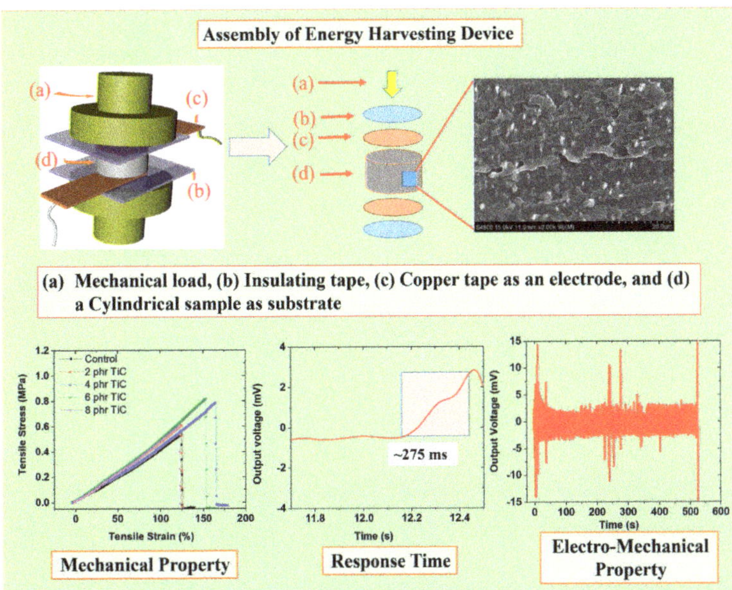

Scheme 1. Assembly of energy harvesting devices and insights into their properties.

3. Results

3.1. Mechanical Properties under Compressive and Tensile Strain

The mechanical properties of composite materials are fundamentally important for various engineering applications. These mechanical properties help us to understand a composite material's behavior under external forces or loads [25]. Moreover, the testing of these properties is very important depending on the nature of the strain under which they are tested. Most often, these strains are compressive or tensile strains, and different mechanical parameters, like the compressive modulus, tensile strength, or elongation at break, are tested [26]. The behavior of the composite materials under a mechanical load is often described through stress–strain curves. Moreover, understanding the mechanical properties is essential in optimizing the type of material, the fabrication process, and the prediction of the overall performance [27]. Thus, the mechanical properties provide the basis for the analysis and design of materials as per the application demands.

Figure 1a,b present the stress–strain curves for the prepared composites. The results show that the stress was higher for increasing strain until the compressive strength. Understanding this behavior involves examining how the material deforms under applied forces or compressive strain. The results show that the TiC and MoS_2 act as reinforcing agents and their addition leads to an increase in the mechanical properties. Here, the rubber matrix is a continuous phase that holds these reinforcing fillers within the composite material. These reinforcing properties of the composites are mainly proposed to be due to filler–rubber interaction [28,29]. Such reinforcing effects of TiC and MoS_2 contribute to improved mechanical properties and thus make them useful for high-load-bearing applications. The main mechanism of these reinforcing properties is to transfer the load from the matrix to the stronger, stiffer reinforcement [30]. These reinforcements are TiC and MoS_2 as the fillers and SR as the rubber matrix. The efficiency of load transfer depends on the interface between the matrix and the reinforcement. The further reinforcement aspects are presented in Figure 1c through the compressive moduli of the composites. All values are reported in Table 2 below. The results show that the reinforcing effects of the MoS_2- and TiC-filled composites are close and nearly the same under compressive strain. However, the control sample without fillers has the lowest compressive modulus. The good compressive moduli of the MoS_2- and TiC-filled composites are due to their improved reinforcing effect in the rubber matrix. These reinforcing effects can help in arresting crack propagation within the matrix [31]. The TiC and MoS_2 particles are proposed to deflect cracks, increasing the composite's toughness to that of the control sample without fillers.

Table 2. Summary of mechanical properties.

Sample Details	Compressive Modulus (MPa)	Tensile Modulus (MPa)	Tensile Strength (MPa)	Fracture Strain (%)
Control	1.55 ± 0.08	0.56 ± 0.03	0.54 ± 0.04	133 ± 7
TiC—2 phr	1.68 ± 0.09	0.66 ± 0.05	0.61 ± 0.05	125 ± 6
TiC—4 phr	1.79 ± 0.08	0.71 ± 0.07	0.78 ± 0.05	165 ± 11
TiC—6 phr	1.95 ± 0.07	0.65 ± 0.05	0.82 ± 0.07	153 ± 9
TiC—8 phr	1.79 ± 0.07	0.61 ± 0.05	0.61 ± 0.06	123 ± 6
MoS_2—2 phr	1.83 ± 0.09	0.53 ± 0.04	0.52 ± 0.04	139 ± 8
MoS_2—4 phr	1.85 ± 0.08	0.68 ± 0.05	0.74 ± 0.07	144 ± 8
MoS_2—6 phr	2.02 ± 0.09	0.66 ± 0.04	0.85 ± 0.09	165 ± 12
MoS_2—8 phr	1.8 ± 0.07	0.68 ± 0.04	0.69 ± 0.07	129 ± 7

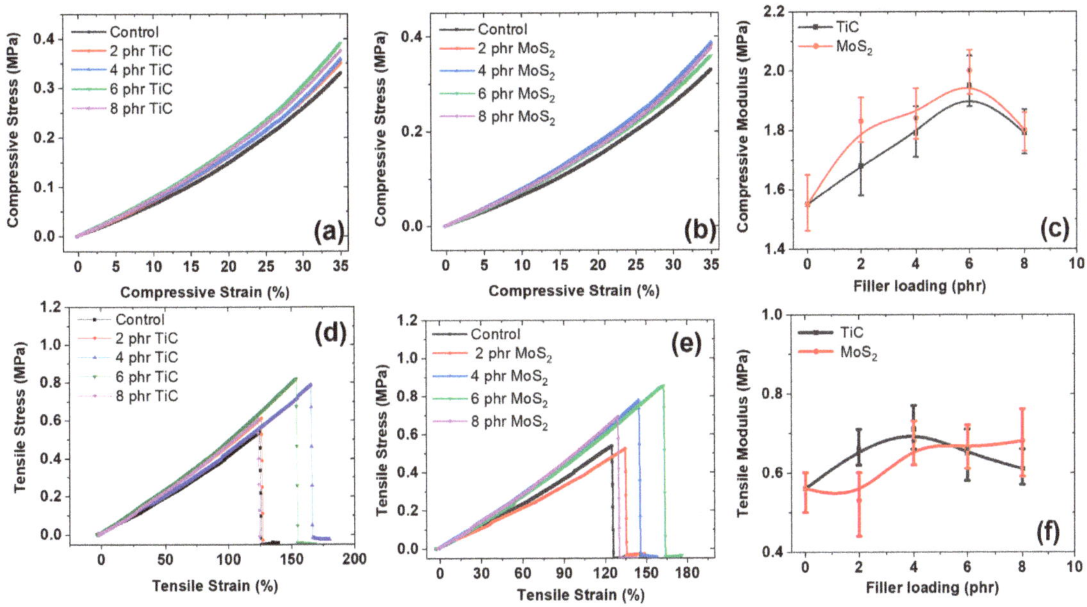

Figure 1. (**a**,**b**) Compressive stress–strain curves of TiC and MoS_2 composites; (**c**) compressive moduli of TiC and MoS_2 composites; (**d**,**e**) tensile stress–strain of TiC and MoS_2 composites; (**f**) tensile moduli of TiC and MoS_2 composites.

The mechanical properties can be further studied through stress–strain curves under tensile strain. Figure 1d,e show that the composites filled with MoS_2 and TiC help in enhancing both the tensile strength and elongation at break as compared to the control sample without fillers. At nearly 6 phr, both fillers exhibit the best tensile strength and elongation at break and fall at 8 phr in the rubber composites. Thus, the inclusion of TiC and MoS_2 significantly increases the mechanical properties. The degree of improvement in the mechanical properties depends on the type, orientation, and volume fraction of the fillers in the rubber matrix [32]. The higher reinforcing effect of MoS_2 and TiC makes the composites stiff, rendering them less prone to deformation under a load. Moreover, these reinforcing agents further improve the ability of a composite to absorb energy and resist fracture, thereby making the composites tough. These tougher composites help to enhance the mechanical properties through mechanisms like crack bridging and deflection [33,34].

The stiff and tougher composites are also examined for their tensile moduli, as presented in Figure 1f. The results agree with those of the compressive modulus, in which the filled composites exhibit higher tensile moduli than the control sample. All values are reported in Table 2 below. Fillers like TiC and MoS_2 can help to redistribute the stress concentrations within the matrix, reducing the likelihood of failure under applied loads. Thus, TiC and MoS_2 are expected to exhibit higher reinforcing effects than the control composite without fillers. However, the tensile modulus falls after 6 phr for TiC and is almost the same for the MoS_2-filled composites. This is expected due to the partial aggregation of TiC, while there is almost a saturation effect with the MoS_2-filled composites. Therefore, the reinforcing properties of composites are integral to their performance and versatility in rubber composites [35]. Here, the reinforcing effect of fillers like TiC and MoS_2 leads to promising mechanical properties. Moreover, understanding and optimizing the interaction between the matrix and reinforcing fillers is important. Thus, engineers can design composites tailored to meet specific demands across various industries [36]. Moreover, under long cycling conditions, some cracks are developed, thereby leading to poorer properties. The mechanism of such crack propagation configurations for these composites is provided in Scheme 2.

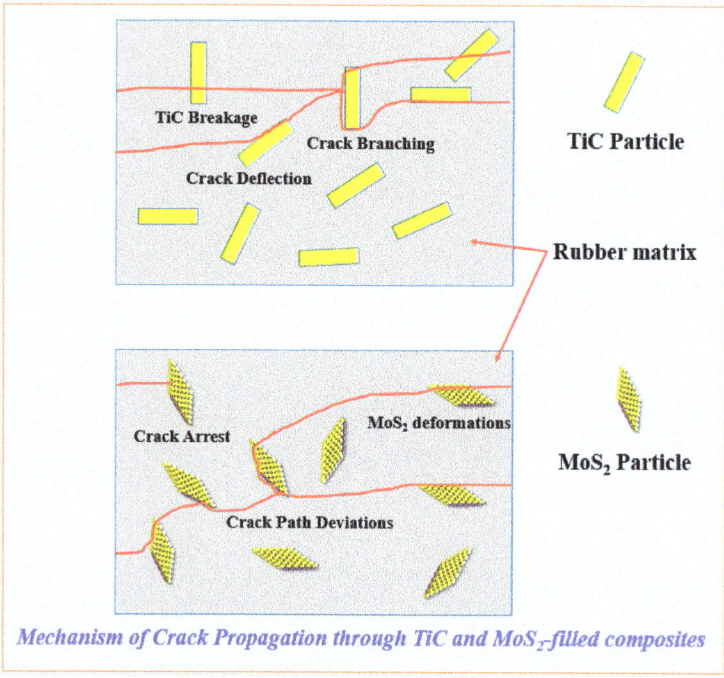

Scheme 2. Crack propagation configurations for different composites.

3.2. Mechanical Properties under Compressive Cyclic Loading

Compressive cyclic loading is very useful in studying fatigue properties and involves applying repeated compressive forces to a composite rubber material. This type of cyclic loading can lead to material fatigue, which is progressive and localized structural damage. The fatigue life is the number of cycles that a material can withstand before failure occurs under cyclic loading [37,38]. It is influenced by several factors, including the magnitude of the stress, the stress range, and the frequency of the cycles. This fatigue behavior generally occurs when a material is subjected to cyclic loading [39]. This type of compressive loading is common in many engineering applications, including structural components and automotive parts, including vibrational loads. Thus, in addition to the static tests described in Figure 1, cyclic loading tests are also important to understand the mechanical stability of the composites. Here, Figure 2a–c show the compressive load at the different strain magnitudes for 6 phr filler loading in the TiC-reinforced composites under compressive cyclic loading. The compressive load increases with increasing compressive strain for both fillers. For example, the compressive load is 40 ± 3 N (10%-TiC), 87 ± 5 N (20%-TiC) and 224 ± 16 N (30%-TiC). Similarly, the compressive load is 36 ± 3 N (10%-MoS_2), 84 ± 6 N (20%-MoS_2) and 170 ± 14 N (30%-MoS_2). This compressive process involves the application of forces or strains that act to compress or shorten a material. This process is opposite to tensile loading, which pulls and stretches the material [40,41].

Moreover, cyclic loading refers to the repeated application and removal of loads over time and it can be symmetric or asymmetric. The results show that the compressive load is higher with increasing strain from 10% to 30%. The high load at higher strain is related to the packing fraction of filler particles in the rubber matrix [42]. For example, the packing is lower at 10%, higher at 20%, and highest for 30% strain. Moreover, a significant increase in compressive load by an order of magnitude was witnessed when the compressive strain increased from 20% to 30%. This can be postulated as the attainment of the optimum packing fraction of filler particles at 30% compressive strain. The results also show a higher compressive load at the first cycle and it then stabilizes in successive cycles. Therefore, the stabilized successive cycles suggest stable mechanical behavior and are beneficial for the durability and fatigue properties under a continuous compressive cyclic process [39,43].

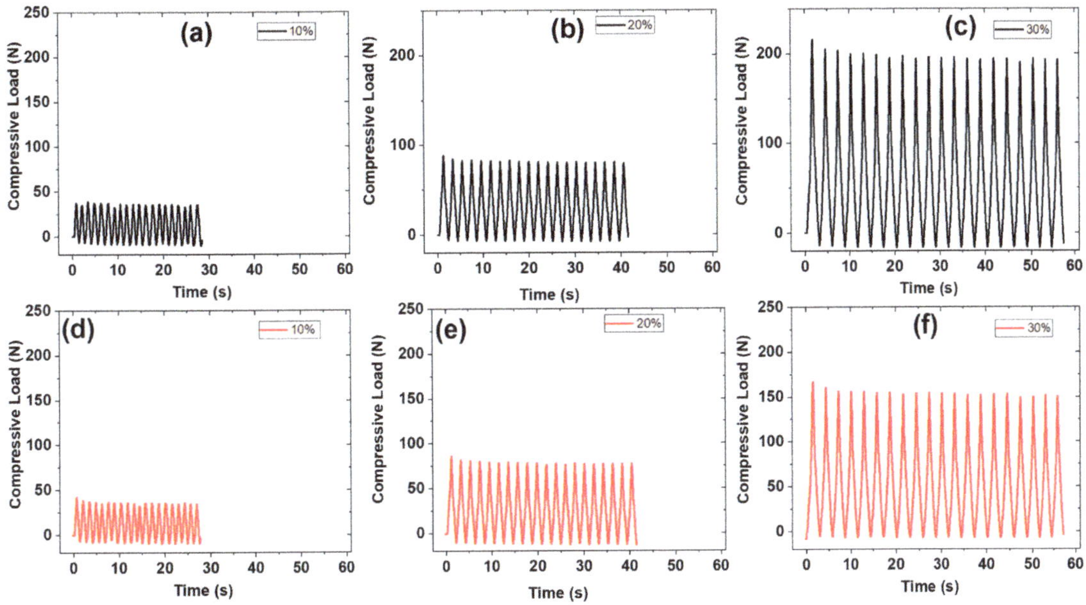

Figure 2. Mechanical properties under compressive cyclic strain at increasing strain from 10 to 30% for (**a**–**c**) TiC-based composites; (**d**–**f**) MoS_2-based composites.

Similarly, Figure 2d–f show the behavior of the compressive load with increasing compressive cyclic strain at 10–30% for the MoS_2-reinforced composites. The results show that the compressive load was higher with increasing compressive strain. These results are proposed to be due to the influence of the packing fractions of filler particles in the rubber composites [42]. Moreover, as discussed above, the compressive load was significantly higher when it increased from 20 to 30% strain. These features are proposed to be due to the favorable packing, which results in a higher compressive load at 30% compressive strain. This behavior is especially important in applications where materials experience fluctuating loads, such as in bridges, buildings, and various mechanical components. Therefore, understanding the mechanical properties under compressive cyclic loading is essential [44,45]. These tests are useful in designing and maintaining components that experience repeated stress. It involves a complex interplay among the material properties, microstructural changes, and load conditions.

3.3. Electromechanical Properties under Compressive Cyclic Loading

As discussed already, compressive cyclic loading involves applying repeated compressive forces to a composite material or device. This type of loading can affect both the mechanical and electrical properties of electromechanical materials. The electromechanical properties in composite materials are related to both electrical and mechanical behaviors [46,47]. These properties are crucial in understanding how these rubber composite materials respond to repeated compressive loads. These electromechanical behaviors refer to the interaction between electrical and mechanical states in a composite material or device. The electroactive materials that exhibit this coupling include piezoelectric materials, electrostrictive materials, and magnetostrictive materials [48,49]. These properties are particularly significant in applications such as sensors, actuators, and various electronic components that experience cyclic mechanical stress.

With the above points in mind, Figure 3a–c present the behavior of the output voltage under increasing compressive cyclic strain from 10% to 30% for the 6 phr TiC. The results show that the output voltage was higher for higher compressive loads up to 20% strain and then decreases at 30% and above. This electromechanical behavior of TiC was slightly different from that of the compressive loads in Figure 2. This difference is proposed to be due to the electrical response, which causes the output voltage to be different from the mechanical behavior [50]. This electrical response is optimum at 20% compressive strain. The testing also involves applying repeated compressive loads to measure changes in the mechanical and electrical properties over time. Thus, both electrical and mechanical stability are important for electromechanical tests. Therefore, understanding the electromechanical properties under compressive cyclic loading is crucial in designing and maintaining reliable electromechanical devices [51,52].

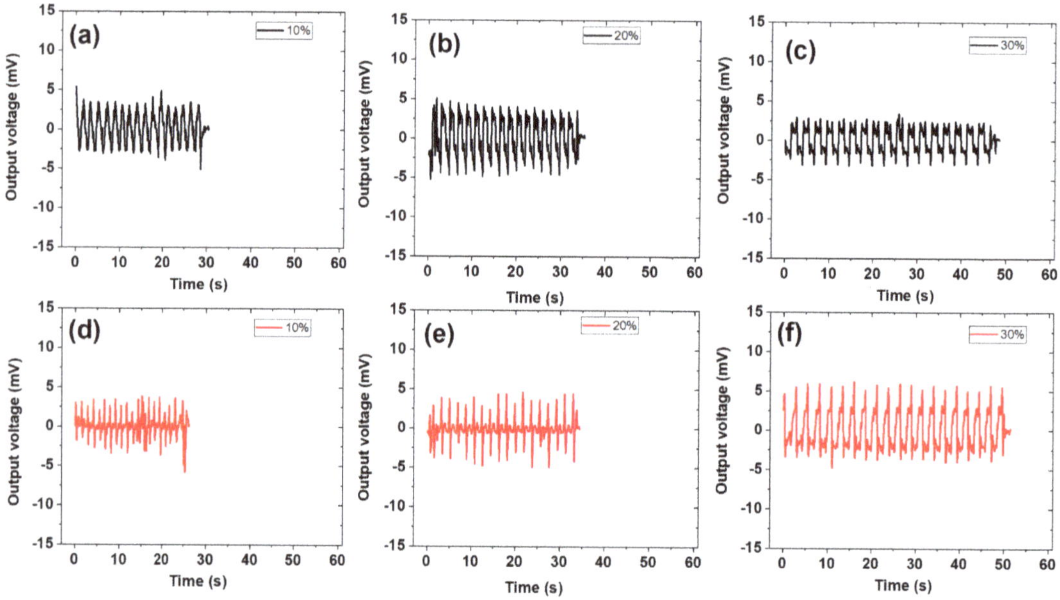

Figure 3. Electromechanical properties under compressive cyclic strain with increasing strain from 10 to 30% for (**a–c**) TiC-based composites; (**d–f**) MoS_2-based composites.

Similarly, Figure 3d–f show the electromechanical behavior for the MoS_2-filled composites at 6 phr. These tests were performed under compressive cyclic loading by increasing the strain from 10 to 30%. Therefore, a higher output voltage at 30% compressive strain was witnessed. These results are proposed to be due to the optimum packing fraction of the MoS_2-filled composite at 30% strain, which results in an improved electrical and mechanical response. The results agree well with Figure 2, in which the mechanical behavior increases with increasing compressive strain. Overall, the electromechanical properties of materials under compressive cyclic loading are essential in achieving good performance that integrates electrical and mechanical properties [53]. The results also show that the selection of the appropriate filler and rubber materials and their optimum processing techniques are critical to obtaining robust electromechanical performance. The usefulness of these studies for various industrial applications extends to piezoelectric sensors, actuators, and energy harvesting devices [54]. The uses of these materials also include medical and sensing monitoring, where they can experience repeating mechanical stress. Finally, piezoelectrical energy harvesting in these composites involves the conversion of a mechanical load into a useful output voltage under compressive cyclic loading [55].

3.4. Response Time of Composites

The addition of fillers in the rubber matrix adds various functionalities to these composite materials. One of the critical functionalities of these rubber composites is their response time [56]. The response time refers to how quickly a material reacts to an applied force, deformation, or external stimulus, which is crucial in applications requiring timely and precise reactions. This response can be apparent in various ways depending on the type of application, such as deformation, vibration damping, or electrical signal generation [57]. The key factors influencing the response time include the material composition, method of fabrication, filler dispersion, and load parameters or type of deformation.

Among them, the type of mechanical response refers to the immediate, reversible deformation to external stimuli. Moreover, filled rubber composites generally exhibit viscoelastic behavior [58]. This behavior results from time-dependent deformation that includes both elastic and viscous components, leading to hysteresis and energy dissipation. However, the response rate studied in this work is related to electromechanical features, i.e., the influence of the output voltage against external stimuli [59]. Therefore, Figure 4a–c show the response times of composites filled with 6 phr of TiC under increasing strain of 10–30%. The results agree with the electromechanical results in Figure 3. For example, the response time was shortest at 20% compressive strain, as expected. This feature is proposed to be due to the optimum packing fraction of the filler particles at this strain [42]. These values further show that our results are consistent with each other. The mechanism behind these electromechanical responses refers to the generation of an electric charge in response to mechanical deformation [59]. Moreover, the process also involves the ability of the composite material to store or dissipate electrical energy in the form of an output voltage. Various factors affect the response time, and the type of filler is one of the most crucial factors [60]. The material type is well known to alter the response time by changing the composite's mechanical and electrical properties.

Figure 4d–f present the response times for the 6 phr of MoS_2-based composites under 10–30%. These results also show the shortest response time at 20% strain. These results could be influenced by the viscoelasticity behavior, which can influence the mechanical response. Moreover, the quality of filler dispersion, which influences the mechanical and electrical response, is important [61]. These factors are influenced by the magnitude of the compressive strain, which correlates with the optimum packing fraction of filler particles at 20% strain. Such features indicate a short response time at 20% strain. It is also proposed that larger or more rapidly applied loads typically result in quicker, more pronounced responses. However, in our case, we test the response time under the same magnetic load under a 1 kN load cell and a similar strain rate. This assists us in understanding the influence of the filler type and strain on the response behavior. The response time is an important factor for various applications [16]. For example, the response time is critical for effective energy dissipation and noise reduction. Moreover, the response time is beneficial for sensor applications, where a rapid electrical response to mechanical stimuli is critical. Therefore, the response time is essential in developing efficient and reliable rubber-composite-based multifunctional products [62].

Experiments were performed to obtain the optimum packing fraction for the filler particles in the composites. The results showed that the optimum packing fraction was 30% for the compressive cyclic mechanical tests, 20% for the response time in the electromechanical tests, and 30% for the electromechanical stability tests. The role of the packing fraction is related to filler networking and percolation. At low filler content and strain, the packing fraction is low, and, at higher filler content, the filler starts to aggregate. This aggregation leads to a reduction in the mechanical and electromechanical properties. However, at certain filler content and strain, the properties are the best, which refers to the optimum packing fraction or filler percolation at this filler content and strain.

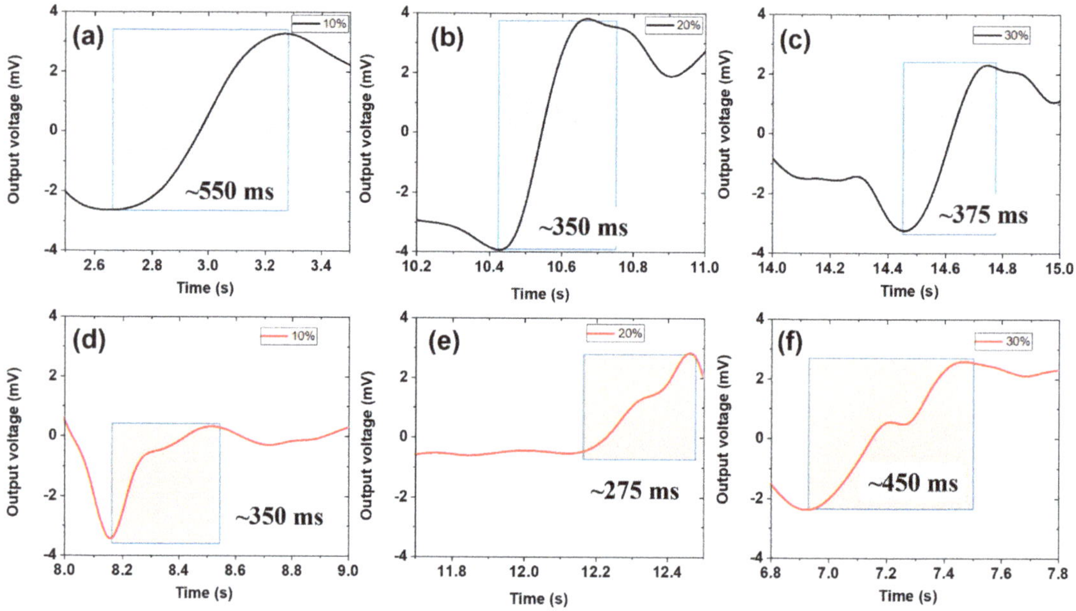

Figure 4. Response times of the composites at 6 phr filler loading, derived from electromechanical properties at different strains from 10 to 30%, for (**a–c**) TiC-based composites; (**d–f**) MoS_2-based composites.

3.5. Mechanical and Electromechanical Stability Tests

Achieving mechanical and electromechanical stability is critical in terms of using the composites in sensors and energy harvesting applications. Such tests help to assess how these rubber composite materials respond to mechanical forces, electrical inputs, and environmental changes over time [63]. These tests ensure that these composite materials can withstand mechanical stresses and strains without failure. These tests are also useful for safety and reliability in various applications, like structural integrity for automotives and as a source of energy harvesting [64]. Moreover, stability tests ensure the consistent performance of these composites under mechanical stress and electrical inputs for crucial precision applications. Most importantly, stability tests ensure the reliability of energy production due to these composites' high dielectric and piezoelectric properties [65].

With the above points in mind, mechanical and electromechanical stability tests were performed and are presented in Figure 5. Cylindrical samples were subjected to 30% compressive strain at 6 phr for the TiC- and MoS_2-based composites. The results show that the mechanical and electromechanical stability is good, and this makes these composites useful for multifunctional applications. The key aspects of mechanical stability and electromechanical stability include fatigue resistance [66]. For example, it ensures the ability to endure repeated loading cycles without significant degradation. The results in Figure 5a–d show that the composites have robust degradation resistance. For example, their initial and final cycles show negligible degradation under continuous compressive cycling [67]. Moreover, the electromechanical properties include the interaction between mechanical deformation and electrical behavior (Figure 5b,d). Various factors influence the fatigue resistance of composites. These factors are the nature of the filler and rubber matrix used, the fabrication method, and the type of strain. Other notable factors are the piezoelectric effect, dielectric properties, and electrical conductivity [68]. Among them, the dielectric properties, such as the ability of a composite material to store and dissipate electric energy, are most critical. Similarly, the mechanical stability is mainly influenced by the filler dispersion and its orientation, the type of polymer matrix, the filler concentration, interfacial bonding, and the method of processing [69]. Almost all of these factors are critical in influencing the mechanical stability. Hence, the favorable mechanical stability depicted in Figure 5a,c shows that the composites prepared in this work have a balance of these factors. Overall, optimizing these properties can be achieved through appropriate material selection, design, and testing. Through these factors, engineers can create advanced composites tailored to meet specific requirements.

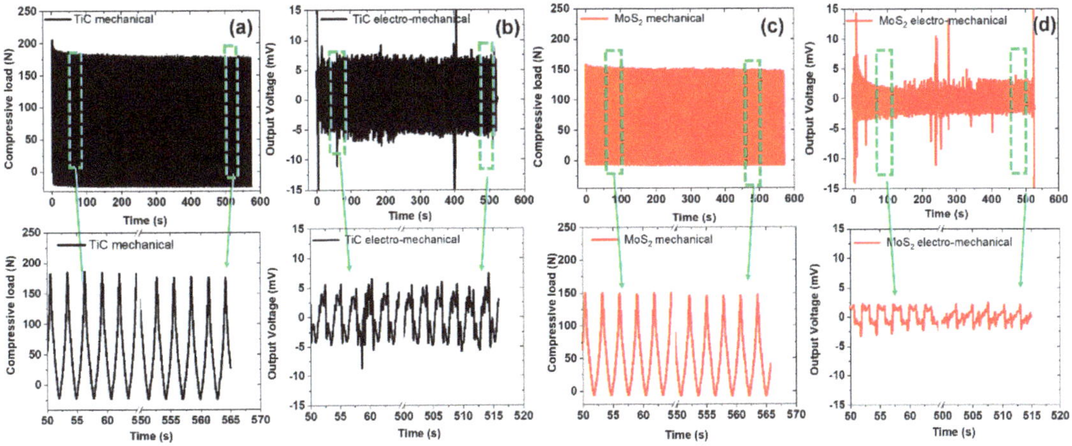

Figure 5. Mechanical and electromechanical stability at 30% compressive cyclic strain for (**a**,**b**) TiC-based composites; (**c**,**d**) MoS_2-based composites.

The main mechanism involved in ensuring mechanical and electromechanical stability is the high fracture toughness of the fillers used, as reported by Alam et al. [70]. Higher fracture toughness impacts the composites by providing high crack resistance, high durability, and safety. Most often, higher crack resistance leads to better mechanical stability. Moreover, this property is essential in maintaining the structural integrity under mechanical loads. In addition, improved fracture toughness allows for the better distribution of mechanical and electrical stresses within the material, reducing the likelihood of localized failure points and enhancing the overall stability. Thus, the composites studied in this work exhibit high mechanical and electromechanical stability, as shown in Figure 5.

3.6. SEM for Study of Filler Dispersion

It is well established that SEM is a powerful tool for the examination of the surface morphology and filler dispersion in composites. The fillers are added to the rubber matrix to improve its mechanical and electrical properties. Moreover, the dispersion of these fillers within the composite matrix significantly affects such properties [61]. Uniform filler dispersion leads to improved performance and consistency. However, the agglomeration or poor dispersion of filler particles can result in weak points and reduced effectiveness [71]. Moreover, SEM helps in assessing the uniformity of dispersion at the micro- and nanoscale, which is vital in achieving the desired electrical and mechanical properties.

With these points in mind, Figure 6a–c provide the SEM micrographs of the control sample without any filler. SEM is performed on these unfilled samples at both low and high resolutions. The study of the microstructure and surface characteristics of unfilled composites is crucial. It helps in understanding their mechanical properties, failure mechanisms, and performance under different conditions [72]. SEM is also used in studying the topological features of composites. SEM shows that the surface topology is smooth and influencing properties such as friction, wear resistance, and adhesion are absent. Moreover, the SEM of the unfilled composites shows the absence of surface defects, cracks, and voids, which might affect the composite material's performance. Similarly, Figure 6d–f show the dispersion of TiC particles at 6 phr in the SR matrix. Generally, the filled composites are engineered to combine the beneficial properties of the matrix material with those of the fillers. The dispersion, distribution, and interaction of the fillers within the matrix critically influence the overall performance of the composite [73]. SEM analysis helps in understanding these factors, which are crucial in optimizing the composite material's properties for specific applications. The low-resolution images show that the TiC particles are uniformly distributed within the rubber matrix. These features result in the improved mechanical and electromechanical properties seen in Figures 1–5 in the above sections. Moreover, in the high-resolution images, the interfacial adhesion between TiC and the rubber matrix was analyzed. The SEM images show good interfacial adhesion and therefore good filler–rubber compatibility. Moreover, there was no evidence of surface defects, cracks, or voids in the filled composites. The SEM images also show that no fracture surfaces were found, providing insights into how TiC influences crack propagation and the failure mechanisms. These features indicate robust resistance against fatigue. These results further support the good mechanical and electromechanical stability of the composites under continuous compressive mechanical deformation, as shown in Figures 3 and 5 above.

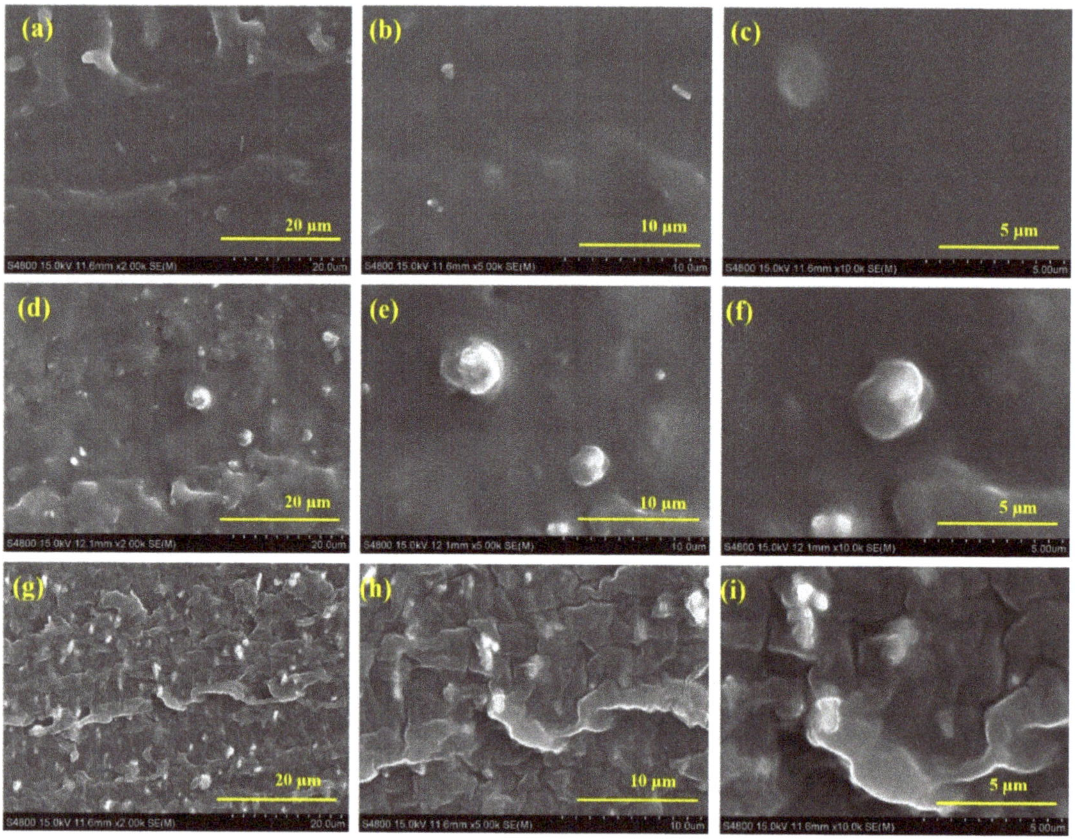

Figure 6. Filler dispersion of different composites at different magnifications: (**a**–**c**) control sample; (**d**–**f**) TiC-based composites, (**g**–**i**) MoS2-based composites.

Similarly, Figure 6g–i show the SEM of the MoS_2-filled composites at 6 phr loading. The results show that the MoS_2 particles are uniformly distributed within the rubber matrix. They therefore support the good properties and stability against fatigue of the tested samples. The results also show how MoS_2 influences the stress within the composite, contributing to the overall toughness or brittleness of the composite [9]. Moreover, the high-resolution image justifies the good interfacial interactions between the MoS_2 particles in the rubber matrix. Moreover, there was no evidence of surface defects, cracks, or voids within the filled composites. Therefore, these results further validate the good filler–rubber compatibility that results in robust performance.

3.7. Prediction of Modulus through Theoretical Modeling

The validation of experimental data through theoretical models is well documented in research. The validation of the modulus of a filled rubber composite is crucial in designing materials with specific mechanical characteristics for diverse applications like energy harvesting. The prediction of the modulus through different models is useful for several reasons [74]. These are as follows: (a) it helps in designing composites with tailored mechanical properties for specific applications of interest, (b) it ensures that the composite will perform as expected under operational conditions, and finally (c) it ensures cost-effectiveness by reducing the need for extensive experimental testing by providing reliable theoretical predictions. Various theoretical models are used to predict the modulus. Among them, Guth-Gold–Smallwood equations and the Halpin–Tsai model are highly useful for filled rubber composites [75,76]. These models strictly depend on the aspect ratio of the filler, the modulus of the control sample, and finally the filler volume fraction in the composites. For example, the equation of the Guth-Gold–Smallwood model [75] is

$$E_{TiC} = E_o\ (1 + 0.67 f_{TiC} \phi_{TiC}) \tag{1}$$

$$EMoS_2 = E_o\ (1 + 0.67 fMoS_2 \phi MoS_2) \tag{2}$$

Similarly, the Halpin–Tsai model [76] that is useful for prediction is

$$E_{TiC} = E_o\ [(1 + 2\ f_{TiC} \phi_{TiC})/(1 - \phi_{TiC}) \tag{3}$$

$$E\ MoS_2 = Eo\ [(1 + 2\ f\ MoS_2 \phi\ MoS_2)/(1 - \phi\ MoS_2) \tag{4}$$

Here, "E" is the predicted modulus of the composite, "E_o" is the modulus of the control sample, "f" is the aspect ratio of the filler, and "ϕ" is the volume fraction of the composite. The results presented in Figure 7a,b show that the models agree well with the literature up to 6 phr and then deviate from the experimental values. This deviation could be due to the partial aggregation of the filler particles in the experimental data. Due to aggregation, the properties fall after 6 phr loading for both the TiC- and MoS_2-filled composites. Several factors influence accurate modulus prediction. These are filler dispersion, the filler morphology, the properties of the rubber matrix, the volume fraction of the fillers, and interfacial bonding [77]. Among them, the filler properties include the modulus, shape, size, and aspect ratio of the filler considered. Similarly, Figure 7c,d show that the model agrees well up to 6 phr for TiC and 8 phr for the MoS_2-filled composites. The results indicate that the experimental data are influenced by uniform filler dispersion up to 6 phr for TiC and then slight aggregation occurs at 8 phr for the TiC-filled composite. Moreover, the MoS_2-filled composites show fair agreement with the predicted models until 8 phr loading. However, the model assumes uniform filler dispersion and perfect interfacial bonding.

These features are difficult to achieve experimentally at higher loadings like 8 phr of TiC, which results in deviations in the experiments from the predicted models. Filler dispersion requires uniform filler distribution with minimum aggregates to achieve good predictions. Finally, the strong interfacial bonding between the matrix and fillers ensures efficient stress transfer, enhancing the composite's modulus [78]. Overall, predicting the moduli of rubber composites through theoretical modeling is essential in designing materials with the desired mechanical properties. By understanding and applying various models, researchers and engineers can develop composites that meet specific performance requirements.

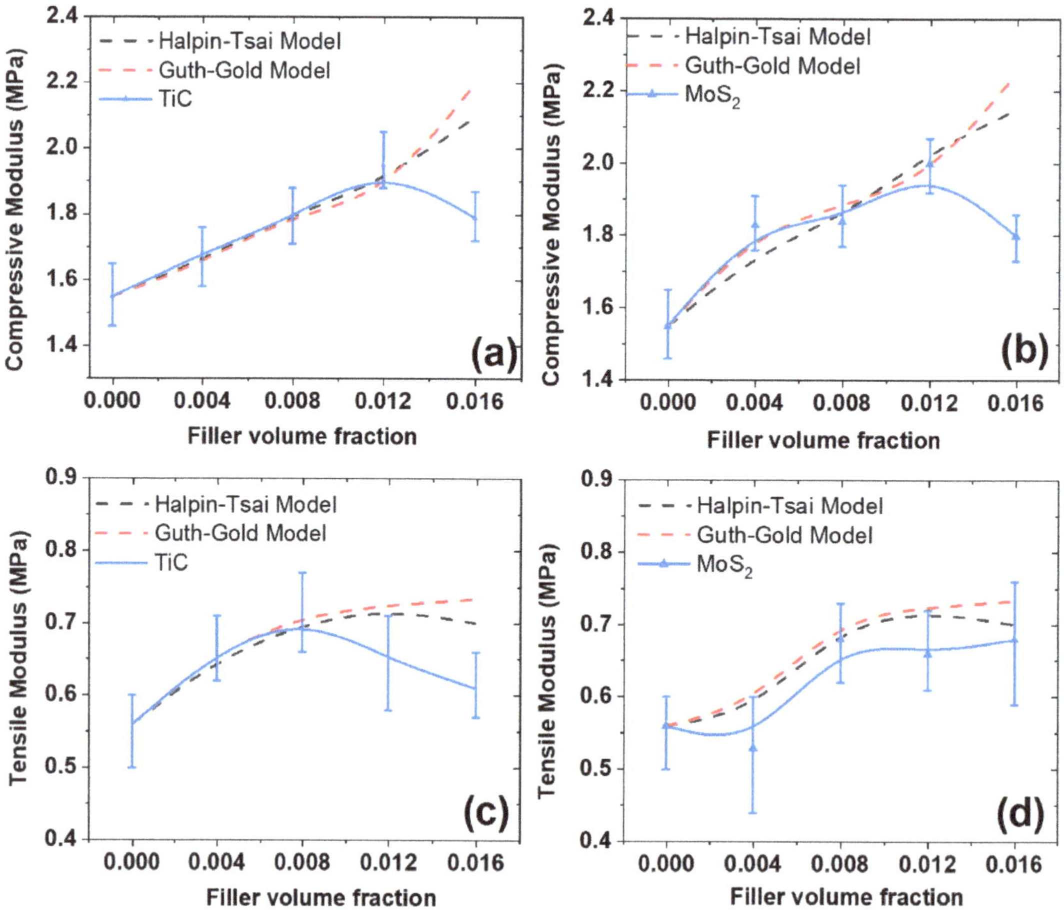

Figure 7. Theoretical models for prediction of composites: (**a**,**c**) TiC-based composites; (**b**–**d**) MoS_2-based composites.

3.8. Calculation of Reinforcing Factor and Reinforcing Efficiency

The reinforcing factor and reinforcing efficiency measure the increase in the stiffness or modulus due to the addition of fillers in rubber composites. The reinforcing factor is typically expressed as the ratio of the composite's modulus to that of the pure rubber matrix [79]. For example,

$$\text{R.F.} = \frac{\text{EF}}{\text{Eo}} \tag{5}$$

Here, R.F. is the reinforcing factor, E_F is the modulus of the filled composite, and E_o is the modulus of the unfilled rubber. Figure 8a,b show the behavior of the R.F. of the TiC and MoS_2-based composites. The results indicate that the R.F. increases with increasing filler content until 6 phr and then falls. The increasing R.F. indicates that the addition of the filler provides reinforcement in the composite. The fall in the R.F. after 6 phr could be due to the partial aggregation of the filler particles in the composite. Various factors affect the R.F. in composites, such as the type of filler and rubber matrix and morphological features like the shape, size, and aspect ratio of the filler [28]. Filler dispersion is also proposed to influence the R.F. of composites. The main mechanism behind the reinforcing effect of fillers is filler–rubber bonding, which enhances the stress transfer from the rubber matrix to the fillers. The calculation of the R.F. is useful as it helps in designing composites with tailored mechanical properties for specific applications of interest. However, there are some challenges [80], such as (a) developing more accurate models to predict the reinforcing factors considering complex interactions between the fillers and the matrix; (b) exploring new types of fillers and their hybrid systems to obtain high reinforcement; (c) obtaining sustainable and eco-friendly fillers that reduce the environmental impact without compromising the performance; and (d) developing a new method of fabrication to obtain better filler dispersion and improved interfacial adhesion and compatibility between the filler particles and rubber matrix.

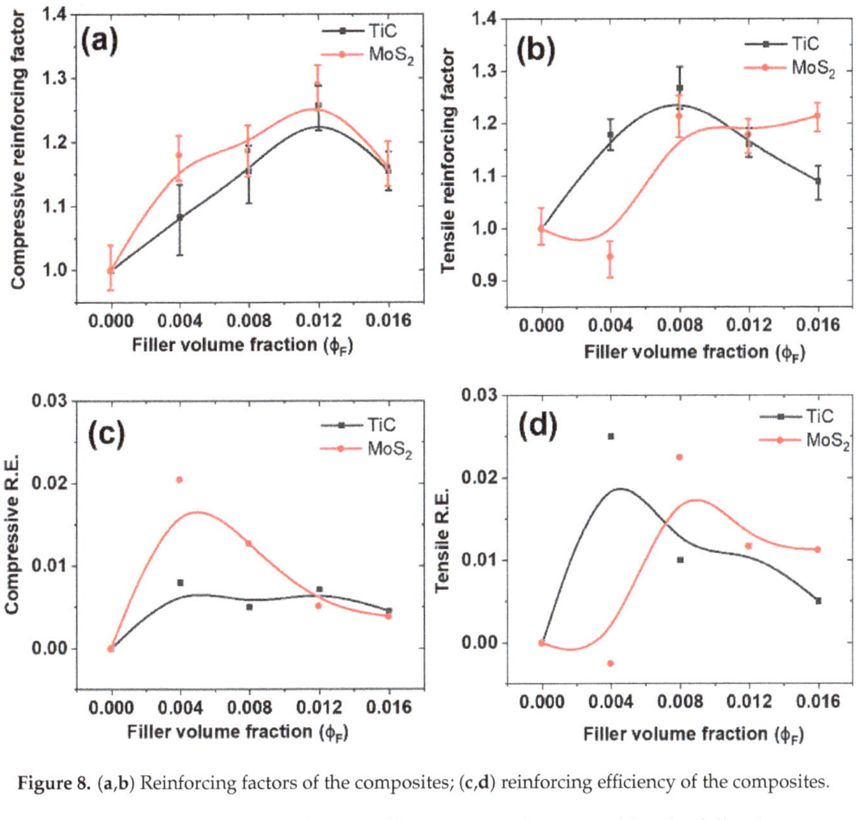

Figure 8. (**a**,**b**) Reinforcing factors of the composites; (**c**,**d**) reinforcing efficiency of the composites.

In the same way, the reinforcing efficiency was determined by the following equations [81]:

$$\text{R.E.at compressive strain} = \frac{\sigma(30\%)\text{filled} - \sigma(30\%)\text{unfilled}}{\text{wt\% of filler}} \tag{6}$$

$$\text{R.E.at tensile strain} = \frac{\sigma(100\%)\text{filled} - \sigma(100\%)\text{unfilled}}{\text{wt\% of filler}} \tag{7}$$

Here, R.E. is the reinforcing efficiency, σ (30%) is the stress of the filled composite at 30% compressive strain, σ (100%) is the stress of the filled composite at 100% tensile strain, and wt% is the weight percentage of the filled rubber. Hence, Figure 8c,d shows the R.E. of the filler particles in the composite. The results indicate that the R.E. depends upon the type of filler and the filler loading in the composites. For example, a higher R.E. indicates that the fillers are more effective in enhancing the modulus of the composite. Generally, a filler with a high reinforcing effect exhibits a higher R.E. Moreover, increasing filler content generally degrades the R.E. This is proposed to be due to the inverse proportionality of the filler content to the R.E. The R.E. quantifies the effectiveness of fillers in enhancing the mechanical properties of a composite [82]. As stated, it considers both the volume fraction of fillers and the resulting improvement in the modulus. Moreover, the R.E. provides insights into how well the fillers contribute to the composite's stiffness relative to their concentration. Therefore, the R.E. is a vital metric in evaluating and optimizing the performance of rubber composites. By understanding and calculating these parameters, we can design composites with enhanced mechanical properties tailored to specific applications.

3.9. Real-Time Monitoring of Human Motion

The real-time monitoring of human motion involves monitoring mechanical deformation through employing smart technology. When coupled with energy harvesting technologies, this monitoring can capture and convert kinetic energy from human activity into electrical energy. This technology can be applied in various fields, such as sports training, physical therapy, or fitness practices [83]. The main aim of such technology is to enhance performance, prevent injuries before they occur, and promote correct movement patterns.

The various applications for real-time monitoring technology include sports monitoring, health monitoring, fitness tracking, and physical therapy. With the above aspects in mind, Figure 9a–f present the behavior of energy generation through different types of human motion, like thumb pressing and finger pressing. The results show that thumb pressing is more efficient than finger pressing for all composites. This is due to the large contact area of the thumb, which provides a larger magnitude of strain than finger pressing [84].

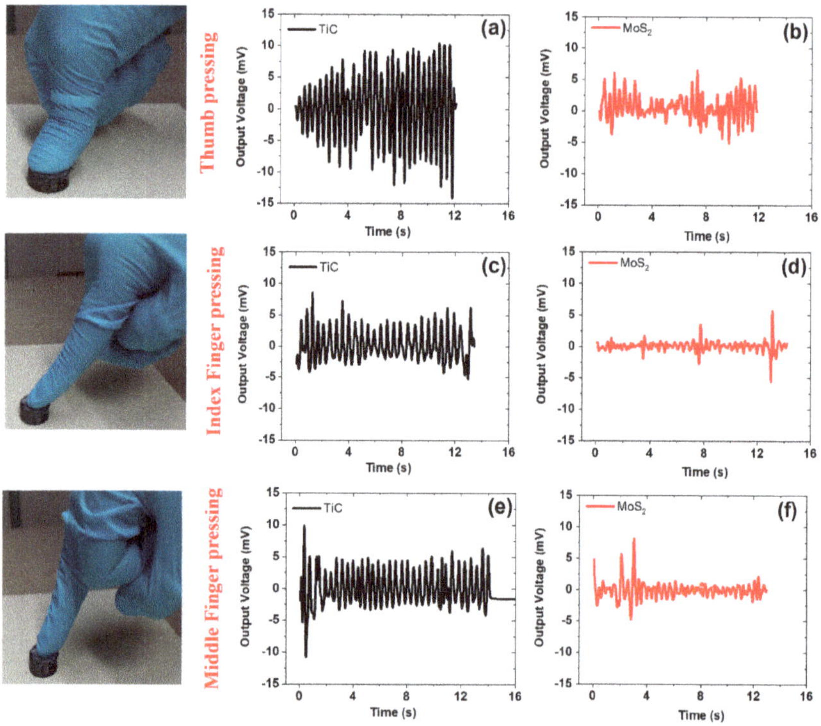

Figure 9. Real-time monitoring of the composites: (**a**,**b**) thumb pressing; (**c**,**d**) index finger pressing; (**e**,**f**) middle finger pressing.

Moreover, the TiC-based composites exhibit a higher output voltage than the MoS_2-filled composites. These results agree with the electromechanical properties in Figures 3 and 5 above. The main mechanism behind this technology involves piezoelectricity. For example, these composite materials generate an electrical charge in response to mechanical stress. Common applications include smart wristbands, thumb pressing, smart rings, and smart textiles. This technology has wide-ranging usefulness. For example, it reduces the dependence on batteries and external power sources, promoting sustainable energy solutions. Moreover, it enables continuous health monitoring, thereby providing valuable data for medical diagnosis and fitness tracking [85]. However, there are some challenges with this technology, such as poor efficiency, data accuracy, and durability. Overall, this technology represents a promising route for wearable technology and sustainable energy solutions. By capturing and converting kinetic energy from daily activities, these systems can power a wide range of devices, enhancing their functionality and sustainability [86].

4. Conclusions

Composites based on TiC, MoS_2, and silicone rubber were successfully fabricated through solution mixing and their multifunctional aspects are detailed below.

(1) TiC and MoS_2 act as reinforcing agents, thereby improving the mechanical properties of the rubber composites. For example, the compressive modulus was 1.55 MPa (control) and increased to 1.95 MPa at 6 phr of TiC and 2.04 MPa at 6 phr of MoS_2. In the same way, the stretchability was 125% (control) and increased to 153% at 6 phr of TiC and 165% at 6 phr of MoS_2.

(2) The electromechanical properties were also greatly improved when using these composites. For example, at 30% strain, the output voltage was 3.5 mV at 6 phr of TiC and 6.7 mV at 6 phr of MoS_2.

(3) The models employed in the present work agree well until 6 phr and then deviate from the experimental values. This deviation could be due to the partial aggregation of the filler particles in the experimental data. The reinforcing factor and reinforcing efficiency show that TiC and MoS_2 as fillers exhibit outstanding reinforcing effects on the silicone rubber matrix, thereby improving the mechanical properties of the composites.

(4) Real-time monitoring reveals the biomechanical properties. The tests show that the output voltage is the highest for thumb pressing. For example, thumb pressing results in an output voltage of ~10 mV at 6 phr of TiC and 5.6 mV at 6 phr of MoS_2.

Author Contributions: Conceptualization, V.K. and M.N.A.; methodology, V.K.; software, V.K.; validation, V.K., M.N.A., M.A.Y. and S.-S.P.; formal analysis, V.K.; investigation, V.K.; resources, S.-S.P.; data curation, V.K.; writing—original draft preparation, V.K.; writing—review and editing, M.N.A., M.A.Y. and S.-S.P.; visualization, M.N.A., M.A.Y. and S.-S.P.; supervision, S.-S.P.; project administration, S.-S.P.; funding acquisition, not applicable. All authors have read and agreed to the published version of the manuscript.

Funding: This research received no external funding.

Institutional Review Board Statement: Not applicable.

Data Availability Statement: The original contributions presented in the study are included in the article, further inquiries can be directed to the corresponding author.

Conflicts of Interest: The authors declare no conflicts of interest.

References

1. Strommer, B.; Battig, A.; Frasca, D.; Schulze, D.; Huth, C.; Bohning, M.; Schartel, B. Multifunctional property improvements by combining graphene and conventional fillers in chlorosulfonated polyethylene rub er composites. *ACS Appl. Polym. Mater.* **2022**, *4*, 1021–1034. [CrossRef]
2. Rakesh, M.; Babu, B.R.N.; Prakash, A.P.G.; Prema, N.S.; Gowda, A.C.; Madhukar, B.S.; Kashimatt, M.G.V.; Pradeep, T.M.; Kumar, B.V.S.; Madhusudan, P. Fabrication of lead zirconate titanate-based polyvinylidene fluoride polymer nano-composites: Microcrystalline, morphological and electrical studies. *J. Mater. Sci. Mater. Electron.* **2023**, *34*, 372. [CrossRef]

3. Zhang, C.; Sun, H.; Zhu, Q. Preparation and property enhancement of poly (Vinylidene Fluoride)(PVDF)/lead zirconate titanate (PZT) composite piezoelectric films. *J. Electron. Mater.* **2021**, *50*, 6426–6437. [CrossRef]
4. Li, T.; Lee, P.S. Piezoelectric energy harvesting technology: From materials, structures, to applications. *Small Struct.* **2022**, *3*, 2100128. [CrossRef]
5. Park, D.; Hong, J.-H.; Choi, D.; Kim, D.; Jung, W.H.; Yoon, S.S.; Kim, K.H.; An, S. Biocompatible and mechanically-reinforced tribopositive nanofiber mat for wearable and antifungal human kinetic-energy harvester based on wood-derived natural product. *Nano Energy* **2022**, *96*, 107091. [CrossRef]
6. Manikkavel, A.; Kumar, V.; Park, S.S. Multifunctionality of MWCNT and TiC hybrid filler silicone composite for energy harvesting and strain sensing. *Polym. Compos.* **2024**, *45*, 6629–6643. [CrossRef]
7. Kumar, V.; Alam, M.N.; Yewale, M.A.; Lee, D.J.; Park, S.S. Mimicking Self-Powered Piezoelectric Energy-Generating Behavior in Silicone Rubber Composites under Compressive and Tensile Strains. *ACS Appl. Electron. Mater.* **2024**, *6*, 1638–1650. [CrossRef]
8. Kathavate, V.S.; Prasad, K.E.; Kiran, M.S.; Zhu, Y. Mechanical characterization of piezoelectric materials: A perspective on deformation behavior across different microstructural length scales. *J. Appl. Phys.* **2022**, *132*, 121103. [CrossRef]
9. Irez, A.B.; Bayraktar, E.; Miskioglu, I. Fracture toughness analysis of epoxy-recycled rubber-based composite reinforced with graphene nanoplatelets for structural applications in automotive and aeronautics. *Polymers* **2020**, *12*, 448. [CrossRef]
10. Fang, S.; Li, F.; Liu, J.; Zhang, L.; Wang, D.; Liu, B.; Wu, S.; Tang, Z.; Guo, B. Rubber-reinforced rubbers toward the combination of high reinforcement and low energy loss. *Nano Energy* **2021**, *83*, 105822. [CrossRef]
11. Zhang, X.; Cai, L.; He, A.; Ma, H.; Li, Y.; Hu, Y.; Zhang, X.; Liu, L. Facile strategies for green tire tread with enhanced filler-matrix interfacial interactions and dynamic mechanical properties. *Compos. Sci. Technol.* **2021**, *203*, 108601. [CrossRef]
12. Ruckdashel, R.R.; Khadse, N.; Park, J.H. Smart E-textiles: Overview of components and outlook. *Sensors* **2022**, *22*, 6055. [CrossRef] [PubMed]
13. Ramegowda, P.C.; Ishihara, D.; Takata, R.; Niho, T.; Horie, T. Finite element analysis of a thin piezoelectric bimorph with a metal shim using solid direct-piezoelectric and shell inverse-piezoelectric coupling with pseudo direct-piezoelectric evaluation. *Compos. Struct.* **2020**, *245*, 112284. [CrossRef]
14. Li, M.; Chen, L.; Li, Y.; Dai, X.; Jin, Z.; Zhang, Y.; Feng, W.; Yan, L.-T.; Cao, Y.; Wang, C. Superstretchable, yet stiff, fatigue-resistant ligament-like elastomers. *Nat. Commun.* **2022**, *13*, 2279. [CrossRef] [PubMed]
15. Chen, L.; Liu, H.; Qi, H.; Chen, J. High-electromechanical performance for high-power piezoelectric applications: Fundamental, progress, and perspective. *Prog. Mater. Sci.* **2022**, *127*, 100944. [CrossRef]
16. Chang, B.P.; Gupta, A.; Muthuraj, R.; Mekonnen, T.H. Bioresourced fillers for rubber composite sustainability: Current development and future opportunities. *Green Chem.* **2021**, *23*, 5337–5378. [CrossRef]
17. Qin, H.; Lu, H.H.; Shen, X.Q.; Xin, Z.J.; Yang, B. Design, preparation and electromechanical characteristics analysis of piezoelectric 1-3-type composites with sandwich polymer structures. *Sens. Actuators A Phys.* **2024**, *366*, 115024. [CrossRef]
18. Javadinasab Hormozabad, S.; Gutierrez Soto, M.; Adeli, H. Integrating structural control, health monitoring, and energy harvesting for smart cities. *Expert Syst.* **2021**, *38*, e12845. [CrossRef]
19. Liu, Y.; Zhao, L.; Wang, L.; Zheng, H.; Li, D.; Avila, R.; Lai, K.W.C.; Wang, Z.; Xie, Z.; Zi, Y.; et al. Skin-Integrated Graphene-Embedded Lead Zirconate Titanate Rubber for Energy Harvesting and Mechanical Sensing. *Adv. Mater. Technol.* **2019**, *4*, 1900744. [CrossRef]
20. Kumar, V.; Kumar, A.; Wu, R.-R.; Lee, D.-J. Room-temperature vulcanized silicone rubber/barium titanate–based high-performance nanocomposite for energy harvesting. *Mater. Today Chem.* **2020**, *16*, 100232. [CrossRef]
21. Deng, H.-T.; Zhang, X.-R.; Wang, Z.-Y.; Wen, D.-L.; Ba, Y.-Y.; Kim, B.; Han, M.-D.; Zhang, H.-X.; Zhang, X.-S. Super-stretchable multi-sensing triboelectric nanogenerator based on liquid conductive composite. *Nano Energy* **2021**, *83*, 105823. [CrossRef]
22. KChung, Y.; Xu, B.; Li, Z.; Liu, Y.; Han, J. Bioinspired ultra-stretchable dual-carbon conductive functional polymer fiber materials for health monitoring, energy harvesting and self-powered sensing. *Chem. Eng. J.* **2023**, *454*, 140384. [CrossRef]
23. Kumar, V.; Kumar, A.; Chhatra, R.K.; Le, D.J. Studies on high performance rubber composites by incorporating titanium dioxide particles with different surface area and particle size. *Nanofabrication* **2022**, *7*, 104–115. [CrossRef]
24. *DIN 53 504*; Testing of Rubber—Determination of Tensile Strength at Break, Tensile Stress at Yield, Elongation at Break and Stress Values in a Tensile Test. Deutsches Institut Fur Normung E.V.: Berlin, Germany, 2017.
25. Mazzanti, V.; Malagutti, L.; Santoni, A.; Sbardella, F.; Calzolari, A.; Sarasini, F.; Mollica, F. Correlation between mechanical properties and processing conditions in rubber-toughened wood polymer composites. *Polymers* **2020**, *12*, 1170. [CrossRef]
26. Han, R.; Wu, Y.; Quan, X.; Niu, K. Effects of crosslinking densities on mechanical properties of nitrile rubber composites in thermal oxidative aging environment. *J. Appl. Polym. Sci.* **2020**, *137*, 49076. [CrossRef]
27. Roy, K.; Debnath, S.C.; Pongwisuthiruchte, A.; Potiyaraj, P. Recent advances of natural fibers based green rubber composites: Properties, current status, and future perspectives. *J. Appl. Polym. Sci.* **2021**, *138*, 50866. [CrossRef]
28. Roshanaei, H.; Khodkar, F.; Alimardani, M. Contribution of filler–filler interaction and filler aspect ratio in rubber reinforcement by silica and mica. *Iran. Polym. J.* **2020**, *29*, 901–909. [CrossRef]
29. Alam, M.N.; Kumar, V.; Jo, C.R.; Ryu, S.R.; Lee, D.J.; Park, S.S. Mechanical and magneto-mechanical properties of styrene-butadiene-rubber-based magnetorheological elastomers conferred by novel filler-polymer interactions. *Compos. Sci. Technol.* **2022**, *229*, 109669. [CrossRef]

30. Maurya, P.; Kota, N.; Gibmeier, J.; Wanner, A.; Roy, S. Review on study of internal load transfer in metal matrix composites using diffraction techniques. *Mater. Sci. Eng. A* **2022**, *840*, 142973. [CrossRef]
31. Tunnicliffe, L.B. Fatigue crack growth behavior of carbon black–reinforced natural rubber. *Rubber Chem. Technol.* **2021**, *94*, 494–514. [CrossRef]
32. Bokobza, L. Elastomer Nanocomposites: Effect of Filler–Matrix and Filler–Filler Interactions. *Polymers* **2023**, *15*, 2900. [CrossRef]
33. Fallahi, H.; Kaynan, O.; Asadi, A. Insights into the effect of fiber–matrix interphase physiochemical-mechanical properties on delamination resistance and fracture toughness of hybrid composites. *Compos. Part A Appl. Sci. Manuf.* **2023**, *166*, 107390. [CrossRef]
34. Białkowska, A.; Bakar, M.; Kucharczyk, W.; Zarzyka, I. Hybrid epoxy nanocomposites: Improvement in mechanical properties and toughening mechanisms—A review. *Polymers* **2023**, *15*, 1398. [CrossRef]
35. Rajak, D.K.; Pagar, D.D.; Menezes, P.L.; Linul, E. Fiber-reinforced polymer composites: Manufacturing, properties, and applications. *Polymers* **2019**, *11*, 1667. [CrossRef]
36. Ramesh, M.; Rajeshkumar, L.N.; Srinivasan, N.; Kumar, D.V.; Balaji, D. Influence of filler material on properties of fiber-reinforced polymer composites: A review. *e-Polymers* **2022**, *22*, 898–916. [CrossRef]
37. Pan, Z.; Lai, Y.; Wang, Y.; Duan, W.; Qiao, Y.; Liu, Y.; Song, C. Fatigue life prediction and effects of cerium oxide-filled vulcanized natural rubber on fatigue life under multiaxial loading. *Fatigue Fract. Eng. Mater. Struct.* **2021**, *44*, 3349–3362. [CrossRef]
38. Parente, J.M.; Santos, P.; Valvez, S.; Silva, M.P.; Reis, P.N.B. Fatigue behaviour of graphene composites: An overview. *Procedia Struct. Integr.* **2020**, *25*, 282–293. [CrossRef]
39. Kuteneva, S.V.; Gladkovsky, S.V.; Vichuzhanin, D.I.; Nedzvetsky, P.D. Microstructure and properties of layered metal/rubber composites subjected to cyclic and impact loading. *Compos. Struct.* **2022**, *285*, 115078. [CrossRef]
40. Manikkavel, A.; Kumar, V.; Alam, M.N.; Kim, U.; Park, S.S. The electro-mechanical energy harvesting configurations in different modes from machine to self-powered wearable electronics. *ACS Appl. Electron. Mater.* **2023**, *5*, 5537–5554. [CrossRef]
41. Kumar, V.; Lee, G.; Choi, J.; Lee, D.J. Studies on composites based on HTV and RTV silicone rubber and carbon nanotubes for sensors and actuators. *Polymer* **2020**, *190*, 122221. [CrossRef]
42. Sriring, M.; Nimpaiboon, A.; Kumarn, S.; Higaki, K.; Higaki, Y.; Kojio, K.; Takahara, A.; Ho, C.C.; Sakdapipanich, J. Film formation process of natural rubber latex particles: Roles of the particle size and distribution of non-rubber species on film microstructure. *Colloids Surf. A Physicochem. Eng. Asp.* **2020**, *592*, 124571. [CrossRef]
43. Yang, L.; Wang, L.; Guo, H.; Du, A. Compressive fatigue behavior of gum and filled sbr vulcanizates. *Polymers* **2021**, *13*, 1497. [CrossRef] [PubMed]
44. Poddaeva, O.; Fedosova, A. Damping capacity of materials and its effect on the dynamic behavior of structures. Review. *Energy Rep.* **2021**, *7*, 299–307. [CrossRef]
45. Zhou, Y.; Zhang, S.; Liu, X.; Yang, J. Monitoring bridge bearings based on rubber-based composite materials. *Mater. Express* **2024**, *14*, 869–876. [CrossRef]
46. Kumar, A.; Ahmad, D.; Patra, K.; Hossain, M. Enhancement of electromechanical properties of natural rubber by adding barium titanate filler: An electro-mechanical study. *J. Appl. Polym. Sci.* **2021**, *138*, 50991. [CrossRef]
47. Yang, D.; Yu, L.; Liang, Y.; Wei, Q.; Ni, Y.; Zhang, L. Enhanced electromechanical performance of natural rubber composites via constructing strawberry-like dielectric nanoparticles. *ACS Appl. Polym. Mater.* **2020**, *2*, 5621–5629. [CrossRef]
48. Xu, X. Intelligent composite materials for use as sensors and actuators. In *Composite Materials*; Elsevier: Amsterdam, The Netherlands, 2021; pp. 465–487.
49. Bocchetta, G.; Fiori, G.; Sciuto, S.A.; Scorza, A. Performance of smart materials-based instrumentation for force measurements in biomedical applications: A methodological review. *Actuators* **2023**, *12*, 261. [CrossRef]
50. Huang, Z.; Yuan, Z.; Sun, G.; Sun, Y.; Hu, J.; He, J. Thickness effect on threshold electric field of ZnO microvaristors/silicone rubber composite with nonlinear conductivity. *Compos. Part A Appl. Sci. Manuf.* **2022**, *158*, 106969. [CrossRef]
51. Azam, S.; Kumar, V.; Park, S.S. Electro-bio-mechanical behavior of graphite nanoplatelets–silicone rubber-based composites for wearable electronic systems. *J. Appl. Polym. Sci.* **2024**, *141*, e55387. [CrossRef]
52. Dubey, K.A.; Mondal, R.K.; Bhardwaj, Y.K. Graphene assisted enhancement in the cyclic electromechanical properties of polyolefin based multiphasic conducting nano carbon black nanocomposites. *Radiat. Phys. Chem.* **2024**, *214*, 111308. [CrossRef]
53. Kumar, V.; Alam, M.N.; Yewale, M.A.; Park, S.S. Modulating the Configurations of "Gel-Type" Soft Silicone Rubber for Electro-Mechanical Energy Generation Behavior in Wearable Electronics. *Gels* **2023**, *9*, 686. [CrossRef] [PubMed]
54. Zabek, D.; Pullins, R.; Pearson, M.; Grzebielec, A.; Skoczkowski, T. Piezoelectric-silicone structure for vibration energy harvesting: Experimental testing and modelling. *Smart Mater. Struct.* **2021**, *30*, 035002. [CrossRef]
55. Chen, W.; Zheng, Q.; Lv, Y.A.; Chen, Y.; Fan, Q.; Zhou, X.; Li, H.; Yu, Q.; Liu, H. Piezoelectric energy harvesting and dissipating behaviors of polymer-based piezoelectric composites for nanogenerators and dampers. *Chem. Eng. J.* **2023**, *465*, 142755. [CrossRef]
56. Kurian, A.S.; Souri, H.; Mohan, V.B.; Bhattacharyya, D. Highly stretchable strain sensors based on polypyrrole-silicone rubber composites for human motion detection. *Sens. Actuators A Phys.* **2020**, *312*, 112131. [CrossRef]
57. Dolui, T.; Natarajan, T.S.; Aiswarya, S.; Chanda, J.; Ghosh, P.; Mukhopadhyay, R.; Wießner, S.; Heinrich, G.; Das, A.; Banerjee, S.S. Stimuli–responsive Mechanoadaptive elastomeric composite materials: Challenges, opportunities, and new approaches. *Adv. Eng. Mater.* **2023**, *25*, 2300584. [CrossRef]

58. Wang, L.; Du, M.; Shan, G.; Lu, Q.; Zuo, M.; Song, Y.; Zheng, Q. Nonlinear viscoelastic behavior and their time-temperature superposition for filled styrene butadiene rubber compounds and vulcanizates. *Compos. Sci. Technol.* **2022**, *230*, 109739. [CrossRef]
59. Wan, B.; Yang, Y.; Guo, R.; Fan, Z.; Deng, P.; Zhang, S. Effect of Vulcanization on the Electro-Mechanical Sensing Characteristics of Multi-Walled Carbon Nanotube/Silicone Rubber Composites. *Polymers* **2023**, *15*, 1412. [CrossRef] [PubMed]
60. Krainoi, A.; Kummerlöwe, C.; Nakaramontri, Y.; Wisunthorn, S.; Vennemann, N.; Pichaiyut, S.; Kiatkamjornwong, S.; Nakason, C. Novel natural rubber composites based on silver nanoparticles and carbon nanotubes hybrid filler. *Polym. Compos.* **2020**, *41*, 443–458. [CrossRef]
61. Kumar, V.; Alam, M.N.; Manikkavel, A.; Song, M.; Lee, D.J.; Park, S.S. Silicone rubber composites reinforced by carbon nanofillers and their hybrids for various applications: A review. *Polymers* **2021**, *13*, 2322. [CrossRef]
62. Banerjee, P.S.; Rana, D.K.; Banerjee, S.S. Influence of microstructural alterations of liquid metal and its interfacial interactions with rubber on multifunctional properties of soft composite materials. *Adv. Colloid Interface Sci.* **2022**, *308*, 102752. [CrossRef]
63. Wang, J.; Wang, W.; Zhang, C.; Yu, W. The electro-mechanical behavior of conductive filler reinforced polymer composite undergone large deformation: A combined numerical-analytical study. *Compos. Part B Eng.* **2018**, *133*, 185–192. [CrossRef]
64. Barrera, C.S.; Tardiff, J.L. Static and dynamic properties of eggshell filled natural rubber composites for potential application in automotive vibration isolation and damping. *J. Clean. Prod.* **2022**, *353*, 131656. [CrossRef]
65. Kumar, V.; Kumar, A.; Han, S.S.; Park, S.S. RTV silicone rubber composites reinforced with carbon nanotubes, titanium-di-oxide and their hybrid: Mechanical and piezoelectric actuation performance. *Nano Mater. Sci.* **2021**, *3*, 233–240. [CrossRef]
66. Zhong, J.; Luo, Z.; Zhang, H.; Ji, Y.; Wang, X.; Gong, W. Influence of interfacial property param of short fiber/rubber composites on fatigue behavior. *Int. J. Fatigue* **2023**, *167*, 107312. [CrossRef]
67. Wang, J.; Zhang, Y.; Ding, J.; Xu, Z.; Zhang, J.; He, Q. Preparation strategy and evaluation method of durable superhydrophobic rubber composites. *Adv. Colloid Interface Sci.* **2022**, *299*, 102549. [CrossRef] [PubMed]
68. Qian, M.; Zou, B.; Shi, Y.; Zhang, Y.; Wang, X.; Huang, W.; Zhu, Y. Enhanced mechanical and dielectric properties of natural rubber using sustainable natural hybrid filler. *Appl. Surf. Sci. Adv.* **2021**, *6*, 100171. [CrossRef]
69. Shimizu, T.; Kishi, R.; Kobashi, K.; Morimoto, T.; Okazaki, T.; Yamada, T.; Hata, K. Improved thermal stability of silicone rubber nanocomposites with low filler content, achieved by well-dispersed carbon nanotubes. *Compos. Commun.* **2020**, *22*, 100482. [CrossRef]
70. Alam, M.N.; Kumar, V.; Lee, D.J.; Choi, J. Synergistically toughened silicone rubber nanocomposites using carbon nanotubes and molybdenum disulfide for stretchable strain sensors. *Compos. Part B Eng.* **2023**, *259*, 110759. [CrossRef]
71. Tian, Q.; Zhang, C.; Tang, Y.; Liu, Y.; Niu, L.; Ding, T.; Li, X.; Zhang, Z. Preparation of hexamethyl disilazane-surface functionalized nano-silica by controlling surface chemistry and its "agglomeration-collapse" behavior in solution polymerized styrene butadiene rubber/butadiene rubber composites. *Compos. Sci. Technol.* **2021**, *201*, 108482. [CrossRef]
72. Chen, J.; Liu, J.; Peng, Z.; Yao, Y.; Chen, S. The microscopic mechanism of size effect in silica-particle reinforced silicone rubber composites. *Eng. Fract. Mech.* **2021**, *255*, 107945. [CrossRef]
73. Liu, H.; Yang, L.; Liu, X.; Cao, J.P.; Zhang, J.; Luo, Z.; Gao, Z. Silicon dioxide nanoparticle decorated graphene with excellent dispersibility in natural rubber composites via physical mixing for application in green tires. *Compos. Part B Eng.* **2023**, *258*, 110700. [CrossRef]
74. Ho, N.X.; Le, T.T.; Le, M.V. Development of artificial intelligence based model for the prediction of Young's modulus of polymer/carbon-nanotubes composites. *Mech. Adv. Mater. Struct.* **2022**, *29*, 5965–5978. [CrossRef]
75. Fukahori, Y.; Hon, A.A.; Jha, V.; Busfield, J.J.C. Modified guth–gold equation for carbon black–filled rubbers. *Rubber Chem. Technol.* **2013**, *86*, 218–232. [CrossRef]
76. Wu, Y.P.; Jia, Q.X.; Yu, D.S.; Zhang, L.Q. Modeling Young's modulus of rubber–clay nanocomposites using composite theories. *Polym. Test.* **2004**, *23*, 903–909. [CrossRef]
77. Alimardani, M.; Razzaghi-Kashani, M.; Ghoreishy, M.H.R. Prediction of mechanical and fracture properties of rubber composites by microstructural modeling of polymer-filler interfacial effects. *Mater. Des.* **2017**, *115*, 348–354. [CrossRef]
78. Han, Y.; Zheng, H.; Liu, Y.; Wang, M.; Wang, J.; Xie, Q.; Jing, S.; Qin, X.; Zhang, L. Synergistic Development of Natural Rubber/Butyl Rubber Composites for Improved Interfacial Bonding and Enhanced Shock-Absorbing Capabilities. *ACS Omega* **2024**, *9*, 13897–13905. [CrossRef] [PubMed]
79. Low, D.Y.S.; Supramaniam, J.; Soottitantawat, A.; Charinpanitkul, T.; Tanthapanichakoon, W.; Tan, K.W.; Tang, S.Y. Recent developments in nanocellulose-reinforced rubber matrix composites: A review. *Polymers* **2021**, *13*, 550. [CrossRef]
80. Kaliyathan, A.V.; Rane, A.V.; Huskic, M.; Kunaver, M.; Kalarikkal, N.; Rouxel, D.; Thomas, S. Carbon black distribution in natural rubber/butadiene rubber blend composites: Distribution driven by morphology. *Compos. Sci. Technol.* **2020**, *200*, 108484. [CrossRef]
81. Kapgate, B.P.; Das, C. Reinforcing efficiency and compatibilizing effect of sol–gel derived in situ silica for natural rubber/chloroprene rubber blends. *RSC Adv.* **2014**, *4*, 58816–58825. [CrossRef]
82. Martone, A.; Formicola, C.; Giordano, M.; Zarrelli, M. Reinforcement efficiency of multi-walled carbon nanotube/epoxy nano composites. *Compos. Sci. Technol.* **2010**, *70*, 1154–1160. [CrossRef]
83. He, M.; Du, W.; Feng, Y.; Li, S.; Wang, W.; Zhang, X.; Yu, A.; Wan, L.; Zhai, J. Flexible and stretchable triboelectric nanogenerator fabric for biomechanical energy harvesting and self-powered dual-mode human motion monitoring. *Nano Energy* **2021**, *86*, 106058. [CrossRef]

84. Liu, Y.; Khanbareh, H.; Halim, M.A.; Feeney, A.; Zhang, X.; Heidari, H.; Ghannam, R. Piezoelectric energy harvesting for self-powered wearable upper limb applications. *Nano Sel.* **2021**, *2*, 1459–1479. [CrossRef]
85. Yu, H.; Li, N.; Zhao, N. How far are we from achieving self-powered flexible health monitoring systems: An energy perspective. *Adv. Energy Mater.* **2021**, *11*, 2002646. [CrossRef]
86. Pan, H.; Qi, L.; Zhang, Z.; Yan, J. Kinetic energy harvesting technologies for applications in land transportation: A comprehensive review. *Appl. Energy* **2021**, *286*, 116518. [CrossRef]

polymers

MDPI

Review

Recent Advancements in Rubber Composites for Physical Activity Monitoring Sensors: A Critical Review

Vineet Kumar [1], Md Najib Alam [1], Gaurav Manik [2] and Sang-Shin Park [1,*]

[1] School of Mechanical Engineering, Yeungnam University, 280 Daehak-Ro, Gyeongsan 38541, Gyeongbuk, Republic of Korea; vineetfri@gmail.com (V.K.); mdnajib.alam3@gmail.com (M.N.A.)
[2] Department of Polymer and Process Engineering, Indian Institute of Technology Roorkee, Saharanpur Campus, Saharanpur 247001, Uttar Pradesh, India; gaurav.manik@pe.iitr.ac.in
* Correspondence: pss@ynu.ac.kr

Abstract: This review provides the latest insight (2020 to 2025) for composite-based physical activity monitoring sensors. These composite materials are based on carbon-reinforced silicone rubber. These composites feature the use of composite materials, thereby allowing the creation of new generation non-invasive sensors for monitoring of sports activity. These physical sports activities include running, cycling, or swimming. The review describes a brief overview of carbon nanomaterials and silicone rubber-based composites. Then, the prospects of such sensors in terms of mechanical and electrical properties are described. Here, a special focus on electrical properties like resistance change, response time, and gauge factor are reported. Finally, the review reports a brief overview of the industrial uses of these sensors. Some aspects are sports activities like boxing or physical activities like walking, squatting, or running. Lastly, the main aspect of fracture toughness for obtaining high sensor durability is reviewed. Finally, the key challenges in material stability, scalability, and integration of multifunctional aspects of these composite sensors are addressed. Moreover, the future research prospects are described for these composite-based sensors, along with their advantages and limitations.

Keywords: graphene; carbon nanotubes; sensors; sports monitoring; durability

1. Introduction

The physical activity monitoring sensors are a hot research topic due to their ability to track and measure various physical activities. These physical activities are running, cycling, or swimming in living beings [1,2]. These sensors have various gadgets like temperature sensors, global positioning systems (GPS), or strain gauges. Here, the temperature sensors provide insights about external weather conditions during sports activity. The GPS sensors use satellite signals to obtain geographical locations [3]. They are often used to track remote outdoor sports activities during hiking or cycling. Moreover, the strain gauges assist in measuring the change in strain or mechanical deformation of wearable and flexible materials. For example, smart clothing monitoring or intense sports activity [4,5]. In some cases, the biosensors are helpful in studying the electric impedance of biological systems. These systems can provide sweating, hydration levels, and muscle fatigue or injuries during sporting. Finally, the pressure sensors help to measure the change in air pressure during hiking or other change in altitude during climbing [6]. Therefore, these composite based sensors can be useful by configuring various monitoring activities that test fitness goals, and other optimized training routines [7].

 https://doi.org/10.3390/polym17081085

These physical activity sensors have numerous advantages for living beings, including humans, aquatic animals, and pulse monitoring in plants [8–10]. For example, the composite-based sensors provide quantitative data about the sports activity involved. Here, the athlete wears the sensor to track progress and make informative decisions about their fitness and health. These wearable sensors offer real-time monitoring, thereby allowing the users to receive immediate feedback and control the biological fatigue. Overall, these sensors help in studying the long-term monitoring of sports activity and the well-being of the athlete [11,12]. It, therefore, provides valuable information on exercise monitoring and helps in identifying risk factors. However, despite the multifunctional uses of these sensors, they have limitations as well. Some of the major limitations are that these sensors have issues like accuracy and reliability. These factors depend on various factors like sensor type, calibration, or environmental factors [13]. Another limitation includes their comfort, breathability, and ability to wear for long time. This causes discomfort to wear, or skin irritation, and finally, the final cost of the sensor is an issue [14]. Therefore, these limitations need further attention in terms of their commercialization. The overview of the present work is summarized in Scheme 1 below.

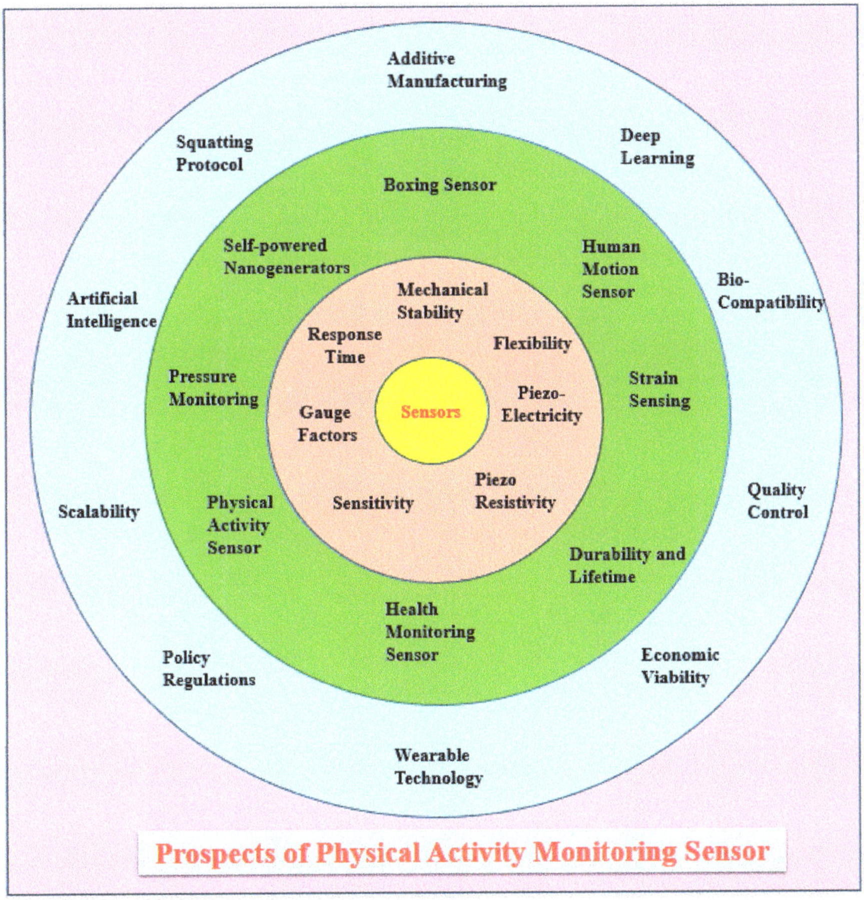

Scheme 1. Overview of present work. Inner ring summarizes properties of explored sensors, middle ring explores their applications, and outermost ring explores the prospects of this work for different commercial uses.

Despite various configurations of physical activity sensors, their performance is strongly influenced by the nature of materials used. These sensors are widely fabricated through stretchable and flexible elastomers like silicone rubber [15]. These silicone rubbers are useful due to their ease of processing and ease of vulcanization. Silicone rubber is a versatile material that is widely used as a synthetic elastomer. It consists of silicon, carbon, hydrogen, and oxygen. Due to its exceptional properties, it is widely used in various industrial applications [16]. Based on vulcanizations, there can be room-temperature, low-temperature, and high-temperature silicone rubbers [17,18]. Other flexible substrates like polyimide can withstand repeated bending and stretching without compromising sensor performance [19]. The key features of silicone rubber are thermal stability, flexibility, elasticity, chemical resistance, and electrical insulation [20].

These features can be understood more efficiently. For example, the silicone rubber is stable under a large temperature range, that is, from −60 °C to 250 °C. Thus, it can withstand a large range of temperatures without degrading. This also makes it useful for prolonged exposure to heat [21]. Silicone rubber has excellent flexibility in a large range of temperatures. These aspects make it useful for flexible substrates in stretchable sensors. Moreover, it has high elastic recovery, thereby maintaining its shape under mechanical loads. The silicone rubber is also highly stable to chemicals, including water, oils, and organic solvents. The used solvents make them useful for challenging environments [22]. Moreover, medical-grade silicone rubber can be useful for various implant medical applications. Finally, silicone rubber has excellent electrical insulating properties, making it useful for electrically insulating coatings and electrical cables [23]. Overall, silicone rubber has high weather and UV resistance, making it useful for various indoor applications.

There are various applications for silicone rubber, such as automotive, medical and healthcare, electronics, and construction applications [24]. For example, silicone rubber is extensively used in automobile applications like gaskets, seals, hoses, and thermal insulations. For electronic purposes, it can be helpful for electronic insulators such as cable components [25]. Finally, silicone rubber is useful for making baby products and personal care items due to its non-toxic nature. Overall, the unique properties of silicone rubber make it an indispensable material in modern manufacturing and technology [26]. With continuous research on silicone rubber, it will likely find even more applications, further demonstrating its versatility. Despite this, the sensor should have an electrode with high electrical conductivity and thermal stability. For this, inorganic nanomaterials like carbon black, graphene, carbon nanotubes, carbon nanofibers, and fullerene are often added [27–29]. They act as a reinforcing filler material in the rubber matrix and are described in detail in the coming sections.

[i] Carbon black: Carbon black is a fine black product obtained from the incomplete combustion of heavy petroleum. It is widely acknowledged that carbon black can reinforce rubber matrix and has thus been used as filler for more than a century [30]. It is widely used in tires and strain sensors due to its high reinforcing and electrical properties. The main properties of carbon black are small particle size, high surface area, and favorable surface chemistry. These properties can be detailed as the particle size of carbon black ranges from 10 to 500 nm [31]. The smaller the particle size, the better the reinforcement offered to the rubber matrix. Similarly, the larger surface area provides better reinforcing properties by improving the filler–rubber interfacial area. Similarly, the higher structure carbon blacks have more branches and voids that result in improved properties like reinforcing properties and electrical and thermal properties [32]. Moreover, the oxygen-carrying functional groups simulate its surface activity. This activity is enhanced by interfacial interactions between filler and rubber matrix. Carbon black has various applications as a reinforcing agent in tires, strain

sensors, and automotive components like vibration-damping components [33,34]. Overall, carbon black plays a crucial role in enhancing the reinforcing properties of the rubber matrix, making it useful for various applications like automotive or strain sensing.

[ii] Graphene: Graphene is a single layer of sp^2 hybridized carbon atoms arranged in a two-dimensional honeycomb lattice. Graphene is also known as the mother of all carbon allotropic forms like carbon nanotube and graphite [35]. It is well known for its wonderful mechanical, electrical, and thermal properties. These unique properties make it an ideal filler candidate to make the rubber matrix useful for strain-sensing applications [36]. For example, the addition of graphene in rubber matrix composites makes them ultra-strong and able to withstand high mechanical deformation. Their high electrical properties also make them useful for superior electrical conductivity. This is due to high electron mobility of graphene [37]. For sensor applications, the piezoresistive effect and short response time of graphene enhance the strain sensitivity. This allows them to detect even minor stimuli. Graphene provides fatigue resistance and promotes repeated bending and stretching without performance degradation [38]. Moreover, graphene also reduces noise and improves the accuracy of strain measurements, thereby making the strain sensor robust and exhibit high performance. This noise can be mechanical, electrical, material-inherent, or contact noise. Here, the mechanical noise can be sourced from mechanical vibrations and shocks, thereby leading to unstable energy outputs. Then, electrical noise is due to internal parasitic components in the sensor, and it distorts output signal integrity. Then, the material-inherent noise is sourced from non-uniform filler dispersion for composite-based sensors. This type of noise can lead to inconsistent signal generation or random charging during loading/de-loading cycles. Finally, the contact noise originates from a variable contact area, pressure, or surface property under cyclic mechanical deformations. These composites have various applications, such as health monitoring, wearable electronics, and robotics [39]. Therefore, adding graphene to strain sensors paves the way for innovative applications in various fields, including structural health monitoring.

[iii] Carbon nanotubes: Carbon nanotubes with cylindrical morphology are also known as robust fillers used in reinforcing rubber matrix. They can be single-wall (SWCNT) or multi-wall carbon nanotubes (MWCNT). When added to a rubber matrix, the carbon nanotubes offer a great combination of mechanical, electrical, and thermal properties [40]. Therefore, the carbon nanotubes enhance the performance and sensitivity of strain sensors, enabling their use in various advanced applications. Moreover, these carbon nanotubes exhibit a high aspect ratio, i.e., high length-to-diameter ratio. This high value helps enhance interaction with the polymer matrix and improves load transfer [41]. The role of carbon nanotubes used in strain sensing involves improved sensitivity. This is based on short response time, high durability, signal stability and high flexibility. These factors can be understood as the networks of carbon nanotubes in rubber matrix altering under mechanical strain, altering the resistance and improving the detection of even small deformations [42]. The carbon nanotubes improve mechanical properties and ensure that sensors can withstand repeated bending and stretching cycles that are crucial for robust wearable applications.

Finally, the carbon nanotube networks assist in achieving stable electrical signals, thereby reducing noise and improving output signals more accurately [43]. These carbon nanotubes added in the rubber matrix are useful for various applications. These are health monitoring, wearable electronics, and medical uses such as implants. Overall, the ability of carbon nanotubes assists in improving sensitivity, durability, and signal stability [44].

Moreover, these composites help in enabling the development of flexible and miniaturized sensors. These features make them a vital component in advancing strain sensor technology.

[iv] Carbon nanofibers: Carbon nanofibers are also cylindrical in morphology like carbon nanotubes, with a diameter ranging from 10–100 nm and lengths up to several micrometers. Therefore, they have a large aspect ratio with the combined properties of carbon nanotubes and traditional carbon fibers [45]. Their properties include high tensile strength, thereby contributing strong mechanical reinforcement to the rubber matrix. Moreover, they offer an optimum balance to strength, flexibility, and stretchability. These are critical for strain-sensing measurements [46]. Carbon nanofibers also offer high electrical conductivity, thereby contributing a strong piezoresistive effect that is crucial for strain sensing. Thus, the change in resistance is critical under the change in mechanical strain, enabling high sensitivity for strain sensors [47]. Moreover, the high thermal conductivity of carbon nanofibers makes them ideal for heat dissipation in rubber composite under mechanical strain [48]. Therefore, carbon nanofibers make the composites thermally stable under large mechanical deformations. As discussed already, carbon fibers have a great impact on strain sensitivity when used in sensing applications. Some other crucial prospects are enhanced piezoresistive sensitivity, high durability, and electrical signal stability [49]. Therefore, they are important for applications such as structural health monitoring, wearable electronics, and robotics.

[v] Fullerenes: Fullerenes are a well-known form of carbon in which the carbon atoms are arranged in spherical, ellipsoidal, or tubular structure. The most commonly studied fullerene is C60, a molecule that is composed of 60 carbon atoms arranged in a hollow sphere [50]. Fullerenes offer properties that make them useful for various advanced applications, such as strain sensors. For example, a fullerene with a robust structure helps composites achieve high mechanical strength due to their stable and symmetrical structure [51]. Like other carbon forms, fullerenes offer a good piezoresistive effect, making them suitable for sensing applications. In other cases, fullerenes can be chemically modified to improve their compatibility with different polymer matrices. This process helps in improving their dispersion and interaction within the composite material [52]. From a sensing perspective, fullerenes offer large strain detection that occurs from resistance change under mechanical strain. Moreover, they offer good mechanical reinforcement, flexibility, mechanical stability against fatigue, and noise reduction. The noise reduction and consistent signals help in obtaining error-free and more accurate data [53]. Therefore, their ability to improve sensitivity, durability, and signal stability. Also, they enable the development of flexible and miniaturized sensors, making them a valuable component in advancing strain sensor technology.

The addition of these materials makes the electrode electrically conductive and can offer a quick response to external stimuli. Moreover, these composite materials need to be biocompatible, non-toxic, and lightweight. They also need to have high fatigue resistance to be useful for sensing prototypes [54]. Thus, combining these robust composite materials during fabrication helps in achieving diverse and innovative physical activity monitoring sensors for in demand applications [55,56]. Therefore, this review paper describes the prospects of the recent literature on physical activity monitoring sensors. This review study reports the sensing of various physical activities like sports activities of human beings, including boxing, tennis playing, running, squatting, etc. [57,58]. It also summarizes the prospects of sports sensors on finger pressing or finger bending, etc.

The first section includes a brief introduction about these sensors and the material used during their fabrication. Finally, the advantages and challenges of these sensors are discussed in view of commercialization. Then, the next step was to describe the fabrication procedure of such sensors from assembly to full real-time monitoring. After discussing their fabrication, some critical mechanical and electrical properties were discussed. These mechanical properties are useful for long-term durability and resistance to mechanical failure or fatigue under continuous mechanical deformation. However, the electrical properties report the understanding of the resistance change, response time, and gauge factor in physical activity monitoring sensors. Then, the final part of the review is focused on the industrial applications of these sensors. This part focuses on the different functionalities of various types of sensors for monitoring various types of physical activities. The studied sensors include a review study on physical monitoring sensors for various uses. For example, we report the sensing of sports activities like boxing or physical activities like walking, squatting, or running. The other parts of this review include important investigations on energy generation for self-powered devices under different mechanical loads. Finally, the main aspect of fracture toughness for higher durability in sensors was reviewed. The study shows that the use of materials with fracture toughness is ideal for the use of composites for fabricating sensors with high durability.

Critical Overview of Rubber Composite-Based Sensors

The rubber composite-based sensors have emerged as a vital class of flexible, stretchable, and wearable sensing materials. These sensors are gaining increasing attention in the fields of soft robotics, healthcare monitoring, and human–machine interfaces. These sensors leverage the mechanical resilience and elasticity of rubber matrices. These matrices are combined with the electrical conductivity of embedded fillers such as carbon black, graphene, carbon nanotubes, or metal nanoparticles. The synergy between mechanical and electrical properties enables rubber composites to exhibit external stimuli. These are mechanical strain, external pressure, or temperature and result in measurable electrical signals. Despite their promising features, the development of rubber composite-based sensors faces several challenges. Among these challenges, the filler dispersion, long-term stability, sensitivity, and hysteresis are in focus. Moreover, repeatability remains critical in determining sensor performance, and the scalability and cost-effectiveness of fabrication techniques continue to impact their commercial viability.

This review aims to evaluate the current advancements critically about the key material systems and fabrication strategies. Moreover, the present review discusses the sensing mechanisms and application domains of rubber composite-based sensors. It also addresses the limitations and open research questions, providing insights into potential pathways for future development in this rapidly evolving field. These sensors have found applications in wearable health monitoring, such as monitoring sports injuries. The other applications include heart-beat monitoring, soft robotics, prosthetics, and electronic skins. Moreover, with the advancements in nanotechnology, 3D printing and self-healing materials are expected to improve the performance of these sensors. There is also a growing interest in multifunctional composites that can sense multiple parameters simultaneously. These parameters are pressure, temperature, and systems with wireless or self-powered functionalities. Overall, these rubber composite-based sensors are at the forefront of flexible and wearable electronics. This is due to their unique combination of mechanical flexibility and tunable electrical properties. However, addressing issues of long-term stability and environmental robustness is essential. Fixing these issues through self-healing and choosing the right material during fabrication is important. Finally, their transition from laboratory prototypes to real-world applications will need to overcome challenges. These challenges are filler dispersion, filler rubber compatibility, and the great interface between filler and rubber matrix.

2. Filler Characterizations

2.1. Structural Features of the Carbon Nanomaterials

The morphology of the carbon nanomaterials, namely carbon black, carbon nanotubes, and graphene, plays a crucial role in influencing the properties of the composites. Therefore, understanding their morphological features is important to obtain properties of interest [59]. Starting with carbon black, Figure 1a shows that carbon black particles consist of quasi-spherical primary particles with diameters typically ranging from 10 to 500 nm. These primary particles fuse together to form larger aggregates, which are the basic morphological units of carbon black [60]. These aggregates can further cluster into loosely bound agglomerates, which can be broken down during processing. The carbon black particles can be useful as a source of reinforcement and electrically conductive additives in rubber composites [61].

Figure 1a shows the morphological features of carbon nanotubes. As described already, the carbon nanotubes are cylindrical tubes composed of rolled-up sheets of graphene. These carbon nanotubes can be single-walled or multi-walled, depending upon the number of layers rolled into a tube [62]. Typically, single-walled CNT has an approximate diameter of 1 nm, while multi-walled CNT have diameters from a few nm up to 10 nm. In addition, carbon nanotubes exhibit a high length-to-diameter ratio, which is the aspect ratio, sometimes exceeding 1000 [63]. This high aspect ratio contributes to the exceptional mechanical and electrical properties of composites. Carbon nanotubes also have various commercial uses such as reinforcing agents, electronic devices, energy storage gadgets like supercapacitors, and strain sensing [64].

Similarly, Figure 1a shows the morphological features of graphene. As already stated, graphene is a two-dimensional single sheet of carbon atoms that are arranged in hexagonal lattices. Graphene has an extremely high surface area, and this feature makes it useful as a source of reinforcement in rubber matrices [65]. In some cases, the functionalization of graphene sheets is performed to make it compatible with the rubber matrix, and improved interfacial interactions are expected. The improved interface makes the composite robust, and high reinforcing properties are expected [66]. Therefore, graphene can be useful in flexible electronics, sensors, and energy storage applications. Overall, the unique morphologies of these carbon nanomaterials are fundamental to their diverse use. For example, they are useful in improving properties for a wide range of applications [67]. Also, carbon black's aggregated structure, carbon nanotubes' high aspect ratio, and tubular form contribute to their respective roles in enhancing materials and devices. Moreover, graphene's two-dimensional planar structure assists in achieving high properties for the application of interest.

Raman spectra are well known as a powerful tool in studying carbon nanomaterials. The Raman spectra assist in providing insights into structural and electronic properties [68]. These properties are described by analyzing the vibrational modes of the carbon atoms. Raman spectra in Figure 1b show that the Raman spectra of carbon black exhibit a D-band at 1350 cm^{-1} [64]. This band originated from breathing modes of sp^2 carbon atoms in rings and indicates defects and disorder in the structure of carbon black. The intensity of the D-band increases with the amount of disorder and defects in the structure of carbon black [69]. Another characteristic G-band located at 1580 cm^{-1} corresponds to the in-plane vibration of sp^2 carbon atoms in the graphitic lattice. The intensity of the G-band provides the degree of graphitization; a higher intensity suggests more graphitic or crystalline regions [70]. Similarly, Figure 1b shows the Raman spectra of carbon nanotubes. The D-band is positioned at 1350 cm^{-1}, like other carbon materials. It indicates the presence of defects, disorder, and amorphous carbon in the carbon nanotubes [71]. Similarly, the G-band position is at 1580 cm^{-1} and provides information about the electronic structure and the degree of graphitization. Moreover, there is another characteristic 2D band at 2700 cm^{-1}. It signifies sensitivity to the electronic properties and can be used to distinguish between single-walled and multi-walled CNTs [72]. Similarly, Raman spectra of graphene were reported in Figure 1b. The results show that D-band and G-band are like other carbon materials. Moreover, the 2D band was also noticed at 2700 cm^{-1}. It is a characteristic peak for the number of graphene layers stacked in the graphene flake [73]. For example, a sharp 2D band is characteristic of single-layer graphene.

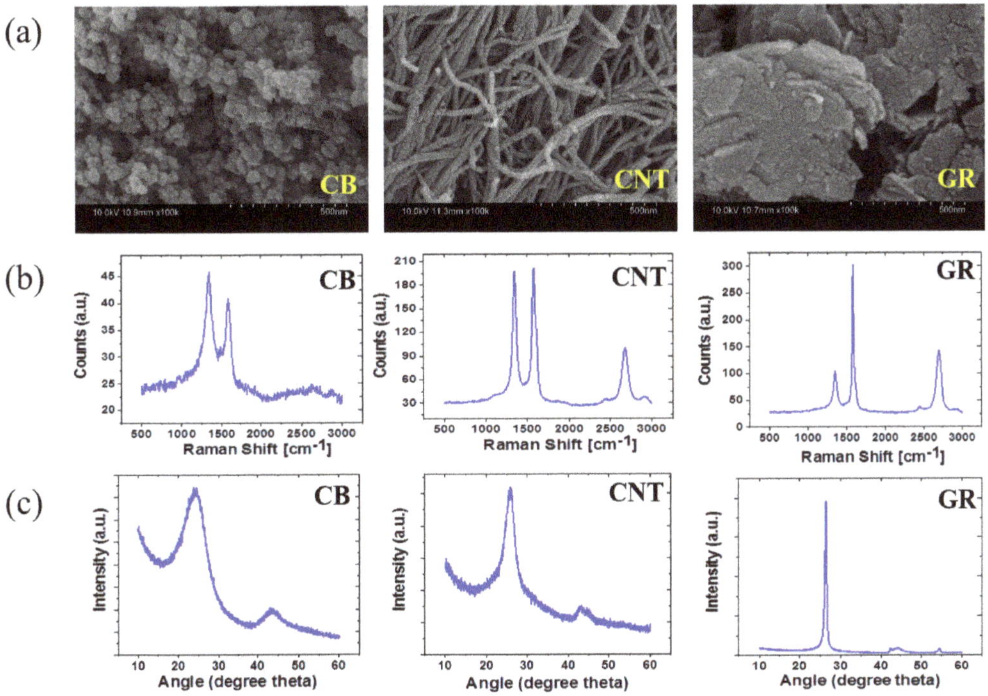

Figure 1. (**a**) SEM of carbon nanomaterials such as CB, CNT, and GR. Here, CB shows oval-shaped morphology, CNT shows tube-like morphology, and GR shows sheet-like morphology; (**b**) Raman spectra of carbon nanomaterials, showing a D-band at 1350 cm^{-1}, G-band at 1580 cm^{-1}, and 2D band at 2700 cm^{-1}. However, the 2D band is absent for CB and the 2D band in GR changes with a change in the number of graphene layers stacked; (**c**) XRD of carbon nanomaterials showing a broad peak around 2θ of 24° to 26° corresponds to the (002) plane of graphite. Reproduced with permission from Elsevier [64].

XRD is well known as a powerful tool that provides the crystallography structure and physical properties of a material. Therefore, it is frequently used to characterize the crystal structure of carbon nanomaterials. Figure 1c provides the XRD of carbon black, and the pattern shows that it is amorphous [64]. It can be stated that the XRD pattern is broad and presents featureless peaks. This broad peak around 2θ of 24° corresponds to the (002) plane of graphitic materials. This feature indicates a short-range order but without a long-range crystalline order [74]. Similarly, the XRD of carbon nanotubes exhibits a prominent peak around 2θ of 26°, corresponding to the (002) plane of graphite. Carbon nanotubes can be briefly useful in electronics, reinforcing agents, and sensors. Finally, the XRD of graphene shows a noticeable peak around 2θ of 26° that relates to the (002) plane of graphite [75]. However, the difference in peak width and intensity is related to the number of graphene layers stacked in the crystalline domain.

2.2. Structural Characterizations

FTIR is regarded as a powerful tool for obtaining the molecular composition and structure of the materials. FTIR is used for studying carbon nanomaterials like graphene or silicone rubber, as studied in the present review. Figure 2a shows the FTIR spectra of graphene [76]. The results show C=C stretching vibrations around 1580 cm^{-1}, justifying the presence of a graphitic structure. It is known to have originated from the stretching vibrations of sp^2 hybridized carbon atoms in the graphene lattice [77]. Further, the O-H stretching was located around 3500 cm^{-1} and proposed from functionalization with hydroxyl groups. It can enhance the hydrophobicity and result in higher surface activity of graphene.

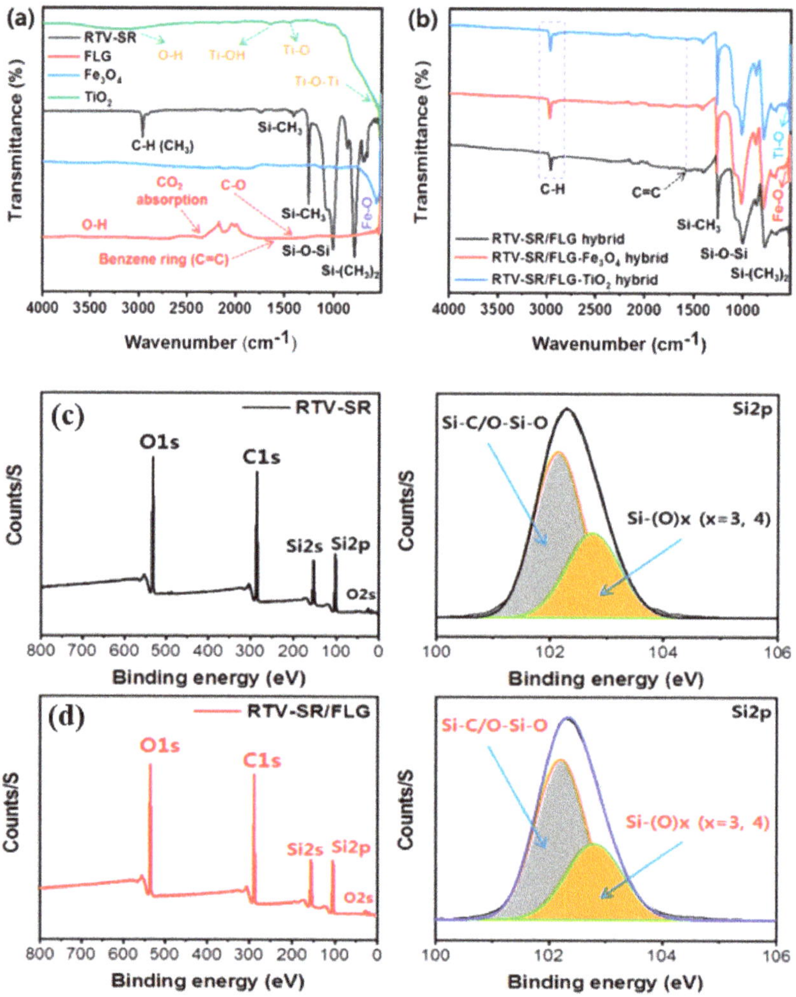

Figure 2. (**a**,**b**) FTIR of carbon nanomaterials and their composites. FTIR of silicone rubber shows functional groups like C-H, Si-CH$_3$, Si-O-Si; FTIR of FLG shows functional groups like O-H, C=C, C-O and adsorption of CO$_2$; FTIR of iron oxide shows functionalities like Fe-O; FTIR of TiO$_2$ shows

functional groups like O-H, Ti-OH, Ti-O, and Ti-O-Ti; finally, FTIR of composite sample shows functionalities like C-H, C=C, Si-CH_3, Si-O-Si; (**c**,**d**) XPS of unfilled and graphene-filled silicone rubber compote samples. Results reveal that silicone rubber contains Si-O-Si and Si-C bonds, indicating silicone polymer networks. Deconvolution results show interactions between graphene and silicone rubber. These interactions can be physical adsorptions or chemical bonding, such as Si-O-Si linkages Reproduced from open access article [76].

C=C stretching is located around 1700 cm^{-1} and indicates the oxidation of graphene [78]. It is proposed to originate from the synthesis method or functionalization process. Similarly, Figure 2b reveals the FTIR of silicone rubber and their composites with respective characteristic absorption bands. They are proposed to originate from the functional groups in silicone rubber. The spectra reveal Si-O-Si stretching vibrations around 1100 cm^{-1}. They are originated from stretching vibrations of the siloxane backbone [79]. Another Si-CH_3 stretching vibration at around 1270 cm^{-1} is known to have originated from the methyl group attached to silicon in the rubber. The C-H band at around 2970 cm^{-1} was proposed to originate from C-H bonds in methyl groups. Thus, FTIR is proposed to be useful for analyzing and validating the chemical composition and structure of both graphene and silicone rubber [80].

XPS is a powerful tool for investigating the surface chemistry of the materials. Analyzing carbon nanomaterials like graphene in silicone rubber composites provides elemental composition. XPS also provide chemical states, interactions, and knowledge about additives [81]. Figure 2c,d show the XPS results of silicone rubber and graphene-filled silicone rubber composites. The results show that the elemental compositions identified are carbon, silicone, and oxygen for both samples. The XPS results further show the presence of functional groups like hydroxyl and carbonyl groups on the surface of graphene [82]. These interactions can be physical adsorptions or chemical bonding, such as Si-O-Si linkages [83]. The results further validate the composite processing's absence of contaminants or residual solvents.

3. Properties

3.1. Mechanical Properties

Understanding the mechanical properties of physical activity monitoring sensors is very important. The composites based on elastomers and electrically conducting fillers are often the best candidates [84]. They exhibit good mechanical properties suitable for fabricating physical activity monitoring sensors. These composites are stretchable and flexible and exhibit high sensitivity and reliability [85]. Moreover, they have robust durability in various industrial applications like physical activity monitoring sensors. The other useful mechanical parameters are elasticity, modulus, fracture strain, and hysteresis losses. [86,87]. Moreover, a balance of these mechanical properties is critical to obtaining robust performance for physical activity monitoring sensors. For example, these monitoring sensors exhibit good stretchability and can help in accurately detecting and measuring small changes [88]. They also exhibit an optimum compressive modulus, which helps in exhibiting greater sensitivity to small changes in strain. Their higher tensile strength is critical for enabling the reliability and durability of strain sensors under various loading conditions [89]. Moreover, their lower hysteresis losses as heat dissipation losses help achieve accurate and repeatable measurements. Finally, the composites with high fatigue resistance ensure long-term reliability and durability [90]. Therefore, mechanical properties are critical for physical activity monitoring sensors.

Figure 3A presents stress–strain curves for different composites to be useful for sensors [91]. The results show two types of behavior. One is an increase in stress with increasing compressive strain. Another is an increasing stress at all percentage strains with increasing MWCNT content in the composite. Firstly, the higher stress resulting from higher strain could be because of higher packing fractions of the ingredients of the composite at higher compressive strain [92,93]. Moreover, the stress increases almost linearly up to 10%, then increases exponentially until a compressive strain limit of 35%. The exponential increase after 10% is due to optimum strain, after which there are higher packing fraction effects. In addition, the higher stress by increasing MWCNT content could be due to higher inter-interfacial interactions [94]. These interactions resist the composites under external mechanical force under the loading phase of strain. During the unloading phase, the interactions re-establish and achieve equilibrium during cyclic mechanical deformations [95].

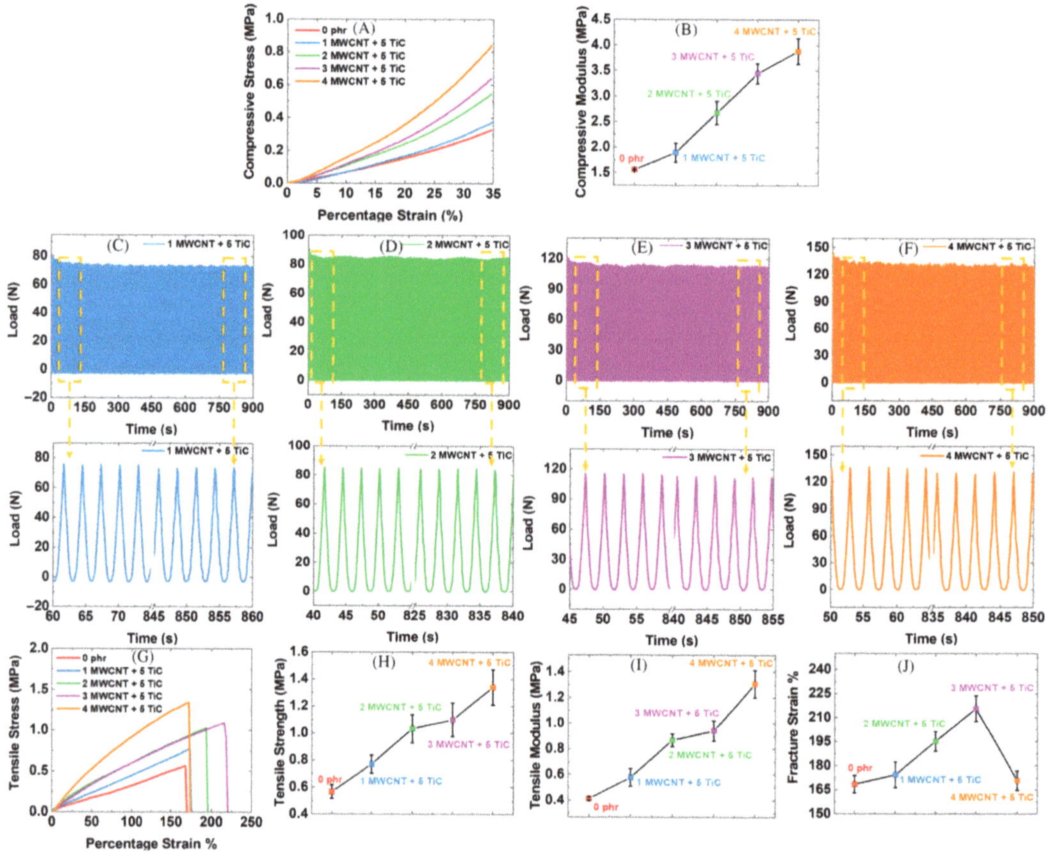

Figure 3. (**A**) Stress–strain profiles under compressive strain. Results show that compressive stress increases with increasing MWCNT + TiC content. Moreover, stress increases almost linearly up to 10%, then increases exponentially until a compressive strain limit of 35%. (**B**) Compressive modulus of different composites. Results show that the compressive modulus increases with increasing content of MWCNT + TiC. (**C–F**) Different samples under cyclic compressive loads. Output mechanical load was stable with negligible losses from the initial to final cycles. (**G**) Stress–strain under tensile strain. Stress was higher at higher strain, reaching a maximum before fracture. (**H–J**) Tensile strength, tensile modulus, and fracture strain of different composites. These mechanical properties increase with increasing MWCNT + TiC content, reaching the maximum at 4 phr of MWCNT + TiC. However, fracture strain decreases after 3 phr of MWCNT. Reproduced with permission from reference [91].

Similarly, the compressive modulus in Figure 3B increases with increasing MWCNT content, reaching a maximum at 4 phr MWCNT + 5 phr TiC sample. This can be influenced by a higher "reinforcing effect" of MWCNT in the composite. Moreover, the higher modulus is due to efficient and robust interfacial interactions in the composites [96]. Finally, the high properties are due to optimum fabrication methodology, dispersion, orientation, and high aspect ratio of MWCNT [97]. These factors strongly influence the improved compressive modulus with increasing MWCNT content in composites. These aspects are important for achieving high performance for physical activity monitoring sensors for mechanical stability and durability [98,99]. Among the various main mechanical factors, mechanical stability is critical for physical activity monitoring sensors. Thus, Figure 3C–F shows the mechanical stability under compressive cycling for different composites. The results reported by Manikkavel et al. [91] show that under continuous compressive cyclic strain, the output mechanical load was stable with negligible losses from the initial to final cycles. This behavior supports their use of physical activity monitoring sensors [100]. These sensors require higher mechanical stability and critical fatigue resistance against mechanical failure. This can be correlated to better load transfer, thereby making the composite more resistant to mechanical deformation [101]. Moreover, the desired modulus and sensitivity can obtain better performance for physical activity monitoring sensors.

Figure 3G reports stress–strain curves under tensile strain. Like previous results in Figure 3A, the stress was higher at higher strain, reaching a maximum before fracture. This could be a result of the breakdown of interfacial interactions in in-rubber structures. These features are also due to other interactions within the composites under increasing tensile strain [102]. Other parameters, like tensile modulus, tensile strength, and fracture strain, can be derived from these stress–strain curves and reported in Figure 3H–J. Most of the results agree well with the stress strain in which the modulus and tensile strength increase while fracture strain falls after 3 phr MWCNT 5 phr TiC. The improved mechanical properties with filler addition are well known, and there are several reasons for such behavior. For example, the higher "reinforcing effect" of MWCNT in composite helps in achieving higher mechanical properties [103]. Higher interfacial interactions with the addition of MWCNT in the composites assist in achieving higher mechanical properties. Finally, good fabrication technology, filler dispersion, and the high aspect ratio of MWCNT in composite help in achieving higher mechanical properties [104].

Finally, the fall in fracture strain after 3 phr MWCNT + 5 phr TiC in Figure 3J could be due to the partial aggregation of filler particles in composite materials. It is essential to understand the modulus of the composites for their usefulness as physical activity monitoring sensors [105]. Moreover, the tensile strength is crucial for composite materials. It provides important information about the quantity of stress a composite can withstand before mechanical failure or fatigue [106]. Moreover, the fatigue behavior is important for understanding the reliability and durability of sensing devices. Finally, the fracture strain is useful to understand the ultimate elongation the composite can withstand before failure. This is one of the most important aspects of the use of composites for various physical activity sensing applications [107].

3.2. Electrical Properties

The physical activity monitoring sensors are generally based on elastomers that are lightweight and stretchable under external mechanical deformations [108]. These physical activity monitoring sensors have key factors like resistance change, response time, and gauge factors while evaluating their performance [109]. These factors are crucial for understanding the performance of a strain-sensing device. The strain sensors work on the principle of piezoresistance in which the resistance change happens in response to applied strain [110]. Such a resistance change is usually proportional to the applied strain and depends upon the material's piezoresistive coefficient. Then, the response time refers to the ability of the strain sensor device to respond to external stimuli [111]. Generally, a faster response time is required for dynamic applications. Therefore, the response time is related to mechanical properties, sensor design, and the electronics of the strain sensor. Finally, the gauge factors measure the sensitivity and quantify the resistance change against the strain [112]. Technically, it is defined as the ratio of relative change in resistance ($\Delta R/R$) to the applied strain ($\Delta\varepsilon$). Thus, these above factors are useful in understanding the performance of strain sensors. For example, using materials with higher piezoresistive coefficients can enhance the resistance change per unit strain, thereby influencing the sensitivity. Overall, researchers need to develop a strain sensor with optimum performance, such as desirable resistance change, faster response time, and optimum gauge factor [113].

Considering these aspects, Park et al. [91] fabricated composites by adding different propositions of MWCNT and fixed 5 phr of TiC into silicone rubber. Finally, these composites were studied for their use in strain sensing by reporting resistance change, response time, and gauge factors [114]. Figure 4A–C report the resistance change by increasing the MWCNT content in the SR matrix. The piezoresistive effect can be understood by composites based on SR matrix and MWCNT and TiC. This helps in achieving higher change in electrical resistance in response to mechanical strain [115]. This can be understood in more detail as the formation of electrically conductive pathways in composite by MWCNT and TiC into SR matrix. Thus, as the composite undergoes compressive mechanical deformation, there is a change in resistance [116]. Finally, it results in enhancing the piezoresistive effect and amplifying the change in resistance for a given level of strain. Another mechanism for achieving higher piezoresistivity is related to filler percolation phenomena [117]. For example, increasing the MWCNT content beyond the threshold helps the composite material transition to a more conductive state. Finally, this assists in making the composite more sensitive to changes in strain-induced resistance [118]. Moreover, the with change in resistance, the sensitivity of the composite can be adjusted by changing the content of MWCNT in the SR matrix. This can be adjusted by tuning MWCNT content for desirable performance. This tuning allows for optimization factors like sensitivity, linearity, and mechanical properties of the final composite [119,120]. Overall, the mechanical stability is important in affecting the fatigue and durability properties of these composites. For example, the mechanical stability of these composites helps to withstand mechanical strain without failure, ensuring reliable performance as a strain sensor over extended usage [121].

In addition to change in resistance, the response time is of utmost importance to study strain sensors. Figure 4D–F present the influence of increasing MWCNT content on the response time of the composites. The results show that the response time decreases with increasing MWCNT content. This behavior is proposed because of the higher electrical conductivity of the composites with an increase in MWCNT content [122]. The transmission of electrical signals or the sensing of changes in electrical properties results in a smaller response time. This can be understood more efficiently as the higher electrical conductivity helps in faster transmission of electrical signals [123]. For example, the resulting electrical response under mechanical deformations can be detected more quickly in materials with higher conductivity. Moreover, the sensing performance is improved with composite with higher electrical conductivity [124]. For example, the time taken to detect changes in resistance due to mechanical deformation can be reduced in more conductive materials. In many cases, the higher electrical conductivity helps in reducing electrical noise or interference in composite materials [125]. It helps to ensure faster and more reliable signal processing and data transmission. It is useful where accurate and timely detection of signals is critical, such as in medical devices [126].

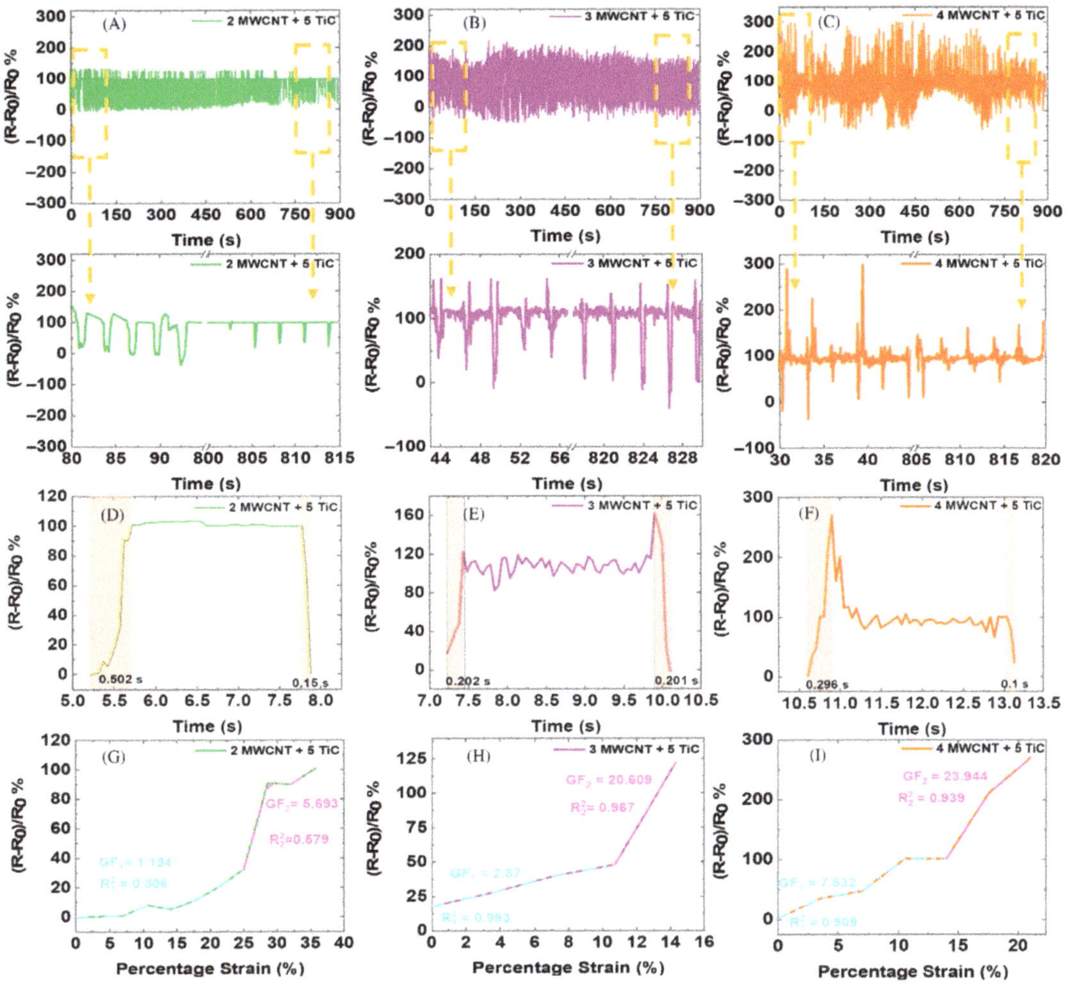

Figure 4. (**A–C**) Relative resistance change for different samples. Results show a higher change in resistance by increasing MWCNT + TiC in composites. Effective piezoresistive exhibit can be understood by composites based on SR matrix and MWCNT and TiC. (**D–F**) Response time for different composites. Response time decreases with increasing MWCNT content. For example, response time of loading phase at 2 phr MWCNT + 5 phr TiC was around 0.502 s, and it decreased to 0.202 s at 3 phr MWCNT + 5 phr TiC. (**G–I**) Gauge factor of different samples. Gauge factor was 1.19 at 25% strain and increased to 5.6 by increasing compressive strain to 35%. Reproduced with permission from reference [91].

Finally, the gauge factor and linearity of the composites are understood by the results reported in Figure 4G–I. It is well known that the gauge factor depends critically on the magnitude of strain and electrical conductivity of the composite studied. This is expected due to an increase in the packing fraction of the additives in the composite during fabrications. The higher packing fraction influences the resistance change, and thereby the higher gauge factor was witnessed [127,128]. In another case, with increasing MWCNT content, the gauge factor and linearity are also better. It is also due to higher electrical conductivity by increasing MWCNT content in the composite. The higher electrical conductivity improves the charge carriers within the composite and allows more efficient transmission of electrical signals [129].

Thus, the process finally influences the change in resistance, thereby leading to better linearity and higher gauge factors. Moreover, higher electrical conductivity helps in enhancing sensitivity to external stimuli. Finally, it enhances the accuracy and reliability of strain-sensing applications [130]. In another case, these strain sensors should exhibit a linear response, where the resistance change is directly proportional to the applied strain over a wide range of strain values. Here, the higher electrical conductivity can contribute to improved linearity by minimizing non-linear effects such as mechanical hysteresis losses [131]. For example, Park et al. [91] show that the linearity was 0.81 for 1 phr MWCNT with 5 phr TiC, and it increases to 0.91 for 3 phr MWCNT with 5 phr TiC. Therefore, electrical conductivity plays a crucial role in enhancing the gauge factor and linearity of composites used in strain-sensing applications.

4. Applications

4.1. Physical Activity Monitoring Sensors Under Tensile and Compressive Strain

In composite materials that are used in strain sensing, the change in resistance can vary based on composition, structure, and electrically conductive properties [132]. For example, under compression strain, the distance between conductive pathways of filler, such as MWCNT, changes in the composite. This leads to a change in resistance and, thus, different sensing performance [133,134]. This case is similar to graphene-based composites, in which the electrically conductive pathways became denser, resulting in resistance change and thus better-sensing properties. Similarly, the change in resistance for CNT and graphene-based composites is higher, thereby resulting in resistance change and lower sensing capacity [135]. Therefore, specific resistance change is based on the type of strain primarily driven by changes in the conductive pathways within the material. Park et al. [136] report a strain sensor fabrication by solution mixing of silicone rubber with carbon nanotube. This mixing was performed for 10 min following the addition of the vulcanizing agent and kept at room temperature for 24 h for vulcanization. After vulcanization, the final composites were tested for properties under tensile and compressive strain.

Figure 5a shows the outline of the wearable sensor in the human body. The real-time sensing responses can be performed under thumb pressing, finger pressing, finger bending, and wrist bending [137]. For strain sensors utilizing composite materials, those used in wearable devices can induce varying responses in terms of resistance change [138, 139]. Figure 5b shows the relaxation and loading under tensile strain from a universal testing machine. These strain types can be understood more extensively as thumb or finger pressing states, as when pressure is applied with the thumb, it results in localized compression on the sensor surface. This compression can result in a decrease in resistance change, thereby increasing conductive pathways in composite elements [140]. Then, the finger bending or wrist bending results in inducing complex strain on the sensor surface. The dimensions of cylindrical and dumbbell samples are presented in Figure 5c. Finally, Figure 5d shows relaxation and loading under compressive strain. Generally, the bending process results in variable compressive or tensile strain across the sensor, resulting in varied response in resistance.

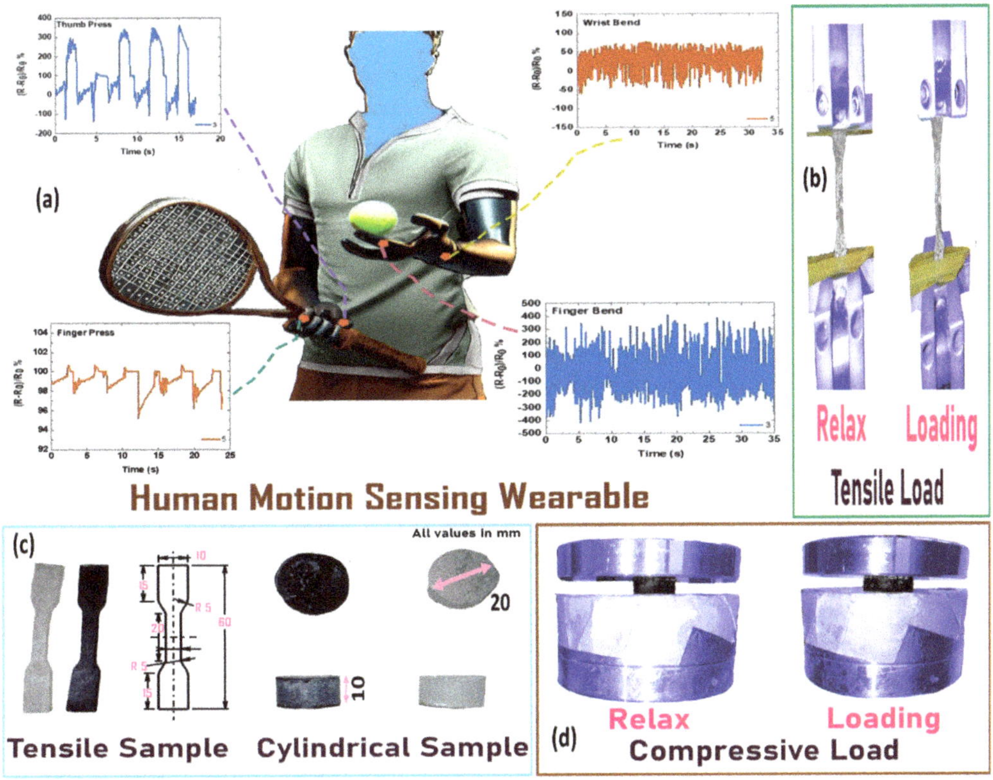

Figure 5. (**a**) Relative resistance change for various human motion sensing. Change in resistance was higher during thumb pressing, mimicking compression strain on cylindrical sample. However, it was highest for finger bending that mimics tensile strain. (**b**) Sample subjected to tensile load. These researchers performed strain sensing using cylindrical samples for compressive tests and dumbbell samples for tensile testing. (**c**) Details of sample geometry. Cylindrical samples were 10 mm thick and 20 mm wide. Tensile test's dumbbell sample has a thickness of 2 mm and a gauge length of 20 mm. (**d**) Sample subjected to compressive load. Relaxation and loading mechanisms are reported. Reproduced with permission from Wiley [136].

4.2. Sensing of Physical Activity Like Boxing

4.2.1. Overview of Deep Learning and Data Analytics for These Sensors

This review introduces the use of deep learning for use in physical activity monitoring sensors. Here, the integration of deep learning and data analytics with composite-based sensors has opened new frontiers in intelligent sensing systems. These sensors use flexible and use multifunctional materials such as rubber composites and electrically conducting fillers. They generate rich, complex data streams in response to various physical stimuli, including pressure, strain, temperature, and motion. Moreover, exploitation of the data from physical activity like walking, running, etc. is performed for accurate sensing. The other functionalities involve the recognition, decision-making, and advanced data-driven approaches. Here, deep learning, a subfield of machine learning based on artificial neural networks, can be explored. It has proven especially powerful in handling large, high-dimensional datasets. It can automatically extract hierarchical features from raw sensor signals. Then, it enables robust classification, regression, and anomaly detection without manual feature engineering. This is particularly valuable for wearable and soft sensors, where signal patterns may be non-linear, noisy, and highly dynamic. Finally, complementing this, data analytics techniques are used. These are statistical modeling, signal processing, and real-time data fusion. They enhance sensor functionality by improving accuracy and enabling sensor calibration. Finally, they help in supporting multi-sensor integration. These technologies transform composite-based sensors from passive signal generators into smart systems. For example, deep learning is capable of adaptive learning, real-time feedback, and predictive analytics. This integration is crucial for applications such as human motion analysis. The other applications are health monitoring, soft robotics, and structural health diagnostics.

4.2.2. For Real-Time Monitoring for Boxing Sensor

The real-time sensing of physical activity involves the integration of smart materials as sensors to monitor physical sports activities such as human motion [141]. They are useful for various industrial applications, such as sports performance analysis and healthcare monitoring [142–144]. Thus, integrating smart materials that are active in piezoelectricity [145], piezoresistivity [146], or capacitive elements [147] within the sensor is useful for sports activity monitoring. These sensors are useful for measuring changes in relative resistance under mechancial stimuli and transmit data through connections for real-time analysis [148]. These sensors are wearable in clothing, footwear, and hand gloves as reported in Figure 6a [149]. The author reports the monitoring of human motions or sports activities like boxing for biomechanical analysis. The viscosity, modulus, and tensile strength can be improved by mixing SiO_2 micro-particles into liquid metals [149]. These aspects allow the sensor used for sports monitoring with high sensitivity with increasing gauge factors. Here, the gauge factors increase from 5.72 at 100% strain to 23.91 at 300% strain. After printing, the sensor was fabricated onto the textile glove for real-time monitoring of the prototype. These sensory arrays are integrated with tactile gloves for evaluating boxing motions like jab, swing, uppercut, etc. [149]. This can be further understood in Figure 6b. Such analysis involves an analysis of sports performance optimization or injury prevention mechanisms. Finally, the real-time data can be obtained from WiFi, Bluetooth, or cloud-based platforms for analysis and visualization. Finally, the signal processing algorithms are employed to understand sensor data. The real-time feedback from physical activity and performance metrics has an accuracy of up to 90.5% [149].

A physical activity like boxing can be monitored with a physical activity sensor with an accuracy of >90% [149]. These sensors represent a powerful tool for promoting advanced research for sports like monitoring jabs, uppercuts, or swings in boxing. These sensors can harness sensor technology, data analytics, and health monitoring solutions [150]. Moreover, these sensors are equipped to sense various advanced sports systems, mechanical motions, or physiological parameters of human health [151]. Moreover, such sensors can capture data on movements, behaviors, and inertial applications. The inertial sensors include accelerometers, global positioning systems, and health monitoring like changes in heartbeat during sports activity [152,153]. Therefore, Xiao et al. [149] report the monitoring of physical activity like boxing through such sensors. Figure 7a shows the monitoring of change in relative resistance during jab, uppercut, and swing operations during boxing using the composite-based physical activity sensor. The reported real-time monitoring results show that the change in relative resistance was higher for the uppercut, medium for the jab, and lower for the swing. The higher resistance change of the uppercut could be proposed due to the higher magnitude of strain that results from higher mechanical deformation and, thus, higher resistance change [154]. For example, a smart tactile glove can be used to evaluate the recognition of a particular activity through an algorithm, as seen in Figure 7b. The final punch was divided into a training set (90%) and a testing set (10%) by adopting the Convolutional Neural Network (CNN) model. Figure 7c reports the real-time monitoring of resistance change against various types of hand motions like finger bending. These reported results show that the resistance to change was highest for the middle finger and lowest for thumb bending, as suggested by Zhang et al. [155]. The results are consistent and reproducible concerning the type of bending.

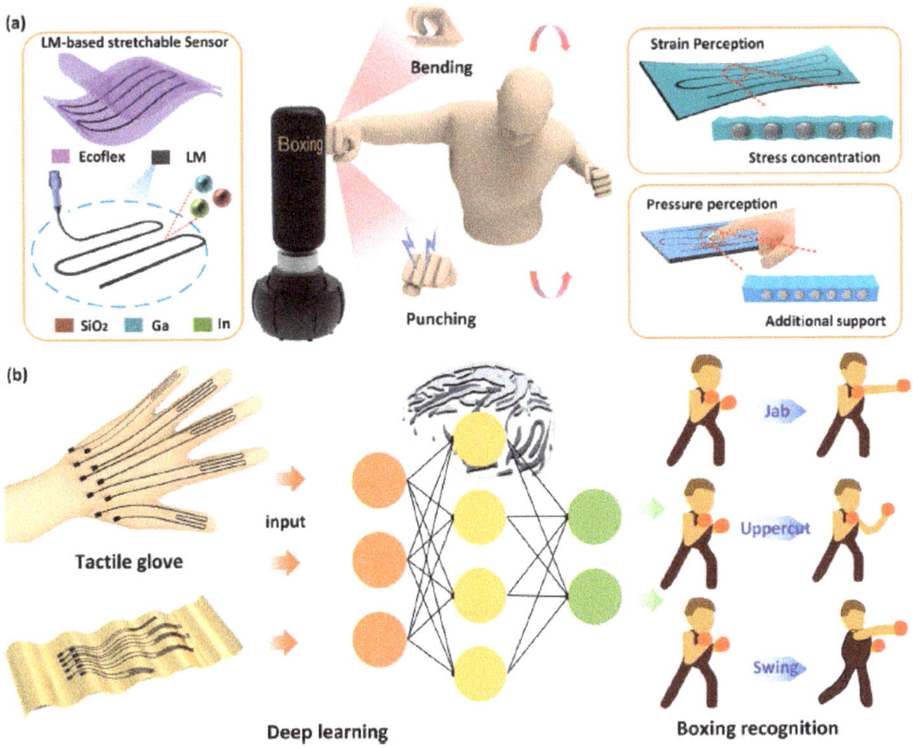

Figure 6. (**a**) Fabrication of sensor and their prospects. Device was developed by 3D printing of liquid metals on silicone rubber with the brand name "Ecoflex". (**b**) Wearable sensor, deep learning, and real-time monitoring of boxing. With future advancements, deep learning and data analytics will play an increasingly central role in unlocking the full potential of composite-based sensor systems. Reproduced with permission from open access article [149].

It is well known that the resistance changes sensors often rely on piezoresistive mechanisms. For example, under mechanical motions of bending or pressing, the electrical resistance changes proportionally depending upon the magnitude of mechanical stimuli [156]. This change in resistance can be measured and correlated with specific human motions like finger pressing or bending during boxing [149]. The wearable sensors embedded in gloves can detect changes in resistance corresponding to repetitive motions like jabs, uppercuts, or swings in boxing. Figure 7d shows the signal processing steps that include data collection. The collected data can be transmitted through the wireless platform, processing scaling transformation of data, and final validation of the results. Finally, Figure 7e shows the optical image of real-time boxing, and Figure 7f shows the confusion map of the classified results. By capturing and quantifying changes in resistance, these sensors can provide valuable insights into movement patterns and sports activities like boxing. Moreover, the change in resistance is the fundamental principle of motion sensing [157]. It, therefore, allows for the detection, quantification, and interpretation of various sports activities like boxing.

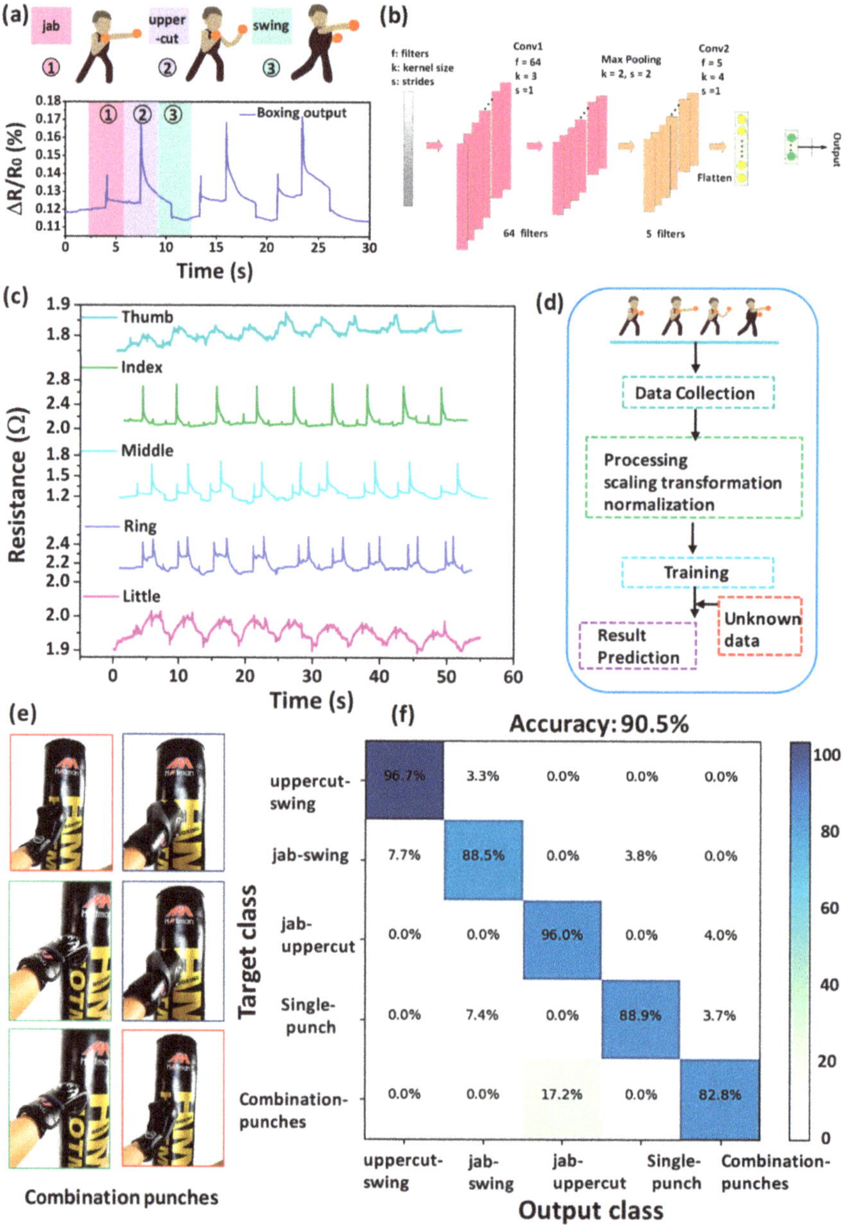

Figure 7. (**a**) Different configurations in boxing. This sensor was fabricated by 3D printing liquid metals on silicone rubber. Mechanical properties like modulus and tensile strength were improved by mixing SiO_2 and micro-particles into liquid metals as reported. (**b**) Filtering process of data. Whole data sample was used in 150 sets, and 30 sets for each punch were reported. (**c**) Resistance change of different human motions during boxing activity. (**d**) Data collection and result prediction process. (**e**) Target class and accuracy of the system. Reported results indicate that boxing indication shows an accuracy of >90%. Overall, resistance to change during different human motions enables real-time monitoring of physical activities. (**f**) Classification test confusion matrix of boxing recognition derived from the tactile glove. Reproduced with permission from open access article [149].

4.3. Monitoring Human Motions Like Squatting, Walking, or Running

Physical activity can be monitored by enabling the real-time tracking of human motions like squatting, walking, or running. The sensor involved helps in providing insights into biomechanics and performance metrics and is often helpful to avoid possible injury [158]. As discussed already, the sensors used in physical activity monitoring involve various useful technologies. These technologies are like accelerometers, gyroscopes, or in some cases equipped with global positioning systems. These sensors are generally integrated with various wearable devices like smartwatches or fitness trackers [159]. The tracker's examinations of parameters like joint angles, acceleration, velocity, and force distribution are useful for potential improvements during monitoring [160]. Yeo et al. [161] reported human motion monitoring like squatting, walking, or running. The detailed protocols of the motions are presented in Figure 8a. The figures detail the squatting, walking, and running protocols for a season and respective timings during data collection. These movements are like squatting, walking, or running [162]. The raw ECG quality, as shown in Figure 8b, shows the overall output from the different human motion protocols. The total protocols of one set of the activities involved are reported in Figure 8c. The final results can be monitored by filtering the data to remove noise-related errors and analyzed to target applications like squatting, walking, or running. Moreover, the sensors can provide feedback on movement patterns and ensure they adhere to prescribed exercises [163]. One key advantage of these sensors is their ability to provide real-time feedback to users. This quality helps the user to immediately correct the feedback and helps to adjust and achieve their goals more effectively. Irrespective of the many advantages of such sensors for monitoring human motions, they also have some limitations. For example, accuracy issues, data interpretation complexities, and user comfort concerns. Moreover, ensuring data privacy and security is crucial, especially when dealing with monitoring [164,165].

4.4. Self-Powered Energy Harvesting Under Different Mechanical Loads

This self-powered green energy harvesting is promising and hoped for a polluted environment and thus studied extensively. There are various ways of harvesting mechanical energy like vibrations, motion-induced, strain-based, or under different mechanical loads through piezoelectricity [166]. These can be understood more efficiently as when the composites were subjected to external vibration mechanical loads from sources like machinery, vehicles, or natural sources that can generate energy through piezoelectric or electromagnetic mechanisms [167,168]. These piezoelectric materials can be lead zirconate or polyvinylidene fluoride that generate energy in response to external mechanical stimuli or vibration sources. However, electromagnetic composites that contain electromagnetic materials generate energy through the motion-induced variation of magnetic fields within the composite structure [169]. Motion-induced energy harvesting involves the generation of voltage through human or vehicle motions. The energy generation in such cases uses mechanisms like piezoelectric, triboelectric, or electromagnetic induction for energy harvesting [170]. The triboelectric mechanisms involve the use of the contact and separation of dissimilar materials within the composite structure. This prototype finally assists in generating electrical charges through the triboelectric effect [171]. The electromagnetic induction mechanisms involve energy generation when composite material is subjected to different magnetic fields due to relative motions, as described by Lai et al. [172].

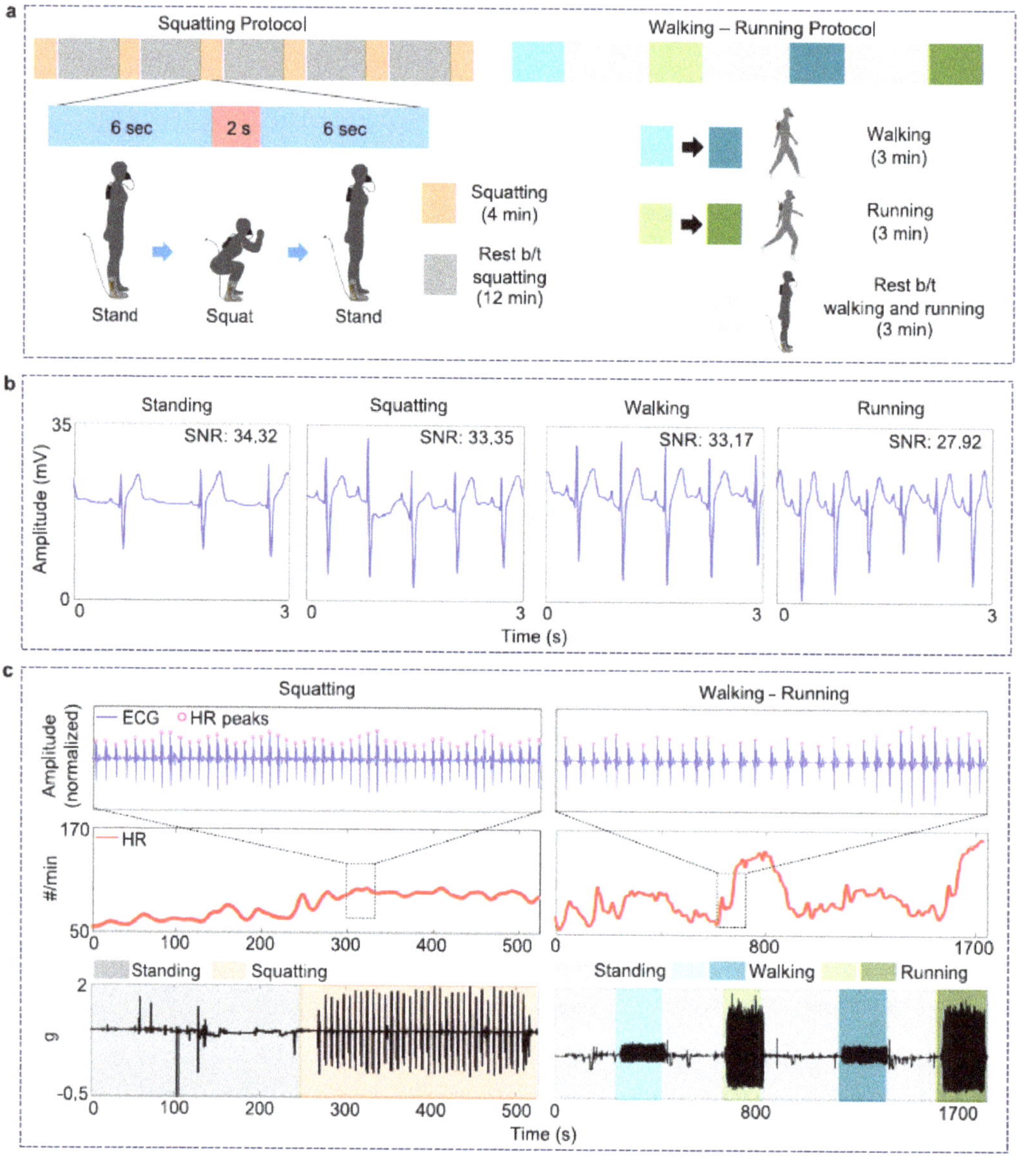

Figure 8. (**a**) Squatting, walking and running protocols. Respective data, like electro-cardio graphs (ECG) obtained through these sensors from physical activities, can be collected by involving algorithms. These algorithms analyze sensor data to recognize and quantify specific movements. (**b**) Output of ECG from different human motions. (**c**) Total protocols of one set of the activities involved. Normalized ECG peaks provide respective respiration rate from protocols. Final results evaluate mechanical stability of sensor during real-time monitoring with reduced noise and remarkable ECG signal quality. Reproduced with permission from open access article [161].

In the present review, energy harvesting through piezoelectric principles is reported. The main mechanism of energy harvesting through the piezoelectric principle involves generating output voltage under the mechanical deformation of these composites. This involves the addition of piezoelectric materials in an elastomeric matrix [173]. The composites were then deformed under mechanical loads on the prototype used for energy generation. Then, the dipoles in the substrate containing these piezoelectric composites are polarized. Thereby, the further mechanical deformation helps further opposite change separation and finally migrated to electrodes through capacitance [174–176]. Finally, this charge can be harvested from the electrode, and the output voltage in the form of electrical energy can be obtained. Park et al. [176] report the setup for harvesting energy under different mechanical loads. Figure 9A reports the setup for obtaining output voltage through piezoelectric materials. The different magnitudes of mechanical loads range from 0.5 N to 4.5 N provided through a universal testing machine as shown in Figure 9A. Figure 9B reports the output voltage under different loads, and Figure 9C shows the maximum voltage generated under particular compressive force.

The relationship between the output voltage and mechanical force can be linear up to a certain mechanical force, like up to 3.5 N, or non-linear after a certain mechanical force like 4.5 N [176]. This change in output voltage under different mechanical forces depends critically on the type of materials used during fabrication, the composition of the ingredients, and the magnitude of mechanical force [177,178]. However, the composite material can exhibit piezoelectric or piezoresistive properties as per need. For example, they can generate an output electrical response under mechanical stress or strain as suggested by Kumar et al. [179]. Finally, Figure 9D shows the electro-mechanical behavior under cyclic mechanical force of 2.5 N and durability tests in Figure 9E. The results show that the output voltage was stable during the initial cycles, like in Figure 9D, but started degrading after ~500 cycles, as reported in Figure 9E in durability tests. There are various reasons for the degradation of the output voltage after 500 cycles. The main reason is crack formation in the composite electrode and substrate. These cracks can lead to changes in the material properties, such as reduced piezoelectric or piezoresistive coefficients like those reported by Manikkavel et al. [180].

This results finally results in a degraded output voltage. The fatigue leading to microstructural damage in mechanical or electrical components. Under repetitive loading-unloading cycles, the fatigue-related degradation can result in a decrease in the output voltage generated [181]. Here, the voids or fabrication defects in the composites and electrical contact degradations at the substrate-electrode interface region [182]. Therefore, proper materials with high fracture toughness must be selected to obtain high durability of the composites for obtaining promising performance and overcoming mechanical failure of the composites [183]. Thus, the role of improving fracture toughness in composites for obtaining good durability with minimum losses is described in Figure 10 in the coming section.

4.5. Current–Voltage Curves, Real-Time Monitoring, and Durability of Composites

It is well known that improving fracture toughness is crucial for obtaining good electro-mechanical stability and mechanical deformation. The high fracture toughness of the composite enables the composite material to resist crack development and propagation [184]. Such resistance to crack propagation helps in improving mechanical integrity under cyclic loading and thus good durability. Moreover, if materials with high fracture toughness are used, they will prevent mechanical failure under mechanical cyclic loading [185]. This will be especially useful for accurate and continuous monitoring of the composites for strain-sensing applications. Higher fracture toughness helps to ensure the accuracy and consistency of strain measurements over time [186]. This is because the mechanical degradation of the strain sensor is expected to be negligible, thereby maintaining the accuracy of the durability tests under constant cyclic loading.

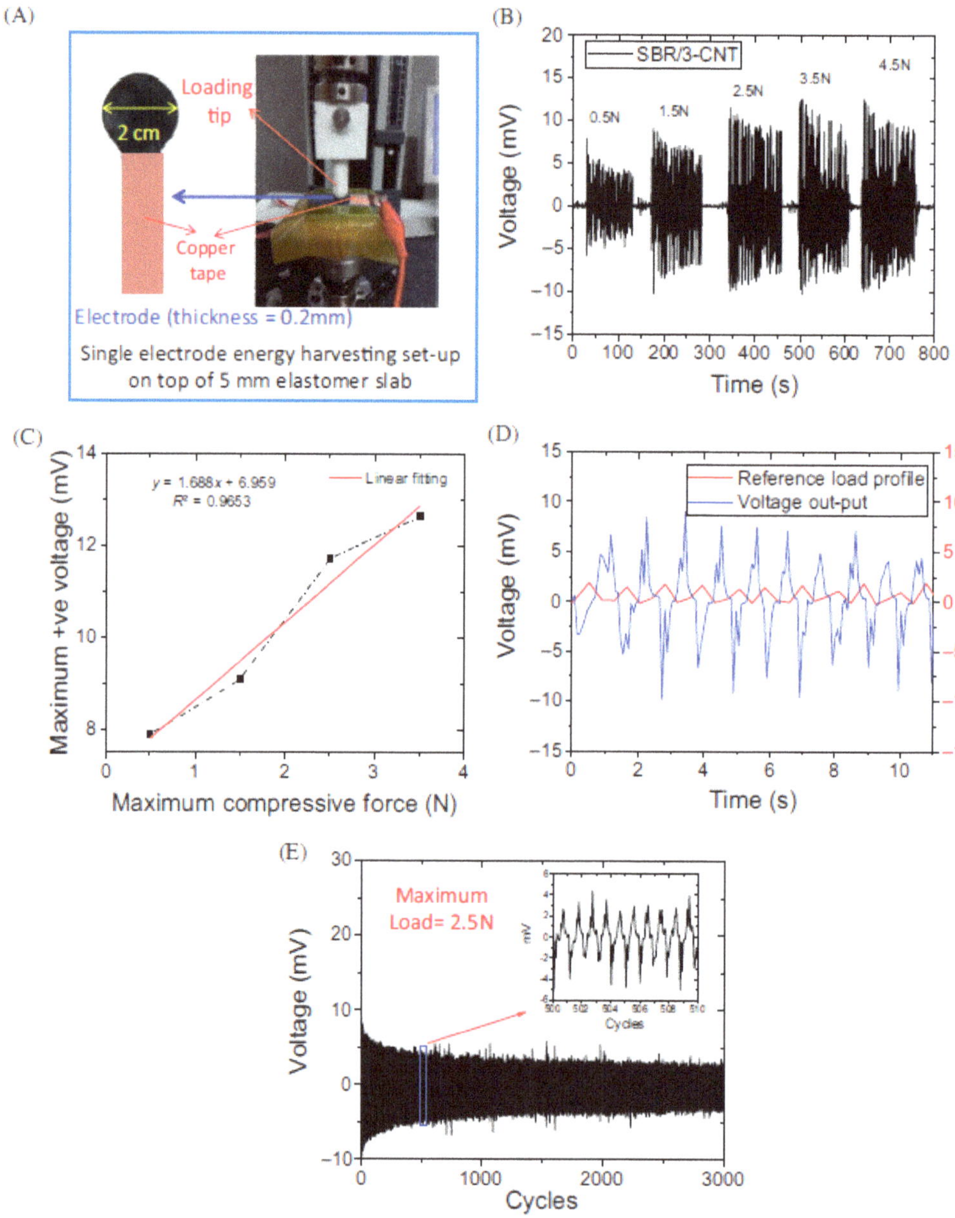

Figure 9. (**A**) Optical image of set-up. Loading tip was around 2 cm in diameter, and copper electrodes were attached to the flexible electrode for collecting the output voltage through a multimeter. (**B**,**C**) Output voltage under different forces. These reported results show that output voltage keeps increasing with increasing mechanical force up to 3.5 N before lowering down at 4.5 N force. (**D**) Output voltage under cyclic loading. (**E**) Durability examinations at a load of 2.5 N. Voltage was stable during the initial cycles but started to degrade after 500 cycles. This degradation is due to crack formation in the electrode and at filler–rubber interface failure. Reproduced with permission from Wiley [132].

Here, Alam et al. [187] report the piezoresistive behavior of the composites with higher fracture toughness in Figure 10. Figure 10a reports the behavior of the current–voltage (I–V) curve for the composite used in strain sensing. It can be understood as variations in electrical resistance or piezoresistance with a change in the magnitude of the applied strain [188]. These I-V curves depict the correlation between applied strain and resulting electric current in the composite sample. The other prospects in I-V curves involve their linear nature at small strain magnitude following Ohm's law. However, at higher strains, the I-V nature is non-linear and leads to a change in slope and threshold voltage [189]. These I-V curves are useful as they can assist in estimating strain sensitivity, gauge factor, hysteresis, and linearity in strain-sensing applications. Figure 10b,c show the change in relative resistance concerning unidirectional and cyclic strain. Moreover, the relative resistance was higher at higher cyclic strains. This can be understood through an increase in inter-particle distance between electrically conductive fillers at higher tensile strain [190, 191]. Moreover, as such piezoresistive materials exhibit an enhanced sensitivity to a higher magnitude of tensile strain. For example, as the material undergoes more significant stretching, its electrical resistance changes more noticeably at a given strain [192]. The main mechanism behind such effects involves the re-orientation of electrically conductive pathways within the composite material. This leads to a more pronounced change in relative resistance at higher levels of tensile strain [187].

The importance of fracture toughness was further validated by Alam et al. [187] by real-time monitoring of finger bending at different angles in Figure 10d. The main mechanism involves transferring mechanical deformation into relative resistance change with finger bending [193,194].

For example, as the finger bends, the sensor undergoes strain, causing changes in its electrical properties, such as resistance. The results show that the relative resistance changes with an increase in finger bending. Such changes can be studied and converted into meaningful data to represent the extent of finger bending. Finally, the durability of the strain sensor prototype is reported in Figure 10e. There are various aspects of improving durability by tuning mechanical stiffness. For example, the composite materials used in fabricating strain sensors are subjected to external mechanical load to obtain a stable piezoresistive response. However, it is subjected to mechanical failure after a certain number of cycles. Thus, Alam et al. [187] report the concept of improving fracture toughness to obtain stable and long durability with minimum voltage losses. For that, the MWCNT was added to the electrode in hybrid to MoS_2 to obtain highly stable durability cycles [187]. The main mechanism behind this higher mechanical stability involves the reduction in the curing process of silicone rubber. It results in the development of uniform crosslink density and thus uniform and better mechanical properties [187]. The literature summary about the sensors and their properties are summarized in Table 1.

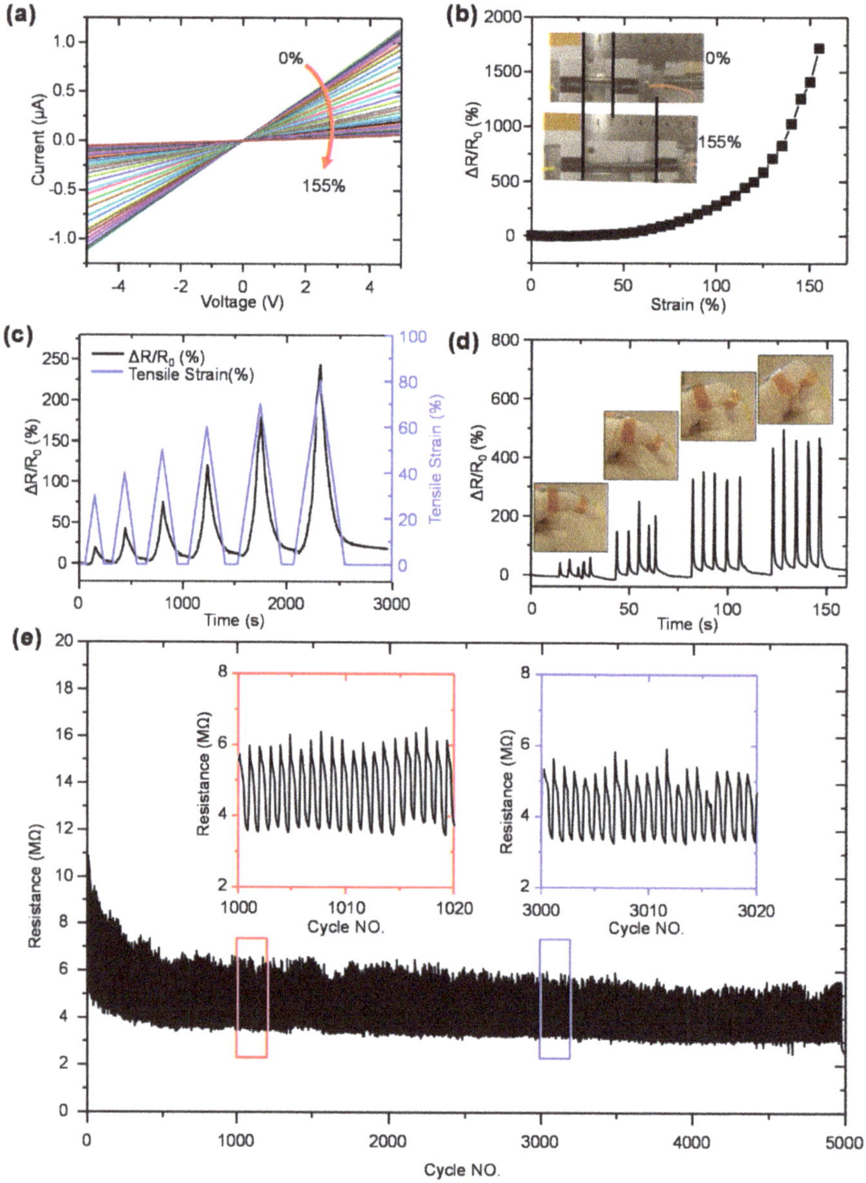

Figure 10. (**a**) Current–voltage (I-V) profiles of the samples under 0–155% strain. These I-V curves provide crucial insights into the nature of electrical behavior under change in mechanical loading. (**b**) Resistance change under uniaxial tensile strain. Results show that resistance to change increases to as high as 1719% at 155% strain. (**c**) Relative resistance change under tensile strain. (**d**) Real-time monitoring of sample by finger bending at different angles. These piezoresistive sensors can detect changes in shape or curvature as the finger is bent. (**e**) Durability cycle of sample. Study shows that adding MoS_2 in hybrid with MWCNT improves fracture toughness and thereby excellent durability with negligible losses for up to 3000 cycles. Reproduced with permission from Elsevier [187].

Table 1. Literature summary on different aspects of composite-based sensors.

Materials Used During Fabrication *	Modulus	Elongation at Break	Electrical Properties	Response Time	Gauge Factor	Reference
MWCNT, SR	6.5 MPa	210%	0.5 kΩ	-	~1.4 at 125%	[18]
LIG, PI, PDMS	-	-	250 Ω	-	75 at 5%	[19]
NCB, PDMS				15 ms	20.97 at 35%	[27]
CNF, PDMS	-	-	10 kΩ/sq	~100 ms	11.11 at 50%	[46]
CNF, PDMS	-	-	-	~50 ms	417 at 100%	[47]
CNT, GNP, SR	11 MPa	-	0.7 kΩ	-	-	[60]
CNT, GNP, CB, SR	7.5 MPa	250%	-	-	7.8 at 100%	[64]
CB, f-GNP, SR	4 MPa	170%	5 kΩ	-	-	[87]
MWCNT, TiC, SR	3.87 MPa	210%	-	100–300 ms	23.9 at 22%	[91]
MWCNT, f-PDMS	-	-	-	12.88 ms	22.64 at 225%	[98]
CNS, TPU	-	200%	30 kΩ	-	57.94 at 100%	[107]
CFM, GO/PAN	-	50%	-	<300 ms	10,688 at 15%	[109]
PVA, Bn, PEI, MXene	-	-	-	120 ms	-	[111]
SnS_2, PDMS	-	-	-	-	3933 at 1.2%	[112]
CNT, UHPC,	166.3 MPa	-	12.95 kΩ	-	28	[117]
MWCNT, Ecoflex	-	1300%	-	-	6.86 at 100%	[120]
Alumina, Kapton	-	120%	-	-	2.6×10^8 at 70%	[128]
MWCNT, SBR	12 MPa	200%	-	-	48.62 at 187%	[132]
MWCNT, SR	3.5 MPa	220%	-	200 and 125 ms	3.7 at 40%	[136]
MWCNT, GNP, SR	9 MPa	175%	-		32.04 at 100%	[139]
PAA-PVA hydrogels	-	30%	-	-	2.03 at 15%	[141]
CIP, CB, PDMS		80%		96 and 50 ms	7.74 at 80%	[156]
MWCNT, MoS_2, SR	0.9 MPa	325%			25.95 at 155%	[187]
SNW, TPU		360%			>16,000 at 360%	[190]
CCB, PVDF, ENR		60%	4 MΩ		1560 at 40%	[191]
CCB, Graphene, SR	7.7 MPa	200%			1.89×10^4 at 44%	[195]
MWCNT, Thinner, SR	1.14 MPa	200%	-	540–110 ms	0.94 at 40%	[196]

* Abbreviations: MWCNT = multi-wall carbon nanotubes, SR = silicone rubber, LIG = laser-induced graphene, PI = polyimide, PDMS = polydimethylsiloxane, NCB = nano carbon black, CNF = carbon nano fibers, CNT = carbon nanotubes, GNP = graphene nanoplatelets, CB = carbon black, f-GNP = functionalized graphene nanoplatelets, TiC = titanium carbide, CNS = carbon nano structures, TPU = thermoplastic polyurethane, GO/PAN = graphene oxide polyanaline, PVA = polyvinyl alcohol, PEI = polyetherimide, UHPC = ultra-high-performance concrete, SBR = styrene butadiene rubber, CIP = carbonyl iron particles, MoS_2 = molybdenum disulfide, CCB = conducting carbon black, PVDF = polyvinylidene fluoride, ENR = epoxied natural rubber.

5. Conclusions

The present review article reports the prospects of carbon nanomaterials-reinforced silicone rubber composites. The use of these composites for industrial applications like physical activity monitoring sensors is reviewed for the years 2020 to 2025. The literature shows that the carbon-based material can be very promising for adding to silicone rubber during composite fabrications. For example, they improve the mechanical, electrical, and thermal properties of silicone rubber when used. These improved properties make these composites useful for industrial applications like sensors. The review provides further insight that the sensors studied through the literature survey show that these carbon materials not only make the composites based on them robust but also feasible for their multi-functionality. Moreover, silicone rubber acts as a great supporting rubber matrix because of its easy processing and electro-active properties. Finally, the literature survey shows that sensors based on silicone rubber and carbon materials exhibit high sensitivity, optimum relative resistance, high gauge factor, and, finally, robust durability. Therefore, the present review article allows the reader to understand the advantages of using these composites for various industrial uses as sensors or energy harvesting.

5.1. Advantages and Challenges

This review presents the robustness of the sensing technique and the role of filler, rubber matrix, and their properties in influencing the performance output. The review further insights that the use of frequently used carbon nanomaterials is novel for achieving high performance. These frequent materials are carbon black, carbon nanotubes, and graphene. Moreover, silicone rubber arises as the best rubber matrix to achieve such high performance. The advantages and challenges of these composites are discussed briefly below.

For carbon black prospects, there are various advantages to using carbon black, such as its cost-effectiveness and high surface area. Thus, it is traditionally used as a filler to reinforce the rubber matrix, as reported by Xu et al. [197]. These high mechanical properties make it useful in sensing with high performance. However, there are some challenges, such as the aggregation of carbon black particles when mixed with a silicone rubber matrix. Another challenge is its limited sensitivity, which leads to a poor gauge factor compared to carbon nanotubes and graphene, as reported by Song et al. [195]. Similarly, for the carbon nanotubes prospects, there are various advantages of using carbon nanotubes in sensing applications. They include high sensitivity and high gauge factors due to excellent electrical properties and high aspect ratio as reported by Pei et al. [198]. However, there are some challenges, such as achieving uniform dispersion. However, the functionalization of carbon nanotubes can be used to achieve good filler–rubber compatibility. Therefore, it helps to achieve uniform dispersion, but the process is challenging and involves acids and other hazardous materials like those presented by Peng et al. [199]. Moreover, it is relatively expensive compared to traditional fillers like carbon black. Finally, obtaining the proper orientation of carbon nanotubes is critical for optimum mechanical performance.

Then, for graphene prospects, the graphene exhibits various extraordinary mechanical and electrical properties, making it ideal for sensing applications. Its high reinforcing properties enhance durability and performance under mechanical strain, as suggested by Guo et al. [200]. However, there are many challenges, such as production and scalability, integration, and re-staking among the graphene layers due to van der Waals forces [201]. Therefore, new methods need to be developed to synthesize large-scale graphene production to exploit their outstanding properties and robust sensing. Silicone rubber is extensively used in sensing applications due to its excellent flexibility, biocompatibility, and high durability. It also serves as an excellent matrix material for fillers like carbon nanomaterials and creates robust sensors, as reported by Kumar et al. [196]. However, there are some challenges, like the dispersion of fillers, balance between conductivity and flexibility etc. Some other challenges are its low sensitivity, hysteresis losses, mechanical stability, and integration in sensing devices, as noted by Chen et al. [202]. Moreover, the medical silicone rubber grade exhibits toxicity that is added by the curing agents and fillers used in reinforcing the rubber matrix. Therefore, the right strategy must be followed to obtain balance properties like optimum stiffness, low toxicity, and high gauge factor or sensitivity for sensing applications, as indicated by Rahmani et al. [203].

5.2. Future Prospects of the Composites

The carbon nanomaterial-reinforced silicone rubber-based composites hold promise to revolutionize the sensing field. These applications include healthcare, environmental monitoring, structural health monitoring. Moreover, wearable technology is also promising perspective of these sensors, as suggested by Haghi et al. [204]. The unique mechanical and electrical properties of these composites provide a promising route for obtaining robust composites with high sensing capacity. For carbon nanotube prospects, carbon nanotube-based composites develop more sensitive and accurate wearable healthcare devices. These devices are useful for tracking vital signs such as heart rate and blood pressure [205]. Moreover, the advanced integrated system of fitness trackers can be useful to monitor physical activity, strain, and stress on muscles and joints. These composites are also useful for designing flexible electronics. For example, electronic skins can mimic human touch and pressure sensitivity for use in robotics, as reported by Palumbo et al. [206]. Finally, these composites are useful for environmental monitoring, including air and water quality. For example, the creation of more sensitive air and water quality sensors capable of detecting trace amounts of pollutants in air and water, as reported by Nasture et al. [207]. Graphene holds promising insights to develop high-performance sensors. These sensors include strain sensors, pressure sensors, chemical sensors, and biological sensors, as suggested by Li et al. [208]. In addition, graphene is promising to create self-powered sensing systems like energy harvesting or energy storage like batteries. The use of graphene also helps to achieve transparent and flexible electronics [209]. For example, the development of touchscreens, displays, and flexible circuits is available using graphene-based composites.

For carbon black prospects, carbon black holds promising solutions for cost-effective gadgets for sensing and reinforcing aspects of composites. For example, by using it in conductive rubber composites, flexible and stretchable sensors with improved mechanical properties can be created [210]. Moreover, carbon black can be useful in creating sensors for automotive applications, such as tire pressure monitoring and strain detection in vehicle components. Therefore, these composites are useful for multifunctional aspects in a wide range of applications, such as electronics, tires, and health monitoring sensing [211]. Silicone rubber reinforced with carbon nanomaterials also holds promising prospects. The first use of silicone rubber was in medical implants. This involves the development of biocompatible sensors for long-term monitoring of physiological parameters within the body, as reported by Kim et al. [212]. Moreover, they can be useful for wearable health monitoring. For example, it can achieve enhanced capability for real-time data collection and analysis for proactive health management based on silicone rubber. In addition, silicone rubber is also useful for environment sensing, structural health monitoring, and finally, robotics, as reported by Horne et al. [213]. Therefore, the composite reviewed in this paper has vast and promising prospects. The combination of these materials offers significant improvements in sensitivity, flexibility, durability, and biocompatibility, as reported by Han et al. [214]. These features articulate the way for advanced sensors in healthcare, environmental monitoring, and structural health monitoring.

5.3. Overview of Cost-Effectiveness, Biocompatibility, and Scalability for These Sensors

As discussed in this review paper, the composite-based sensors hold a promising front for high-performance sensors. These aspects are due to their flexibility, adaptability, and low mechanical modulus. However, their practical adoption requires overcoming factors such as cost-effectiveness, biocompatibility, and scalability. These aspects are crucial for commercial viability, particularly in fields like wearable electronics.

(a) **Cost-effectiveness**: For the sensors reviewed in this paper, a key advantage of rubber composite sensors is their potential for low-cost production. Generally, most of the materials, such as silicone rubber and polyurethane, are relatively inexpensive and abundant in nature. This makes the fabrication of these sensors at large scale feasible. Moreover, many commonly used conductive fillers (e.g., carbon black or graphite) offer a cost-effective option. These cheap fillers are alternatives to high-cost nanomaterials like graphene or metallic nanoparticles. However, cost challenges can arise when advanced fillers such as monolayer graphene or silver nanowires are used to achieve high performance. Another challenge is the use of complex fabrication techniques such as laser patterning and 3D printing. Overall, balancing performance with economic feasibility remains a central focus, especially for applications like single-use sensors.

(b) **Biocompatibility**: This review summarizes the importance of biocompatibility for the use of these sensors for wearable applications. For example, biocompatibility is essential for applications involving direct contact with human skin or implantation. Many rubber matrices, such as medical-grade silicone and thermoplastic polyurethanes (TPUs), are inherently biocompatible and useful with low-to-no toxicity. Moreover, they have a long history of use in biomedical devices. The choice of fillers, however, significantly affects biocompatibility. For example, the carbon-based fillers like CNTs and graphene in higher concentrations could make the composite toxic. Low content of fillers generally shows acceptable biocompatibility when embedded in a stable matrix. However, concerns remain about potential cytotoxicity if the fillers are released due to material degradation. To overcome this challenge, encapsulation layers and surface treatments are often employed to mitigate for long-term safety.

(c) **Scalability**: This review work gives insight into the prospects of scalability for these sensors. Here, the scalability is a major factor influencing the transition from research prototypes to market-ready products. Recently, simple processing techniques such as solution casting, screen printing, and roll-to-roll manufacturing have had a promising effect on scalability. For example, these techniques enabled the scalable production of rubber composite films and sensors. Therefore, these methods are well-suited for producing large-area or batch-fabricated devices. Nonetheless, challenges in maintaining uniform filler dispersion and ensuring integrating sensors into complex devices. These devices are useful as wearables or smart textiles and can limit scalability. Additionally, multi-step processes or post-fabrication treatments can increase complexity and reduce throughput.

5.4. Overview of Composite Sensors Compared to Alternatives Like Textile-Based or Capacitive Sensors

(a) **Composite sensor**: The present review describes the use of composite-based nano-invasive sensors and their use to monitor physical activities. These composite sensors are made by combining two or more different materials. These materials contain polymers and electrically conductive fillers like carbon nanotubes, graphene, or metal particles. This review paper further insight that these composites-based sensors can sense physical changes. These physical activities can be running, jumping, or walking via the piezoresistive principle. For example, the pressure, strain, and temperature originate from physical activities, and the sensor performs the change through changes in electrical properties. The key features of these sensors include high flexibility, light weight, tunable sensitivity, and multimodal sensing. For example, the tuning of the sensitivity of these sensors can be performed through filler content. Moreover, their capacity for multimodal sensing involves their sensitivity to changes in pressure, strain, or temperature. Finally, their light weight makes them useful for soft and wearable applications. However, there are some limitations to these sensors, like uniform filler dispersion, signal draft, or hysteresis losses, and finally, the durability is a great issue.

(b) **Textile-based sensors**: Textile sensors are greatly useful and made from conductive threads or coatings integrated into the fabrics. These sensors are also piezoresistive, where sensing is achieved through a change in resistance. The key features of these sensors are their seamless integration with garments, their light weight, and their ease of wearing. Due to these features, these sensors are comfortable to wear, scalable, washable, and washable. Moreover, they have high flexibility and are easy to fabricate. However, some challenges involve lower durability, higher sensitivity, and sensitivity to the environment. The sensitivity to the environment includes the change in temperature or humidity in the surrounding environment, which results in a change in resistance. Moreover, these sensors have limited multifunctionality and have less ability to detect multiple stimuli.

(c) **Capacitive sensors**: Capacitive sensors work by studying change in capacitance caused by mechanical deformations. These types of sensprs are frequently used in touchscreens and pressure sensors. The key features of these sensors include fast response time, high sensitivity, and moderate flexibility. The advantages of using these sensors include high precision, non-contact sensing, and low power consumption. Due to their low power consumption, these sensors are frequently useful for military applications. However, these sensors have some limitations, such as sensitivity to humidity, limited flexibility, and sensitivity to external noise. Moreover, these sensors have limited multimodal sensing, moderate durability, and moderate-to-tough fabrication processes.

Funding: This work received no external funding.

Institutional Review Board Statement: Not applicable.

Data Availability Statement: Not applicable.

Conflicts of Interest: The authors declare no conflicts of interest.

References

1. Javed, A.R.; Faheem, R.; Asim, M.; Baker, T.; Beg, M.O. A smartphone sensors-based personalized human activity recognition system for sustainable smart cities. *Sustain. Cities Soc.* **2021**, *71*, 102970. [CrossRef]
2. Sousa Lima, W.; Souto, E.; El-Khatib, K.; Jalali, R.; Gama, J. Human activity recognition using inertial sensors in a smartphone: An overview. *Sensors* **2019**, *19*, 3213. [CrossRef] [PubMed]
3. Zekavat, S.; Buehrer, R.M.; Durgin, G.D.; Lovisolo, L.; Wang, Z.; Goh, S.T.; Ghasemi, A. An overview on position location: Past, present, future. *Int. J. Wirel. Inf. Netw.* **2021**, *28*, 45–76. [CrossRef]
4. Muhammad Sayem, A.S.; Hon Teay, S.; Shahariar, H.; Luise Fink, P.; Albarbar, A. Review on smart electro-clothing systems (SeCSs). *Sensors* **2020**, *20*, 587. [CrossRef]
5. Ahsan, M.; Hon, S.T.; Albarbar, A. Development of novel big data analytics framework for smart clothing. *IEEE Access* **2020**, *8*, 146376–146394. [CrossRef]
6. Meng, K.; Xiao, X.; Liu, Z.; Shen, S.; Tat, T.; Wang, Z.; Chen, J. Kirigami-inspired pressure sensors for wearable dynamic cardiovascular monitoring. *Adv. Mater.* **2022**, *34*, 2202478. [CrossRef]
7. Taylor, J.L.; Bonikowske, A.R.; Olson, T.P. Optimizing outcomes in cardiac rehabilitation: The importance of exercise intensity. *Front. Cardiovasc. Med.* **2021**, *8*, 734278. [CrossRef]
8. Qi, J.; Yang, P.; Newcombe, L.; Peng, X.; Yang, Y.; Zhao, Z. An overview of data fusion techniques for Internet of Things enabled physical activity recognition and measure. *Inf. Fusion* **2020**, *55*, 269–280. [CrossRef]

9. Zhang, C.; Zhang, C.; Wu, X.; Ping, J.; Ying, Y. An integrated and robust plant pulse monitoring system based on biomimetic wearable sensor. *NPJ Flex. Electron.* **2022**, *6*, 43. [CrossRef]
10. Tan, Y.; Liu, X.; Tang, W.; Chen, J.; Zhu, Z.; Li, L.; Li, H. Flexible pressure sensors based on bionic microstructures: From plants to animals. *Adv. Mater. Interfaces* **2022**, *9*, 2101312. [CrossRef]
11. Nascimento, L.M.S.D.; Bonfati, L.V.; Freitas, M.L.B.; Mendes Junior, J.J.A.; Siqueira, H.V.; Stevan, S.L., Jr. Sensors and systems for physical rehabilitation and health monitoring—A review. *Sensors* **2020**, *20*, 4063. [CrossRef] [PubMed]
12. Ye, S.; Feng, S.; Huang, L.; Bian, S. Recent progress in wearable biosensors: From healthcare monitoring to sports analytics. *Biosensors* **2020**, *10*, 205. [CrossRef] [PubMed]
13. Giordano, M.R.; Malings, C.; Pandis, S.N.; Presto, A.A.; McNeill, V.; Westervelt, D.M.; Beekmann, M.; Subramanian, R. From low-cost sensors to high-quality data: A summary of challenges and best practices for effectively calibrating low-cost particulate matter mass sensors. *J. Aerosol Sci.* **2021**, *158*, 105833. [CrossRef]
14. Ling, Y.; An, T.; Yap, L.W.; Zhu, B.; Gong, S.; Cheng, W. Disruptive, soft, wearable sensors. *Adv. Mater.* **2020**, *32*, 1904664. [CrossRef]
15. Hu, P.; Madsen, J.; Skov, A.L. One reaction to make highly stretchable or extremely soft silicone elastomers from easily available materials. *Nat. Commun.* **2022**, *13*, 370. [CrossRef]
16. Yang, W.; Lin, Y.; Zhu, Y.; Zhen, C.; Tao, W.; Luo, Y.; Wang, X. Synergistic effect of dual-modification strategy on thermal conductivity and thermal stability of h-BN/silicone rubber composites: Experiments and simulations. *Int. Commun. Heat Mass Transf.* **2025**, *163*, 108716. [CrossRef]
17. Kosmela, P.; Sałasińska, K.; Kowalkowska-Zedler, D.; Barczewski, M.; Piasecki, A.; Saeb, M.R.; Hejna, A. Fire-Retardant Flexible Foamed Polyurethane (PU)-Based Composites: Armed and Charmed Ground Tire Rubber (GTR) Particles. *Polymers* **2024**, *16*, 656. [CrossRef]
18. Kumar, V.; Lee, G.; Choi, J.; Lee, D.J. Studies on composites based on HTV and RTV silicone rubber and carbon nanotubes for sensors and actuators. *Polymer* **2020**, *190*, 122221. [CrossRef]
19. Wang, H.; Zhao, Z.; Liu, P.; Guo, X. A soft and stretchable electronics using laser-induced graphene on polyimide/PDMS composite substrate. *npj Flex. Electron.* **2022**, *6*, 26. [CrossRef]
20. Guo, Y.; Qiu, H.; Ruan, K.; Wang, S.; Zhang, Y.; Gu, J. Flexible and insulating silicone rubber composites with sandwich structure for thermal management and electromagnetic interference shielding. *Compos. Sci. Technol.* **2022**, *219*, 109253. [CrossRef]
21. Ullah, R.; Akbar, M.; Amin, S. Measuring electrical, thermal and mechanical properties of DC-stressed HTV silicone rubber loaded with nano/micro-fillers exposed to long-term aging. *Appl. Nanosci.* **2020**, *10*, 2101–2111. [CrossRef]
22. Brook, M.A. Functional silicone oils and elastomers: New routes lead to new properties. *Chem. Commun.* **2023**, *59*, 12813–12829. [CrossRef]
23. Haghshenas, N.; Nejat, A.; Seyedmehdi, S.A.; Ou, J.; Amirfazli, A.; Chini, S.F. Adhesion aspects of silicone rubber coatings for high voltage insulators: A critical review. *Rev. Adhes. Adhes.* **2021**, *9*, 434–480.
24. Kumar, V.; Alam, M.N.; Park, S.S. Review of Recent Progress on Silicone Rubber Composites for Multifunctional Sensor Systems. *Polymers* **2024**, *16*, 1841. [CrossRef] [PubMed]
25. Kumar, V.; Alam, M.N.; Park, S.S. Review on functionalized CNTs reinforced silicone rubber composites for potential wearable applications. *Polym. Compos.* **2024**, *45*, 12503–12529. [CrossRef]
26. Wang, G.; Li, A.; Zhao, W.; Xu, Z.; Ma, Y.; Zhang, F.; He, Q. A review on fabrication methods and research progress of superhydrophobic silicone rubber materials. *Adv. Mater. Interfaces* **2021**, *8*, 2001460. [CrossRef]
27. Hu, J.; Yu, J.; Li, Y.; Liao, X.; Yan, X.; Li, L. Nano carbon black-based high performance wearable pressure sensors. *Nanomaterials* **2020**, *10*, 664. [CrossRef]
28. You, R.; Liu, Y.Q.; Hao, Y.L.; Han, D.D.; Zhang, Y.L.; You, Z. Laser fabrication of graphene-based flexible electronics. *Adv. Mater.* **2020**, *32*, 1901981. [CrossRef]
29. Yan, T.; Wu, Y.; Yi, W.; Pan, Z. Recent progress on fabrication of carbon nanotube-based flexible conductive networks for resistive-type strain sensors. *Sens. Actuators A Phys.* **2021**, *327*, 112755. [CrossRef]
30. Fan, Y.; Fowler, G.D.; Zhao, M. The past, present and future of carbon black as a rubber reinforcing filler–A review. *J. Clean. Prod.* **2020**, *247*, 119115. [CrossRef]
31. Khodabakhshi, S.; Fulvio, P.F.; Andreoli, E. Carbon black reborn: Structure and chemistry for renewable energy harnessing. *Carbon* **2020**, *162*, 604–649. [CrossRef]
32. Verma, A.; Baurai, K.; Sanjay, M.R.; Siengchin, S. Mechanical, microstructural, and thermal characterization insights of pyrolyzed carbon black from waste tires reinforced epoxy nanocomposites for coating application. *Polym. Compos.* **2020**, *41*, 338–349. [CrossRef]
33. Dwivedi, C.; Manjare, S.; Rajan, S.K. Recycling of waste tire by pyrolysis to recover carbon black: Alternative & environment-friendly reinforcing filler for natural rubber compounds. *Compos. Part B Eng.* **2020**, *200*, 108346.

34. Milone, G.; Vlachakis, C.; Tulliani, J.M.; Al-Tabbaa, A. Strain Monitoring of Concrete Using Carbon Black-Based Smart Coatings. *Materials* **2024**, *17*, 1577. [CrossRef]
35. Kumar, V.; Alam, M.N.; Manikkavel, A.; Song, M.; Lee, D.J.; Park, S.S. Silicone rubber composites reinforced by carbon nanofillers and their hybrids for various applications: A review. *Polymers* **2021**, *13*, 2322. [CrossRef]
36. Irani, F.S.; Shafaghi, A.H.; Tasdelen, M.C.; Delipinar, T.; Kaya, C.E.; Yapici, G.G.; Yapici, M.K. Graphene as a piezoresistive material in strain sensing applications. *Micromachines* **2022**, *13*, 119. [CrossRef]
37. Chen, L.; Li, N.; Yu, X.; Zhang, S.; Liu, C.; Song, Y.; Wang, Z. A general way to manipulate electrical conductivity of graphene. *Chem. Eng. J.* **2023**, *462*, 142139. [CrossRef]
38. Jen, Y.M.; Ni, W.L. Effect of dispersing multiwalled carbon nanotubes and graphene nanoplatelets hybrids in the matrix on the flexural fatigue properties of carbon/epoxy composites. *Polymers* **2022**, *14*, 918. [CrossRef]
39. Farooq, A.S.; Zhang, P. A comprehensive review on the prospects of next-generation wearable electronics for individualized health monitoring, assistive robotics, and communication. *Sens. Actuators A Phys.* **2022**, *344*, 113715. [CrossRef]
40. Nurazzi, N.M.; Sabaruddin, F.A.; Harussani, M.M.; Kamarudin, S.H.; Rayung, M.; Asyraf, M.R.M.; Aisyah, H.A.; Norrrahim, M.N.F.; Ilyas, R.A.; Abdullah, N.; et al. Mechanical performance and applications of cnts reinforced polymer composites—A review. *Nanomaterials* **2021**, *11*, 2186. [CrossRef]
41. Banerjee, S.; Kar, K.K. Characteristics of carbon nanotubes. In *Handbook of Nanocomposite Supercapacitor Materials I: Characteristics*; Springer: Berlin/Heidelberg, Germany, 2020; pp. 179–214.
42. Dong, W.; Peng, S.; Wang, K.; Huang, Y.; Shi, L.; Wu, F.; Li, W. Integrated triboelectric self-powering and piezoresistive self-sensing cementitious composites for intelligent civil infrastructure. *Nano Energy* **2025**, *135*, 110656. [CrossRef]
43. Cardenas, J.A.; Andrews, J.B.; Noyce, S.G.; Franklin, A.D. Carbon nanotube electronics for IoT sensors. *Nano Futures* **2020**, *4*, 012001. [CrossRef]
44. Deshmukh, M.A.; Jeon, J.Y.; Ha, T.J. Carbon nanotubes: An effective platform for biomedical electronics. *Biosens. Bioelectron.* **2020**, *150*, 111919. [CrossRef] [PubMed]
45. Sharma, R.; Kar, K.K. Characteristics of carbon nanofibers. In *Handbook of Nanocomposite Supercapacitor Materials I: Characteristics*; Springer: Berlin/Heidelberg, Germany, 2020; pp. 215–245.
46. Sengupta, D.; Chen, S.H.; Michael, A.; Kwok, C.Y.; Lim, S.; Pei, Y.; Kottapalli, A.G.P. Single and bundled carbon nanofibers as ultralightweight and flexible piezoresistive sensors. *npj Flex. Electron.* **2020**, *4*, 9. [CrossRef]
47. Peng, S.; Wu, S.; Yu, Y.; Sha, Z.; Li, G.; Hoang, T.T.; Thai, M.T.; Do, T.N.; Chu, D.; Wang, C.H. Carbon nanofiber-reinforced strain sensors with high breathability and anisotropic sensitivity. *J. Mater. Chem. A* **2021**, *9*, 26788–26799. [CrossRef]
48. Shi, S.; Mou, P.; Wang, D.; Li, X.; Teng, S.; Zhou, M.; Wang, G. Co/carbon nanofiber with adjustable size and content of Co nanoparticles for tunable microwave absorption and thermal conductivity. *J. Mater.* **2024**, *10*, 124–133. [CrossRef]
49. Wang, B.; Yin, X.; Cheng, R.; Li, J.; Ying, G.; Chen, K. Compressible, superelastic and fatigue resistant carbon nanofiber aerogels derived from bacterial cellulose for multifunctional piezoresistive sensors. *Carbon* **2022**, *199*, 318–328. [CrossRef]
50. Pan, Y.; Liu, X.; Zhang, W.; Liu, Z.; Zeng, G.; Shao, B.; Chen, M. Advances in photocatalysis based on fullerene C60 and its derivatives: Properties, mechanism, synthesis, and applications. *Appl. Catal. B Environ.* **2020**, *265*, 118579. [CrossRef]
51. Kausar, A. Fullerene nanofiller reinforced epoxy nanocomposites—Developments, progress and challenges. *Mater. Res. Innov.* **2021**, *25*, 175–185. [CrossRef]
52. Kausar, A. Potential of polymer/fullerene nanocomposites for anticorrosion applications in the biomedical field. *J. Compos. Sci.* **2022**, *6*, 394. [CrossRef]
53. Tavakkolnia, I.; Jagadamma, L.K.; Bian, R.; Manousiadis, P.P.; Videv, S.; Turnbull, G.A.; Samuel, I.D.W.; Haas, H. Organic photovoltaics for simultaneous energy harvesting and high-speed MIMO optical wireless communications. *Light Sci. Appl.* **2021**, *10*, 41. [CrossRef] [PubMed]
54. Andrew, J.J.; Dhakal, H.N. Sustainable biobased composites for advanced applications: Recent trends and future opportunities–A critical review. *Compos. Part C Open Access* **2022**, *7*, 100220. [CrossRef]
55. Luo, Y.; Abidian, M.R.; Ahn, J.-H.; Akinwande, D.; Andrews, A.M.; Antonietti, M.; Bao, Z.; Berggren, M.; Berkey, C.A.; Bettinger, C.J.; et al. Technology roadmap for flexible sensors. *ACS Nano* **2023**, *17*, 5211–5295. [CrossRef] [PubMed]
56. Demrozi, F.; Pravadelli, G.; Bihorac, A.; Rashidi, P. Human activity recognition using inertial, physiological and environmental sensors: A comprehensive survey. *IEEE Access* **2020**, *8*, 210816–210836. [CrossRef]
57. Sun, F.; Zhu, Y.; Jia, C.; Zhao, T.; Chu, L.; Mao, Y. Advances in self-powered sports monitoring sensors based on triboelectric nanogenerators. *J. Energy Chem.* **2023**, *79*, 477–488. [CrossRef]
58. Sahu, M.; Hajra, S.; Panda, S.; Rajaitha, M.; Panigrahi, B.K.; Rubahn, H.-G.; Mishra, Y.K.; Kim, H.J. Waste textiles as the versatile triboelectric energy-harvesting platform for self-powered applications in sports and athletics. *Nano Energy* **2022**, *97*, 107208. [CrossRef]
59. Dong, C.; Zhao, R.; Yao, L.; Ran, Y.; Zhang, X.; Wang, Y. A review on WO3 based gas sensors: Morphology control and enhanced sensing properties. *J. Alloys Compd.* **2020**, *820*, 153194. [CrossRef]

60. Kumar, V.; Wu, R.R.; Lee, D.J. Morphological aspects of carbon nanofillers and their hybrids for actuators and sensors. *Polym. Compos.* **2019**, *40*, E373–E382. [CrossRef]
61. Alves, A.M.; Cavalcanti, S.N.; da Silva, M.P.; Freitas, D.M.; Agrawal, P.; de Mélo, T.J. Electrical, rheological, and mechanical properties copolymer/carbon black composites. *J. Vinyl Addit. Technol.* **2021**, *27*, 445–458. [CrossRef]
62. Kumar, V.; Lee, D.J. Rubber nanocomposites reinforced with single-wall and multiwall carbon nanotubes for industrial applications. *Rubber Chem. Technol.* **2020**, *93*, 157–171. [CrossRef]
63. Thiruvengadam, M.; Rajakumar, G.; Swetha, V.; Ansari, M.A.; Alghamdi, S.; Almehmadi, M.; Halawi, M.; Kungumadevi, L.; Raja, V.; Sarbudeen, S.S.; et al. Recent insights and multifactorial applications of carbon nanotubes. *Micromachines* **2021**, *12*, 1502. [CrossRef] [PubMed]
64. Kumar, V.; Lee, G.; Singh, K.; Choi, J.; Lee, D.J. Structure-property relationship in silicone rubber nanocomposites reinforced with carbon nanomaterials for sensors and actuators. *Sens. Actuators A Phys.* **2020**, *303*, 111712. [CrossRef]
65. Kumar, V.; Tang, X. New horizons in nanofiller-based polymer composites II. *Polymers* **2023**, *15*, 4259. [CrossRef]
66. Sun, X.; Huang, C.; Wang, L.; Liang, L.; Cheng, Y.; Fei, W.; Li, Y. Recent progress in graphene/polymer nanocomposites. *Adv. Mater.* **2021**, *33*, 2001105. [CrossRef] [PubMed]
67. Cao, C.-F.; Yu, B.; Guo, B.-F.; Hu, W.-J.; Sun, F.-N.; Zhang, Z.-H.; Li, S.-N.; Wu, W.; Tang, L.-C.; Song, P.; et al. Bio-inspired, sustainable and mechanically robust graphene oxide-based hybrid networks for efficient fire protection and warning. *Chem. Eng. J.* **2022**, *439*, 134516. [CrossRef]
68. Wang, X.; Huang, S.C.; Hu, S.; Yan, S.; Ren, B. Fundamental understanding and applications of plasmon-enhanced Raman spectroscopy. *Nat. Rev. Phys.* **2020**, *2*, 253–271. [CrossRef]
69. Li, Z.; Deng, L.; Kinloch, I.A.; Young, R.J. Raman spectroscopy of carbon materials and their composites: Graphene, nanotubes and fibres. *Prog. Mater. Sci.* **2023**, *135*, 101089. [CrossRef]
70. Moseenkov, S.I.; Kuznetsov, V.L.; Zolotarev, N.A.; Kolesov, B.A.; Prosvirin, I.P.; Ishchenko, A.V.; Zavorin, A.V. Investigation of amorphous carbon in nanostructured carbon materials (A Comparative Study by TEM, XPS, Raman Spectroscopy and XRD). *Materials* **2023**, *16*, 1112. [CrossRef]
71. Thapliyal, V.; Alabdulkarim, M.E.; Whelan, D.R.; Mainali, B.; Maxwell, J.L. A concise review of the Raman spectra of carbon allotropes. *Diam. Relat. Mater.* **2022**, *127*, 109180. [CrossRef]
72. Roy, D.; Kanojia, S.; Mukhopadhyay, K.; Eswara Prasad, N. Analysis of carbon-based nanomaterials using Raman spectroscopy: Principles and case studies. *Bull. Mater. Sci.* **2021**, *44*, 31. [CrossRef]
73. Kumar, V.; Kumar, A.; Lee, D.J.; Park, S.S. Estimation of number of graphene layers using different methods: A focused review. *Materials* **2021**, *14*, 4590. [CrossRef] [PubMed]
74. Low, I.M.; Albetran, H.M.; Degiorgio, M. Structural characterization of commercial graphite and graphene materials. *J. Nanotechnol. Nanomater.* **2020**, *1*, 23–30.
75. Saleem, H.; Haneef, M.; Abbasi, H.Y. Synthesis route of reduced graphene oxide via thermal reduction of chemically exfoliated graphene oxide. *Mater. Chem. Phys.* **2018**, *204*, 1–7. [CrossRef]
76. Kumar, V.; Kumar, A.; Song, M.; Lee, D.J.; Han, S.S.; Park, S.S. Properties of silicone rubber-based composites reinforced with few-layer graphene and iron oxide or titanium dioxide. *Polymers* **2021**, *13*, 1550. [CrossRef]
77. Zhang, W.; Chai, C.; Fan, Q.; Song, Y.; Yang, Y. PBCF-Graphene: A 2D Sp2 Hybridized Honeycomb Carbon Allotrope with a Direct Band Gap. *ChemNanoMat* **2020**, *6*, 139–147. [CrossRef]
78. Kumar, P.; Divya, N.; Ratan, J.K. Study on the physico-chemical properties of reduced graphene oxide with different degrees of reduction temperature. *J. Iran. Chem. Soc.* **2021**, *18*, 201–211. [CrossRef]
79. Liang, T.; Zhang, Z.; Ma, X.; Xiang, Y.; Huang, H.; Jiang, X. Characteristics of powdered layer on silicone rubber surface. *J. Mater. Res. Technol.* **2021**, *14*, 36–46. [CrossRef]
80. Sarath, P.S.; Moni, G.; George, J.J.; Haponiuk, J.T.; Thomas, S.; George, S.C. A study on the influence of reduced graphene oxide on the mechanical, dynamic mechanical and tribological properties of silicone rubber nanocomposites. *J. Compos. Mater.* **2021**, *55*, 2011–2024. [CrossRef]
81. Krishna, D.N.G.; Philip, J. Review on surface-characterization applications of X-ray photoelectron spectroscopy (XPS): Recent developments and challenges. *Appl. Surf. Sci. Adv.* **2022**, *12*, 100332. [CrossRef]
82. Peng, C.; Zhang, X. Chemical functionalization of graphene nanoplatelets with hydroxyl, amino, and carboxylic terminal groups. *Chemistry* **2021**, *3*, 873–888. [CrossRef]
83. Sun, F.; Yang, S.; Wang, Q. Selective decomposition process and mechanism of Si–O–Si cross-linking bonds in silane cross-linked polyethylene by solid-state shear milling. *Ind. Eng. Chem. Res.* **2020**, *59*, 12896–12905. [CrossRef]
84. Chen, T.; Xie, Y.; Wang, Z.; Lou, J.; Liu, D.; Xu, R.; Xiao, X. Recent advances of flexible strain sensors based on conductive fillers and thermoplastic polyurethane matrixes. *ACS Appl. Polym. Mater.* **2021**, *3*, 5317–5338. [CrossRef]
85. Zhou, J.; Guo, X.; Xu, Z.; Wu, Q.; Chen, J.; Wu, J.; Huang, Z. Highly sensitive and stretchable strain sensors based on serpentine-shaped composite films for flexible electronic skin applications. *Compos. Sci. Technol.* **2020**, *197*, 108215. [CrossRef]

86. Kumar, V.; Lee, D.J.; Park, S.S. Multi-functionality prospects in functionalized and pristine graphene nanosheets reinforced silicone rubber composites: A focused review. *FlatChem* **2023**, *41*, 100535. [CrossRef]
87. Kumar, V.; Parvin, N.; Manikkavel, A.; Lee, D.Y.; Mandal, T.K.; Joo, S.W.; Park, S.S. Improved interfacial mechanical strength and synergy in properties of nano-carbon black reinforced rubber composites containing functionalized graphite nanoplatelets. *Surf. Interfaces* **2023**, *39*, 102941. [CrossRef]
88. Yang, R.; Zhang, W.; Tiwari, N.; Yan, H.; Li, T.; Cheng, H. Multimodal sensors with decoupled sensing mechanisms. *Adv. Sci.* **2022**, *9*, 2202470. [CrossRef]
89. Wu, S.; Peng, S.; Yu, Y.; Wang, C.H. Strategies for designing stretchable strain sensors and conductors. *Adv. Mater. Technol.* **2020**, *5*, 1900908. [CrossRef]
90. Kumar, V.; Alam, M.N.; Yewale, M.A.; Park, S.S. Stretchable rubber composites with lower hysteresis losses, improved magnetic effect, and a robust magnetic sensitivity. *Sens. Actuators A Phys.* **2024**, *373*, 115425. [CrossRef]
91. Manikkavel, A.; Kumar, V.; Park, S.S. Multifunctionality of MWCNT and TiC hybrid filler silicone composite for energy harvesting and strain sensing. *Polym. Compos.* **2024**, *45*, 6629–6643. [CrossRef]
92. Duan, Y.; Zhang, L.; Tian, D.; Liao, S.; Wang, Y.; Xu, Y.; Hu, Y. Modulus matching strategy of ultra-soft electrically conductive silicone composites for high performance electromagnetic interference shielding. *Chem. Eng. J.* **2023**, *472*, 144934. [CrossRef]
93. Schäfer, K.; Braun, T.; Riegg, S.; Musekamp, J.; Gutfleisch, O. Polymer-bonded magnets produced by laser powder bed fusion: Influence of powder morphology, filler fraction and energy input on the magnetic and mechanical properties. *Mater. Res. Bull.* **2023**, *158*, 112051. [CrossRef]
94. Zhang, L.; Zhou, J.; Yang, H.; Wang, X.; Cai, Z.; Zhu, M. Effect of modulation of interfacial properties on the tribological properties of viscoelastic epoxy resin damping coatings. *Polym. Test.* **2021**, *100*, 107229. [CrossRef]
95. Kumar, V.; Alam, M.N.; Manikkavel, A.; Park, S.S. Efficacy of graphitic allotrope's surface area in silicone rubber based composites for reverse piezo-electric actuation: Nano effect. *Sens. Actuators A Phys.* **2022**, *336*, 113411. [CrossRef]
96. Chen, B.; Kondoh, K.; Li, J.S.; Qian, M.J.C.P.B.E. Extraordinary reinforcing effect of carbon nanotubes in aluminium matrix composites assisted by in-situ alumina nanoparticles. *Compos. Part B Eng.* **2020**, *183*, 107691. [CrossRef]
97. Amoroso, L.; Heeley, E.L.; Ramadas, S.N.; McNally, T. Crystallisation behaviour of composites of HDPE and MWCNTs: The effect of nanotube dispersion, orientation and polymer deformation. *Polymer* **2020**, *201*, 122587. [CrossRef]
98. Wang, P.; Wei, W.; Li, Z.; Duan, W.; Han, H.; Xie, Q. A superhydrophobic fluorinated PDMS composite as a wearable strain sensor with excellent mechanical robustness and liquid impalement resistance. *J. Mater. Chem. A* **2020**, *8*, 3509–3516. [CrossRef]
99. Zhu, P.; Ou, H.; Kuang, Y.; Hao, L.; Diao, J.; Chen, G. Cellulose nanofiber/carbon nanotube dual network-enabled humidity sensor with high sensitivity and durability. *ACS Appl. Mater. Interfaces* **2020**, *12*, 33229–33238. [CrossRef]
100. Kumar, V.; Alam, M.N.; Azam, S.; Manikkavel, A.; Park, S.S. The tough and multi-functional stretchable device based on silicone rubber composites. *Polym. Adv. Technol.* **2023**, *34*, 2167–2178. [CrossRef]
101. Li, Z.; Young, R.J.; Kinloch, I.A. Interfacial stress transfer in graphene oxide nanocomposites. *ACS Appl. Mater. Interfaces* **2013**, *5*, 456–463. [CrossRef]
102. Jiao, B.; Chen, J.; Gong, M. Investigation of interfacial degradation mechanism and debonding behavior for degraded rubber asphalt-aggregate molecular system using a molecular dynamics method. *J. Clean. Prod.* **2024**, *451*, 142123. [CrossRef]
103. Yang, B.; Wang, S.; Song, Z.; Liu, L.; Li, H.; Li, Y. Molecular dynamics study on the reinforcing effect of incorporation of graphene/carbon nanotubes on the mechanical properties of swelling rubber. *Polym. Test.* **2021**, *102*, 107337. [CrossRef]
104. Mohd Nurazzi, N.; Muhammad Asyraf, M.R.; Khalina, A.; Abdullah, N.; Sabaruddin, F.A.; Kamarudin, S.H.; Ahmad, S.; Mahat, A.M.; Lee, C.L.; Aisyah, H.A.; et al. Fabrication, functionalization, and application of carbon nanotube-reinforced polymer composite: An overview. *Polymers* **2021**, *13*, 1047. [CrossRef] [PubMed]
105. Ruth, S.R.A.; Beker, L.; Tran, H.; Feig, V.R.; Matsuhisa, N.; Bao, Z. Rational design of capacitive pressure sensors based on pyramidal microstructures for specialized monitoring of biosignals. *Adv. Funct. Mater.* **2020**, *30*, 1903100. [CrossRef]
106. Hassanifard, S.; Behdinan, K. Fatigue response of multiscale extrusion-based additively manufactured acrylonitrile butadiene styrene-graphene nanoplatelets composites. *Compos. Part B Eng.* **2024**, *279*, 111464. [CrossRef]
107. Chen, J.; Liu, J.; Yao, Y.; Chen, S. Effect of microstructural damage on the mechanical properties of silica nanoparticle-reinforced silicone rubber composites. *Eng. Fract. Mech.* **2020**, *235*, 107195. [CrossRef]
108. Georgopoulou, A.; Clemens, F. Piezoresistive elastomer-based composite strain sensors and their applications. *ACS Appl. Electron. Mater.* **2020**, *2*, 1826–1842. [CrossRef]
109. Wu, Y.; Yan, T.; Zhang, K.; Pan, Z. Flexible and anisotropic strain sensors based on highly aligned carbon fiber membrane for exercise monitoring. *Adv. Mater. Technol.* **2021**, *6*, 2100643. [CrossRef]
110. Lincon, M.I.; Chalivendra, V. High strain rate damage sensing in intra-ply hybrid composites under dynamic shear loading. *Int. J. Impact Eng.* **2023**, *173*, 104439. [CrossRef]
111. Peng, W.; Han, L.; Huang, H.; Xuan, X.; Pan, G.; Wan, L.; Pan, L. A direction-aware and ultrafast self-healing dual network hydrogel for a flexible electronic skin strain sensor. *J. Mater. Chem. A* **2020**, *8*, 26109–26118. [CrossRef]

112. Yan, W.; Fuh, H.-R.; Lv, Y.; Chen, K.-Q.; Tsai, T.-Y.; Wu, Y.-R.; Shieh, T.-H.; Hung, K.-M.; Li, J.; Zhang, D.; et al. Giant gauge factor of Van der Waals material based strain sensors. *Nat. Commun.* **2021**, *12*, 2018. [CrossRef]
113. Duan, L.; D'hooge, D.R.; Cardon, L. Recent progress on flexible and stretchable piezoresistive strain sensors: From design to application. *Prog. Mater. Sci.* **2020**, *114*, 100617. [CrossRef]
114. Wan, C.; Zhang, L.; Yong, K.T.; Li, J.; Wu, Y. Recent progress in flexible nanocellulosic structures for wearable piezoresistive strain sensors. *J. Mater. Chem. C* **2021**, *9*, 11001–11029. [CrossRef]
115. Gilanizadehdizaj, G.; Aw, K.C.; Stringer, J.; Bhattacharyya, D. Facile fabrication of flexible piezo-resistive pressure sensor array using reduced graphene oxide foam and silicone elastomer. *Sens. Actuators A Phys.* **2022**, *340*, 113549. [CrossRef]
116. Wang, Z.; Cong, Y.; Fu, J. Stretchable and tough conductive hydrogels for flexible pressure and strain sensors. *J. Mater. Chem. B* **2020**, *8*, 3437–3459. [CrossRef]
117. Song, F.; Chen, Q.; Zhang, M.; Jiang, Z.; Ding, W.; Yan, Z.; Zhu, H. Exploring the piezoresistive sensing behaviour of ultra-high performance concrete: Strategies for multiphase and multiscale functional additives and influence of electrical percolation. *Compos. Part B: Eng.* **2023**, *267*, 111042. [CrossRef]
118. Tian, K.; Pan, Q.; Deng, H.; Fu, Q. Shear induced formation and destruction behavior of conductive networks in nickel/polyurethane composites during strain sensing. *Compos. Part A Appl. Sci. Manuf.* **2020**, *130*, 105757. [CrossRef]
119. Lin, Y.; Yin, Q.; Wang, J.; Jia, H.; Yuan, G.; Wang, J. Sensitivity enhanced, highly stretchable, and mechanically robust strain sensors based on reduced graphene oxide-aramid nanofibers hybrid fillers. *Chem. Eng. J.* **2022**, *443*, 136468. [CrossRef]
120. Zhang, Y.; Zhu, X.; Liu, Y.; Liu, L.; Xu, Q.; Liu, H.; Chen, L. Ultra-stretchable monofilament flexible sensor with low hysteresis and linearity based on MWCNTs/Ecoflex composite materials. *Macromol. Mater. Eng.* **2021**, *306*, 2100113. [CrossRef]
121. Guo, H.; Jerrams, S.; Xu, Z.; Zhou, Y.; Jiang, L.; Zhang, L.; Wen, S. Enhanced fatigue and durability of carbon black/natural rubber composites reinforced with graphene oxide and carbon nanotubes. *Eng. Fract. Mech.* **2020**, *223*, 106764. [CrossRef]
122. Zhou, K.; Dai, K.; Liu, C.; Shen, C. Flexible conductive polymer composites for smart wearable strain sensors. *SmartMat* **2020**, *1*, e1010. [CrossRef]
123. Li, Y.; Hong, J.; Zhang, J.; Yang, H.; Wang, H.; Ye, L.; Li, Y. Synergistically enhanced thermal conductivity, electrical insulation, and mechanical toughness of polymer composites with carbon nanofibers segregated by alumina nanoparticles. *Compos. Part B Eng.* **2025**, *289*, 111935. [CrossRef]
124. Wang, Y.; Liu, A.; Han, Y.; Li, T. Sensors based on conductive polymers and their composites: A review. *Polym. Int.* **2020**, *69*, 7–17. [CrossRef]
125. Wang, J.; Li, Q.; Li, K.; Sun, X.; Wang, Y.; Zhuang, T.; Wang, H. Ultra-high electrical conductivity in filler-free polymeric hydrogels toward thermoelectrics and electromagnetic interference shielding. *Adv. Mater.* **2022**, *34*, 2109904. [CrossRef]
126. Tan, L.; Yu, K.; Bashir, A.K.; Cheng, X.; Ming, F.; Zhao, L.; Zhou, X. Toward real-time and efficient cardiovascular monitoring for COVID-19 patients by 5G-enabled wearable medical devices: A deep learning approach. *Neural Comput. Appl.* **2021**, *35*, 13921–13934. [CrossRef]
127. Averardi, A.; Cola, C.; Zeltmann, S.E.; Gupta, N. Effect of particle size distribution on the packing of powder beds: A critical discussion relevant to additive manufacturing. *Mater. Today Commun.* **2020**, *24*, 100964. [CrossRef]
128. Aslanidis, E.; Skotadis, E.; Tsoukalas, D. Resistive crack-based nanoparticle strain sensors with extreme sensitivity and adjustable gauge factor, made on flexible substrates. *Nanoscale* **2021**, *13*, 3263–3274. [CrossRef]
129. Coetzee, D.; Venkataraman, M.; Militky, J.; Petru, M. Influence of nanoparticles on thermal and electrical conductivity of composites. *Polymers* **2020**, *12*, 742. [CrossRef]
130. Zhang, D.; Ren, B.; Zhang, Y.; Xu, L.; Huang, Q.; He, Y.; Zheng, J. From design to applications of stimuli-responsive hydrogel strain sensors. *J. Mater. Chem. B* **2020**, *8*, 3171–3191. [CrossRef]
131. Penas, R.; Balmes, E.; Gaudin, A. A unified non-linear system model view of hyperelasticity, viscoelasticity and hysteresis exhibited by rubber. *Mech. Syst. Signal Process.* **2022**, *170*, 108793. [CrossRef]
132. Alam, M.N.; Kumar, V.; Lee, D.J.; Park, S.S. Styrene–butadiene rubber-based nanocomposites toughened by carbon nanotubes for wide and linear electromechanical sensing applications. *Polym. Compos.* **2024**, *45*, 2485–2499. [CrossRef]
133. Kumar, V.; Alam, M.N.; Yewale, M.A.; Park, S.S. Great elastomer-based composites for robust and portable wearable sensors: A critical review. *Sens. Actuators A Phys.* **2025**, *389*, 116563. [CrossRef]
134. Kumar, V.; Lee, D.J. Studies of nanocomposites based on carbon nanomaterials and RTV silicone rubber. *J. Appl. Polym. Sci.* **2017**, *134*, 44407. [CrossRef]
135. Lee, J.Y.; Kumar, V.; Tang, X.W.; Lee, D.J. Mechanical and electrical behavior of rubber nanocomposites under static and cyclic strain. *Compos. Sci. Technol.* **2017**, *142*, 1–9. [CrossRef]
136. Manikkavel, A.; Kumar, V.; Park, S.S. MWCNT-reinforced Silicone's Dual Response to Human and Mechanical Motion: Exploring Strain Sensing Behavior and Practical Applications. *Colloids Surf. A Physicochem. Eng. Asp.* **2024**, *688*, 133536. [CrossRef]
137. Kumar, V.; Alam, M.N.; Yewale, M.A.; Lee, D.J.; Park, S.S. Mimicking Self-Powered Piezoelectric Energy-Generating Behavior in Silicone Rubber Composites under Compressive and Tensile Strains. *ACS Appl. Electron. Mater.* **2024**, *6*, 1638–1650. [CrossRef]

138. Kumar, V.; Alam, M.N.; Yewale, M.A.; Park, S.S. Robust performance for composites using silicone rubber with different prospects for wearable electronics. *Polym. Adv. Technol.* **2024**, *35*, e6264. [CrossRef]
139. Kumar, V.; Alam, M.N.; Manikkavel, A.; Choi, J.; Lee, D.J. Investigation of silicone rubber composites reinforced with carbon nanotube, nanographite, their hybrid, and applications for flexible devices. *J. Vinyl Addit. Technol.* **2021**, *27*, 254–263. [CrossRef]
140. Manikkavel, A.; Kumar, V.; Alam, M.N.; Kim, U.; Park, S.S. The electro-mechanical energy harvesting configurations in different modes from machine to self-powered wearable electronics. *ACS Appl. Electron. Mater.* **2023**, *5*, 5537–5554. [CrossRef]
141. Wang, L.; Xu, T.; Fan, C.; Zhang, X. Wearable strain sensor for real-time sweat volume monitoring. *IScience* **2021**, *24*, 102028. [CrossRef]
142. Yadav, A.; Yadav, N.; Wu, Y.; RamaKrishna, S.; Hongyu, Z. Wearable strain sensors: State-of-the-art and future applications. *Mater. Adv.* **2023**, *4*, 1444–1459. [CrossRef]
143. Patel, V.; Chesmore, A.; Legner, C.M.; Pandey, S. Trends in workplace wearable technologies and connected-worker solutions for next-generation occupational safety, health, and productivity. *Adv. Intell. Syst.* **2022**, *4*, 2100099. [CrossRef]
144. Cheng, Y.; Wang, K.; Xu, H.; Li, T.; Jin, Q.; Cui, D. Recent developments in sensors for wearable device applications. *Anal. Bioanal. Chem.* **2021**, *413*, 6037–6057. [CrossRef] [PubMed]
145. Parvin, N.; Kumar, V.; Park, S.S.; Mandal, T.K.; Joo, S.W. Enhanced piezoelectric energy harvesting using hybrid composites of MWCNTs and partially-reduced GO in RTV-SR for stable voltage generation. *Surf. Interfaces* **2024**, *44*, 103681. [CrossRef]
146. Chung, D.D.L. A critical review of piezoresistivity and its application in electrical-resistance-based strain sensing. *J. Mater. Sci.* **2020**, *55*, 15367–15396. [CrossRef]
147. Ma, Z.; Zhang, Y.; Zhang, K.; Deng, H.; Fu, Q. Recent progress in flexible capacitive sensors: Structures and properties. *Nano Mater. Sci.* **2023**, *5*, 265–277. [CrossRef]
148. Javaid, M.; Haleem, A.; Rab, S.; Singh, R.P.; Suman, R. Sensors for daily life: A review. *Sens. Int.* **2021**, *2*, 100121. [CrossRef]
149. Qiu, Y.; Zou, Z.; Zou, Z.; Setiawan, N.K.; Dikshit, K.V.; Whiting, G.; Yang, F.; Zhang, W.; Lu, J.; Zhong, B.; et al. Deep-learning-assisted printed liquid metal sensory system for wearable applications and boxing training. *npj Flex. Electron.* **2023**, *7*, 37. [CrossRef]
150. Malekloo, A.; Ozer, E.; AlHamaydeh, M.; Girolami, M. Machine learning and structural health monitoring overview with emerging technology and high-dimensional data source highlights. *Struct. Health Monit.* **2022**, *21*, 1906–1955. [CrossRef]
151. Luo, J.; Gao, W.; Wang, Z.L. The triboelectric nanogenerator as an innovative technology toward intelligent sports. *Adv. Mater.* **2021**, *33*, 2004178. [CrossRef]
152. El-Sheimy, N.; Youssef, A. Inertial sensors technologies for navigation applications: State of the art and future trends. *Satell. Navig.* **2020**, *1*, 2. [CrossRef]
153. Ponciano, V.; Pires, I.M.; Ribeiro, F.R.; Marques, G.; Villasana, M.V.; Garcia, N.M.; Zdravevski, E.; Spinsante, S. Identification of diseases based on the use of inertial sensors: A systematic review. *Electronics* **2020**, *9*, 778. [CrossRef]
154. Rana, M.; Mittal, V. Wearable sensors for real-time kinematics analysis in sports: A review. *IEEE Sens. J.* **2020**, *21*, 1187–1207. [CrossRef]
155. Zhang, J.; Hu, Y.; Zhang, L.; Zhou, J.; Lu, A. Transparent, ultra-stretching, tough, adhesive carboxyethyl chitin/polyacrylamide hydrogel toward high-performance soft electronics. *Nano-Micro Lett.* **2023**, *15*, 8. [CrossRef]
156. Lei, M.; Feng, K.; Ding, S.; Wang, M.; Dai, Z.; Liu, R.; Zhou, B. Breathable and waterproof electronic skin with three-dimensional architecture for pressure and strain sensing in nonoverlapping mode. *ACS Nano* **2022**, *16*, 12620–12634. [CrossRef]
157. Worsey, M.T.; Espinosa, H.G.; Shepherd, J.B.; Thiel, D.V. An evaluation of wearable inertial sensor configuration and supervised machine learning models for automatic punch classification in boxing. *IoT* **2020**, *1*, 360–381. [CrossRef]
158. McDevitt, S.; Hernandez, H.; Hicks, J.; Lowell, R.; Bentahaikt, H.; Burch, R.; Anderson, B. Wearables for biomechanical performance optimization and risk assessment in industrial and sports applications. *Bioengineering* **2022**, *9*, 33. [CrossRef]
159. Bayoumy, K.; Gaber, M.; Elshafeey, A.; Mhaimeed, O.; Dineen, E.H.; Marvel, F.A.; Martin, S.S.; Muse, E.D.; Turakhia, M.P.; Tarakji, K.G.; et al. Smart wearable devices in cardiovascular care: Where we are and how to move forward. *Nat. Rev. Cardiol.* **2021**, *18*, 581–599. [CrossRef]
160. Ajdaroski, M.; Tadakala, R.; Nichols, L.; Esquivel, A. Validation of a device to measure knee joint angles for a dynamic movement. *Sensors* **2020**, *20*, 1747. [CrossRef]
161. Kim, J.; Kantharaju, P.; Yi, H.; Jacobson, M.; Jeong, H.; Kim, H.; Yeo, W.H. Soft wearable flexible bioelectronics integrated with an ankle-foot exoskeleton for estimation of metabolic costs and physical effort. *npj Flex. Electron.* **2023**, *7*, 3. [CrossRef]
162. Quaid, M.A.K.; Jalal, A. Wearable sensors based human behavioral pattern recognition using statistical features and reweighted genetic algorithm. *Multimed. Tools Appl.* **2020**, *79*, 6061–6083. [CrossRef]
163. Kristoffersson, A.; Lindén, M. A systematic review of wearable sensors for monitoring physical activity. *Sensors* **2022**, *22*, 573. [CrossRef] [PubMed]
164. Elayan, H.; Aloqaily, M.; Guizani, M. Sustainability of healthcare data analysis IoT-based systems using deep federated learning. *IEEE Internet Things J.* **2021**, *9*, 7338–7346. [CrossRef]

165. Thapa, C.; Camtepe, S. Precision health data: Requirements, challenges and existing techniques for data security and privacy. *Comput. Biol. Med.* **2021**, *129*, 104130. [CrossRef] [PubMed]
166. Iqbal, M.; Nauman, M.M.; Khan, F.U.; Abas, P.E.; Cheok, Q.; Iqbal, A.; Aissa, B. Vibration-based piezoelectric, electromagnetic, and hybrid energy harvesters for microsystems applications: A contributed review. *Int. J. Energy Res.* **2021**, *45*, 65–102. [CrossRef]
167. Gao, L.; Xu, X.; Han, H.; Yang, W.; Zhuo, R.; Wei, Q.; Hu, N. A broadband hybrid blue energy nanogenerator for smart ocean IoT network. *Nano Energy* **2024**, 109697. [CrossRef]
168. Rahman, M.T.; Rana, S.S.; Salauddin, M.; Maharjan, P.; Bhatta, T.; Kim, H.; Cho, H.; Park, J.Y. A highly miniaturized freestanding kinetic-impact-based non-resonant hybridized electromagnetic-triboelectric nanogenerator for human induced vibrations harvesting. *Appl. Energy* **2020**, *279*, 115799. [CrossRef]
169. Yuan, F.; Yu, Y.; Wu, J.; Peng, J.; Jiang, C.; Fan, X.; Li, J. RCF crack direction assessment in moving ferromagnetic material by DC electromagnetic NDT technique. *NDT E Int.* **2023**, *138*, 102882. [CrossRef]
170. Li, G.; Wang, J.; Fu, S.; Shan, C.; Wu, H.; An, S.; Hu, C. A nanogenerator enabled by a perfect combination and synergetic utilization of triboelectrification, charge excitation and electromagnetic induction to reach efficient energy conversion. *Adv. Funct. Mater.* **2023**, *33*, 2213893. [CrossRef]
171. Wang, Z.L. From contact electrification to triboelectric nanogenerators. *Rep. Prog. Phys.* **2021**, *84*, 096502. [CrossRef]
172. Lai, Y.; Lu, H.; Wu, H.; Zhang, D.; Yang, J.; Ma, J.; Shamsi, M.; Vallem, V.; Dickey, M.D. Elastic multifunctional liquid–metal fibers for harvesting mechanical and electromagnetic energy and as self-powered sensors. *Adv. Energy Mater.* **2021**, *11*, 2100411. [CrossRef]
173. Parvin, N.; Kumar, V.; Manikkavel, A.; Park, S.S.; Mandal, T.K.; Joo, S.W. Great new generation carbon microsphere-based composites: Facile synthesis, properties and their application in piezo-electric energy harvesting. *Appl. Surf. Sci.* **2023**, *613*, 156078. [CrossRef]
174. Kumar, V.; Mandal, T.K.; Kumar, A.; Alam, M.N.; Parvin, N. Multifunctional properties of chemically reduced graphene oxide and silicone rubber composite: Facile synthesis and application as soft composites for piezoelectric energy harvesting. *Express Polym. Lett.* **2022**, *16*, 978–995. [CrossRef]
175. Kumar, V.; Manikkavel, A.; Kumar, A.; Alam, M.N.; Hwang, G.J.; Park, S.S. Stretchable piezo-electric energy harvesting device with high durability using carbon nanomaterials with different structure and their synergism with molybdenum disulfide. *J. Vinyl Addit. Technol.* **2022**, *28*, 813–827. [CrossRef]
176. Kumar, V.; Lee, D.J.; Park, S.S. Multifunctional composites-based sensors for health monitoring: Latest insights and prospects. *Mater. Today Commun.* **2025**, *44*, 112187. [CrossRef]
177. Manikkavel, A.; Kumar, V.; Kim, J.; Lee, D.J.; Park, S.S. Investigation of high temperature vulcanized and room temperature vulcanized silicone rubber based on flexible piezo-electric energy harvesting applications with multi-walled carbon nanotube reinforced composites. *Polym. Compos.* **2022**, *43*, 1305–1318. [CrossRef]
178. Kumar, V.; Azam, S.; Alam, M.N.; Hong, W.B.; Park, S.S. Novel rubber composites based on copper particles, multi-wall carbon nanotubes and their hybrid for stretchable devices. *Polymers* **2022**, *14*, 3744. [CrossRef]
179. Kumar, V.; Alam, M.N.; Park, S.S. Soft composites filled with iron oxide and graphite nanoplatelets under static and cyclic strain for different industrial applications. *Polymers* **2022**, *14*, 2393. [CrossRef]
180. Manikkavel, A.; Kumar, V.; Lee, D.J. Simple fracture model for an electrode and interfacial crack in a dielectric elastomer under tensile loading. *Theor. Appl. Fract. Mech.* **2020**, *108*, 102626. [CrossRef]
181. Kumar, V.; Kumar, A.; Wu, R.R.; Lee, D.J. Room-temperature vulcanized silicone rubber/barium titanate–based high-performance nanocomposite for energy harvesting. *Mater. Today Chem.* **2020**, *16*, 100232. [CrossRef]
182. Toleukhanova, S.; Shen, T.H.; Chang, C.; Swathilakshmi, S.; Bottinelli Montandon, T.; Tileli, V. Graphene Electrode for Studying CO_2 Electroreduction Nanocatalysts under Realistic Conditions in Microcells. *Adv. Mater.* **2024**, *36*, 2311133. [CrossRef]
183. Lee, C.H.; Khalina, A.; Lee, S.; Liu, M. A comprehensive review on bast fibre retting process for optimal performance in fibre-reinforced polymer composites. *Adv. Mater. Sci. Eng.* **2020**, *2020*, 6074063. [CrossRef]
184. Ritchie, R.O. Toughening materials: Enhancing resistance to fracture. *Philos. Trans. R. Soc. A* **2021**, *379*, 20200437. [CrossRef] [PubMed]
185. Mousavi, S.R.; Estaji, S.; Paydayesh, A.; Arjmand, M.; Jafari, S.H.; Nouranian, S.; Khonakdar, H.A. A review of recent progress in improving the fracture toughness of epoxy-based composites using carbonaceous nanofillers. *Polym. Compos.* **2022**, *43*, 1871–1886. [CrossRef]
186. Rahmatabadi, D.; Ghasemi, I.; Baniassadi, M.; Abrinia, K.; Baghani, M. 3D printing of PLA-TPU with different component ratios: Fracture toughness, mechanical properties, and morphology. *J. Mater. Res. Technol.* **2022**, *21*, 3970–3981. [CrossRef]
187. Alam, M.N.; Kumar, V.; Lee, D.J.; Choi, J. Synergistically toughened silicone rubber nanocomposites using carbon nanotubes and molybdenum disulfide for stretchable strain sensors. *Compos. Part B Eng.* **2023**, *259*, 110759. [CrossRef]

188. Ben Atitallah, B.; Rajendran, D.; Hu, Z.; Ramalingame, R.; Quijano Jose, R.B.; da Veiga Torres, R.; Bouchaala, D.; Derbel, N.; Kanoun, O. Piezo-resistive pressure and strain sensors for biomedical and tele-manipulation applications. *Adv. Sens. Biomed. Appl.* **2021**, *38*, 47–65.
189. Nirosha, R.; Agarwal, R. Characterization and modeling of threshold voltage for organic and amorphous thin-film transistors. *Microelectron. Reliab.* **2023**, *147*, 115054. [CrossRef]
190. Wang, X.; Liu, X.; Ge, X.; Schubert, D.W. Superior sensitive, high-tensile flexible fabric film strain sensor. *Compos. Part A Appl. Sci. Manuf.* **2023**, *172*, 107610. [CrossRef]
191. Salaeh, S.; Das, A.; Stöckelhuber, K.W.; Wießner, S. Fabrication of a strain sensor from a thermoplastic vulcanizate with an embedded interconnected conducting filler network. *Compos. Part A Appl. Sci. Manuf.* **2020**, *130*, 105763. [CrossRef]
192. Al-Bahrani, M.; Bouaissi, A.; Cree, A. The fabrication and testing of a self-sensing MWCNT nanocomposite sensor for oil leak detection. *Int. J. Low-Carbon Technol.* **2022**, *17*, 622–629. [CrossRef]
193. Zhang, R.; Liu, C.; Wei, C.; Wang, Y.; Li, F.; Zhang, Z.; Tang, L. Thermoplastic charge-transfer hydrogels for highly sensitive strain and temperature sensors. *J. Mater. Chem. A* **2023**, *11*, 8320–8329. [CrossRef]
194. Liu, J.; Chen, X.; Sun, B.; Guo, H.; Guo, Y.; Zhang, S.; Tang, J. Stretchable strain sensor of composite hydrogels with high fatigue resistance and low hysteresis. *J. Mater. Chem. A* **2022**, *10*, 25564–25574. [CrossRef]
195. Song, P.; Wang, G.; Zhang, Y. Preparation and performance of graphene/carbon black silicone rubber composites used for highly sensitive and flexible strain sensors. *Sens. Actuators A Phys.* **2021**, *323*, 112659. [CrossRef]
196. Kumar, V.; Manikkavel, A.; Yewale, M.A.; Alam, M.N.; Park, S.S. Lightweight, compressible, stretchable, ultra-soft, and mechanically stable composites for piezo-electric energy generators and strain sensing. *Mater. Res. Bull.* **2024**, *179*, 112962. [CrossRef]
197. Xu, J.; Yu, J.; Xu, J.; Sun, C.; He, W.; Huang, J.; Li, G. High-value utilization of waste tires: A review with focus on modified carbon black from pyrolysis. *Sci. Total Environ.* **2020**, *742*, 140235. [CrossRef]
198. Pei, Z.; Zhang, Q.; Liu, Y.; Zhao, Y.; Dong, X.; Zhang, Y.; Sang, S. A high gauge-factor wearable strain sensor array via 3D printed mold fabrication and size optimization of silver-coated carbon nanotubes. *Nanotechnology* **2020**, *31*, 305501. [CrossRef]
199. Peng, J.; He, Y.; Zhou, C.; Su, S.; Lai, B. The carbon nanotubes-based materials and their applications for organic pollutant removal: A critical review. *Chin. Chem. Lett.* **2021**, *32*, 1626–1636. [CrossRef]
200. Guo, H.; Ji, P.; Halász, I.Z.; Pirityi, D.Z.; Bárány, T.; Xu, Z.; Zheng, L.; Zhang, L.; Liu, L.; Wen, S. Enhanced fatigue and durability properties of natural rubber composites reinforced with carbon nanotubes and graphene oxide. *Materials* **2020**, *13*, 5746. [CrossRef]
201. Md Said, N.H.; Liu, W.W.; Khe, C.S.; Lai, C.W.; Zulkepli, N.N.; Aziz, A. Review of the past and recent developments in functionalization of graphene derivatives for reinforcement of polypropylene nanocomposites. *Polym. Compos.* **2021**, *42*, 1075–1108. [CrossRef]
202. Chen, B.; Dai, J.; Song, T.; Guan, Q. Research and development of high-performance high-damping rubber Materials for high-damping rubber isolation bearings: A review. *Polymers* **2022**, *14*, 2427. [CrossRef]
203. Rahmani, P.; Shojaei, A. A review on the features, performance and potential applications of hydrogel-based wearable strain/pressure sensors. *Adv. Colloid Interface Sci.* **2021**, *298*, 102553. [CrossRef] [PubMed]
204. Haghi, M.; Danyali, S.; Ayasseh, S.; Wang, J.; Aazami, R.; Deserno, T.M. Wearable devices in health monitoring from the environmental towards multiple domains: A survey. *Sensors* **2021**, *21*, 2130. [CrossRef] [PubMed]
205. Kumar, S.; Pavelyev, V.; Tripathi, N.; Platonov, V.; Sharma, P.; Ahmad, R.; Khosla, A. Recent advances in the development of carbon nanotubes based flexible sensors. *J. Electrochem. Soc.* **2020**, *167*, 047506. [CrossRef]
206. Palumbo, A.; Li, Z.; Yang, E.H. Trends on carbon nanotube-based flexible and wearable sensors via electrochemical and mechanical stimuli: A review. *IEEE Sens. J.* **2022**, *22*, 20102–20125. [CrossRef]
207. Nasture, A.M.; Ionete, E.I.; Lungu, F.A.; Spiridon, S.I.; Patularu, L.G. Water quality carbon nanotube-based sensors technological barriers and late research trends: A bibliometric analysis. *Chemosensors* **2022**, *10*, 161. [CrossRef]
208. Li, K.; Yang, W.; Yi, M.; Shen, Z. Graphene-based pressure sensor and strain sensor for detecting human activities. *Smart Mater. Struct.* **2021**, *30*, 085027. [CrossRef]
209. Esteghamat, A.; Akhavan, O. Graphene as the ultra-transparent conductive layer in developing the nanotechnology-based flexible smart touchscreens. *Microelectron. Eng.* **2023**, *267*, 111899. [CrossRef]
210. Zhang, Z.; Xiang, D.; Wu, Y.; Zhang, J.; Li, Y.; Wang, M.; Li, Y. Effect of carbon black on the strain sensing property of 3D printed conductive polymer composites. *Appl. Compos. Mater.* **2022**, *29*, 1235–1248. [CrossRef]
211. Dinesh, A.; Saravanakumar, P.; Prasad, B.R.; Raj, S.K. Carbon black based self-sensing cement composite for structural health monitoring–A review on strength and conductive characteristics. In *Materials Today: Proceedings*; Elsevier: Amsterdam, The Netherlands, 2023.
212. Kim, D.-S.; Choi, Y.W.; Shanmugasundaram, A.; Jeong, Y.-J.; Park, J.; Oyunbaatar, N.-E.; Kim, E.-S.; Choi, M.; Lee, D.-W. Highly durable crack sensor integrated with silicone rubber cantilever for measuring cardiac contractility. *Nat. Commun.* **2020**, *11*, 535. [CrossRef]

213. Horne, J.; McLoughlin, L.; Bury, E.; Koh, A.S.; Wujcik, E.K. Interfacial phenomena of advanced composite materials toward wearable platforms for biological and environmental monitoring sensors, armor, and soft robotics. *Adv. Mater. Interfaces* **2020**, *7*, 1901851. [CrossRef]
214. Han, R.; Li, Y.; Zhu, Q.; Niu, K. Research on the preparation and thermal stability of silicone rubber composites: A review. *Compos. Part C Open Access* **2022**, *8*, 100249. [CrossRef]

polymers

MDPI

Review

Enhancing Polylactic Acid (PLA) Performance: A Review of Additives in Fused Deposition Modelling (FDM) Filaments

Ioan Plamadiala [1], Catalin Croitoru [1,*], Mihai Alin Pop [2] and Ionut Claudiu Roata [1]

1 Materials Engineering and Welding Department, Transilvania University of Brasov, 500036 Brasov, Romania; ioan.plamadiala@unitbv.ro (I.P.); ionut.roata@unitbv.ro (I.C.R.)
2 Department of Materials Science, Transilvania University of Brasov, 500036 Brasov, Romania; mihai.pop@unitbv.ro
* Correspondence: c.croitoru@unitbv.ro

Abstract: This review explores the impact of various additives on the mechanical properties of polylactic acid (PLA) filaments used in Fused Deposition Modeling (FDM) 3D printing. While PLA is favored for its biodegradability and ease of use, its inherent limitations in strength and heat resistance necessitate enhancements through additives. The impact of natural and synthetic fibers, inorganic particles, and nanomaterials on the mechanical properties, printability, and overall functionality of PLA composites was examined, indicating that fiber reinforcements, such as carbon and glass fibers, significantly enhance tensile strength and stiffness, while natural fibers contribute to sustainability but may compromise mechanical stability. Additionally, the inclusion of inorganic particulate fillers like calcium carbonate improves dimensional stability and printability, although larger particles can lead to agglomeration issues. The study highlights the potential for improved performance in specific applications while acknowledging the need for further investigation into optimal formulations and processing conditions.

Keywords: PLA; mechanical properties; additives; carbon fiber; glass fiber; additive manufacturing; graphene; CNT

1. Introduction

The increasing demand for functional prototypes and end-use parts has driven the rapid development of Additive Manufacturing (AM) technologies, with Fused Deposition Modeling (FDM) or Fused Filament Fabrication (FFF) being one of the most popular methods. FDM relies on filaments, which are fed through a heated extruder and deposited layer-by-layer to create 3D objects. Choosing the right filament material is crucial for successful printing and achieving the desired final product properties. Over recent years, there has been growing interest in biopolymers as sustainable FDM alternatives to petroleum-based materials [1].

Among biopolymers, polylactic acid (PLA) stands out for its widespread production from renewable resources (240,000 tons annually) and its biodegradability and biocompatibility [2].

These qualities make PLA attractive for various applications, including commodity goods, medical devices, and FDM 3D printing [2]. PLA can be synthesized through several pathways, including condensation of lactic acid or ring-opening polymerization of lactide (its cyclic dimer) in the presence of tin octanoate or zinc stearate catalysts (Figure 1) [3].

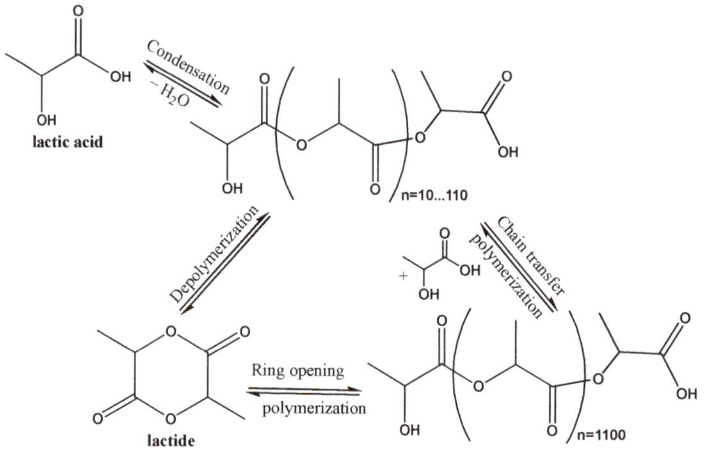

Figure 1. Main reaction pathways for polylactic acid (PLA) synthesis.

There is a noticeable upward trend in the total number of publications on PLA, as extracted from the Clarivate ISI Web of Science database (Figure 2), indicating increasing interest and research activity in this biodegradable polymer. Although the number of publications specifically involving PLA and 3D printing/FDM is less than the total PLA publications, it also shows a rising trend. This suggests that the use of PLA in 3D printing is gaining traction within the research community, especially since 2015. The growing number of publications could reflect advancements in PLA applications, improvements in 3D printing technology, and a broader recognition of the importance of sustainable materials. The same increasing trend could be seen in the number of studies involving 3D printing with custom-made (non-commercial) PLA filaments.

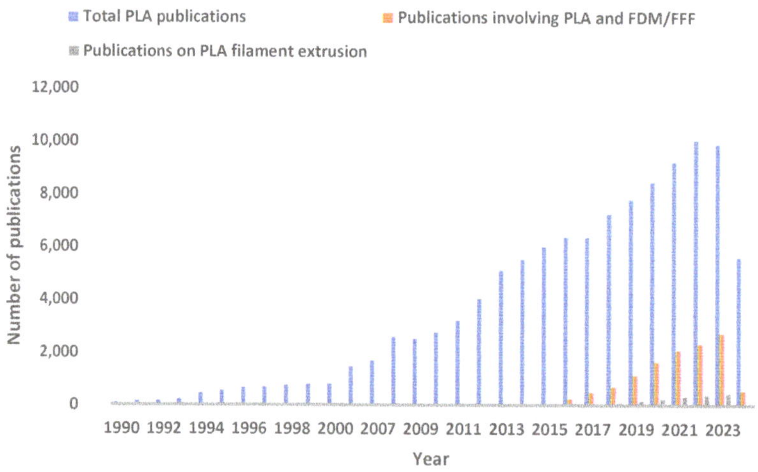

Figure 2. Number of publications involving PLA indexed in the Clarivate ISI Web of Science database (accessed on 2 January 2025).

PLA's popularity in FDM stems from several advantages. First, its low melting temperature (150–180 °C) allows for easy printing and good layer adhesion, enabling intricate designs [4]. However, PLA also has limitations for FDM applications compared to other polymers, including those from the polyester class, such as polyhydroxybutyrate (PHB), polybutylene succinate (PBS), polyethylene terephthalate glycol (PETG) or poly(acrylonitrile-co-butadiene-co-styrene) (ABS), as outlined in Figure 3 [5].

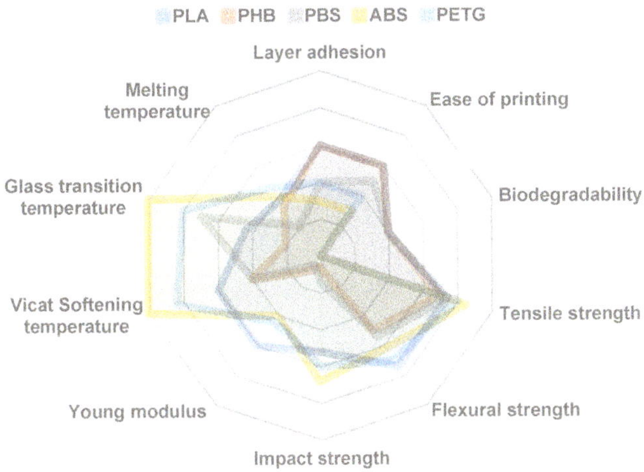

Figure 3. Main properties of commonly used fused deposition modeling (FDM) polymers in pure form (relative comparison) based on data from [6].

Compared to ABS or PETG, PLA has lower strength and heat resistance. Its heat deflection temperature (HDT) is 53–56 °C, meaning PLA parts can deform under elevated temperatures [5]. Additionally, PLA exhibits good tensile strength but lacks flexibility, making it unsuitable for load-bearing applications. Furthermore, PLA's high thermal expansion coefficient can cause warping and deformation during printing.

Additionally, PLA is not as strong or durable as other FDM materials to organic solvents (e.g., acetone, ethyl acetate, toluene, dichloromethane, aliphatic ketones), as the printed parts either dissolve or delaminate when exposed to these compounds. Also, more challenging alkaline/acidic conditions and humid environments can pose a problem, which may affect the longevity and performance of printed objects.

PLA's inherent limitations can be addressed through the incorporation of various fillers and additives into filaments. These can improve mechanical properties, thermal resistance, and printability.

While PLA is often praised for being an eco-friendly material in 3D printing, a deeper examination reveals a more nuanced picture. Typical PLA composite filaments for FDM applications go beyond just the reinforcing materials (like fibers or powders). They also include a surprising array of additional ingredients [7]. These additives, although necessary to achieve desired properties, can raise concerns about the overall environmental impact and efficiency [8–12]. The most important additives used in PLA filaments are the following:

- Coupling Agents: These improve the adhesion between the reinforcing material and the PLA matrix. Examples include silanes and maleic anhydride grafted polymers, with typical loading ratios of 0.5–3 wt. %
- Impact Modifiers: These additives enhance the toughness and impact resistance of the composite. They can be elastomeric polymers (e.g., thermoplastic polyurethane, TPU), core–shell particles, or other toughening agents, usually added in 5–15 wt. % ratios.
- Processing Aids: Lubricants (fatty acids or waxes at 0.5–3 wt. % loading) and plasticizers (e.g., poly (ethylene glycol) (PEG) at typical loadings of 5–10 wt. %) improve the flow and printability of the filament during printing.
- Nucleating Agents and Antioxidants: These additives promote crystallization, improve thermal properties, and prevent degradation of the PLA during processing

and use. Examples include talcum powder, calcium carbonate (1–5 wt. %), phenolic antioxidants, and UV absorbers (0.1–3%).

- Processing, since crystallinity results in poor dimensional stability due to shrinkage and
- Warpage in FDM processes.
- Flame Retardants: While these enhance safety, they often require high loading percentages, impacting printability and potentially adding weight (e.g., aluminum trihydroxide at 10–13 wt. %).
- Colorants: These add desired colors to the filament and printed parts without significantly affecting mechanical properties.

Despite the advancements, challenges remain in using additives with PLA for FDM printing. These challenges include compatibility issues between the additives, the filler material, and the PLA itself. Additionally, optimizing printing parameters and ensuring scalability of the manufacturing process are ongoing areas of research [3]. While these additives can improve mechanical performance, they may also lead to lower adhesion between layers compared to pure PLA. This is because faster solidification and larger voids can occur during printing with additives [13].

Future research should focus on overcoming these compatibility challenges and exploring the potential of PLA formulations with natural components for FDM printing [3]. A key area is achieving better dispersion of additives within the PLA matrix, which is crucial for optimal mechanical properties and avoiding defects in printed parts [3]. It is also important to consider the cost-effectiveness of additives, as some can be expensive and hinder biodegradability or the commercial viability of PLA filaments for certain applications [14].

Despite challenges like uneven printing, clogging, and filler–matrix distribution, natural fiber-reinforced PLA composites hold promise in addressing these issues [15], making the 3D-printed materials suitable for a wider range of applications [5,16]. For instance, adding cellulose-based fibers up to 30 wt. % can significantly increase PLA's stiffness and impact strength without impacting the density and environmental footprint of the material [17,18].

Researchers are pushing the boundaries of PLA's performance by exploring a wider range of fillers and reinforcements beyond natural fibers. This includes incorporating various synthetic fibers and particulate materials like metallic particles, ceramics, polymers or biomass-derived particles (e.g., sawdust). Metallic particles can enhance thermal conductivity, wear resistance, and electrical conductivity of PLA composites [19,20]. Certain iron or cobalt fillers can impart new properties in PLA, such as magnetic properties, which could be useful in sensor-based applications. However, achieving homogeneous dispersion and mitigating potential increases in weight and brittleness depending on the metal used are ongoing challenges. Similarly, ceramics can improve wear resistance, flame retardancy, and stiffness, but ensuring good dispersion and avoiding negative effects on printability and mechanical properties (the stress concentrator effect) require further research [21]. Both polymeric additives and biomass particles can significantly enhance the properties of PLA, making it more suitable for functional applications. Polymeric additives can improve toughness, and could strengthen the adhesion between PLA and reinforcing fibers [22].

Biomass particles, such as wood or micronized plant materials, can significantly increase strength, making PLA composites suitable for demanding applications [23,24].

Nanomaterial fillers offer even more exciting possibilities due to their high surface area-to-volume ratio. These materials have the potential to significantly improve properties like strength, flame retardancy, and barrier properties within PLA composites [25,26]. However, safe handling and proper integration of these materials within the PLA matrix necessitate further investigation.

This paper aims to provide an overview of the potential of PLA filaments filled with various reinforcing agents for FDM/FFF printing. The focus will be on how the incorporation of natural and synthetic fibers (like glass and carbon fibers), inorganic particles (such as calcium carbonate), organic biomass particles, nanomaterials or polymers can enhance the properties of PLA filaments. The advantages and disadvantages of these fillers/additives are discussed, focusing on their impact on mechanical strength, heat resistance, printability, and overall functionality of the printed parts. By examining the current state of research on filled PLA filaments, this paper aims to emphasize future developments in this field and promote the use of PLA for a wider range of FDM/FFF printing applications.

2. Reinforcing Agents for PLA FDM/FFF Applications

2.1. Fibers

In the realm of FDM 3D printing, PLA filament manufacturers have embraced the use of fiber reinforcements to elevate the material's mechanical properties and, in some cases, its environmental footprint. These reinforcements come in two main categories: synthetic and natural fibers.

Synthetic fibers, including carbon fiber, glass fiber, and Kevlar, are popular choices for reinforcing PLA due to their exceptional strength and stiffness. This translates to significantly enhanced mechanical properties in the resulting composites. They typically exhibit improvements in tensile strength, modulus, and impact resistance. For example, carbon fiber reinforcement can lead to a remarkable 150% increase in tensile strength, while glass fiber can boost the modulus by a range of 50–100%. Additionally, these synthetic fibers enhance the thermal stability and heat deflection temperature of PLA. Notably, carbon fiber can significantly improve thermal conductivity, making the composite suitable for high-temperature applications [27].

However, incorporating synthetic fibers presents a challenge in the filament extrusion process. Specialized equipment and techniques become necessary to ensure uniform fiber dispersion and prevent clogging during FDM printing.

Natural fibers like kenaf, hemp, flax, and bamboo are gaining traction as sustainable alternatives to synthetic fibers. Their biodegradability, renewability, and often lower cost align well with the eco-friendly nature of PLA. While not as dramatic as synthetic fibers, natural fibers enhance the tensile strength and modulus of PLA composites. Kenaf and hemp fibers, for instance, can increase tensile strength by 20–50%, with flax and bamboo offering similar improvements.

The sustainability benefits of natural fibers are undeniable. They are renewable, biodegradable, and have a lower environmental impact compared to their synthetic counterparts. Additionally, natural fibers are less expensive, making composites more affordable. They also offer a weight advantage, which can be crucial in applications where weight reduction is a priority.

However, natural fibers come with their own set of challenges. Their hygroscopic nature, meaning they tend to absorb moisture, can affect the mechanical properties and dimensional stability of the composites. This can be a significant drawback in applications where moisture resistance is critical. Furthermore, the properties of natural fibers can vary depending on factors like source, growing conditions, and processing methods. This variability can translate to inconsistencies in the performance of the final composites. Like synthetic fibers, incorporating natural fibers into PLA filaments can pose challenges during the extrusion and FDM printing processes, requiring careful attention to ensure uniform dispersion and prevent clogging. Some strategies that mitigate clogging are as follows:

- Using a large nozzle: Fibers can be larger than standard PLA fillers, so using a larger nozzle (e.g., 0.6 mm or 0.8 mm) can help prevent clogging.
- Printing at a higher temperature: Printing at a slightly higher temperature than standard PLA (around 200–220 °C) can help the fibers flow more easily through the nozzle.
- Reducing retraction: Minimizing retraction distance and speed can help prevent the filament from being pulled back into the nozzle, which can cause clogging.
- Adjusting printing speed: Printing at a slower speed can give the filament more time to melt and flow through the nozzle, reducing the risk of clogging.
- Drying the filament: Especially natural fibers can absorb moisture from the air, which can cause inconsistent extrusion and clogging.

2.1.1. Carbon Fiber

Carbon fiber-reinforced PLA (CFR-PLA) is one of the most popular composites. There are two main types of carbon fiber reinforcement: chopped and continuous. Chopped fibers, the most common option, are short lengths dispersed within the PLA, offering easier printing but with a lower reinforcement potential. Continuous fibers, embedded in a polymer sheath, provide the highest level of strength and stiffness but require specialized printers due to potential clogging and higher processing temperatures. The choice between these options depends on your priorities [28]. Chopped fibers are ideal for affordability and moderate improvements, while continuous fibers offer the best performance but necessitate a more advanced printing setup.

Typical fiber loadings range from 5% to 20% by weight. For instance, a study reported that 15% carbon fiber loading in PLA increased the tensile strength by over 100% compared to pure PLA [29]. Other studies conversely report reduced tensile strength but increased Young's modulus and elongation at break, for example, from 1.04 GPa and 6.26%, to 1.26 GPa and 7.81%, respectively [30,31]. The main weakness in these PLA-CF composites lies in the fiber–matrix interface, with fiber pull-out being the primary failure mode due to interface destruction. This effect is more pronounced at higher fiber loads reported to the PLA matrix [32].

The mechanical properties of 3D-printed carbon fiber PLA composite are significantly influenced by the FDM process parameters. The following parameters have been found to have a significant impact on the mechanical properties of CFR-PLA, as referenced from the literature [32–34]:

- Nozzle geometry: Square nozzle geometry increases tensile strength and reduces void geometry (with up to 7%) in carbon nanofiber-reinforced polylactic acid composites during FDM printing [35]. Also, this geometry enhances the fracture toughness with less inter-bead voids and larger bonded areas in carbon fiber-reinforced PLA composites [36].
- Printing Layer Height and print orientation: The layer height has a significant effect on the mechanical properties of CFR-PLA. A lower layer height (0.1–0.2 mm) can result in improved tensile strength and Young's modulus, while a higher layer height (0.4–0.6 mm) can lead to a decrease in mechanical properties [37]. A study mentions that a print orientation of 45° presents superior mechanical strengths [38]. Optimal process parameters for enhancing the impact strength of carbon fiber-reinforced PLA during FDM printing include layer thickness of 0.04 mm and shell thickness of 1.2 mm [39].
- Extrusion Width: The extrusion width has a significant impact on the mechanical properties of CFR-PLA. A wider extrusion width (0.4–0.6 mm) can result in improved

tensile strength and Young's modulus, while a narrower extrusion width (0.1–0.2 mm) can lead to a decrease in mechanical properties [32].

- Printing Temperature: The printing temperature has a significant effect on the mechanical properties of CFR-PLA. A higher printing temperature (230–250 °C) can result in improved tensile strength and Young's modulus, while a lower printing temperature (200–220 °C) can lead to a decrease in mechanical properties and lower dispersability of the fiber in the polymer melt [40]. Several studies indicate that the maximum tensile properties for carbon fiber-reinforced PLA are attained at a nozzle temperature of 230 °C [41,42]. Higher nozzle temperatures in the FDM process increase the crystallinity of printed PLA, promoting its nucleation, thus enhancing mechanical properties and rheological properties [13].
- Printing Speed: The printing speed has a significant impact on the mechanical properties of CFR-PLA. A slower printing speed (10–20 mm/s) can result in improved tensile strength and Young's modulus, while a faster printing speed (50–100 mm/s) can lead to a decrease in mechanical properties [38].
- Infill Percentage: Increasing the infill ratio in FDM printing increases the mechanical properties of both polylactic acid and carbon fiber-reinforced polylactic acid, with a maximum tensile strength of 35.65 MPa for PLA [43]. A higher infill percentage (50–70%) can result in improved tensile strength and Young's modulus, while a lower infill percentage (20–30%) can lead to a decrease in mechanical properties [44].

2.1.2. Glass Fiber

The addition of glass fibers (GFs) to 3D-printed PLA composites significantly enhances their mechanical and thermal properties. Optimal fiber content and innovative printing methods can further improve these properties, making GF-reinforced PLA composites suitable for a wide range of engineering applications. However, considerations such as moisture sensitivity and layer adhesion need to be addressed to fully leverage the benefits of glass fiber reinforcement.

- Mechanical Strength Enhancement: The addition of glass fibers significantly improves the tensile and flexural strength of PLA composites. For instance, composites with 30% glass fibers show superior mechanical properties compared to neat PLA. Continuous glass fiber reinforcement in PLA composites can achieve high flexural strength and modulus, reaching up to 312 MPa and 21.5 GPa, respectively [45–47]. The mechanical properties of PLA composites are optimized at specific glass fiber contents. For example, tensile strength and flexural strength are maximized at 20–30% glass fiber content, while impact strength peaks at 45% [45,48].
- Thermal Stability: GF addition enhances the thermal stability of PLA composites. Composites with 30% glass fibers exhibit excellent thermal stability, making them suitable for applications where thermal resistance is crucial. The heat deflection temperature and linear-expansion coefficient are significantly improved with the addition of short glass fibers [49]. Glass fiber-reinforced polylactic acid composites with isothermal heat treatment significantly improve mechanical properties and thermal stability, enabling their use in applications like automotive, aerospace, and electronics [50]. A study reported that PLA-GF composite filaments are thermally stable until 311 °C [51].
- Layer Adhesion and Printability: The incorporation of glass fibers can improve the printability and dimensional stability of PLA composites, although it may also lead to faster solidification and larger voids, which can affect layer adhesion [13,46].
- Hygromechanical Properties: Glass fiber-reinforced PLA composites exhibit anisotropic behavior, with better performance in the longitudinal direction compared to the trans-

verse direction. Moisture uptake can lead to a reduction in mechanical properties, particularly in the transverse direction [52].

- Chemical and structural stability: While glass fiber reinforcement in PLA (10–30% loading) significantly improves its strength and stiffness for demanding 3D printing applications, a study revealed interesting effects on weatherability. Compared to neat PLA, glass fiber-reinforced PLA (GF-PLA) showed improved mechanical properties even after exposure to sunlight and moisture (weathering). This is because the PLA itself degrades due to weathering, becoming weaker and developing cracks. Glass fibers, however, are much more stable and help the composite maintain its structural integrity. Even after extensive weathering, GF-PLA retains some mechanical advantages over neat PLA, thanks to the reinforcing effect of the relatively stable glass fibers [53].

Glass fibers often undergo a surface treatment, typically with a silane coupling agent, to improve adhesion with the PLA matrix. This enhances the stress transfer between the fibers and the polymer, leading to better mechanical properties in the final composite. Regarding the GF type and FDM process parameters, the most pronounced effects on the properties of the 3D-printed composites can be summarized as follows:

- The type and orientation of glass fibers in 3D-printed PLA composites significantly influence their mechanical and thermal properties. E-glass fibers are the most common and widely used due to their affordability, good balance of mechanical properties, and compatibility with various resins (including PLA). They offer good tensile strength, stiffness, and dimensional stability when incorporated into PLA filaments. Glass fibers for FDM/FFF applications typically come in diameters ranging from 6 microns to 20 microns. Thinner fibers offer better dispersion within the PLA matrix but may require higher loading percentages to achieve desired strength improvements. Conversely, thicker fibers can provide greater reinforcement but might be more challenging to disperse uniformly and could create printing issues like nozzle clogging.

Continuous glass fibers enhance longitudinal tensile strength but suffer from poor transverse properties due to anisotropic behavior. Short glass fibers (around 250 μm) improve thermal stability, making them ideal for thermally demanding applications. The presence and type of voids formed during 3D printing also play a crucial role in determining the composite's strength. Comparatively, 3D-printed composites may exhibit reduced mechanical performance compared to injection-molded ones due to differences in fiber orientation [46].

- Influence of Layer Thickness and Orientation: Thinner layers and specific orientations (e.g., flat orientation) improve tensile and flexural strengths. On-edge orientation with +45°/−45° raster angles enhance mechanical strength and modulus [44].
- Extrusion rate and Temperature: Lower extrusion rates and higher printing temperatures improve mechanical properties by enhancing layer adhesion and reducing voids. The impact strength of the composite was found to decrease with increasing extrusion rate, from 2.5 kJ/m^2 to 1.5 kJ/m^2 when the extrusion rate was increased from 10 mm/s to 30 mm/s. A higher extrusion rate of 50 mm/s resulted in an impact strength of 1.2 kJ/m^2 [54,55]. The printing temperature has a significant impact on the mechanical properties of GF PLA composites. A study found that increasing the printing temperature from 190 °C to 230 °C resulted in a significant improvement in tensile strength and modulus of the composite. However, excessive printing temperatures can lead to warping and delamination of the composite. One study found that the printing temperature range of 230 °C to 240 °C resulted in a decrease in tensile strength by 1.02% and a decrease in elastic modulus by 3.32% [56]. Bed temperature also plays a

role, with higher bed temperatures (50–70 °C) improving crystallinity and mechanical properties [57].

- Infill Density: The infill density of the composite affects its mechanical properties. A study found that increasing the infill density from 10% to 50% resulted in a significant improvement in tensile strength and modulus of the composite [58]. Another study reported that an infill density of 30% yielded the highest tensile strength and modulus values [59]. A study found that samples with 50% and 75% infill density exhibited higher overall toughness compared to 100% infill density samples [51].

While glass fibers enhance tensile and flexural strength in PLA, they can make the printed parts more brittle, reducing their ability to withstand sudden impacts. Additionally, the abrasive nature of glass fibers can roughen the surface finish and accelerate wear on the printer's nozzle, requiring more frequent replacements [48].

2.1.3. Plant Fibers

Lignocellulosic fibers are derived from plant biomass (such as stems, leaves, seed husks and so on) and mainly consist of cellulose, hemicellulose, and lignin in various proportions. These natural fibers (NFs) are renewable, biodegradable, and have a low environmental impact. The most used lignocellulosic fibers in conjunction with PLA for FDM 3D printing include, for example, wood pulp fibers, cotton (lyocell), bamboo, sisal, jute, flax, sugarcane bagasse, ramie, henequen and hemp [60], although several other leaf-based materials are also considered (e.g., pineapple leaf fibers, microalgae). Typical fiber loadings used in 3D-printed PLA-based composites usually range from 1 to 10 wt. % [61].

NFs such as hemp, sisal, jute, flax, cotton or kenaf exhibit a range of mechanical and thermal properties that make them suitable for diverse applications. Hemp (Cannabis sativa) boasts high tensile strength (550–900 MPa) and good flexural strength (20–30 MPa), along with low moisture absorption (2–5%) and thermal conductivity (0.03–0.06 W/mK). Similarly, sisal (Agave sisalana) exhibits high tensile strength (400–600 MPa) and comparable thermal properties, including low density (1.3–1.5 g/cm^3) and high thermal resistance (220–250 °C). Flax (*Linum usitatissimum*) shares these advantageous properties, with tensile strength reaching 500–700 MPa and extremely low moisture absorption (0.5–1.5%). Kenaf (Hibiscus cannabinus) also demonstrates impressive tensile strength (400–600 MPa) and thermal properties, low conductivity and high resistance up to 250 °C. In contrast, jute (Corchorus capsularis) and cotton (Gossypium spp.) exhibit moderate to low mechanical strength, with jute offering a tensile strength of 300–400 MPa and cotton 200–300 MPa, and both also suffer from high moisture absorption (10. . .20%) and lower thermal stability [62–66].

Lignocellulosic fibers are a natural, renewable, and readily available resource that offers several advantages as a reinforcement material for PLA in FDM 3D printing, such as sustainability, being a biodegradable and eco-friendly alternative to synthetic fibers like glass or carbon fiber, contributing to a more sustainable manufacturing process [67]. They also contribute to the overall composite weight reduction, having a lower density compared to many other reinforcing materials, leading to lighter-weight composites (ranging from 1.1 to 1.2 g/cm^3, depending on the fiber type and content, neat PLA having an average density of 1.24 g/cm^3). However, they also present some challenges, such as moisture absorption; lignocellulosics are hygroscopic, meaning they absorb moisture from the environment. The moisture absorption rate can vary from 1% to 15% by weight. This can negatively impact the dimensional stability and mechanical properties of the composite. The properties of the fibers can vary depending on the type of plant, harvesting time, age of the plant, processing methods, and growing conditions. This variability can affect the consistency of the final composite material [68]. Some strategies for mitigation of these shortcomings are presented in the descriptive flow diagram from Figure 4 [69,70].

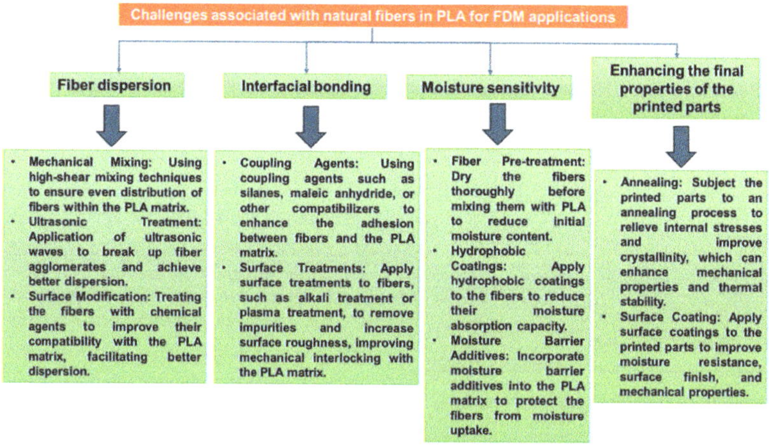

Figure 4. Flow diagram of challenges and mitigation strategies associated with the use of plant fibers in FDM applications.

Inhomogeneous fiber–matrix distribution in natural fiber-reinforced PLA composites for 3D printing remains a major challenge, along with uneven printing and clogging [71]. Plant fibers typically have lower surface energy than synthetic fibers, which can be modified through treatments to improve bonding with PLA. Surface treatments can be applied to lignocellulosic fibers to improve their compatibility with the PLA matrix and enhance the performance of the composite. Common surface treatment methods include [72]:

- Alkali Treatment: This treatment removes impurities, lignin and hemicellulose from the fiber surface, promoting better bonding with the PLA matrix and potentially improving mechanical properties [73].
- Bleaching and pulping: Natural fibers in their raw state harbor various impurities like waxes, oils, and lignin. These substances act as barriers, hindering the formation of strong bonds between the fiber and the PLA matrix. Pulping and bleaching processes effectively remove a significant portion of these impurities, particularly lignin. Lignin, a natural binder in plant cell walls, creates a weak interface with PLA. By eliminating lignin, pulping and bleaching create a cleaner fiber surface, allowing for better adhesion with the PLA matrix. These processes can slightly roughen the surface of the fibers. This increased surface area allows for better mechanical interlocking with the PLA, leading to a stronger bond and improved stress transfer between the fiber and the matrix [74,75]. Pulping and bleaching hemp fibers towards delignified short fibers significantly improves their compatibility with poly(lactic acid) and results in a 52% increase in tensile strength [76].
- Silane Coupling Agents: Silane coupling agents can create a chemical bond between the fibers and the PLA matrix, leading to improved interfacial adhesion and stress transfer, ultimately enhancing the mechanical properties of the composite [77].
- Enzymatic modification: Enzymatic-Assisted Modification of Thermomechanical Pulp Fibers Improves Interfacial Adhesion with Poly(lactic acid) via grafting of less hydrophilic moieties to the lignocellulose fiber surface, e.g., laccase-assisted grafting of octyl gallate (OG) or lauryl gallate (LG) onto the fiber surface [78].

  Studies have shown that the inclusion of NFs in PLA can significantly influence the mechanical, thermal, and physical properties of the resulting composites [77].

- Mechanical Properties: Lignocellulosic fiber reinforcement can improve the tensile strength, modulus, and impact resistance of PLA composites. A study on the tensile behavior of 3D-printed PLA-based composites reinforced with natural henequen fibers found that the tensile strength and modulus increased with increasing fiber loading up to 5 wt. % [32]. Another study on lignocellulose nanofiber/polylactic acid (LCNF/PLA) composites reported that the tensile strength and modulus increased with increasing fiber loading up to 10 wt. % [79]. Research has shown increases in tensile strength by 20–50% and modulus by 30–50% with the incorporation of wood fibers like kenaf and hemp [80]. The addition of natural fibers, such as wood, hemp, or flax, can improve the stiffness and strength of PLA composites. For example, a study found that the addition of 30 wt. % industrial hemp fibers to PLA increased the Young's modulus by 10.9 Gpa and the tensile strength by 82.9 Mpa [73]. Natural fibers can also improve the toughness of PLA composites by increasing their resistance to crack propagation and improving their impact resistance [81]. The strength of 3D-printed kenaf/PLA composites increases to 3 wt. % but decreases significantly to 5 and 7 wt. % due to voids, and extrusion temperature affects the structure of the filaments [82].
- Wear and friction: NF-reinforced PLA composites exhibit reduced wear rates and friction coefficients, enhancing their suitability for tribological applications. Incorporation of natural fiber mats into the PLA matrix significantly improves wear behavior, resulting in a 10–44% reduction in friction coefficient and over 70% reduction in specific wear rate [83].
- Thermal Properties: The thermal stability of PLA composites with NFs can be slightly affected compared to neat PLA. Some studies report a decrease in the glass transition temperature (T_g) due to the presence of the fibers [84]. However, the thermal degradation temperature may not be significantly impacted.
- Physical Properties: NF reinforcement can decrease or increase the density of the composite compared to neat PLA, depending on fiber loading and type. Additionally, the hygroscopic nature of these fibers can affect the moisture absorption of the composite, potentially leading to dimensional changes.
- Chemical and moisture stability: PLA composites are more sensitive to high temperatures than to water. The mechanical properties, such as Young's modulus, decrease significantly at elevated temperatures (80 °C), but the addition of natural fibers helps reduce this decrease [85].

Specifically, the typical mechanical and thermal stability properties of several natural fiber-reinforced PLA printed composites are given in Table 1.

Table 1. Comparison of mechanical properties of PLA composites reinforced with different types of natural lignocellulose fibers [data compiled from [86–88]] *.

Fiber Type	Tensile Strength (MPa)	Tensile Modulus (Gpa)	Flexural Strength (MPa)	Impact Resistance	Thermal Stability
Flax	50	5	60	High	Good
Jute	40	3	40	Medium	Medium
Sisal	30	2	30	Low	Low
Kenaf	45	4	50	Medium	Medium
Ramie	35	3	35	Low	Low
Cotton (lyocell)	25	2	25	Low	Low

* For comparison's sake, the values corresponding to the mechanical properties were chosen as the median value for the respective composite type at the closest to 5 wt. % fiber loading.

Several fibers' structural properties and FDM process parameters play a crucial role in achieving good printability and the desired properties in NF-reinforced PLA composites [8,9]:

- Typically, fibers used in 3D printing are short fibers, often between 0.3 and 3 mm in length. Longer fibers can improve tensile strength but may cause issues with printability and nozzle clogging. The diameter of plant fibers can vary widely depending on the type of fiber. Common ranges are from 10 to 100 µm. Also, higher aspect ratio generally leads to better reinforcement. Typical aspect ratios for fibers used in 3D printing are between 50 and 200 [89].
- Printing Temperature: A higher printing temperature can improve the flowability of the PLA matrix and facilitate better fiber dispersion. However, excessively high temperatures can lead to thermal degradation of the wood fibers. The extrusion temperature for PLA-based composites containing lignocellulose fibers in FDM printing generally ranges between 200 °C and 220 °C. This temperature range plays a critical role in influencing the color, physical attributes, and mechanical properties of composites with heat-treated wood fibers, primarily due to changes in the chemical composition of lignocellulosic materials [77]. Research indicates that the PLA matrix in wood–plastic composite (WPC) parts remains amorphous during the FDM printing process, as the crystallization rate of PLA is too slow to achieve full crystallinity [90].
- Infill Pattern and density: The infill pattern can also affect the properties of the composites. A study on the 3D printing of PLA-based composites reinforced with natural lyocell fibers found that the tensile strength and modulus increased with the use of a concentric infill pattern compared to a grid or tri-hexagonal pattern [91]. Infill density significantly influences the improvement of mechanical strength in 3D-printed PLA composites with natural fiber reinforcement. Good tensile strength was observed at 25% of infill for flax fiber-reinforced 3D-printed composites [86]. Optimal load bearing is reported more often at filament crossing angles of −45/+45° [92].
- Build orientation: Generally, NF-PLA composite parts printed with the fibers aligned along the loading direction (e.g., on-edge orientation) exhibit higher flexural strength and modulus compared to parts printed with fibers perpendicular to the loading direction (e.g., flat orientation). The build orientation also influences the surface finish and the need for post-processing steps. For instance, printing a part with a smooth surface finish on the exterior layers can improve the flexural properties by reducing surface defects that could initiate cracks or failures. Additionally, post-processing techniques like annealing or heat treatment can be applied more effectively when the build orientation is considered, further enhancing the flexural properties of the composite part [93].

Also, Mansingh et al. [94] investigated the potential of fabricating fully biodegradable composites using an innovative 3D printing technique, with a particular focus on applications in food and medical product packaging. The weight percentage (wt. %) of powdered raw Pineapple Leaf Fiber (PALF) and alkali-treated PALF in the 3D-printed green composite presented a notable influence on its mechanical and thermal properties. Notably, the incorporation of 3 wt. % alkali-treated PALF yielded the optimal reinforcement, resulting in the highest tensile strength (42.9 MPa) and flexural strength (51.9 MPa). In contrast, composites reinforced with raw PALF exhibited superior ductility, achieving a maximum elongation at break of 6.89%. Additionally, the density of the 3D-printed composites increased proportionally with the wt. % of PALF content. The crystalline structure and chemical bonding characteristics of the composites were characterized through Fourier Transform Infrared (FTIR) spectroscopy and X-ray Diffraction (XRD) analysis. Microstructural analysis further revealed the presence of impurities, voids, and fiber degradation within the 3D-printed composites. These findings underscore the influence of fiber treatment and composition on the performance and structural integrity of the green composites.

Several other studies involving the usage of fibers and PLA for 3D printing applications are summarized in Table 2.

Table 2. Composite filaments and 3D-printed materials involving PLA and fibers.

Reinforcement/Additive	Description	Properties	Applications	Reference
Cellulose nanofiber (CNF)	The CNF/PLA compounds were obtained by blending a PEG surface-modified CNF aqueous suspension with microscale PLA powder, followed by air-drying using various mass ratios of CNF to PLA, 0.25:99.75, 0.5:99.5, and 1:99, respectively.	Tensile strength 50.7 MPa, Young modulus E, 3.73 MPa	Environmentally friendly method which significantly reduced the use of chemical reagents and shortened the processing time.	[95]
steam-exploded coconut fiber (F) and PLA or PLA/PBS	PBS content: 0 or 20 wt. %; F content: 3, 6 or 10 wt. %	Impact strengths of PLA97/F3 and PLA77/PBS20/F3 were approximately 8.5% and 7.4% higher than those of the original PLA and PLA/PBS, respectively. During the degradation test, both PLA97/F3 and PLA77/PBS20/F3 exhibited higher tensile strengths than the original materials.		[96]
short lyocell fibers	Lyocell fibers (FCP400) with a nominal length of 400 μm, loading reported to PLA: 10%, 20%, 30%	By combining fiber fibrillation, matrix modification, and post-printing annealing, we achieved an excellent balance of tensile strength (85 MPa), Young's modulus (7.2 GPa), and strain at break (3.2%)—the highest reported values for such composites.	performance structures using 100% bio-derived materials	[91]
Lignocellulose nanofiber/polylactic acid (LCNF/PLA) composite	Cellulose raw material used was hardwood bleached kraft pulp (HBKP) with a degree of polymerization (DPv) of ~750; cooking active alkali charges of 10%, 13%, 16%, and 19% (calculated by sodium oxide) with a liquor-to-chip ratio of 1:4.	The flexural strength of the CNF/PLA composite increased from 92.7 MPa to 151.2 MPa by combining 10% CNF (without lignin) with PLA. The flexural strength of LCNF/PLA composite with internal lignin content of 3.7% (0.37% of the total mass) was increased from 151.2 MPa to 234.5 MPa, which is 153.0% higher than that of pure PLA.		[79]
Carbon nanotubes (CNTs) were coated on short glass fibers (SGFs)	CNT-coated SGFs through a one-step flame synthesis technique. SGF- and CNT-SGF-reinforced composite filaments were fabricated with filler loadings of 1, 5 and 10 wt. %	Specimens using SGF- ($\geq$5 wt. %) and CNT-SGF (1–5 wt. %)-reinforced PLA filaments exhibited higher Young's modulus and tensile strength values due to the enhanced interface adhesion. Furthermore, the FDM printing raster angles ($\pm$45° and 0°/90°) did not noticeably affect the tensile properties of the samples made of the same material.		[97]

2.2. PLA Reinforced with Particles

Some common examples of particle reinforcements for Polylactic Acid (PLA) used in FDM 3D printing, along with their properties and typical loading weights, are as follows:

- Calcium Carbonate ($CaCO_3$): This is a widely used mineral filler for PLA. It is inexpensive, abundant, and can improve the stiffness, dimensional stability, and printability of PLA. For example, adding 20–30% by weight of $CaCO_3$ particles can increase the modulus and heat deflection temperature of PLA while also improving layer adhesion. A specific product is Omyacarb TF-100, a fine ground calcium carbonate with a median particle size of 1.3 microns, which can be added at 20–30% by weight. A study in the Journal of Polymers and the Environment found that adding 20% $CaCO_3$ to PLA increased its flexural modulus by 50% and reduced its thermal expansion coefficient by 30%.

- Carbon Black: Carbon black particles can be used to improve the electrical conductivity and mechanical properties of PLA. They are typically added at a loading weight of 1–10%. For example, a research paper in the Composites Science and Technology journal reported that adding 5% carbon black to PLA increased its electrical conductivity by several orders of magnitude and its tensile strength by 20%.
- Talcum (Magnesium Silicate): Talc is a soft, lamellar mineral that can act as a reinforcing agent and nucleating agent in PLA. It improves stiffness, impact strength, and heat resistance of PLA. For instance, adding 10–20% by weight of talcum particles can enhance the dimensional stability and surface finish of printed parts. An example is Luzenac Talc, which has a median particle size of 1.5 microns and can be added at 10–20% by weight.
- Nanoclay: Nanoclays, such as montmorillonite, are commonly used to improve the barrier and mechanical properties of PLA. They act as nano-scale reinforcements, increasing stiffness, strength, and heat resistance. For example, Cloisite 30B, an organic-modified montmorillonite nanoclay, can be added at 5–10% by weight to PLA to improve its gas barrier properties and mechanical strength. For instance, adding 3% nanoclay to PLA increased its tensile strength by 25% and its flexural modulus by 35% [98].
- Silicon dioxide (SiO_2): Silicon dioxide particles, also known as silica, can be used to improve the toughness and impact resistance of PLA. They are typically added at a loading weight of 1–5%. For instance, a research paper in the Journal of Applied Polymer Science reported that adding 3% silica to PLA increased its impact strength by 50% and its elongation at break by 100%.
- Metal Powders: Metal particles, such as copper or brass, can be added to PLA to create metallic-looking prints with improved mechanical properties. These particles increase the weight, density, and stiffness of the printed parts. For example, adding 5–15% by weight of brass powder with a particle size of 35–45 microns can create a brass-infused PLA with a metallic sheen.

2.2.1. PLA with Calcium Carbonate

$CaCO_3$ is a naturally occurring mineral that exists in several crystalline forms, including aragonite, calcite, and vaterite [99,100]. These polymorphs have different crystal habits, particle sizes, and surface areas, which can influence the final properties of the PLA filaments [101]. The use of $CaCO_3$ particles in PLA filaments can improve their mechanical properties, such as tensile strength and flexural strength, by acting as a reinforcing material [101,102]. Additionally, $CaCO_3$ can enhance the thermal stability and resistance to degradation of PLA filaments [101]. Also, this type of composite is researched as a potential biomaterial (ceramic-coated scaffolds), showing potential for further evaluations in bone tissue engineering applications [100]. Regarding the structural and morphological characteristics of calcium carbonate, several parameters influence the properties of the 3D-printed composites [103]:

- Morphology: The morphology of the $CaCO_3$ particles, such as spheroidal (granular) or rod-like (Figure 5), can also impact the mechanical properties of PLA filaments. Rod-like particles, for example, have been found to have a more pronounced influence in improving mechanical properties compared to spherical particles. Particles with higher specific surface areas tend to agglomerate more within the PLA matrix, leading to a decrease in mechanical properties [104]. A critical value for the specific surface of calcium carbonate in correlation to PLA was not determined, but as an orientative value, it is around 7 m^2/g for polyolefin matrices, beyond which increased agglomeration causes a significant decrease in strength and impact resistance [105].

- Amorphous calcium carbonate (ACC) derived from various sources, such as limestone (LS), chalk, or animal origins like white eggshells (WESs), has been explored as a filler material for polylactic acid (PLA) composites. These fillers are particularly beneficial due to their potential to enhance the mechanical properties of PLA-based materials. Notably, the particle size of the ACC plays a significant role in the resulting composite's performance. Studies have shown that fillers with a particle size of 32 μm tend to exhibit superior tensile strength compared to those with a larger particle size of 63 μm. This suggests that smaller particles contribute more effectively to the reinforcement of the PLA matrix, likely due to their higher surface area, which facilitates better interaction with the polymer matrix. In addition to tensile strength, the tensile modulus of PLA composites is significantly influenced by the filler content. For amorphous calcium carbonate-based fillers, the tensile modulus increases as the filler content rises, indicating enhanced stiffness and structural integrity. The highest tensile modulus was observed at a filler content of 20 wt. % for both 32 μm and 63 μm particle sizes, suggesting that this concentration represents an optimal balance between filler loading and the material's mechanical properties. When comparing the toughness of different ACC sources in PLA composites, it was found that limestone (LS)-based fillers provided superior toughness compared to those derived from white eggshells (WESs). This difference in toughness is likely due to variations in the microstructure and the intrinsic properties of the fillers, which influence the energy absorption capacity of the composite materials. Overall, the use of amorphous calcium carbonate fillers, particularly from limestone, shows promising potential in improving the mechanical performance of PLA composites, offering a viable option for enhancing the material's strength and durability for various applications [106].

Figure 5. Collection of observed morphologies of synthesized $CaCO_3$ polymorphs (aragonite (**a**), calcite (**b**), commercially available calcite (**c**), and vaterite (**d**)) observed under SEM [distributed under the terms and conditions of the Creative Commons Attribution (CC BY) license https://creativecommons.org/licenses/by/4.0/, accessed on 13 June 2024, from [104]].

The processing parameters for printing PLA-$CaCO_3$ composites, referenced from the literature, are as follows [107–109]:

- Nozzle Temperature Optimal Range: 190 °C to 210 °C. Higher temperatures can improve the dispersion of $CaCO_3$ particles within the PLA matrix, improving the mechanical properties.
- Printing Speed: 50 to 80 mm/s. Slower speeds can lead to better particle dispersion and improved mechanical properties.
- Infill Density: 50% to 80%. Higher densities can enhance the mechanical properties of the printed part, especially with higher $CaCO_3$ loads.
- Infill Pattern: Use a grid or honeycomb pattern for better stress distribution and to take advantage of the improved stiffness of PLA-$CaCO_3$ composites.
- Layer Height: 0.1 to 0.3 mm. Fine layers can improve the surface finish and mechanical properties, especially with high-quality prints.

A low-permeability polylactic acid (PLA) coating to calcium carbonate ($CaCO_3$) micro-particles was successfully applied, which encapsulated co-precipitated bovine serum albumin-fluorescein isothiocyanate (BSA–FITC), utilizing the solvent/oil/water (S/O/W) emulsion technique (Figure 6) [110,111]. This effectively sealed the micro-particle pores and significantly decreased the shell permeability. The incorporation of an additional polyvinyl alcohol (PVA) layer significantly improved the stability of the coated microparticles and refined the size distribution of the resultant PLA/$CaCO_3$ particles. The formulation of the particles was optimized with a PVA concentration of 2.5%, which was found to be the most effective for achieving a stable and uniform particle size distribution. To gain a deeper understanding of particle formation, the varying mass ratios of $CaCO_3$ microparticles to PLA were systematically examined across different samples. Bovine serum albumin-fluorescein isothiocyanate (BSA–FITC) was observed within the PLA particles across all formulations, with the $CaCO_3$/PLA mass ratio ranging from 0.1 to 1.2. At $CaCO_3$/PLA mass ratios below 0.8, the polymer-coated microparticles exhibited increased resistance to ethylenediaminetetraacetic acid (EDTA) treatment, thereby maintaining the integrity of their bioactive cargo. In contrast, at a $CaCO_3$/PLA mass ratio of 1.2, the PLA-coated particles displayed increased susceptibility to EDTA due to the thinner PLA coating on each $CaCO_3$ particle and the presence of pores in the PLA shell. This compromised the stability of the bioactive content within the particles. Based on these findings, a $CaCO_3$/PLA mass ratio of 0.8 was determined to be optimal, as it provided the highest protein payload while maintaining the stability of the microparticles against dissolution. This ratio offers a favorable balance between bioactive content retention and the structural integrity of the particle coating.

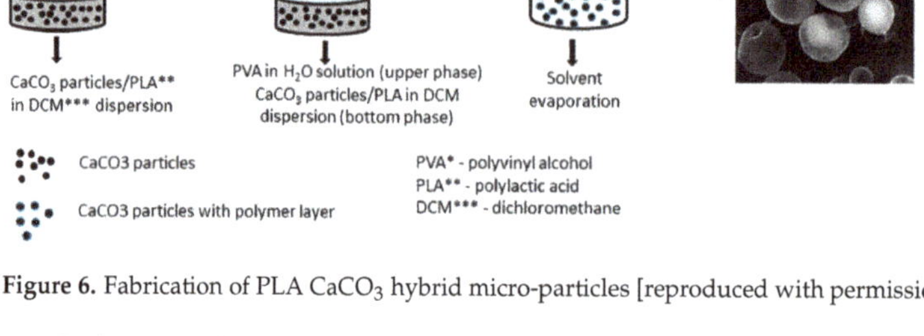

Figure 6. Fabrication of PLA $CaCO_3$ hybrid micro-particles [reproduced with permission from [111].

Both PLA and $CaCO_3$ are biodegradable materials [112], making them highly suitable for biomedical applications. Given that $CaCO_3$ is a natural component of bone, this microcapsule system holds potential for the storage and controlled delivery of bone growth factors or other pharmaceutical agents aimed at treating bone-related diseases. Additionally, the system could provide enhanced mechanical support during cargo delivery. Advanced delivery strategies, such as targeted drug delivery, could be realized by incorporating magnetic nanoparticles into the PLA shell. This modification would enable magnetic navigation, allowing precise delivery of drug-loaded microcapsules to specific target sites. The PLA-coated $CaCO_3$ microparticles, encapsulating bioactive molecules, are anticipated to have diverse applications in biomedicine, particularly as efficient drug storage and delivery systems. Furthermore, the hybrid microcapsules developed in this study could be employed for the encapsulation of a wide range of water-soluble substances, including proteins, insulin, cytostatics, and polysaccharides. This versatility underscores the potential of these microcapsules as a multifunctional platform for various therapeutic and pharmaceutical applications

2.2.2. PLA with Zinc Oxide

Polylactic Acid-Zinc Oxide (PLA-ZnO) nanocomposites have emerged as versatile and multifunctional materials, suitable for a wide array of applications due to their exceptional antibacterial, photocatalytic, and ultraviolet (UV) absorption properties. Their superior antibacterial performance offers a significant advantage over pure polymers by reducing the risk of infection through fomite transmission. Despite these promising attributes, the current utilization of PLA-ZnO nanocomposites, predominantly produced via conventional fabrication methods, remains limited to simple geometries such as films and fibers. To date, few studies have explored the additive manufacturing (AM) of PLA-ZnO nanocomposites, primarily due to challenges in material processing and the production of filaments with properties that can be reliably preserved post-3D printing [113].

As research advances to overcome these challenges, the integration of PLA-ZnO nanocomposites into AM has the potential to expand their application spectrum significantly. AM offers a distinct advantage in fabricating complex geometries and customized components, which could unlock novel uses for these nanocomposites. This intersection of 3D printing with advanced antibacterial materials is garnering increasing interest within the biomedical field, with the goal of creating personalized antibacterial medical devices tailored to individual patient anatomies. Furthermore, this innovation aligns with the priorities of the post-pandemic AM industry, aiming to deliver safer and more hygienic materials for enhanced environmental security. Nevertheless, a critical concern is whether the functional properties of PLA-ZnO nanocomposites, particularly their antibacterial effectiveness, can be retained at sufficient levels after undergoing the AM process. The transition from conventional manufacturing to AM introduces additional processing steps, such as high-temperature extrusion and deposition, which may alter the intrinsic properties of the nanocomposite materials. Understanding the impact of these processes on the structural integrity and functionality of PLA-ZnO nanocomposites remains a key research focus, as it will determine the feasibility of their broader implementation in additive manufacturing [114].

2.2.3. PLA with Silicon Carbide and Graphite

Silicon carbide (SiC) and graphite are chosen as fillers for polylactic acid (PLA), primarily due to their mechanical and thermal attributes. A significant reduction in recovery time is observed for composites with high filler loading. For instance, a composite comprising 50 wt. % carbon and 10 wt. % SiC in a PLA matrix exhibits an 87% decrease in recovery time compared to pure PLA [115]. The evaluation of shape recovery performance is based on three key parameters: recovery ratio, recovery rate, and recovery time. The recovery ratio is defined as the proportion of the initial deformation angle restored after the recovery process. The recovery rate represents the instantaneous time derivative of the recovery ratio, providing insight into the speed of shape restoration. Recovery time is measured as the interval between the initiation of the thermal trigger and the point at which the recovery rate reaches its maximum value. A detailed methodology for the extrusion of printer filaments and the subsequent 3D printing process is thoroughly described to ensure reproducibility and clarity.

By altering the material composition, and consequently its thermal conductivity, the rate of shape memory response can be effectively controlled. The correlation between material composition and shape recovery rate, determined through thermal conductivity analysis, offers a framework for designing structures capable of activating shape memory responses at tailored rates. The advanced capabilities of 3D printing, particularly the use of multiple filaments and print heads, enable the fabrication of structures with highly heterogeneous material compositions. This approach represents a significant departure from conventional methods, such as variations in temperature or structural thickness, that have traditionally been used to modulate the timing of shape memory response activation within structures.

Several mechanical properties (UTS: ultimate tensile strength (MPa); strain at break (%); flexural strength (MPa) and impact strength (kJ/m^2)) of 3D-printed PLA reinforced with various inorganic fillers selected from the literature are given in Figure 7. In this figure, the average value was calculated from the samples sets presented in the indicated studies.

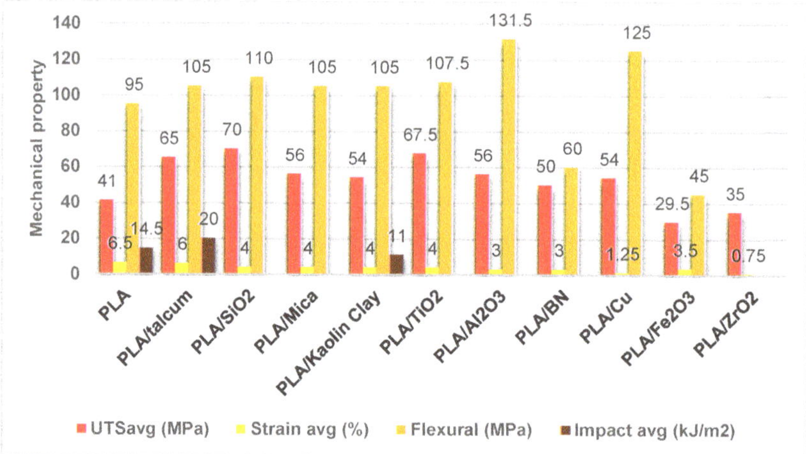

Figure 7. Average values for the mechanical properties of 3D-printed PLA reinforced with inorganic additives: neat PLA [116]; talcum [117,118]; silica (SiO_2) [119,120]; mica [121,122]; kaolin clay [123]; titania (TiO_2) [124,125]; alumina (Al_2O_3) [126]; boron nitride (BN) [127]; copper (Cu) [128]; iron (III) oxide (Fe_2O_3) [129]; zirconia (ZrO_2) [130].

Overall, the inclusion of various fillers, such as, notably, titanium dioxide (PLATiO_2) and copper (PLA/Cu), appears to improve the mechanical properties of the PLA material compared to the baseline PLA without any fillers.

2.3. PLA Reinforced with Carbon-Based Nanomaterials

The most effective nanomaterials for enhancing the mechanical properties of PLA in FFF/FDM 3D printing are [131–133]:

- Carbon nanotubes (CNTs): Several studies have shown that incorporating CNTs into PLA can significantly improve its tensile strength, modulus, and toughness, while reducing elongation at break. The mechanical properties are influenced by interfacial adhesion and CNT dispersion.

- Graphene nanoplatelets (GNPs): Adding GNPs to PLA can enhance its mechanical properties, depending on factors such as raster direction and nanoplatelet size. Larger GNPs and optimizing raster orientation can improve the mechanical capabilities of graphene-reinforced PLA.
- Short carbon fibers (SCFs): Incorporating SCFs into PLA has been found to enhance its tensile and shear modulus in various printing directions. Continuous carbon fibers can also increase the tensile and flexural behavior of PLA, although they may reduce failure strain.
- Combinations of carbon nanofillers: Using multiple carbon nanofillers, such as GNPs and multi-wall carbon nanotubes (MWCNTs), can produce PLA nanocomposites with better mechanical properties compared to single fillers. For example, adding SCFs and graphene to PLA tripled its mechanical strength [134].

The maximum filler content is typically limited to avoid drastically decreasing the elastic strain of the composites and to allow adequate extrusion of the filaments during FFF/FDM printing. Optimizing printing parameters, such as temperature and infill, can also influence the mechanical properties of PLA nanocomposites.

The addition of carbon nanotubes to polylactic acid (PLA) composites during FDM printing can significantly improve the mechanical, thermal, and electrical properties of the resulting materials. CNTs act as reinforcing agents, increasing the mechanical strength and stiffness of the PLA matrix by providing additional load-bearing capacity and resistance to deformation. This aspect can improve the dimensional accuracy of 3D-printed PLA parts by reducing the warpage and shrinkage that can occur during the printing process. The strong interfacial bonding between CNTs and PLA matrix enhances the mechanical properties of the composite by reducing the likelihood of debonding and improving the transfer of stresses between the CNTs and the matrix. These nanofillers can form a network structure within the PLA matrix, which enhances the mechanical properties of the composite by providing additional mechanical reinforcement and improving the resistance to deformation [135,136].

The CNT content has a significant influence on the PLA/CNT composite, with the optimal CNT loading depending on the specific application and desired properties. The mechanical properties improve by increasing CNT content up to a certain threshold, beyond which the benefits may plateau or even diminish due to issues like CNT agglomeration [137]. Studies comparing CNTs with other reinforcements like cellulose nanocrystals (CNCs) also show that CNTs provide superior improvements in mechanical properties [138].

- Effect of CNT Content on Thermal Properties: The addition of CNTs can improve the thermal conductivity of the PLA matrix, which can lead to improved mechanical properties by reducing thermal gradients and stress concentrations within the composite. The presence of CNTs can accelerate the crystallization of PLA, leading to improved mechanical properties by increasing the degree of crystallinity and reducing the likelihood of defects and imperfections within the matrix. The presence of CNTs and other nanofillers like nanoclay increases the thermal stability and storage modulus of PLA composites, making them more robust at higher temperatures [139].
- Effect of CNT Content on Electrical Properties: The electrical conductivity of PLA/CNT composites has also been investigated, with the addition of CNTs being shown to significantly improve the electrical conductivity of the materials. The CNT content has a significant influence on the electrical properties of PLA/CNT composites during FDM printing. The electrical resistivity varied from approximately $1 \times 10^{12}\ \Omega/m^2$ to $1 \times 10^{2}\ \Omega/m^2$ for CNT contents ranging from 0 wt. % to 8 wt. % [139].

- Effect of CNT on the mechanical properties: The optimal carbon nanotube (CNT) content for achieving the best mechanical properties in PLA/CNT composites is still a topic of ongoing research. However, studies have shown that a CNT content of around 5–10 wt. % can result in significant improvements in mechanical properties (10–60% in tensile strength, up to 30% for flexural strength) compared to neat PLA [140]. It has been found that a 6% CNT content led to a substantial increase in tensile and flexural strength, as well as improved electrical conductivity [139]. However, in other studies, it has been found that while CNT inclusion increased the Young's modulus by 30% at 5% CNT loading, it reduced tensile strength and overall toughness [137].

CNT structure, as well as the FDM process parameters such as extrusion temperature, building direction, infill percentage and geometry, and layer height were key factors in determining the properties of the printed materials [33].

- CNT structure and dimensions: The type (single-walled carbon nanotubes, SWCNTs, multi-walled carbon nanotubes, MWCNTs) and length of CNTs play crucial roles in determining the performance of 3D-printed parts. Longer carbon nanotubes (typically 0.65–1.3 mm) result in higher thermal and electrical conductivities, but no appreciable change in the mechanical properties of PLA/CNT composites during FDM printing [141]. SWCNTs have a more significant impact on the material properties than MWCNTs, but they are generally more expensive. Long MWCNT inclusion determines better performance at lower concentrations [142].
- Nozzle temperature: The optimal printing temperature range for PLA-CNT composites is between 180 °C and 220 °C, with an optimal printing temperature of around 200 °C for tensile strength and 220 °C for flexural strength. A higher printing temperature can lead to improved bonding between the PLA matrix and CNTs, resulting in improved mechanical properties, but may also reduce the impact strength of the composite material [143]. Higher extrusion temperatures can reduce the void fraction in the printed parts but may also result in less alignment of CNTs due to radial flow and fusion between adjacent layers [137]. For better electrical and thermal conductivity, a lower temperature of 200 °C is preferred [139].
- Layer height and infill geometry: The specimens printed at 0.100 mm layer height, gyroid-type infill geometry and number of perimeters of 6 have maximum tensile strength [144]. For better thermal and electrical conductivity, greater layer heights are more desirable [139].
- Build direction: The mechanical properties of 3D-printed PLA composites exhibit significant anisotropy based on build orientation. Tensile strength and Young's modulus are higher when the material is printed in a flat orientation compared to an upright orientation [93].

Functionalized MWCNTs can significantly enhance the mechanical properties of polymer composites. The functionalization of carbon nanotubes (CNTs) through mild acid treatment enhances their nucleating effect during polylactic acid (PLA) crystallization and promotes better dispersion of the CNTs within the PLA matrix. The PLA/f-CNT nanocomposite exhibits higher degrees of crystallinity, as expected, because f-CNTs act more efficiently as nucleating agents than c-CNTs.

Differential Scanning Calorimetry (DSC) analysis revealed no thermal property gradients in 3D-printed specimens across all analyzed samples, including pure PLA, PLA/c-CNT, and PLA/f-CNT. This uniformity in thermal properties, encompassing transition temperatures and degrees of crystallinity, was consistent across the top, central, and base layers of the printed specimens. Furthermore, the 3D printing process did not alter the melting temperature of the materials, indicating that the thermal stability of the composites remained unaffected. The surface functionalization of carbon nanotubes (CNTs) significantly improved the composite's mechanical performance. Specifically, the PLA/f-CNT samples exhibited a 43% higher storage modulus at body temperature (37 °C) compared to PLA/c-CNT samples, indicating that functionalized CNTs act as more effective reinforcing agents. Scanning Electron Microscopy (SEM) imaging of the fracture interfaces further revealed enhanced interlayer adhesion in PLA/f-CNT samples (Figure 8). This improvement is attributed to the better dispersion of functionalized CNTs within the PLA matrix, resulting in a more uniform material flow during the printing process. Under tensile stress, PLA/f-CNT samples demonstrated markedly superior mechanical properties, with layer-to-layer adhesion contributing to a significant increase in tensile strength (from 29.4 ± 0.7 MPa in PLA/c-CNT to 41.6 ± 1.4 MPa in PLA/f-CNT). These findings highlight the efficacy of CNT surface modification in enhancing the thermal and mechanical properties of 3D-printed PLA/CNT nanocomposites. In particular, the study underscores the role of mild acid treatment (HNO_3, 5 mol L^{-1}) in functionalizing CNTs and optimizing their reinforcing potential [145].

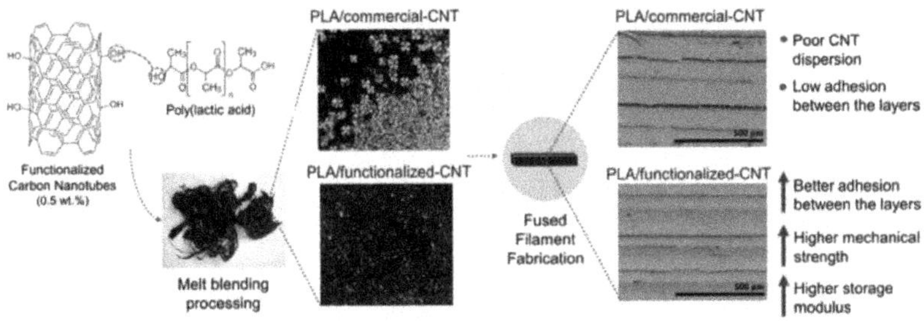

Figure 8. Differences between PLA/commercial-CNT and PLA/functionalized-CNT [reproduced with permission from [145].

2.4. *Natural and Synthetic Polymer Blends with Biomass Fillers*

2.4.1. Hydroxypropyl-Methylcellulose-Reinforced PLA

Biocompatible hydroxypropyl methylcellulose (HPMC)/polylactic acid (PLA) composites were successfully fabricated as filaments and 3D-printed parts using fused deposition modeling (FDM). This study examined the effects of HPMC content on various properties, including microstructure, chemical structure, thermal behavior, mechanical performance, and water contact angle (CA). The results demonstrated a uniform distribution and dispersion of HPMC within the PLA matrix. Incorporating HPMC up to a concentration of 7% did not affect the chemical structure of PLA during the melt blending and filament extrusion processes. However, HPMC incorporation significantly influenced the thermal properties of the composites, leading to an increased glass transition temperature and a reduced cold crystallization temperature [146]. While the tensile strengths of the composite filaments were comparable to those of neat PLA, the tensile and impact strengths of the 3D-printed samples decreased as the HPMC content increased. This reduction was attributed to increased porosity induced by the presence of HPMC. Additionally, the incorporation of HPMC resulted in a lower water contact angle, which indicates enhanced hydrophilicity. This characteristic could be particularly advantageous for applications in the biomedical field, where surface wettability plays a critical role.

2.4.2. PLA with Lignin and Polymerized-Lignin-Treated NFC

Composite PLA filaments containing unmodified and lignin/polymerized lignin surface-modified nanofibrillated cellulose (NFC) were analyzed to assess their mechanical, thermal, and structural properties. The addition of NFC was found to reduce the stretchability, toughness, breaking tenacity, bending stiffness, and compression resistance of the filaments, while slightly enhancing their initial modulus. Despite a marginal toughening effect imparted by NFC, the improvement was limited due to poor interfacial interactions between NFC and the PLA matrix [147].

The mechanical performance of the composite filaments was closely associated with the NFC's surface modification and concentration within the PLA matrix. When the NFC content was increased to 5 wt. %, the mechanical properties showed the most significant deterioration. However, incorporating lignin/polymerized lignin surface-modified NFC mitigated the decline in mechanical properties and resulted in an increase in storage modulus. The observed limitations in toughness improvement were attributed to poor interfacial adhesion and the presence of NFC agglomerates. These agglomerates, characterized by irregular plate-like shapes and uneven dispersion, acted as structural weak points, thereby impairing the mechanical integrity of the filaments. Scanning Electron Microscopy (SEM) images revealed the structural inconsistencies, while Fourier Transform Infrared (FTIR) analysis confirmed interactions between lignin-modified NFC and the PLA matrix. Dynamic Mechanical Analysis (DMA) further demonstrated a notable increase in elasticity with the addition of modified NFC. Although no significant differences in thermal stability were observed between filaments containing unmodified and modified NFC, the composite filaments exhibited an enhanced crystallization rate compared to neat PLA [148].

Despite only minor variations in mechanical properties, the incorporation of NFC improved the thermal stability of the composites. Lignin, known for its antioxidant properties and UV protection capability, enhanced the thermal oxidative resistance of the filaments, thereby extending the lifespan of 3D-printed objects. Additionally, the light-brown and dark-brown hues imparted by lignin/polymerized lignin surface-modified NFC conferred a unique and aesthetically appealing appearance to the filaments and their corresponding 3D-printed constructs [149].

2.4.3. PLA with Acetylated Tannin

This study investigated the properties and 3D printability of a composite material composed of Polylactic Acid (PLA) and Acetylated Tannin (AT). The acetylation of tannin enhanced its dispersion within the PLA matrix, enabling the successful fabrication of PLA/AT composite filaments using a twin-screw extrusion process. The inclusion of 20 wt. % AT in the PLA matrix did not significantly compromise the tensile properties of the composite. Thermogravimetric analysis indicated that the PLA/AT composites could be safely processed at 3D printing temperatures below 250 °C to avoid thermal decomposition. While acetylated tannin improved the overall printability of PLA systems, optimal printability was achieved at temperatures below 220 °C to prevent phase separation and aggregation of AT, especially at higher loading levels [150].

The incorporation of acetylated tannin at varying concentrations did not substantially affect the melting and glass transition temperatures of PLA, as it had minimal impact on the intermolecular interactions or chain flexibility of the PLA polymer. However, a reduction in the crystallinity of PLA/AT composites was observed, which accelerated their degradation in aquatic environments, particularly under alkaline conditions. This accelerated degradation could be advantageous for short-term biomedical applications, such as implantable devices, where controlled degradation is a desirable attribute [151].

2.4.4. PLA with Cork

Biodegradable composites were synthesized, composed of cork and Polylactic Acid (PLA) with varying proportions of cork. A decline was observed in the tensile mechanical attributes of the composites as the proportion of cork was augmented. The impact resistance demonstrated an initial decrease with the incorporation of cork; however, it exhibited an increase with further escalation in the cork content. The viscoelastic characteristics of the composites displayed a diminishing trend with the increase in cork content. In contrast, the specific modulus and specific strength of the composites increased with higher cork content, highlighting the potential of cork–PLA composites for lightweight and impact-resistant 3D-printed structures. This improvement in specific mechanical properties underscores their suitability for advanced engineering applications. A filament composed of a 5% *w*/*w* cork–PLA composite was successfully developed, demonstrating compatibility with fused deposition modeling (FDM) processes. The 3D-printed cork–PLA composite exhibited slightly reduced tensile mechanical properties compared to its compression-molded counterpart. However, an exception was observed in elongation at break, where the 3D-printed composite displayed superior ductility, emphasizing its potential for applications requiring enhanced flexibility and resilience [152].

2.4.5. PLA/Wood Flour

Composite filaments of Wood Fiber (WF) and Polylactic Acid (PLA) were synthesized, and their properties were evaluated in this study. Additionally, 3D specimens were fabricated utilizing the Fused Deposition Modelling (FDM) technique. The study tells us that WF/PLA composite filament demonstrated compatibility with the FDM process, indicating its suitability for 3D printing. Moreover, the incorporation of WF resulted in alterations in the microstructure of the PLA fracture surface. The interfaces between the WF and PLA were distinctly visible and the resistance to initial deformation of the composite was observed to be enhanced upon the addition of WF, in comparison to pure PLA. Also, the onset temperature for thermal degradation of the composites exhibited a slight decrease, while the final residual ratio post-thermal decomposition of the composites showed an increase and the addition of WF at a concentration of 5 wt. % did not influence the melting temperature of the PLA. [153]

Parameters of 0.2 mm layer height, 0.7 mm nozzle diameter, 75% fill density, and 35 mm/s velocity maximize the flexural strength of wood–PLA composite parts built through FFF. [154]

2.4.6. PLA/HA with PCL Addition

The fabrication of bespoke materials, composed of biocompatible and biodegradable polymers either independently or in conjunction with mineral components, resulted in printable substances possessing chemical stability and mechanical attributes conducive to bone regeneration. The mechanical characteristics were validated to meet the mechanical threshold requisite for trabecular bone applications. The tailoring of degradation rates exhibited a strong correlation with the material composition, indicating an enhancement through the integration of mineral phases, such as hydroxyapatite, which further accelerated the degradation of PLA–PCL amalgamations. This, in conjunction with the introduction of regulated porosity and scaffold architecture, could potentially facilitate the control and equilibrium of biomaterial resorption and neo-osteogenesis as required [155].

2.4.7. PLA with TPS and ESO

This research is driven by the current market demand for cost-effective alternatives to Polylactic Acid (PLA). In this study, novel compositions based on PLA, modified by Thermoplastic Starch (TPS) and Epoxidized Soybean Oil (ESO), were developed to enhance the ductility of PLA and reduce the cost of the products without compromising their biodegradability [156]. The efficacy of the proposed compositions was validated by studying their rheological, mechanical, and thermal properties, water resistance, and compostability. An increase in the Melt Flow Rate (MFR) value resulted in enhanced adaptability to injection and blow molding processes, thereby expanding the range of short-life application products from a technological perspective. To evaluate the suitability of the obtained PLA/TPS (thermoplastic starch) blends for various packaging industry applications, their mechanical properties were thoroughly analyzed. The incorporation of epoxidized soybean oil (ESO) resulted in materials with enhanced softness, improved impact strength (up to 16.69 kJ/m^2), and increased tensile and ductile properties, including an elongation at break of approximately 8.8%, compared to native PLA. Scanning Electron Microscopy (SEM) of fractured surfaces confirmed these improvements. Additionally, the presence of ESO in the blends was observed to delay water diffusion into the matrix, thereby potentially improving the dimensional stability of products subjected to short-term water exposure. The modified TPS, prepared with ESO, was shown to be processable using standard PLA machinery, facilitating its integration into existing production systems. Furthermore, the ability to substitute up to 25% of PLA with ESO-modified TPS offers a cost-effective alternative while maintaining comparable properties and compostability to that of pure PLA, underscoring its potential as a sustainable material for packaging applications [157].

Several other studies involving the usage of different fillers and PLA for 3D printing applications are summarized in Table 3.

Table 3. Composite filaments and 3D-printed materials involving PLA and various natural and synthetic fillers.

Type	Reinforcement/Aditive	Description	Properties	Applications	Reference
Particle and fiber hybrid-reinforced PLA	MXene ($Ti_3C_2T_x$) and recycled carbon fiber (rCF)	MrCF (fiber mass fraction 10 wt. % with 1 wt. % $Ti_3C_2T_x$) and 178 g of dried PLA	Compared to pure PLA, the modified PLA displayed significant improvements: 15.6% in toughness, 112.1% in flexural strength, and 31.8% in notched impact strength.	The modified PLA demonstrated superior electromagnetic shielding performance due to the absorption properties of the composite material.	[158]
Particle-based composite	Ferronickel slag (FNS)	composites with FNS contents of 2.0, 6.0, 10.0, and 14.0 wt. %	PLA/5.0 wt. % FNS composite exhibited the most significant improvement in mechanical properties, with a roughly 18% increase in both tensile and flexural strength compared to unfilled PLA thermoplastic		[159]
	Magnesium	1% Mg reported to PLA	Among the 3D-printed samples with different infill orientations ($\theta = 0°, 45°, \pm45°, 90°$), the highest UTS was measured in specimens with 0° infill orientation (UTS = 43 MPa), whereas the lowest UTS was measured for the 90° infill orientation (UTS = 26 MPa). Mg particles also reduce the mechanical properties at all infill orientations.		[160]
	Parawood (*Hevea brasiliensis*)	Four different parawood powder weight ratios were blended with PLA, encompassing 0%, 5%, 10%, and 15% (*w*/*w*)	The maximum tensile strength was attained when the parawood powder content was 15% *w*/*w*.	Large-scale screw-extrusion 3D printing process for producing furniture parts	[161]

Table 3. *Cont.*

Type	Reinforcement/Aditive	Description	Properties	Applications	Reference
	Pyrolyzed HSC (hazelnut shell carbon)	2 % HSC was added to the PLA matrix	The 3D-printed PLA-HSC parts demonstrated excellent photothermal performance with a light absorption intensity of around 93%	Low-cost, compostable, high-efficiency photothermal conversion materials with shaping freedom	[162]
	Salix alba sawdust	1%, 2%, 3%, and 4% (weight %) Salix alba were loaded in the PLA matrix.		PLA-salix alba-based composite feedstock has the controlled MFI and may be used for scaffold preparation in biomedical applications.	[163]
	PLA/Olive wood waste	olive wood scraps (with an amount ranging between 10 and 20% by weight)	The use of wooden scraps in place of virgin PLA granulate reveals a rise in the impact in the extrusion (ca. 27%) and 3D printing (11%) stages	Environmental benefits resulting from the inclusion of wood scraps in PLA filaments (20% of wood corresponds to 10% environmental impact reduction) supporting further research in this area.	[164]
	Cu_2O nanoparticles (50-100 nm diameter)	PLA-3 wt. % Cu_2O mixture	Superior mechanical performance of the 3D-printed nanocomposite at a 0° raster angle, while the mechanical properties gradually decreased for raster angles of 45° and 90°.	This composite can be utilized for the fabrication of various prostheses featuring specific groove geometries, bio parts in medical devices, robotics surgical systems, and implants	[165]
	PLA/$MgAl_2O_4$:Sm^{3+} composite filaments	Composite filaments of PLA with different amounts of phosphors (0, 1, 2, 3 and 4 phr)	Luminescence intensity of the composites regularly increased with the inorganic phosphor contents	Luminescent bioplastic composite filaments. Filaments and samples of the composites emitted pink-orange light under UV light	[166]
Sheet composite	graphene oxide (GO)	different content of GO (0.4, 0.8, 1.6, 2.4 and 4.0 wt. %) reported to PLA	PLA/GO scrolled fibers and printed fibers with graphene content of 0.4 wt. % exhibit outstanding mechanical properties with strength increases of about 32.7% and 35.2%.		[167]
Polymer-based composite	PLA-TPU	50%, 70% and 90% TPU reported to PLA	The UTS values for PLA50, PLA70, and PLA90 were achieved at 27.27 MPa, 40.91 MPa, and 54.18 MPa, respectively. Also, the fracture toughness results were consistent with the mechanical properties. The PLA90 had 1.69 and 2.36 times higher fracture toughness than the PLA70 and PLA50 compounds, respectively.	By increasing the amount of TPU, the printability decreased due to higher melt strength and viscosity, incomplete melting, feeding problems caused by buckling, and incomplete integration between adjacent rasters and layers.	[168]

Table 3. *Cont.*

Type	Reinforcement/Aditive	Description	Properties	Applications	Reference
	thermoplastic starch (TPS)/poly(lactic acid) (PLA)/poly(butylene adipate-co-terephthalate) (PBAT) composite	Ratio of TPS:PLA:PBAT was fixed at 50:40:10 wt. %.	113% increase in elongation at break and the 190% rise in impact strength compared to PLA	highly renewable filaments for 3D printing	[169]
	PLA with biodegradable elastomer poly (butyleneadipate-co-terephthalate) (PBAT) and poly (methyl methacrylate) (PMMA)		Tensile strength and breaking elongation were 94.8% and 3650%, respectively, of that of 3D-printed pure SC-PLA	Controlled compatibilization effect of PMMA endows PBAT with good dispersibility in the quaternary system, without affecting the hierarchical crystallization of enantiomeric PLA matrices for complete stereo-complexation	[170]
	Chitosan	PLA and 10 wt. % chitosan	The acidic environment caused by the degradation of PLA can be counteracted by chitosan, probably due to its protonation.	Biocomposite materials	[171]

3. Conclusions

In the dynamic landscape of additive manufacturing, the quest for high-performance materials continues, and our exploration of reinforced PLA composites has revealed promising innovations and practical applications. The integration of natural fibers into PLA enhances strength and stiffness, offering an eco-friendly alternative while maintaining mechanical integrity. By incorporating metallic particles, researchers have pushed the boundaries of PLA's capabilities, improving thermal and electrical properties and expanding their utility in functional parts. PLA composites reinforced with cellulose fibers demonstrate excellent printability and mechanical performance, with the synergy between PLA and cellulose opening doors to sustainable materials for various industries. Carbon fibers provide significant improvements in tensile strength, sometimes exceeding a 100% increase when loaded at 15%, though these materials require careful handling to prevent clogging and ensure uniform dispersion. Glass fibers enhance flexural strength and thermal stability but tend to make composites brittle, reducing impact resistance. Natural fibers, such as kenaf and hemp, improve sustainability and weight reduction but are limited by moisture absorption and mechanical stability, which can affect layer adhesion. Among particle reinforcements, calcium carbonate improves stiffness, dimensional stability, and printability, though large particles can cause agglomeration and reduce mechanical performance if not properly managed. Carbon nanotubes (CNTs) show potential in improving mechanical, thermal, and electrical properties, though their effectiveness depends on proper dispersion and bonding within the PLA, with surface treatments enhancing their effects. However, challenges like clogging, dispersion issues, and moisture absorption remain common for both synthetic and natural fibers in FDM printing, requiring adjustments in nozzle size, temperature, and other parameters to facilitate smoother flow. Continuous carbon fiber composites stand out for high-strength applications but require specialized equipment and processing, while natural fibers align with eco-friendly goals but underperform compared with synthetic fibers like carbon and glass, which offer superior properties. Among particle-based systems, calcium carbonate is particularly promising for improving mechanical properties and printability without significant processing challenges. Suitable printing parameters, including higher temperatures (230–250 °C) for particles, lower temperatures (below 220 °C) for natural fibers, increased infill density (50–70 %), slower printing speeds (10–20 mm/s), and lower layer heights (0.1–0.2 mm), are critical for maximizing performance, particularly in fiber-reinforced composites. While each reinforcement system presents unique benefits and challenges, continuous carbon fibers and calcium carbonate-based composites appear to offer the most promising results depending on the application. The optimization of printing parameters, such as nozzle size, temperature, layer height, and infill density, is crucial for achieving the best outcomes. Reinforced PLA composites align with global efforts to reduce reliance on fossil-based polymers, and the journey toward optimized PLA composites involves collaboration across disciplines, including materials science, engineering, and design. By bridging the gap between theory and practice, reinforced PLA composites hold immense promise for a greener, more resilient additive manufacturing industry.

Funding: This research received no external funding.

Conflicts of Interest: The authors declare no conflict of interest.

Abbreviations

PLA	polylactic acid
FDM	fused deposition modeling
ABS	poly (acrylonitrile-co-butadine-co-styrene)
PETG	polyethylene-terephthalate glycol
FFF	fused filament fabrication
PHB	polyhydroxy butyrate
PBS	polybutylene succinate
CPA-PLA	carbon fiber-reinforced polylactic acid
GFs	glass fiber
OG	octylgallate
LG	lauryl gallate
PALF	pineapple leaf fiber
FTIR	Fourier transform infrared
XRD	X-ray diffraction
HBICP	hardwood bleached kraft pulp
LCNF	lignocellose nanofiber
SGFs	Short glass fibers
LS	limestone
WES	white eggshell
BSA-FITC	Bovine serum albumin -fluorescein isothiocyanate
S/O/W	solvent/oil/water
DCM	dichloro methane
UTS	Ultimate tensile strength
BN	Boron nitride
SCFs	short carbon fibers
MWCNTs	Multiwalled carbon nanotubes
SEM	scanning electron microscopy
AT	acetylated tannin
DMA	dynamic mechanical analysis
PCL	polycaprolactone
ESO	Epoxidized soybean oil

TPS	Thermoplastic starch
MFR	Melt flow rate
FNS	Ferronickel slab
PBAT	polybutylene adipate-co-terephthalate
PMMA	poly(methyl methacrylate)
f-CNTs	functionalized carbon nanotubes
phr	parts per hundred resin (polymer)

References

1. Hottle, T.; Bilec, M.; Landis, A. Sustainability assessments of bio-based polymers. *Polym. Degrad. Stab.* **2013**, *98*, 1898–1907. [CrossRef]
2. Li, G.; Zhao, M.; Xu, F.; Yang, B.; Li, X.; Meng, X.; Teng, L.; Sun, F.; Li, Y. Synthesis and Biological Application of Polylactic Acid. *Molecules* **2020**, *25*, 5023. [CrossRef] [PubMed]
3. Caminero, M.; Chacón, J.; García-Plaza, E.; Núñez, P.; Reverte, J.; Bécar, J. Additive Manufacturing of PLA-Based Composites Using Fused Filament Fabrication: Effect of Graphene Nanoplatelet Reinforcement on Mechanical Properties, Dimensional Accuracy and Texture. *Polymers* **2019**, *11*, 799. [CrossRef] [PubMed]
4. Kaczmarek, H.; Nowicki, M.; Vuković-Kwiatkowska, I.; Nowakowska, S. Crosslinked blends of poly (lactic acid) and polyacrylates: AFM, DSC and XRD studies. *J. Polym. Res.* **2013**, *20*, 91. [CrossRef]
5. Ilyas, R.A.; Sapuan, S.M.; Harussani, M.M.; Hakimi, M.Y.A.Y.; Haziq, M.Z.M.; Atikah, M.S.N.; Asyraf, M.R.M.; Ishak, M.R.; Razman, M.R.; Nurazzi, N.M.; et al. Polylactic Acid (PLA) Biocomposite: Processing, Additive Manufacturing and Advanced Applications. *Polymers* **2021**, *13*, 1326. [CrossRef]
6. Ncube, L.K.; Ude, A.U.; Ogunmuyiwa, E.N.; Zulkifli, R.; Beas, I.N. Environmental Impact of Food Packaging Materials: A Review of Contemporary Development from Conventional Plastics to Polylactic Acid Based Materials. *Materials* **2020**, *13*, 4994. [CrossRef]
7. Gu, J.; Wensing, M.; Uhde, E.; Salthammer, T. Characterization of particulate and gaseous pollutants emitted during operation of a desktop 3D printer. *Environ. Int.* **2019**, *123*, 476–485. [CrossRef]
8. Potter, P.M.; Al-Abed, S.R.; Hasan, F.; Lomnicki, S.M. Influence of polymer additives on gas-phase emissions from 3D printer filaments. *Chemosphere* **2021**, *279*, 130543. [CrossRef]
9. Alberts, E.; Ballentine, M.; Barnes, E.; Kennedy, A. Impact of metal additives on particle emission profiles from a fused filament fabrication 3D printer. *Atmos. Environ.* **2021**, *244*, 117956. [CrossRef]
10. Bhagia, S.; Bornani, K.; Agrawal, R.; Satlewal, A.; Ďurkovič, J.; Lagaňa, R.; Bhagia, M.; Yoo, C.G.; Zhao, X.; Kunc, V.; et al. Critical review of FDM 3D printing of PLA biocomposites filled with biomass resources, characterization, biodegradability, upcycling and opportunities for biorefineries. *Appl. Mater. Today* **2021**, *24*, 101078. [CrossRef]
11. Cojocaru, V.; Frunzaverde, D.; Miclosina, C.-O.; Marginean, G. The Influence of the Process Parameters on the Mechanical Properties of PLA Specimens Produced by Fused Filament Fabrication—A Review. *Polymers* **2022**, *14*, 886. [CrossRef] [PubMed]
12. Weber, R.; Zhang, Q.; Wong, J.; Davis, A.; Black, M. Fine particulate and chemical emissions from desktop 3D printers. *NIP Digit. Fabr. Conf.* **2016**, *32*, 121–123. [CrossRef]
13. Thumsorn, S.; Prasong, W.; Kurose, T.; Ishigami, A.; Kobayashi, Y.; Ito, H. Rheological Behavior and Dynamic Mechanical Properties for Interpretation of Layer Adhesion in FDM 3D Printing. *Polymers* **2022**, *14*, 2721. [CrossRef]
14. Rodríguez-Panes, A.; Claver, J.; Camacho, A.M. The Influence of Manufacturing Parameters on the Mechanical Behaviour of PLA and ABS Pieces Manufactured by FDM: A Comparative Analysis. *Materials* **2018**, *11*, 1333. [CrossRef] [PubMed]
15. Tao, Y.; Li, P.; Zhang, J.; Wang, S.; Shi, S.Q.; Kong, F. A review of fused filament fabrication of continuous natural fiber reinforced thermoplastic composites: Techniques and materials. *Polym. Compos.* **2023**, *44*, 8200–8222. [CrossRef]
16. Błędzki, A.; Gassan, J. Composites reinforced with cellulose based fibres. *Prog. Polym. Sci.* **1999**, *24*, 221–274. [CrossRef]
17. Yusri, M.A.H.M.; Zuhri, M.Y.M.; Ishak, M.R.; Azman, M.A. The Capabilities of Honeycomb Core Structures Made of Kenaf/Polylactic Acid Composite under Compression Loading. *Polymers* **2023**, *15*, 2179. [CrossRef]
18. Brząkalski, D.; Sztorch, B.; Frydrych, M.; Pakuła, D.; Dydek, K.; Kozera, R.; Boczkowska, A.; Marciniec, B.; Przekop, R.E. Limonene Derivative of Spherosilicate as a Polylactide Modifier for Applications in 3D Printing Technology. *Molecules* **2020**, *25*, 5882. [CrossRef]
19. Buj-Corral, I.; Sivatte-Adroer, M. An Experimental Investigation about the Dimensional Accuracy and the Porosity of Copper-Filled PLA Fused Filament Fabrication Parts. *Metals* **2023**, *13*, 1608. [CrossRef]
20. Sztorch, B.; Brząkalski, D.; Pakuła, D.; Frydrych, M.; Špitalský, Z.; Przekop, R.E. Natural and Synthetic Polymer Fillers for Applications in 3D Printing—FDM Technology Area. *Solids* **2022**, *3*, 508–548. [CrossRef]
21. Kantaros, A.; Soulis, E.; Petrescu, F.I.T.; Ganetsos, T. Advanced Composite Materials Utilized in FDM/FFF 3D Printing Manufacturing Processes: The Case of Filled Filaments. *Materials* **2023**, *16*, 6210. [CrossRef] [PubMed]

22. Fernández, V.; García, A.; Camacho, A.M.; Claver, J.; Rodríguez, A.; Sebastián, M.A. Methodology to estimate the modulus of elasticity of parts manufactured by FFF/FDM combining finite element simulations and experimental tests. *IOP Conf. Ser. Mater. Sci. Eng.* **2021**, *1193*, 012103. [CrossRef]
23. Anwajler, B.; Zdybel, E.; Tomaszewska-Ciosk, E. Innovative Polymer Composites with Natural Fillers Produced by Additive Manufacturing (3D Printing)—A Literature Review. *Polymers* **2023**, *15*, 3534. [CrossRef] [PubMed]
24. Pal, A.K.; Mohanty, A.K.; Misra, M. Additive manufacturing technology of polymeric materials for customized products: Recent developments and future prospective. *RSC Adv.* **2021**, *11*, 36398–36438. [CrossRef]
25. Rafiee, R.; Amini, M.; Zehtabzadeh, H. Examining Tensile Properties in Nanocomposite Samples Fabricated Through Extrusion-Based Additive Manufacturing. *Arab. J. Sci. Eng.* **2023**, *49*, 10913–10926. [CrossRef]
26. Ryan, K.; Down, M.; Hurst, N.; Keefe, E.; Banks, C. Additive manufacturing (3D printing) of electrically conductive polymers and polymer nanocomposites and their applications. *eScience* **2022**, *2*, 365–381. [CrossRef]
27. Hu, C.; Hau, W.N.J.; Chen, W.; Qin, Q.-H. The fabrication of long carbon fiber reinforced polylactic acid composites via fused deposition modelling: Experimental analysis and machine learning. *J. Compos. Mater.* **2021**, *55*, 1459–1472. [CrossRef]
28. Do, T.; Le, M.; Nguyen, T.; Le Thai, H. Effect of Infill Density and Printing Patterns on Compressive Strength of ABS, PLA, PLA-CF Materials for FDM 3D Printing. *Mater. Sci. Forum* **2022**, *1068*, 19–27. [CrossRef]
29. Valvez, S.; Santos, P.; Parente, J.M.; Silva, M.P.; Reis, P.N.B. 3D printed continuous carbon fiber reinforced PLA composites: A short review. *Procedia Struct. Integr.* **2020**, *25*, 394–399. [CrossRef]
30. Cao, M.; Cui, T.; Yue, Y.; Li, C.; Guo, X.; Jia, X.; Wang, B. Investigation of Carbon Fiber on the Tensile Property of FDM-Produced PLA Specimen. *Polymers* **2022**, *14*, 5230. [CrossRef]
31. Shah, A.K.; Jain, A. Microstructure and mechanical properties of filament and fused deposition modelling printed polylactic-acid and carbon-fiber reinforced polylactic-acid. *J. Reinf. Plast. Compos.* **2024**, *43*, 516–531. [CrossRef]
32. Dou, H.; Cheng, Y.; Ye, W.; Zhang, D.; Li, J.; Miao, Z.; Rudykh, S. Effect of Process Parameters on Tensile Mechanical Properties of 3D Printing Continuous Carbon Fiber-Reinforced PLA Composites. *Materials* **2020**, *13*, 3850. [CrossRef] [PubMed]
33. Kamaal, M.; Anas, M.; Rastogi, H.; Bhardwaj, N.; Rahaman, A. Effect of FDM process parameters on mechanical properties of 3D-printed carbon fibre–PLA composite. *Prog. Addit. Manuf.* **2021**, *6*, 63–69. [CrossRef]
34. Hu, Y.; Ladani, R.B.; Brandt, M.; Li, Y.; Mouritz, A.P. Carbon fibre damage during 3D printing of polymer matrix laminates using the FDM process. *Mater. Des.* **2021**, *205*, 109679. [CrossRef]
35. Papon, E.A.; Haque, A. Tensile properties, void contents, dispersion and fracture behaviour of 3D printed carbon nanofiber reinforced composites. *J. Reinf. Plast. Compos.* **2018**, *37*, 381–395. [CrossRef]
36. Papon, E.A.; Haque, A. Fracture toughness of additively manufactured carbon fiber reinforced composites. *Addit. Manuf.* **2019**, *26*, 41–52. [CrossRef]
37. Krajangsawasdi, N.; Blok, L.G.; Hamerton, I.; Longana, M.L.; Woods, B.K.S.; Ivanov, D.S. Fused Deposition Modelling of Fibre Reinforced Polymer Composites: A Parametric Review. *J. Compos. Sci.* **2021**, *5*, 29. [CrossRef]
38. Lee, D.; Wu, G.-Y. Parameters Affecting the Mechanical Properties of Three-Dimensional (3D) Printed Carbon Fiber-Reinforced Polylactide Composites. *Polymers* **2020**, *12*, 2456. [CrossRef]
39. Selvam, A.; Mayilswamy, S.; Whenish, R.; Velu, R.; Subramanian, B. Preparation and Evaluation of the Tensile Characteristics of Carbon Fiber Rod Reinforced 3D Printed Thermoplastic Composites. *J. Compos. Sci.* **2021**, *5*, 8. [CrossRef]
40. Peng, X.; Zhang, M.; Guo, Z.; Sang, L.; Hou, W. Investigation of processing parameters on tensile performance for FDM-printed carbon fiber reinforced polyamide 6 composites. *Compos. Commun.* **2020**, *22*, 100478. [CrossRef]
41. Magri, A.E.; El Mabrouk, K.; Vaudreuil, S.; Touhami, M.E. Mechanical properties of CF-reinforced PLA parts manufactured by fused deposition modeling. *J. Thermoplast. Compos. Mater.* **2021**, *34*, 581–595. [CrossRef]
42. Rao, V.D.P.; Rajiv, P.; Geethika, V.N. Effect of fused deposition modelling (FDM) process parameters on tensile strength of carbon fibre PLA. *Mater. Today Proc.* **2019**, *18 Pt 6*, 2012–2018. [CrossRef]
43. Öteyaka, M.; Aybar, K.; Öteyaka, H.C. Effect of Infill Ratio on the Tensile and Flexural Properties of Unreinforced and Carbon Fiber-Reinforced Polylactic Acid Manufactured by Fused Deposition Modeling. *J. Mater. Eng. Perform.* **2021**, *30*, 5203–5215. [CrossRef]
44. Liu, Z.; Lei, Q.; Xing, S. Mechanical characteristics of wood, ceramic, metal and carbon fiber-based PLA composites fabricated by FDM. *J. Mater. Res. Technol.* **2019**, *8*, 3741–3751. [CrossRef]
45. Chen, K.; Yu, L.; Cui, Y.; Jia, M.; Pan, K. Optimization of printing parameters of 3D-printed continuous glass fiber reinforced polylactic acid composites. *Thin-Walled Struct.* **2021**, *164*, 107717. [CrossRef]
46. Vinyas, M.; Athul, S.; Harursampath, D.; Thoi, T.N. Experimental evaluation of the mechanical and thermal properties of 3-D printed PLA and its composites. *Mater. Res. Express* **2019**, *6*, 115301. [CrossRef]
47. Ismail, K.I.; Pang, R.; Ahmed, R.; Yap, T.C. Tensile Properties of In Situ 3D Printed Glass Fiber-Reinforced PLA. *Polymers* **2023**, *15*, 3436. [CrossRef]

48. Dickson, A.N.; Barry, J.N.; McDonnell, K.A.; Dowling, D.P. Dowling, Fabrication of continuous carbon, glass and Kevlar fibre reinforced polymer composites using additive manufacturing. *Addit. Manuf.* **2017**, *16*, 146–152. [CrossRef]
49. Arai, S.; Tsunoda, S.; Yamaguchi, A.; Ougizawa, T. Effects of short-glass-fiber content on material and part properties of poly(butylene terephthalate) processed by selective laser sintering. *Addit. Manuf.* **2018**, *21*, 683–693. [CrossRef]
50. Wang, G.; Zhang, D.; Li, B.; Wan, G.; Zhao, G.; Zhang, A. Strong and thermal-resistance glass fiber-reinforced polylactic acid (PLA) composites enabled by heat treatment. *Int. J. Biol. Macromol.* **2019**, *129*, 448–459. [CrossRef]
51. Chicos, L.-A.; Pop, M.A.; Zaharia, S.-M.; Lancea, C.; Buican, G.R.; Pascariu, I.S.; Stamate, V.-M. Fused Filament Fabrication of Short Glass Fiber-Reinforced Polylactic Acid Composites: Infill Density Influence on Mechanical and Thermal Properties. *Polymers* **2022**, *14*, 4988. [CrossRef] [PubMed]
52. Wang, K.; Chen, Y.; Long, H.; Baghani, M.; Rao, Y.; Peng, Y. Hygrothermal aging effects on the mechanical properties of 3D printed composites with different stacking sequence of continuous glass fiber layers. *Polym. Test.* **2021**, *100*, 107242. [CrossRef]
53. Varsavas, S.; Kaynak, C. Weathering degradation performance of PLA and its glass fiber reinforced composite. *Mater. Today Commun.* **2017**, *15*, 344–353. [CrossRef]
54. Wang, G.; Zhang, D.; Wan, G.; Li, B.; Zhao, G. Glass fiber reinforced PLA composite with enhanced mechanical properties, thermal behavior, and foaming ability. *Polymer* **2019**, *181*, 121803. [CrossRef]
55. Tümer, E.H.; Erbil, H.Y. Extrusion-Based 3D Printing Applications of PLA Composites: A Review. *Coatings* **2021**, *11*, 390. [CrossRef]
56. Tang, C.; Liu, J.; Yang, Y.; Liu, Y.; Jiang, S.; Hao, W. Effect of process parameters on mechanical properties of 3D printed PLA lattice structures. *Compos. Part C Open Access* **2020**, *3*, 100076. [CrossRef]
57. Yang, Y.; Yang, B.; Chang, Z.; Duan, J.; Chen, W. Research Status of and Prospects for 3D Printing for Continuous Fiber-Reinforced Thermoplastic Composites. *Polymers* **2023**, *15*, 3653. [CrossRef]
58. Gunasekaran, K.; Aravinth, V.; Kumaran, C.M.; Madhankumar, K.; Kumar, S.P. Pradeep Kumar, Investigation of mechanical properties of PLA printed materials under varying infill density. *Mater. Today Proc.* **2021**, *45 Pt 2*, 1849–1856. [CrossRef]
59. Ambati, S.S.; Ambatipudi, R. Effect of infill density and infill pattern on the mechanical properties of 3D printed PLA parts. *Mater. Today Proc.* **2022**, *64 Pt 1*, 804–807. [CrossRef]
60. Shavandi, A.; Hosseini, S.; Okoro, O.V.; Nie, L.; Babadi, F.E.; Melchels, F. 3D Bioprinting of Lignocellulosic Biomaterials. *Adv. Healthc. Mater.* **2020**, *9*, 2001472. [CrossRef]
61. Dickson, A.N.; Abourayana, H.M.; Dowling, D.P. 3D Printing of Fibre-Reinforced Thermoplastic Composites Using Fused Filament Fabrication—A Review. *Polymers* **2020**, *12*, 2188. [CrossRef] [PubMed]
62. Mazzanti, V.; Malagutti, L.; Mollica, F. FDM 3D Printing of Polymers Containing Natural Fillers: A Review of their Mechanical Properties. *Polymers* **2019**, *11*, 1094. [CrossRef] [PubMed]
63. Tarrés, Q.; Melbø, J.; Delgado-Aguilar, M.; Espinach, F.; Mutjé, P.; Chinga-Carrasco, G. Chinga-Carrasco, Bio-polyethylene reinforced with thermomechanical pulp fibers: Mechanical and micromechanical characterization and its application in 3D-printing by fused deposition modelling. *Compos. Part B Eng.* **2018**, *153*, 70–77. [CrossRef]
64. Le Duigou, A.; Barbé, A.; Guillou, E.; Castro, M. 3D printing of continuous flax fibre reinforced biocomposites for structural applications. *Mater. Des.* **2019**, *180*, 107884. [CrossRef]
65. Savic, A.; Antonijevic, D.; Jelic, I.; Zakic, D. Thermomechanical behavior of bio-fiber composite thermal insulation panels. *Energy Build.* **2020**, *229*, 110511. [CrossRef]
66. Asim, M.; Paridah, M.T.; Chandrasekar, M.; Shahroze, R.M.; Jawaid, M.; Nasir, M.; Siakeng, R. Thermal stability of natural fibers and their polymer composites. *Iran. Polym. J.* **2020**, *29*, 625–648. [CrossRef]
67. Estakhrianhaghighi, E.; Mirabolghasemi, A.; Zhang, Y.; Lessard, L.; Akbarzadeh, A. 3D-Printed Wood-Fiber Reinforced Architected Cellular Composites 2020. *Adv. Eng. Mater.* **2020**, *22*, 2000565. [CrossRef]
68. Le Duigou, A.; Correa, D.; Ueda, M.; Matsuzaki, R.; Castro, M. A review of 3D and 4D printing of natural fibre biocomposites. *Mater. Des.* **2020**, *194*, 108911. [CrossRef]
69. Rajendran Royan, N.R.; Leong, J.S.; Chan, W.N.; Tan, J.R.; Shamsuddin, Z.S.B. Current State and Challenges of Natural Fibre-Reinforced Polymer Composites as Feeder in FDM-Based 3D Printing. *Polymers* **2021**, *13*, 2289. [CrossRef]
70. Parandoush, P.; Lin, D. A review on additive manufacturing of polymer-fiber composites. *Compos. Struct.* **2017**, *182*, 36–53. [CrossRef]
71. Lee, C.H.; Padzil, F.N.B.M.; Lee, S.H.; Ainun, Z.M.A.; Abdullah, L.C. Potential for Natural Fiber Reinforcement in PLA Polymer Filaments for Fused Deposition Modeling (FDM) Additive Manufacturing: A Review. *Polymers* **2021**, *13*, 1407. [CrossRef] [PubMed]
72. Bledzki, A.K.; Jaszkiewicz, A.; Scherzer, D. Mechanical properties of PLA composites with man-made cellulose and abaca fibres. *Compos. Part A Appl. Sci. Manuf.* **2009**, *40*, 404–412. [CrossRef]

73. Rajeshkumar, G.; Seshadri, S.A.; Devnani, G.; Sanjay, M.; Siengchin, S.; Maran, J.P.; Al-Dhabi, N.A.; Karuppiah, P.; Mariadhas, V.A.; Sivarajasekar, N.; et al. Environment friendly, renewable and sustainable poly lactic acid (PLA) based natural fiber reinforced composites—A comprehensive review. *J. Clean. Prod.* **2021**, *310*, 127483. [CrossRef]
74. Goriparthi, B.K.; Suman, K.N.S.; Rao, N.M. Effect of fiber surface treatments on mechanical and abrasive wear performance of polylactide/jute composites. *Compos. Part A Appl. Sci. Manuf.* **2012**, *43*, 1800–1808. [CrossRef]
75. Vilaseca, F.; Mendez, J.; Pèlach, A.; Llop, M.; Cañigueral, N.; Gironès, J.; Turon, X.; Mutjé, P. Composite materials derived from biodegradable starch polymer and jute strands. *Process Biochem.* **2007**, *42*, 329–334. [CrossRef]
76. Aguado, R.J.; Espinach, F.X.; Julián, F.; Tarrés, Q.; Delgado-Aguilar, M.; Mutjé, P. Tensile Strength of Poly(lactic acid)/Bleached Short Hemp Fiber Fully Green Composites as Replacement for Polypropylene/Glass Fiber. *Polymers* **2023**, *15*, 146. [CrossRef]
77. Deb, D.; Jafferson, J. Natural fibers reinforced FDM 3D printing filaments. *Mater. Today Proc.* **2021**, *46 Pt 2*, 1308–1318. [CrossRef]
78. Filgueira, D.; Holmen, S.; Melbø, J.K.; Moldes, D.; Echtermeyer, A.T.; Chinga-Carrasco, G. Enzymatic-Assisted Modification of Thermomechanical Pulp Fibers To Improve the Interfacial Adhesion with Poly(lactic acid) for 3D Printing. *ACS Sustain. Chem. Eng.* **2017**, *5*, 9338–9346. [CrossRef]
79. Zhang, Q.; Ma, L.; Zhang, X.; Zhang, L.; Wang, Z. Lignocellulose nanofiber/polylactic acid (LCNF/PLA) composite with internal lignin for enhanced performance as 3D printable filament. *Ind. Crops Prod.* **2022**, *178*, 114590. [CrossRef]
80. Ahmed, W.; Alnajjar, F.; Zaneldin, E.; Al-Marzouqi, A.H.; Gochoo, M.; Khalid, S. Implementing FDM 3D Printing Strategies Using Natural Fibers to Produce Biomass Composite. *Materials* **2020**, *13*, 4065. [CrossRef]
81. Billings, C.; Siddique, R.; Sherwood, B.; Hall, J.; Liu, Y. Additive Manufacturing and Characterization of Sustainable Wood Fiber-Reinforced Green Composites. *J. Compos. Sci.* **2023**, *7*, 489. [CrossRef]
82. Shahar, F.S.; Hameed Sultan, M.T.; Safri, S.N.A.; Jawaid, M.; Abu Talib, A.R.; Basri, A.A.; Md Shah, A.U. Physical, thermal and tensile behaviour of 3D printed kenaf/PLA to suggest its usability for ankle–foot orthosis—A preliminary study. *Rapid Prototyp. J.* **2022**, *28*, 1573–1588. [CrossRef]
83. Bajpai, P.K.; Singh, I.; Madaan, J. Tribological behavior of natural fiber reinforced PLA composites. *Wear* **2013**, *297*, 829–840. [CrossRef]
84. Gupta, B.; Revagade, N.; Hilborn, J. Poly(lactic acid) fiber: An overview. *Prog. Polym. Sci.* **2007**, *32*, 455–482. [CrossRef]
85. Mazur, K.E.; Borucka, A.; Kaczor, P.; Gądek, S.; Kuciel, S. Mechanical Performance Under Various Conditions of 3D Printed Polylactide Composites with Natural Fibers. *Preprint* **2021**. Available online: https://www.researchsquare.com/article/rs-517727/v1 (accessed on 10 December 2024). [CrossRef]
86. Paulo, A.; Santos, J.; da Rocha, J.; Lima, R.; Ribeiro, J. Mechanical Properties of PLA Specimens Obtained by Additive Manufacturing Process Reinforced with Flax Fibers. *J. Compos. Sci.* **2023**, *7*, 27. [CrossRef]
87. Aliotta, L.; Gigante, V.; Coltelli, M.B.; Cinelli, P.; Lazzeri, A. Evaluation of Mechanical and Interfacial Properties of Bio-Composites Based on Poly(Lactic Acid) with Natural Cellulose Fibers. *Int. J. Mol. Sci.* **2019**, *20*, 960. [CrossRef]
88. Graupner, N.; Herrmann, A.S.; Müssig, J. Natural and man-made cellulose fibre-reinforced poly(lactic acid) (PLA) composites: An overview about mechanical characteristics and application areas. *Compos. Part A Appl. Sci. Manuf.* **2009**, *40*, 810–821. [CrossRef]
89. Dogru, A.; Sozen, A.; Neser, G.; Seydibeyoglu, M.O. Effects of Aging and Infill Pattern on Mechanical Properties of Hemp Reinforced PLA Composite Produced by Fused Filament Fabrication (FFF). *Appl. Sci. Eng. Prog.* **2021**, *14*, 651–660. [CrossRef]
90. Yang, T.-C.; Yeh, C.-H. Morphology and Mechanical Properties of 3D Printed Wood Fiber/Polylactic Acid Composite Parts Using Fused Deposition Modeling (FDM): The Effects of Printing Speed. *Polymers* **2020**, *12*, 1334. [CrossRef]
91. Gauss, C.; Pickering, K.L.; Graupner, N.; Müssig, J. 3D-printed polylactide composites reinforced with short lyocell fibres—Enhanced mechanical properties based on bio-inspired fibre fibrillation and post-print annealing. *Addit. Manuf.* **2023**, *77*, 103806. [CrossRef]
92. Belarbi, Y.E.; Guessasma, S.; Belhabib, S.; Benmahiddine, F.; Hamami, A.E.A. Effect of Printing Parameters on Mechanical Behaviour of PLA-Flax Printed Structures by Fused Deposition Modelling. *Materials* **2021**, *14*, 5883. [CrossRef] [PubMed]
93. Gonabadi, H.; Yadav, A.; Bull, S.J. The effect of processing parameters on the mechanical characteristics of PLA produced by a 3D FFF printer. *Int. J. Adv. Manuf. Technol.* **2020**, *111*, 695–709. [CrossRef]
94. Mansingh, B.B.; Binoj, J.S.; Tan, Z.Q.; Eugene, W.W.L.; Amornsakchai, T.; Abu Hassan, S.; Goh, K.L. Comprehensive characterization of raw and treated pineapple leaf fiber/polylactic acid green composites manufactured by 3D printing technique. *Polym. Compos.* **2022**, *43*, 6051–6061. [CrossRef]
95. Zhang, Z.; Wang, W.; Li, Y.; Fu, K.; Tong, X.; Cao, B.; Chen, B. 3D printing of cellulose nanofiber/polylactic acid composites via an efficient dispersion method. *Compos. Commun.* **2023**, *43*, 101731. [CrossRef]
96. Inseemeesak, B.; Siripaiboon, C.; Somkeattikul, K.; Attasophonwattana, P.; Kiatiwat, T.; Punsuvon, V.; Areeprasert, C. Biocomposite fabrication from pilot-scale steam-exploded coconut fiber and PLA/PBS with mechanical and thermal characterizations. *J. Clean. Prod.* **2022**, *379 Pt 1*, 134517. [CrossRef]
97. Zhao, G.; Liu, H.-Y.; Cui, X.; Du, X.; Zhou, H.; Mai, Y.-W.; Jia, Y.-Y.; Yan, W. Tensile properties of 3D-printed CNT-SGF reinforced PLA composites. *Compos. Sci. Technol.* **2022**, *230 Pt 2*, 109333. [CrossRef]

98. Tenn, N.; Follain, N.; Soulestin, J.; Crétois, R.; Bourbigot, S.; Marais, S. Effect of Nanoclay Hydration on Barrier Properties of PLA/Montmorillonite Based Nanocomposites. *J. Phys. Chem. C* **2012**, *117*, 12117–12135. [CrossRef]
99. Lenshin, A.S.; Frolova, V.E.; Kannykin, S.V.; Domashevskaya, E.P. Microstructural and Hydrophilic Properties of Polylactide Polymer Samples with Various 3D Printing Patterns. *Polymers* **2024**, *16*, 1281. [CrossRef]
100. Ghiasi, M.; Abdollahy, M.; Khalesi, M.R.; Ghiasi, E. Control of morphology, specific surface area and agglomeration of precipitated calcium carbonate crystals through a multiphase carbonation process. *CrystEngComm* **2020**, *22*, 1970–1984. [CrossRef]
101. Faba, S.; Agüero, Á.; Arrieta, M.P.; Martínez, S.; Romero, J.; Torres, A.; Galotto, M.J. Foaming of 3D-Printed PLA/$CaCO_3$ Composites by Supercritical CO_2 Process for Sustainable Food Contact Materials. *Polymers* **2024**, *16*, 798. [CrossRef]
102. Varga, P.; Lorinczy, D.; Tóth, L.; Pentek, A.; Nyitrai, M.; Maroti, P. Novel PLA-CaCO3 composites in additive manufacturing of upper limb casts and orthotics—A feasibility study. *Mater. Res. Express* **2019**, *6*, 045317. [CrossRef]
103. Donate, R.; Paz, R.; Quintana, Á.; Bordón, P.; Monzón, M. Calcium Carbonate Coating of 3D-Printed PLA Scaffolds Intended for Biomedical Applications. *Polymers* **2023**, *15*, 2506. [CrossRef] [PubMed]
104. Zárybnická, L.; Ševčík, R.; Pokorný, J.; Machová, D.; Stránská, E.; Šál, J. $CaCO_3$ Polymorphs Used as Additives in Filament Production for 3D Printing. *Polymers* **2022**, *14*, 199. [CrossRef] [PubMed]
105. Zhu, Y.; Allen, G.; Jones, P.; Adams, J.; Gittins, D.; Heard, P.; Skuse, D. Dispersion characterisation of CaCO3 particles in PP/CaCO3 composites. *Compos. Part A Appl. Sci. Manuf.* **2014**, *60*, 38–43. [CrossRef]
106. Betancourt, N.G.; Cree, D.E. Mechanical Properties of Poly (lactic acid) Composites Reinforced with $CaCO_3$ Eggshell Based Fillers. *MRS Adv.* **2017**, *2*, 2545–2550. [CrossRef]
107. Pavon, C.; Aldas, M.; Samper, M.D.; Motoc, D.L.; Ferrandiz, S.; López-Martínez, J. Mechanical, Dynamic-Mechanical, Thermal and Decomposition Behavior of 3D-Printed PLA Reinforced with $CaCO_3$ Fillers from Natural Resources. *Polymers* **2022**, *14*, 2646. [CrossRef]
108. Aliotta, L.; Cinelli, P.; Coltelli, M.B.; Lazzeri, A. Rigid filler toughening in PLA-Calcium Carbonate composites: Effect of particle surface treatment and matrix plasticization. *Eur. Polym. J.* **2019**, *113*, 78–88. [CrossRef]
109. Shi, X.; Zhang, G.; Siligardi, C.; Ori, G.; Lazzeri, A. Comparison of Precipitated Calcium Carbonate/Polylactic Acid and Halloysite/Polylactic Acid Nanocomposites. *J. Nanomater.* **2015**, *2015*, 905210. [CrossRef]
110. Zhi, Z.; Liu, R.; Wang, W.; Dewettinck, K.; Bockstaele, F. Recent progress in oil-in-water-in-oil (O/W/O) double emulsions. *Crit. Rev. Food Sci. Nutr.* **2022**, *63*, 6196–6207. [CrossRef]
111. Kudryavtseva, V.L.; Zhao, L.; Tverdokhlebov, S.I.; Sukhorukov, G.B. Fabrication of PLA/CaCO3 hybrid micro-particles as carriers for water-soluble bioactive molecules. *Colloids Surf. B Biointerfaces* **2017**, *157*, 481–489. [CrossRef]
112. Aframehr, W.; Molki, B.; Heidarian, P.; Behzad, T.; Sadeghi, M.; Bagheri, R. Effect of calcium carbonate nanoparticles on barrier properties and biodegradability of polylactic acid. *Fibers Polym.* **2017**, *18*, 2041–2048. [CrossRef]
113. Murariu, M.; Dubois, P. PLA composites: From production to properties. *Adv. Drug Deliv. Rev.* **2016**, *107*, 17–46. [CrossRef] [PubMed]
114. Chong, W.J.; Shen, S.; Li, Y.; Trinchi, A.; Pejak, D.; Kyratzis, I.; Sola, A.; Wen, C. Additive manufacturing of antibacterial PLA-ZnO nanocomposites: Benefits, limitations and open challenges. *J. Mater. Sci. Technol.* **2022**, *111*, 120–151. [CrossRef]
115. Liu, W.; Wu, N.; Pochiraju, K. Shape recovery characteristics of SiC/C/PLA composite filaments and 3D printed parts. *Compos. Part A-Appl. Sci. Manuf.* **2018**, *108*, 1–11. [CrossRef]
116. Darnal, A.; Shahid, Z.; Deshpande, H.; Kim, J.; Muliana, A. Tuning mechanical properties of 3D printed composites with PLA:TPU programmable filaments. *Compos. Struct.* **2023**, *318*, 117075. [CrossRef]
117. Yu, F.; Liu, T.; Zhao, X.; Yu, X.; Lu, A.; Wang, J. Effects of talc on the mechanical and thermal properties of polylactide. *J. Appl. Polym. Sci.* **2012**, *125*, E99–E109. [CrossRef]
118. Yu, W.; Wang, X.; Ferraris, E.; Zhang, J. Melt crystallization of PLA/Talc in fused filament fabrication. *Mater. Des.* **2019**, *182*, 108013. [CrossRef]
119. Goldschmidt, G.; de Moraes, E.; de Oliveira, A.N.; Hotza, D. Production and characterization of 3D-printed silica-based cellular structures. *Open Ceram.* **2022**, *9*, 100225. [CrossRef]
120. Ahmed, W.; Siraj, S.; Al-Marzouqi, A.H. 3D Printing PLA Waste to Produce Ceramic Based Particulate Reinforced Composite Using Abundant Silica-Sand: Mechanical Properties Characterization. *Polymers* **2020**, *12*, 2579. [CrossRef]
121. Travieso-Rodriguez, J.A.; Jerez-Mesa, R.; Llumà, J.; Traver-Ramos, O.; Gomez-Gras, G.; Roa Rovira, J.J. Mechanical Properties of 3D-Printing Polylactic Acid Parts subjected to Bending Stress and Fatigue Testing. *Materials* **2019**, *12*, 3859. [CrossRef]
122. Monaldo, E.; Ricci, M.; Marfia, S. Mechanical properties of 3D printed polylactic acid elements: Experimental and numerical insights. *Mech. Mater.* **2023**, *177*, 104551. [CrossRef]
123. Coppola, B.; Cappetti, N.; Di Maio, L.; Scarfato, P.; Incarnato, L. Incarnato; Influence of 3D printing parameters on the properties of PLA/clay nanocomposites. *AIP Conf. Proc.* **2018**, *1981*, 020064. [CrossRef]
124. Mofokeng, J.P.; Luyt, A.S. Dynamic mechanical properties of PLA/PHBV, PLA/PCL, PHBV/PCL blends and their nanocomposites with TiO2 as nanofiller. *Thermochim. Acta* **2015**, *613*, 41–53. [CrossRef]

125. Xu, J.; Xu, F.; Gao, G. The Effect of 3D Printing Process Parameters on the Mechanical Properties of PLA Parts. *J. Phys. Conf. Ser.* **2021**, *2133*, 012026. [CrossRef]
126. Subramaniyan, M.; Karuppan, S. Mechanical properties of sandwich products obtained by 3D printing from PLA-PLA/Al2O3. *Polimery* **2024**, *68*, 646–651. [CrossRef]
127. Keshavamurthy, R.; Tambrallimath, V.; Patil, S.; Rajhi, A.A.; Duhduh, A.A.; Khan, T.M.Y. Mechanical and Wear Studies of Boron Nitride-Reinforced Polymer Composites Developed via 3D Printing Technology. *Polymers* **2023**, *15*, 4368. [CrossRef]
128. Ji, Q.; Wang, Z.; Yi, J.; Tang, X. Mechanical Properties and a Constitutive Model of 3D-Printed Copper Powder-Filled PLA Material. *Polymers* **2021**, *13*, 3605. [CrossRef]
129. Kumar, S.; Singh, R.; Singh, T.; Batish, A. On mechanical characterization of 3-D printed PLA-PVC-wood dust-Fe3O4 composite. *J. Thermoplast. Compos. Mater.* **2022**, *35*, 36–53. [CrossRef]
130. Bennett, C.; Sojithamporn, P.; Thanakulwattana, W.; Wattanutchariya, W.; Leksakul, K.; Nakkiew, W.; Jantanasakulwong, K.; Rachtanapun, P.; Suhr, J.; Sawangrat, C. Optimization of 3D Printing Technology for Fabrication of Dental Crown Prototype Using Plastic Powder and Zirconia Materials. *Materials* **2022**, *15*, 8618. [CrossRef]
131. Luo, X.; Cheng, H.; Wu, X. Nanomaterials Reinforced Polymer Filament for Fused Deposition Modeling: A State-of-the-Art Review. *Polymers* **2023**, *15*, 2980. [CrossRef]
132. Amza, C.G.; Zapciu, A.; Constantin, G.; Baciu, F.; Vasile, M.I. Enhancing Mechanical Properties of Polymer 3D Printed Parts. *Polymers* **2021**, *13*, 562. [CrossRef] [PubMed]
133. Camargo, J.C.; Machado, R.; Almeida, E.C.; Silva, E.F.M.S. Mechanical properties of PLA-graphene filament for FDM 3D printing. *Int. J. Adv. Manuf. Technol.* **2019**, *103*, 2423–2443. [CrossRef]
134. Al Abir, A.; Chakrabarti, D.; Trindade, B. Fused Filament Fabricated Poly(lactic acid) Parts Reinforced with Short Carbon Fiber and Graphene Nanoparticles with Improved Tribological Properties 2019. *Polymers* **2023**, *15*, 2451. [CrossRef] [PubMed]
135. Khan, S.; Joshi, K.; Deshmukh, S. A comprehensive review on effect of printing parameters on mechanical properties of FDM printed parts. *Mater. Today Proc.* **2022**, *50 Pt 5*, 2119–2127. [CrossRef]
136. Zhou, X.; Deng, J.; Fang, C.; Lei, W.; Song, Y.; Zhang, Z.; Huang, Z.; Li, Y. Additive manufacturing of CNTs/PLA composites and the correlation between microstructure and functional properties. *J. Mater. Sci. Technol.* **2021**, *60*, 27–34. [CrossRef]
137. Patanwala, H.S.; Hong, D.; Vora, S.R.; Bognet, B.; Ma, A.W.K. The microstructure and mechanical properties of 3D printed carbon nanotube-polylactic acid composites. *Polym. Compos.* **2018**, *39*, E1060–E1071. [CrossRef]
138. Agbakoba, V.C.; Hlangothi, P.; Andrew, J.; John, M.J. Mechanical and Shape Memory Properties of 3D-Printed Cellulose Nanocrystal (CNC)-Reinforced Polylactic Acid Bionanocomposites for Potential 4D Applications. *Sustainability* **2022**, *14*, 12759. [CrossRef]
139. Yang, L.; Li, S.; Zhou, X.; Liu, J.; Li, Y.; Yang, M.; Yuan, Q.; Zhang, W. Effects of carbon nanotube on the thermal, mechanical, and electrical properties of PLA/CNT printed parts in the FDM process. *Synth. Met.* **2019**, *253*, 122–130. [CrossRef]
140. Podsiadły, B.; Matuszewski, P.; Skalski, A.; Słoma, M. Carbon Nanotube-Based Composite Filaments for 3D Printing of Structural and Conductive Elements. *Appl. Sci.* **2021**, *11*, 1272. [CrossRef]
141. Jang, H.G.; Yang, B.; Khil, M.-S.; Kim, S.Y.; Kim, J. Comprehensive study of effects of filler length on mechanical, electrical, and thermal properties of multi-walled carbon nanotube/polyamide 6 composites. *Compos. Part A Appl. Sci. Manuf.* **2019**, *125*, 105542. [CrossRef]
142. Elaskalany, M.; Behdinan, K. Effect of carbon nanotube type and length on the electrical conductivity of carbon nanotube polymer nanocomposites. *Mater. Res. Express* **2023**, *10*, 105010. [CrossRef]
143. Jayanth, N.; Jaswanthraj, K.; Sandeep, S.; Mallaya, N.H.; Siddharth, S.R. Effect of heat treatment on mechanical properties of 3D printed PLA. *J. Mech. Behav. Biomed. Mater.* **2021**, *123*, 104764. [CrossRef] [PubMed]
144. Chokshi, H.; Shah, D.B.; Patel, K.M.; Joshi, S.J. Experimental investigations of process parameters on mechanical properties for PLA during processing in FDM. *Adv. Mater. Process. Technol.* **2021**, *8*, 696–709. [CrossRef]
145. De Bortoli, L.; de Farias, R.; Mezalira, D.; Schabbach, L.; Fredel, M. Functionalized carbon nanotubes for 3D-printed PLA-nanocomposites: Effects on thermal and mechanical properties. *Mater. Today Commun.* **2022**, *31*, 103402. [CrossRef]
146. Jiang, G.; Yang, T.; Xu, J.; Tao, D.; Luo, C.; Wang, C.; Dong, Q.; Wang, Y. Investigation into hydroxypropyl-methylcellulose-reinforced polylactide composites for fused deposition modelling. *Ind. Crops Prod.* **2020**, *146*, 112174. [CrossRef]
147. Makri, S.; Xanthopoulou, E.; Klonos, P.; Grigoropoulos, A.; Kyritsis, A.; Tsachouridis, K.; Anastasiou, A.; Deligkiozi, I.; Nikolaidis, N.; Bikiaris, D. Effect of Micro- and Nano-Lignin on the Thermal, Mechanical, and Antioxidant Properties of Biobased PLA–Lignin Composite Films. *Polymers* **2022**, *14*, 5274. [CrossRef]
148. Gregor-Svetec, D.; Leskovšek, M.; Leskovar, B.; Stanković Elesini, U.; Vrabič-Brodnjak, U. Analysis of PLA Composite Filaments Reinforced with Lignin and Polymerised-Lignin-Treated NFC. *Polymers* **2021**, *13*, 2174. [CrossRef]
149. Domínguez-Robles, J.; Martin, N.; Fong, M.; Stewart, S.; Irwin, N.; Rial-Hermida, M.; Donnelly, R.; Larrañeta, E. Antioxidant PLA Composites Containing Lignin for 3D Printing Applications: A Potential Material for Healthcare Applications. *Pharmaceutics* **2019**, *11*, 165. [CrossRef]

150. Liao, J.; Brosse, N.; Pizzi, A.; Hoppe, S.; Zhou, X.; Du, G. Characterization and 3D printability of poly (lactic acid)/acetylated tannin composites. *Ind. Crops Prod.* **2020**, *149*, 112320. [CrossRef]
151. Liao, J.; Brosse, N.; Hoppe, S.; Zhou, X.; Xi, X.; Du, G.; Pizzi, A. Interfacial improvement of poly (lactic acid)/tannin acetate composites via radical initiated polymerization. *Ind. Crops Prod.* **2021**, *159*, 113068. [CrossRef]
152. Daver, F.; Lee, K.; Brandt, M.; Shanks, R. Cork–PLA composite filaments for fused deposition modelling. *Compos. Sci. Technol.* **2018**, *168*, 230–237. [CrossRef]
153. Tao, Y.; Wang, H.; Li, Z.; Li, P.; Shi, S. Development and Application of Wood Flour-Filled Polylactic Acid Composite Filament for 3D Printing. *Materials* **2017**, *10*, 339. [CrossRef] [PubMed]
154. Zandi, M.D.; Jerez-Mesa, R.; Lluma-Fuentes, J.; Roa, J.J.; Travieso-Rodriguez, J.A. Experimental analysis of manufacturing parameters' effect on the flexural properties of wood-PLA composite parts built through FFF. *Int. J. Adv. Manuf. Technol.* **2020**, *106*, 3985–3998. [CrossRef]
155. Åkerlund, E.; Diez-Escudero, A.; Grzeszczak, A.; Persson, C. The Effect of PCL Addition on 3D-Printable PLA/HA Composite Filaments for the Treatment of Bone Defects. *Polymers* **2022**, *14*, 3305. [CrossRef]
156. Belhassen, R.; Vilaseca, F.; Mutjé, P.; Boufi, S. Thermoplasticized starch modified by reactive blending with epoxidized soybean oil. *Ind. Crops Prod.* **2014**, *53*, 261–267. [CrossRef]
157. Xiong, Z.; Yang, Y.; Feng, J.; Zhang, X.; Zhang, C.; Tang, Z.; Zhu, J. Preparation and characterization of poly(lactic acid)/starch composites toughened with epoxidized soybean oil. *Carbohydr. Polym.* **2013**, *92*, 810–816. [CrossRef]
158. Hu, Y.; Gu, J.; Zhang, Y.; Liu, G.; Yi, X.; Liu, X. FDM 3D printed MXene/Recycled carbon fibre reinforced polylactic acid composites: Interface optimization, toughening and enhanced electromagnetic shielding performance. *Compos. Commun.* **2024**, *48*, 101953. [CrossRef]
159. Vidakis, N.; Kalderis, D.; Michailidis, N.; Papadakis, V.; Mountakis, N.; Argyros, A.; Spiridaki, M.; Moutspoulou, A.; Petousis, M. Environmentally friendly polylactic acid/ferronickel slag composite filaments for material extrusion 3D printing: A comprehensive optimization of the filler content. *Mater. Today Sustain.* **2024**, *27*, 100881. [CrossRef]
160. Zeynivandnejad, M.; Moradi, M.; Sadeghi, A. Mechanical, physical, and degradation properties of 3D printed PLA + Mg composites. *J. Manuf. Process.* **2023**, *101*, 234–244. [CrossRef]
161. Suvanjumrat, C.; Chansoda, K.; Chookaew, W. Additive manufacturing advancement through large-scale screw-extrusion 3D printing for precision parawood powder/PLA furniture production. *Clean. Eng. Technol.* **2024**, *20*, 100753. [CrossRef]
162. Yin, Q.; Kong, F.; Wang, S.; Du, J.; Tao, Y.; Li, P. Hazelnut shell carbon filled polylactic acid composite filaments for 3D printing photothermal structures. *J. Mater. Res. Technol.* **2023**, *27*, 3923–3935. [CrossRef]
163. Ranjan, N.; Kumar, R.; Singh, R.; Ramniwas, S.; Kumar, V. 1.40—On extrusion of PLA-salix alba feedstock filament for scaffolding applications. In *Comprehensive Materials Processing*, 2nd ed.; Saleem, H., Ed.; Elsevier: Amsterdam, The Netherlands, 2024; pp. 367–373. ISBN 9780323960212. [CrossRef]
164. Fico, D.; Rizzo, D.; De Carolis, V.; Montagna, F.; Palumbo, E.; Corcione, C.E. Development and characterization of sustainable PLA/Olive wood waste composites for rehabilitation applications using Fused Filament Fabrication (FFF). *J. Build. Eng.* **2022**, *56*, 104673. [CrossRef]
165. Hatkeposhti, J.K.; Kordani, N.; Pasha, M.A.; Barari, A. Fracture load in double keyhole notch PLA-Cu2O nanocomposites manufactured via compression molding and 3D printing: An experimental and numerical study. *J. Mech. Behav. Biomed. Mater.* **2024**, *153*, 106504. [CrossRef]
166. Rimpongpisarn, T.; Wattanathana, W.; Sukthavorn, K.; Nootsuwan, N.; Hanlumyuang, Y.; Veranitisagul, C.; Laobuthee, A. Novel luminescent PLA/MgAl2O4:Sm3+ composite filaments for 3D printing application. *Mater. Lett.* **2019**, *237*, 270–273. [CrossRef]
167. Qian, Y.; Li, C.; Qi, Y.; Zhong, J. 3D printing of graphene oxide composites with well controlled alignment. *Carbon* **2021**, *171*, 777–784. [CrossRef]
168. Rahmatabadi, D.; Ghasemi, I.; Baniassadi, M.; Abrinia, K.; Baghani, M. 3D printing of PLA-TPU with different component ratios: Fracture toughness, mechanical properties, and morphology. *J. Mater. Res. Technol.* **2022**, *21*, 3970–3981. [CrossRef]
169. Ju, Q.; Tang, Z.; Shi, H.; Zhu, Y.; Shen, Y.; Wang, T. Thermoplastic starch based blends as a highly renewable filament for fused deposition modeling 3D printing. *Int. J. Biol. Macromol.* **2022**, *219*, 175–184. [CrossRef]
170. Yang, J.; Li, W.; Mu, B.; Xu, H.; Hou, X.; Yang, Y. 3D printing of toughened enantiomeric PLA/PBAT/PMMA quaternary system with complete stereo-complexation: Compatibilizer architecture effects. *Polymer* **2022**, *242*, 124590. [CrossRef]
171. Hui, I.; Pasquier, E.; Solberg, A.; Agrenius, K.; Håkansson, J.; Chinga-Carrasco, G. Biocomposites containing poly(lactic acid) and chitosan for 3D printing—Assessment of mechanical, antibacterial and in vitro biodegradability properties. *J. Mech. Behav. Biomed. Mater.* **2023**, *147*, 106136. [CrossRef]

polymers

MDPI

Review

Machine Learning in 3D and 4D Printing of Polymer Composites: A Review

Ivan Malashin [1,*], Igor Masich [1], Vadim Tynchenko [1,*], Andrei Gantimurov [1], Vladimir Nelyub [1,2], Aleksei Borodulin [1], Dmitry Martysyuk [3] and Andrey Galinovsky [3]

1 Artificial Intelligence Technology Scientific and Education Center, Bauman Moscow State Technical University, 105005 Moscow, Russia; imasich@emtc.ru (I.M.); agantimurov@emtc.ru (A.G.)
2 Scientific Department, Far Eastern Federal University, 690922 Vladivostok, Russia
3 Center NTI "Digital Materials Science: New Materials and Substances", Bauman Moscow State Technical University, 105005 Moscow, Russia; dmart9945@mail.ru (D.M.); a_galinovskiy@bmstu.ru (A.G.)
* Correspondence: ivan.p.malashin@gmail.com (I.M.); vadimond@mail.ru (V.T.); Tel.: +7-926-875-7128 (I.M.)

Abstract: The emergence of 3D and 4D printing has transformed the field of polymer composites, facilitating the fabrication of complex structures. As these manufacturing techniques continue to progress, the integration of machine learning (ML) is widely utilized to enhance aspects of these processes. This includes optimizing material properties, refining process parameters, predicting performance outcomes, and enabling real-time monitoring. This paper aims to provide an overview of the recent applications of ML in the 3D and 4D printing of polymer composites. By highlighting the intersection of these technologies, this paper seeks to identify existing trends and challenges, and outline future directions.

Keywords: machine learning; polymer composites; 3D printing; 4D printing; design optimization

1. Introduction

In recent years, 3D and 4D printing of polymer composites have gained attention across various industries, ranging from healthcare [1,2] to aerospace [3,4]. Simultaneously, machine learning (ML) is emerging as a key tool for optimizing additive manufacturing (AM) processes, offering methods to predict material properties, improve printing parameters, and enhance overall efficiency. Despite the growing interest in this intersection, the number of studies combining ML with 4D printing (4DP) of polymer composites remains relatively limited, with most publications focusing on specific applications of ML. According to data from Scopus, there are currently 47 publications that include the terms "machine learning", "polymer composites", and "3D or 4D printing". Notably, the year 2024 has seen the highest output, with 21 publications, suggesting a significant increase in research activity in this area.

Figure 1 presents visualizations of the analysis of publications related to machine learning, polymer composites, and 3D/4D printing based on Scopus data. Figure 1a illustrates the changes in the number of publications on this topic over time, providing insights into the relevance and growing interest in this research area. Figure 1b shows the distribution of these publications across various fields, highlighting the most active research directions. Figure 1c depicts the geographical distribution of publications from different countries, revealing the regions that are actively contributing to this field of study. Finally, Figure 1d presents the types of documents prevalent in this area, whether they are articles, conference papers, or other formats.

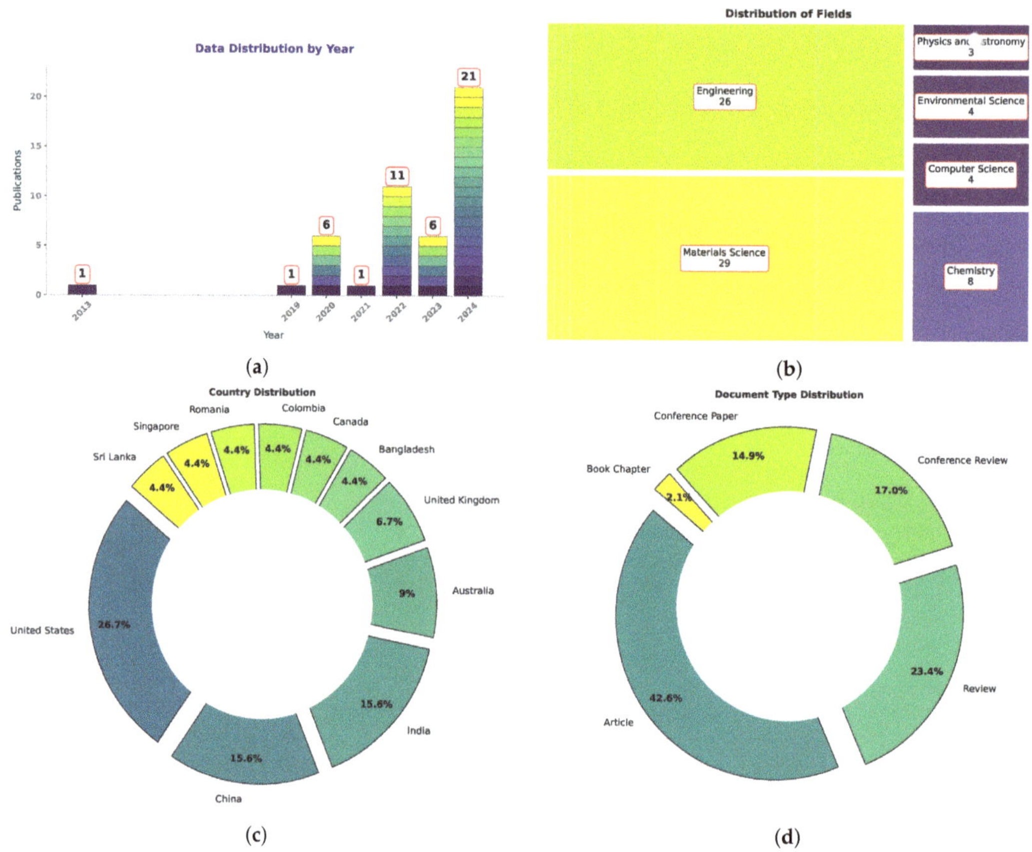

Figure 1. Data visualizations illustrating various distributions: (**a**) annual publication counts over the years, (**b**) distribution of fields in the analyzed dataset, (**c**) geographical distribution of contributions by country, and (**d**) distribution of document types. According to Scopus data.

The aim of this review is to systematically analyze the existing literature, discuss current advancements, and identify key challenges in the application of ML to 3DP and 4DP of polymer composites. Additionally, this review provides recommendations for future research to integrate these technologies. The structure of the paper covers the role of ML in material enhancement, process optimization, property prediction, and quality monitoring, as well as a discussion of the future prospects in this field.

To facilitate the reader's understanding of these topics, the paper presents a structured outline. It begins with an overview of 3D and 4D printing technologies in Section 2, laying the groundwork for understanding various additive manufacturing methods. This includes examining material extrusion (Section 2.1) and stereolithography (Section 2.2), followed by an exploration of selective laser sintering (Section 2.3) and digital light processing (Section 2.4). Next, the paper delves into jetting technologies (Section 2.5) and direct ink writing (Section 2.6), highlighting their unique features and applications. The concept of 4D printing is introduced (Section 2.7), showcasing its potential for creating adaptive materials. The properties of polymer composites used in these technologies are analyzed (Section 3), followed by a discussion on the applications of machine learning in both 3D (Section 4) and 4D printing (Section 5) of polymer composites. This leads into a section on transforming polymer composites into ceramics and other materials (Section 6), addressing relevant processes and implications. The paper also discusses challenges and limitations associated with these technologies (Section 7), providing a critical perspective on their current state. It concludes with future directions for research and development (Section 8), emphasizing the importance of continued innovation in additive manufacturing, before wrapping up with a summary of the key findings (Section 9).

2. 3D and 4D Printing: An Overview

Three-dimensional printing (3DP) is a technology that enables the creation of complex structures by layering materials according to digital models. The most common techniques used in 3DP of polymer composites include fused deposition modeling (FDM), stereolithography (SLA), and selective laser sintering (SLS). These methods allow for high precision and customization in the production of functional parts [5], which makes 3DP suitable for a wide range of applications, from rapid prototyping to the production of end-use products in industries like aerospace, automotive, healthcare, and consumer goods.

Table 1 outlines key 3DP and extrusion techniques, including their processes, common materials, potential applications, advantages, and associated challenges.

Table 1. Summary of 3DP techniques with updated terminology according to ASTM 52900:2021.

Technique	Process	Materials	Applications	Advantages	Challenges
Fused Deposition Modeling (FDM) [6–8]/ Fused Filament Fabrication (FFF) [9–11]	Material Extrusion	PLA, ABS, PETG, nylon, composite filaments (e.g., carbon fiber-reinforced)	Prototypes, consumer products, lightweight structures	Low cost, widely accessible, variety of materials available	Rough surface finish, limited mechanical strength
Stereolithography (SLA) [12–14]	Vat Photopolymerization	Photopolymer resins (tough, flexible, bio-compatible)	High-detail prototypes, medical/dental models	High precision, smooth surface finish, fine details	Brittle materials, requires post-processing (curing)
Selective Laser Sintering (SLS) [15–17]	Powder Bed Fusion	Nylon, polyamide, TPU, composites	Functional prototypes, aerospace/ automotive parts	No support structures needed, strong mechanical properties	Rough surface, more expensive than FDM/SLA
Digital Light Processing (DLP) [18–20]	Vat Photopolymerization	Photopolymer resins	Jewelry, dental devices, high-resolution prototypes	Faster than SLA, high resolution	Relies on photosensitive resins, requires post-curing and washing to ensure full polymerization
Multi Jet Fusion (MJF) [21–23]	Powder Bed Fusion	Nylon, TPU	Functional parts, small batch manufacturing	Excellent mechanical properties, no supports required	Rough surface finish, requires post-processing (dyeing, etc.)
Material Jetting (PolyJet by Stratasys) [24–26]	Material Jetting	Photopolymers (rigid, rubber-like, transparent)	Multi-material prints, medical models, tactile products	High resolution, smooth finish, multi-material printing	Material durability limitations, complex post-processing
Direct Ink Writing (DIW) [27–29]	Material Extrusion	Hydrogels, silicones, composite pastes	Soft robotics, biomedical devices, tissue engineering	Can print functional/ biologically active materials	Limited material types, weaker mechanical properties

Four-dimensional printing extends the capabilities of 3DP by introducing time as a fourth dimension (all techniques are summarized in Table 2). It involves the fabrication of smart materials or structures that can change shape, properties, or behavior over time in response to external stimuli, such as heat, moisture, or light [30].

Table 2. Summary of 4DP techniques.

Technique	Process	Materials	Applications	Advantages	Challenges
Shape Memory Polymer (SMP)-Based 4DP [31–34]	Printed using techniques like FDM, SLA, or SLS, incorporating shape memory polymers that "remember" a programmed shape and return to it when exposed to stimuli (e.g., heat).	Shape memory polymers, composites	Self-assembling structures, biomedical devices (e.g., stents), robotics, adaptive products	Programmable and responsive to stimuli	Limited availability of high-performance SMP materials, complex control over transformations
Hydrogel-Based 4DP [35–39]	Hydrogels are printed using DIW, PolyJet, or SLA, designed to change shape or properties in response to water or humidity.	Hydrogels, stimuli-responsive polymers (e.g., pH-responsive, temperature-responsive)	Tissue scaffolds, drug delivery systems, wearable electronics	Biocompatibility, highly responsive to environmental conditions	Control over swelling, ensuring long-term stability
Stimuli-Responsive Composite-Based 4DP [40–45]	Printed using standard 3DP methods (FDM, SLS, etc.) but with composite materials that react to stimuli such as light, magnetic fields, or heat.	Composites with nanoparticles, liquid crystal elastomers, magnetically active particles	Soft robotics, aerospace components, deployable structures	Tailored responses to specific external stimuli	Complex fabrication processes, challenges in controlling transformations

These dynamic transformations open up possibilities for creating self-assembling systems, adaptive materials, and responsive structures, which are particularly relevant in fields like biomedical devices, robotics, and adaptive architectures. To better understand these aspects, a Venn diagram in Figure 2 illustrates the key components of different AM techniques, including process types, materials used, and applications. Surrounding petals emphasize the main advantages and limitations associated with each technology.

However, both 3D and 4DP of polymer composites come with inherent challenges. The mechanical properties of printed parts can depending on the printing parameters, the type of polymer matrix used, and the incorporation of reinforcing fillers. In 4DP, the complexity increases further, as materials need to exhibit predictable and controllable transformations. The quality and functionality of printed parts are influenced by numerous factors, such as material selection, process control, and post-processing, making it difficult to optimize production efficiently using traditional trial-and-error methods [46,47].

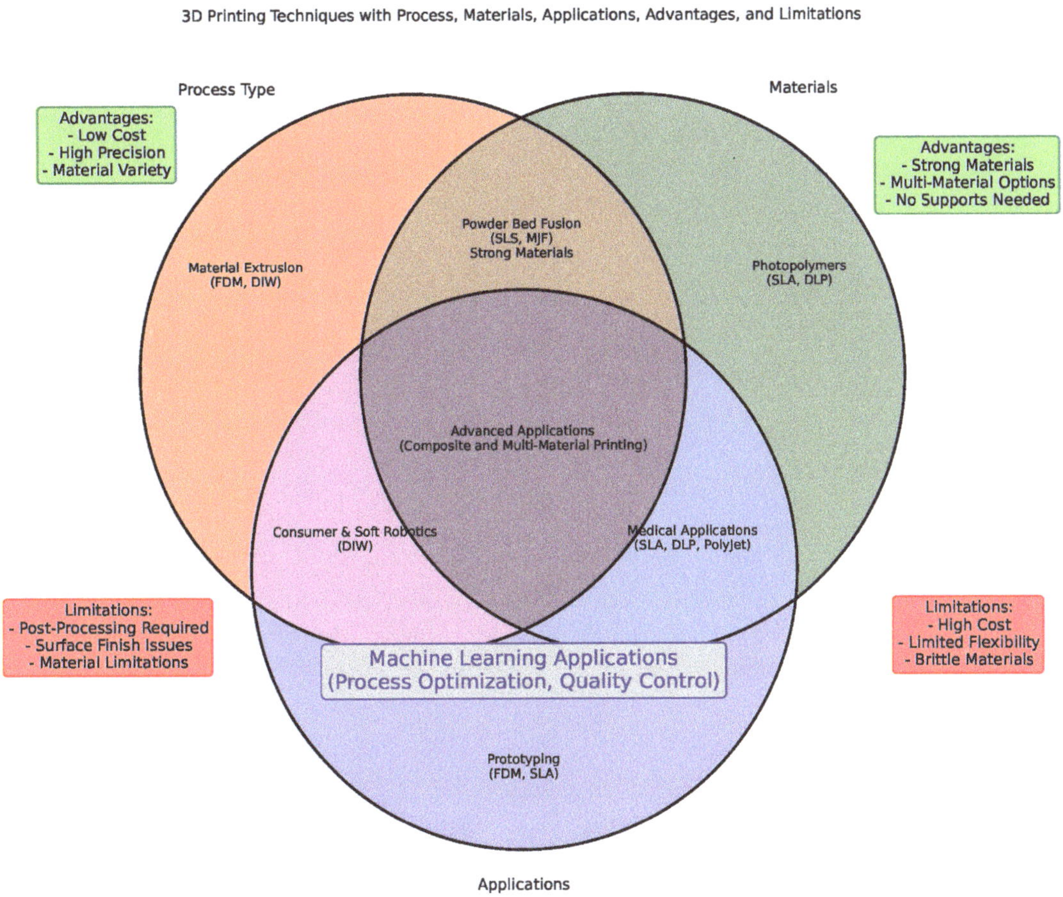

Figure 2. Venn diagram illustrating key aspects of various 3DP techniques.

Figure 3 presents a conceptual graph that illustrates the interconnections among key concepts in this domain. The graph includes elements, such as ML methods (e.g., artificial neural networks, variational autoencoders, and generative adversarial networks), as well as their applications in quality control, defect detection, and process optimization.

This approach helps us to better understand how different components interact with each other and identifies promising directions for further research and development in the fields of ML and AM. ML has the potential to predict the performance of polymer composites in various environmental conditions and assist in real-time monitoring of the printing process, ensuring quality and consistency [48,49]. By leveraging ML, the development of smart materials and structures can become more efficient, reducing experimentation time and enhancing the overall reliability of 3D- and 4D-printed products.

Figure 3. Bibliometric network visualization.

2.1. Material Extrusion

Material extrusion, the FDM process, also known as FFF, is an AM technique in which a 3D object is constructed layer by layer using melted material [50–59]. The process begins with model preparation, where a 3D model is created using CAD (Computer-Aided Design) software. This model is then processed by slicing software, which divides the model into horizontal layers and generates instructions, or G-code, for the 3D printer. A schematic representation of the process is shown in Figure 4.

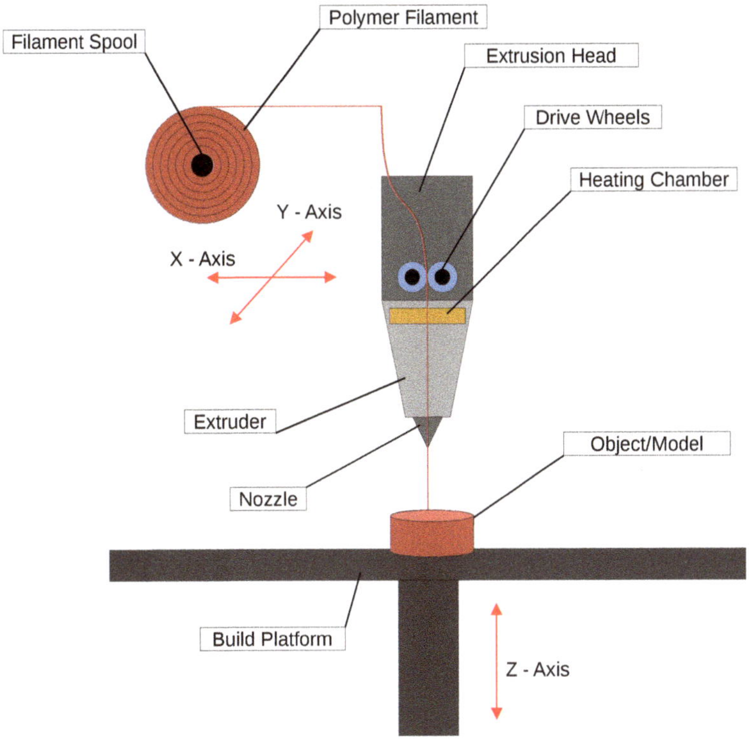

Figure 4. Schematic representation of FDM process.

Next is the material feeding stage. The material, typically a thermoplastic filament, is fed into the printer's extruder, which heats it until it reaches a semi-liquid state [60]. Extrusion and deposition follow, where the melted material is extruded through a nozzle onto the build platform. The print head follows a set path, depositing material layer by layer. As each layer is completed, the nozzle or build platform moves up to prepare for the next layer [61].

The object is then built layer by layer, with the material gradually forming the 3D shape. For complex geometries or overhangs, the printer can deposit a secondary, support material alongside the main material to hold these structures in place [62].

After deposition, each layer cools and solidifies, locking into the layer below. This process repeats until the entire model is complete. In the final stage, the object is removed from the platform, and any support structures are detached, either mechanically or by dissolving if they are water-soluble. Final steps may include surface treatments, such as sanding, polishing, or painting, to improve surface quality [63].

2.1.1. Fused Deposition Modeling

Subramani et al. identify optimal FDM parameters for ABS components; Melentiev et al. improve adhesion in multi-material parts by combining FDM with chemical deposition and electroplating; Bahrami et al. enhance wear resistance in Fe–ABS composites using a GA-ANN optimization model; and Hajjaj compares the mechanical properties of zirconia restorations produced by FDM and CAD/CAM milling, finding FDM-printed zirconia to be mechanically inferior.

Subramani et al. [64] investigate how various FDM parameters, such as infill density, printing speed, platform, and extruder temperature, affect the mechanical properties of ABS components produced on a Creality Ender-3 3D printer [64]. Mechanical properties like tensile strength, yield strength, and elastic modulus were evaluated using a Multi-Criteria Decision-Making (MCDM) [65] approach [66]. The optimal printing settings—35% infill, 0.25 mm layer height, 40 mm/s speed, 75 °C platform, and 210 °C extruder temperature—were identified for manufacturing impellers. Additionally, field emission scanning electron microscopy (FESEM) [67] provided insights into surface defects and material behavior.

Melentiev et al. [68] explore multiprocess additive manufacturing (MPAM) to produce multi-material components, focusing on improving the adhesion strength between metal and polymer interfaces in 3D-printed parts. By combining FDM with chemical deposition and electroplating [69], the research aims to enhance the structural integrity of metalized plastic components, which typically suffer from poor adhesion. The study focused on creating a hierarchically structured surface on ABS parts through 3DP and acid etching copper adhesion [70]. The experiment involved 3DP, surface treatment, copper deposition, electroplating, and adhesion testing, offering insights for industries using MPAM for advanced electronics and multi-material devices.

FDM has limitations in wear resistance [71]. To address this, Fe particles (10%, 20%, and 30%) were added [72] to ABS to create Fe–ABS composite filaments, and parts were printed using varying filling patterns, nozzle temperatures, and layer thicknesses. Wear testing showed that Fe percentage had the greatest effect on wear reduction, followed by filling pattern, while nozzle temperature had the least impact. For optimization, a genetic algorithm–artificial neural network (GA-ANN) [73] model slightly outperformed the response surface methodology (RSM) [74], with results closely matching experimental data at a 0.25% error rate.

Hajjaj [75] compares the mechanical properties of zirconia dental restorations made using 3DP (FDM) and CAD/CAM milling, focusing on the effects of conventional versus speed sintering. A total of 60 bars were tested for flexural strength and modulus, while 40 discs were used for Vickers microhardness testing. Results showed that milled zirconia had higher flexural strength and modulus than FDM-printed zirconia. The sintering cycle did not affect flexural properties, but speed sintering the Vickers microhardness of milled zirconia did have an effect. Overall, FDM-printed zirconia was mechanically inferior to milled zirconia.

2.1.2. Fused Filament Fabrication

This section summarizes recent studies on FFF. Khan reviews how process parameters affect the mechanical properties of lightweight polymers; Kariuki identifies optimal printing parameters for carbon fiber-reinforced polyamide 12; Garcia compares FFF, Metal Injection Molding, and Powder Metallurgy for 17-4 PH stainless steel, noting FFF's superior tribocorrosion resistance; and Kalinke explores sustainable practices in FFF using renewable and recycled materials to enhance environmental sustainability.

Fused filament fabrication (FFF) is a cost-effective 3DP method for lightweight polymer structures. Key mechanical properties like flexural and impact strength are influenced by process parameters and material selection [76]. Filled polymers often perform better, and crystallinity plays a key role in the final properties. Review [76] discusses emerging trends such as topology optimization and polymer recyclability, while highlighting research gaps and proposing directions for further in FFF technology.

Kariuki et al. [77] investigate the flexural behavior of 3D-printed short carbon fiber-reinforced polyamide 12 (PA12-CF) [78] parts produced using fused filament fabrication (FFF). Using an L18 Taguchi design and Gray relational analysis, the optimal printing parameters were identified. Build orientation had the most impact on flexural properties, with a rectilinear infill pattern producing a flexural strength of 119.9 MPa and modulus of 3038 MPa, while a concentric pattern improved strength by 15.8%. This work provides valuable insights into optimizing FFF parameters for enhanced mechanical performance in carbon fiber composites.

Garcia et al. [79] compare the effects of different manufacturing methods— FFF, Metal Injection Molding (MIM), and conventional Powder Metallurgy (PM)—on the properties of 17-4 PH stainless steel. FFF and MIM both produced near-dense parts, but MIM samples showed the highest hardness. Corrosion behavior was similar for FFF and MIM, both outperforming PM. However, FFF parts exhibited superior tribocorrosion resistance, attributed to higher proportions of delta ferrite and retained austenite in their microstructure. These findings highlight the potential of FFF for producing corrosion-resistant, durable components.

Kalinke et al. [80] explore sustainable methods for enhancing the development, treatment, and applications of 3D-printed objects, particularly in FFF. The paper discusses various conductive and non-conductive filaments made from renewable biopolymers [81], bioplasticizers, and recycled materials, detailing how these choices impact material properties. They also highlight alternative strategies for sustainability, including recycling, adjusting printing parameters, and system miniaturization. These approaches aim to reduce environmental impact while producing high-quality, cost-effective 3D-printed products, aligning with Green Chemistry principles and Circular Economy concepts.

2.2. Stereolithography

In SLA printing for polymer composites [44,57–59,82–87], the process is adapted to use specialized resin blends that incorporate composite materials, such as ceramic, carbon, or glass fibers, to enhance the mechanical properties, thermal resistance, or surface finish of the printed object. This approach combines the precision and detail of SLA with the strength and functionality of composite materials, creating parts that are suitable for more demanding applications.

The process begins with model preparation in CAD software, where a 3D model is designed and then sliced into layers. The composite resin is prepared in a vat and often contains finely distributed particles (e.g., glass, ceramic, or carbon fibers) that are suspended within the photopolymer base. This composite resin is carefully formulated to maintain a uniform consistency, ensuring that particles do not settle and are evenly distributed throughout each layer of the print.

In the layer curing phase, a UV laser or projector selectively cures each layer, hardening both the photopolymer and the embedded particles simultaneously. The UV laser follows a precise path to solidify each layer, bonding the particles into a matrix that enhances the overall strength and durability of the printed part [88]. Between each layer, the platform moves incrementally to allow the next thin layer of resin to coat the surface [89].

The layer-by-layer bonding process creates a composite structure, embedding the particles within the cured photopolymer matrix [90]. Support structures are added automatically by the slicing software when needed, especially for overhangs or complex geometries, and are printed in the same composite material [91].

Once the print is completed, post-processing begins. The part is removed from the resin vat, cleaned of any excess resin, and may undergo an additional UV curing process to fully harden the composite [92]. Support structures are then removed, and the object may be further processed through sanding, polishing, or coating, depending on the application's requirements [93].

SLA printing with polymer composites enables the production of parts with improved mechanical properties [94], thermal stability [95], and surface quality [96], making it ideal for engineering prototypes, end-use parts, and high-performance applications in industries such as aerospace, automotive, and medical devices. This technique expands the scope of

SLA by providing a balance between high resolution and enhanced material strength, tailored to meet the specific demands of advanced manufacturing. A schematic representation of the SLA process is shown in Figure 5.

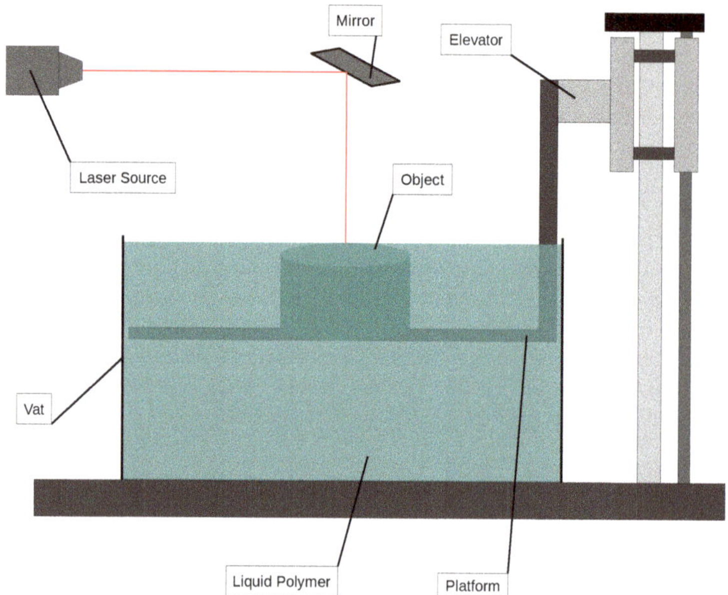

Figure 5. Schematic representation of SLA process.

Hydrogel-based electronics are promising for wearable devices but face challenges like low conductivity and stretchability. Sun et al. [97] present a projection SLA 3DP method to create high-conductive, flexible hydrogel antennas for wireless sensing [98]. The photocurable silver-based hydrogel forms conductive pathways after partial dehydration, achieving a conductivity of 387 S cm^{-1}. Sealed circuits maintain stable resistance under 100% strain for 30 days, with added features like stretchability and shape memory. Custom flexible RFID tags were created, enabling accurate eye movement tracking and passive wireless sensing.

Zhou et al. [99] examine Stereolithography Additive Manufacturing (SLAM) for producing advanced ceramic objects with complex geometries [100], highlighting its resolution and surface quality. It addresses the challenges in achieving the desired performance due to the necessity of thermal debinding (TD) [99] to remove binders, which can lead to defects and prolonged processing times. Key topics covered include the impact of raw materials on photocurable ceramic suspensions, the mechanisms and characterization methods of the TD process, and strategies for designing effective TD profiles. The review concludes with insights into the challenges and future directions for TD in ceramic SLAM, providing a foundational understanding for optimizing TD processes in research and industry.

Kulkarni et al. [101] investigate the use of SLA to print polymer nanocomposite samples of stimuli-responsive spin crossover (SCO) materials with resins DS3000 and PEGDA-250. The analysis showed that incorporating SCO particles improved mechanical properties, with transformation strains of 1.2–1.5% at high loads (13–15 vol.%), enabling thermal expansion peaks. Two SCO complexes were synthesized and characterized, demonstrating their suitability for actuator applications due to favorable spin transition properties. The findings emphasize the importance of effective particle dispersion for optimal performance in SLA-printed composites.

Pharmaceutical 3DP [102–111] is advancing rapidly, offering the potential for highly personalized medicine. SLA is a particularly promising technology due to its high resolution and compatibility with heat-sensitive drugs. However, the lack of specialized excipients for pharmaceutical SLA limits material options. Curti et al. [103] investigate how formulation factors—such as photoinitiator concentration, polymer size, and liquid filler type—affect the print quality of SLA 3D-printed medicines. By screening 156 photopolymer formulations, it highlights how these factors influence print outcomes, providing valuable insights for future development of personalized 3D-printed pharmaceuticals.

2.3. Selective Laser Sintering

In SLS for polymer composites printing [112–121], a powdered composite material—typically a blend of a polymer base like nylon with reinforcing particles such as carbon fiber, glass beads, or ceramic—is used to create strong, high-performance parts. This process leverages the strength and durability of composite materials within the flexibility of polymer-based 3DP, ideal for demanding applications in industries such as aerospace, automotive, and consumer goods. A schematic representation of the SLS process is shown in Figure 6.

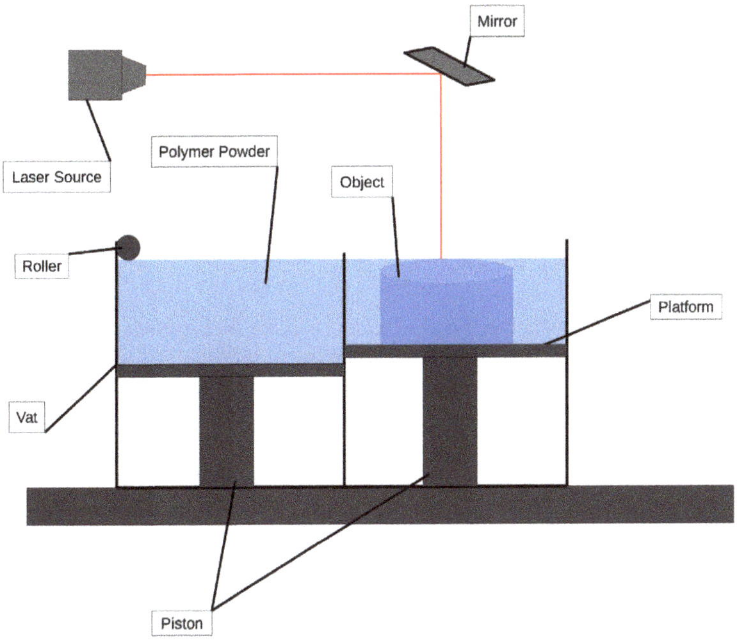

Figure 6. Schematic representation of SLS process.

The process starts with model preparation using CAD software, where a 3D model is designed and then processed through slicing software to divide the model into thin, horizontal layers. These layers are used to generate precise instructions for the SLS printer.

In the powder preparation stage, the composite powder is loaded into the build chamber [122]. This powder must be well-mixed to ensure an even distribution of reinforcement particles, like carbon or glass fibers, throughout the polymer base. A powder bed is formed, and the printer's roller or recoater evenly spreads a thin layer of the composite powder across the build platform [123].

During layer sintering, a laser selectively fuses areas of the powder bed, following the contours of the sliced model layer [124]. The laser heats the polymer particles to their melting point, allowing them to fuse together while also bonding the embedded reinforcement particles within the polymer matrix. Once a layer is completed, the platform lowers slightly, and a new layer of composite powder is spread over the previous one. This layer-by-layer sintering process continues until the entire object is formed, with each new layer bonding to the one beneath it. Because of the self-supporting nature of the powder bed, support structures are typically not required, allowing for more complex geometries without additional material waste.

After printing, the object is surrounded by unsintered powder, which acts as a support and is carefully removed in the post-processing phase. The remaining loose powder is brushed or blown off, often followed by bead-blasting or compressed air cleaning to reveal the printed part. Additional finishing steps such as sanding or coating can further enhance the part's surface quality and durability [125].

SLS printing with polymer composites produces parts with superior mechanical strength, stiffness, and thermal stability compared with traditional polymers, due to the reinforcing particles integrated within the polymer matrix. This method is well-suited for functional prototypes, tooling, and end-use parts that require the combined benefits of both high-performance polymers and reinforcement materials.

Song et al. [126] highlight SLS's role in medical engineering for producing complex biomedical products, particularly implants and prosthetics using biocompatible materials; Azam et al. focus on SLS processing of polymer materials, emphasizing innovations in piezoresistive strain-sensing and the process–structure–property relationships; Han et al. develop a method for creating carbon nanotube-anchored α-ZrP nanohybrids to enhance polyamide 12 composites, resulting in significant improvements in mechanical properties and functional characteristics; and Zhang et al. investigate the effects of process parameters on carbon fiber-reinforced PEEK composites, achieving notable strength and modulus enhancements for industrial applications.

SLS has been particularly successful in creating electrically conductive polymer composites (ECPCs) by forming a segregated filler network along powder boundaries. Azam et al. [127] focus on SLS processing of polymer materials, highlighting the consolidation mechanisms, process parameters, and innovations in piezoresistive strain-sensing materials and self-sensing structures. They also explore the intricate process–structure–property relationships in SLS-printed polymer composites.

Han et al. [128] introduce a simple method to synthesize a carbon nanotube (CNT)-anchored α-ZrP nanohybrid (CNT@α-ZrP) for enhancing polyamide 12 (PA12) composites using ball-milling followed by SLS. CNTs serve dual functions: providing black coloration for efficient heat absorption and reinforcing the PA12 matrix. The α-ZrP nanosheets primarily enhance the mechanical and functional properties of PA12 composites. The resulting PA12/CNT@α-ZrP composites show improvements in Young's modulus (98.9%), tensile strength (33.1%), and impact strength (34.6%), along with better wear resistance, flame retardancy, and reduced smoke production. This method offers an industrial approach to producing robust and functional SLS-based structures.

Carbon fiber-reinforced PEEK (CF/PEEK) composites fabricated via SLS offer excellent mechanical properties and are highly promising for advanced applications. Zhang et al. [120] investigate the impact of process parameters—such as laser power, layer thickness, paving speed, and carbon fiber content—on the microstructure and performance of CF/PEEK composites. Key findings include achieving a failure strength of 117 MPa with a layer thickness of 0.08 mm and an optimal fiber weight fraction of 15%. The highest elastic modulus reached 8400 MPa, surpassing previous works. The study also reveals nonlinear relationships between paving speed and strength, with longer carbon fibers improving strength. The research provides insights into optimizing SLS-CF/PEEK composites for industrial applications.

2.4. Digital Light Processing

In DLP printing for polymer composites [121,129–137], the process utilizes a high-resolution digital projector to cure photopolymer resins that are often blended with composite materials, such as ceramic, carbon fiber, or glass particles. This combination allows for the creation of highly detailed parts with enhanced mechanical properties, making DLP a suitable choice for various advanced manufacturing applications. A schematic representation of the DLP process is shown in Figure 7.

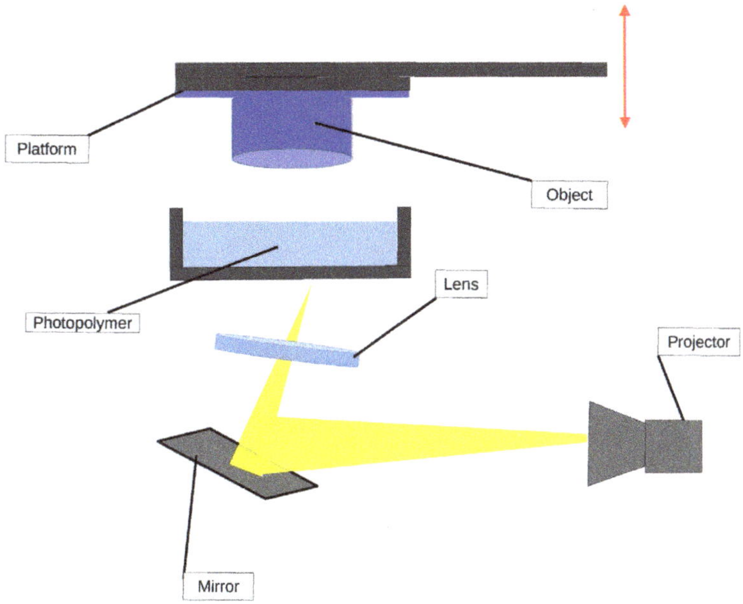

Figure 7. Schematic representation of DLP process.

The process begins with model preparation in CAD software, where a 3D model is designed and optimized for printing. The model is then sliced into thin layers by slicing software, which generates the necessary instructions for the DLP printer.

In the resin preparation phase, a vat is filled with liquid photopolymer resin mixed with composite materials. The resin is specially formulated to ensure uniform dispersion of the composite particles, allowing for consistent curing and material properties throughout the printed part.

During the layer curing phase, the DLP printer uses a digital light projector to expose the surface of the resin to UV light. The projector displays a complete layer of the model at once, curing the resin in a pattern that corresponds to the sliced model [138]. This process allows for rapid curing of an entire layer simultaneously, significantly speeding up the printing process compared with traditional layer-by-layer methods.

Once a layer is cured, the build platform moves upward (or the resin vat moves downward) to allow fresh resin to flow over the cured layer, preparing for the next layer. The layer-by-layer construction continues until the entire object is complete, with each layer bonding to the one beneath it.

For complex geometries, support structures are often generated by the slicing software to prevent deformation during printing. These supports are printed using the same composite resin and can be easily removed after the printing process is finished.

After the printing is complete, the part undergoes post-processing. It is removed from the resin vat and cleaned of any excess uncured resin, typically using isopropyl alcohol. The part may then be subjected to further UV curing to ensure complete hardening. Any support structures are removed, and final finishing processes, such as sanding or coating, can be applied to achieve the desired surface quality.

DLP printing with polymer composites offers several advantages, including high resolution and smooth surface finishes, while also enhancing the mechanical properties of the printed parts. This technology is particularly well-suited for applications in industries such as dental and medical devices, jewelry, and high-performance engineering components, where precision and material strength are paramount.

Melentiev et al. [139] present lithography metal additive manufacturing (LMAM) for high-resolution metal parts with excellent density and tensile strength; Guo et al. enhance photosensitive resin with multi-walled carbon nanotubes, improving mechanical properties; Senthooran et al. incorporate mica into DLP-printed samples, significantly increasing tensile and flexural strength; and Wang et al. create a flexible multistage honeycomb absorber from carbonyl iron and MWCNTs, demonstrating exceptional electromagnetic wave absorption.

DLP is employed in high-resolution AM, enabling the 3DP of complex metallic parts with micrometer precision. Melentiev et al. [139] present lithography metal additive manufacturing (LMAM), a method that utilizes DLP with a photosensitive resin filled with metal powder. The process yields intricate structures with a spatial resolution of 35 μm and surface roughness of 1–2 μm without support structures. Sintered stainless steel parts exhibit 99.3% density and 93% tensile strength relative to annealed 316 L steel. LMAM is ideal for fabricating small, precise devices in fields such as biomedicine, microheat exchangers, and pharmaceutical engineering.

Guo et al. [137] explores the potential of multi-walled carbon nanotubes (MWCNTs) to enhance the structural, mechanical, and electrical properties of materials through AM, specifically focusing on DLP techniques. Despite the growing interest in MWCNT-reinforced composites, there is limited research on their integration into photosensitive resin (PR) systems using DLP, particularly concerning the distribution patterns of MWCNTs. This investigation fabricated MWCNTs-reinforced PR (MWCNTs-PR) and examined how varying MWCNT content affects the microstructure and mechanical properties of the composite. Findings indicate that adding 0.05 wt% MWCNTs enhances the elastic modulus by 25% and the bending strength by 2% compared with pure PR. To achieve a more uniform MWCNT distribution, a combination of ultrasonic treatment and mechanical stirring was employed. The study further developed a multi-material layered 3DP structure, demonstrating that the 10001 structure achieved the highest bending modulus, outperforming the control group by 14.9%. Finally, finite element analysis was utilized to validate the enhanced bending resistance mechanism attributed to the MWCNTs in the PR.

Senthooran et al. [133] explore the enhancement of mechanical and thermal properties in 3D-printed samples using DLP by incorporating mica as an inorganic filler at 5%, 10%, and 15% concentrations, along with a KH570 silane coupling agent for better dispersion. The results show improvements: tensile strength increased by 85% and flexural strength by 132% with mica addition. Thermogravimetric analysis (TGA) and scanning electron microscopy (SEM) were used for thermal and morphological evaluations. The findings highlight advancements in AM technology through DLP techniques.

Research on wideband electromagnetic (EM) absorbers in the 75–110 GHz range is limited, hindering millimeter-wave technology advancements. Wang et al. [140] introduce a novel flexible multistage honeycomb structure absorber (FMHSA) made from carbonyl iron (CIP), multi-walled carbon nanotubes (MWCNTs), and flexible photopolymer resin (FPR), fabricated via DLP 3DP. The FMHSA achieves exceptional EM wave absorption with a bandwidth of 35 GHz at a 150° bending angle and a minimum reflection loss of −37.04 dB. Its notable properties include flexibility, recoverability, and lightweight design, paving the way for improved wearable absorbers.

2.5. Jetting 3D Technologies

Jetting 3D technologies include MJF and PJP, both of which can utilize polymer composites to create high-performance parts with enhanced mechanical properties.

In Multi Jet Fusion [141–150], a layer of polymer composite powder, often a blend of nylon and reinforcing materials like carbon fiber or glass beads, is spread across the build platform. Inkjet print heads selectively apply a fusing agent to specific areas of the powder bed, allowing for controlled heating and fusion when exposed to infrared light. This layer-by-layer process continues until the part is fully formed, with each layer bonding to the previous one. After cooling, excess un-fused powder is removed and can be recycled. MJF produces parts that exhibit superior strength and durability, making it ideal for functional prototypes and end-use applications in industries such as aerospace and automotive. A schematic representation of the MJM process is shown in Figure 8.

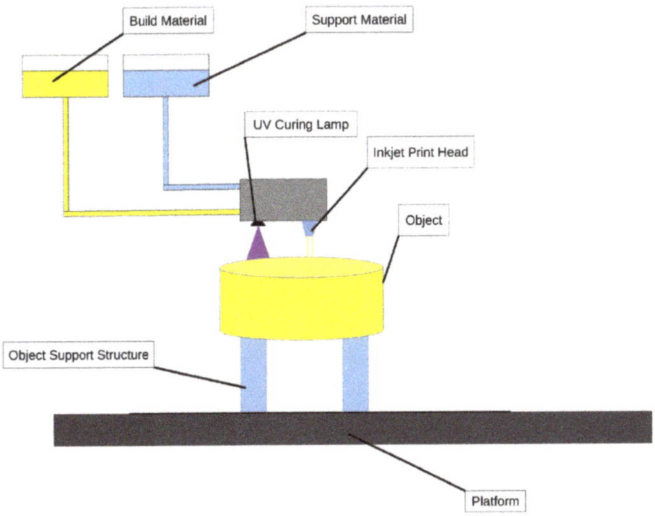

Figure 8. Schematic representation of MJM process.

PolyJet printing [151–160] utilizes a different approach by jetting ultra-thin layers of liquid photopolymer resin, often enhanced with composite materials. The print heads spray the resin, which can include reinforcing particles, and immediately cure it with UV light. This method allows for the incorporation of various materials in a single print job, enabling the creation of parts with tailored mechanical properties and surface finishes. After printing, support structures are easily removed, resulting in high-resolution parts. PolyJet is particularly well-suited for applications requiring intricate details and multi-material capabilities, such as dental devices and intricate consumer products.

2.5.1. Multi Jet Fusion

Alomarah et al. find that MJF outperforms FFF in producing stronger auxetic structures; Tan et al. develop a framework showing how fiber weight affects porosity in fiber-reinforced composites; Kafi et al. explore the impact of build height and orientation on the mechanical properties of MJF-printed polypropylene; and Conway et al. assess the geometric accuracy of MJF surgical guides, achieving precise distortion predictions using machine learning.

MJF and FFF are explored for fabricating a hybrid auxetic structure in AM. Alomarah et al. [143] find that MJF produces robust specimens with high dimensional accuracy, while FFF suffers from large pores in connecting areas, indicating lower print quality. MJF specimens exhibit plateau stress with high peaks when compressed along the Y-axis, whereas FFF specimens display a smooth plateau stress. MJF specimens achieve the highest specific energy absorption (SEA) at 2.1 and 2.5 J g^{-1}, exceeding the 0.495 and 0.480 J g^{-1} of FFF specimens. Additionally, the auxetic features (negative Poisson's ratio) remain unaffected by the manufacturing methods. This research underscores the influence of fabrication techniques on the mechanical properties and energy absorption capabilities of cellular materials.

AM of fiber-reinforced polymer composites is gaining attention for its ability to create lightweight, functional products. However, pore defects remain a concern, necessitating a better understanding of pore formation. Tan et al. [117] present a powder-scale multi-physics framework to simulate the printing process of fiber-reinforced polymer composites in powder bed fusion in MJF. The framework incorporates various phenomena, including particle flow dynamics, laser–particle interaction, heat transfer, and multiphase fluid flow. The melt depths of glass fiber-reinforced polyamide 12 parts fabricated via selective laser sintering were measured to validate the model. Results indicate that increasing the fiber weight fraction leads to a lower densification rate, larger porosity, and reduced pore sphericity in the composites.

Kafi et al. [142] validate the absorption phenomena in MJF-printed polypropylene (PP) using Laser Flash and Corrected Porosity methods. It investigates how build height and orientation affect tensile properties, crystallinity, porosity, and thermophysical attributes in MJF-printed PP coupons. Results indicate that crystallinity and tensile performance are consistent across orientations, but Z-oriented samples exhibit 35% lower strain and increased porosity compared with XY samples. Micro-CT scans revealed that horizontal positioning improved contrast for porosity analysis. A correlation was established between Laser Flash half-time and porosity when corrections were applied, indicating that lower absorption occurs in less dense Z samples. The findings highlight the importance of accurately determining porosity to understand absorption in MJF-printed PP, offering insights into predicting mechanical properties and enhancing the overall quality of MJF-produced parts.

Conway et al. [161] examine the repeatability and geometric accuracy of AM surgical guides for personalized knee surgery. A total of 258 unique guide designs were created, and 2100 parts were produced using MJF AM. An automated measurement technique gathered 8400 individual feature dimensions, revealing standard deviations in feature size ranging from 0.076 to 0.173 mm and consistent deviations from target dimensions of −0.308 to 0.017 mm. ML models were developed to predict these geometric distortions, achieving accuracy within 0.033 to 0.075 mm, allowing for effective predictions across various part sizes.

2.5.2. PolyJet Printing

This section highlights advancements in PolyJet printing. Azpiazu et al. assess how thermocycling and surface finishing impact the strength and hardness of dental prostheses, finding certain finishes yield better performance. Krause et al. examine material choice and print orientation effects on microfluidic channel accuracy, noting optimal results for wider channels. Aberdeen et al. explore the interface design and mechanical failure of bi-material coupons, emphasizing the need for further research on geometric designs to strengthen material interfaces.

PolyJet 3DP is an advanced AM technology that deposits photopolymeric materials in micron-sized droplets, curing them with ultraviolet (UV) light. It excels in creating complex, multi-material structures with exceptional precision, achieving layer thicknesses as fine as 16 microns [162]. Its versatility allows for a wide range of materials, including rigid, flexible, and transparent options, enabling the production of components with tailored mechanical and optical properties. While widely used in industries like aerospace and healthcare, challenges remain in material performance and print optimization, necessitating ongoing research to enhance interfacial bonding and mechanical properties.

Azpiazu et al. [163] evaluate the effects of thermocycling and different surface finishing protocols on the flexural strength and surface hardness of a novel photopolymer designed for monolithic polychromatic dental prostheses made via PolyJet 3DP. A total of 90 specimens were divided into three groups based on finishing protocols: Pumice + Moldent, Pumice + Optiglaze, and Polycril + Moldent. Results showed that thermocycling reduced the flexural strength across all groups, with the Optiglaze group demonstrating the highest strength after thermocycling. The analysis also revealed an interaction between thermocycling and finishing protocols concerning surface hardness, with the Optiglaze group exhibiting the highest hardness values.

Krause et al. [164] investigate the impact of materials and print orientations on the 3DP of microfluidic channels as negative features using PolyJet technology and the Stratasys Objet500 printer. Two sets of chips, each containing channel pairs made from a high-contrast reference material and a sacrificial material embedded in clear photopolymer resin, were printed. The planned channel widths ranged from 64 to 992 µm, and the channels were printed either parallel or perpendicular to the jetting head's movement. The findings indicate that reproducibility and accuracy were optimal for channels with a width of 600 µm or greater, with the best channel morphology achieved when the printer head moved parallel to the channel's longitudinal axis.

Aberdeen et al. [165] explore the interface design and mechanical failure dynamics of PolyJet-printed bi-material coupons using material jetting technology, specifically PolyJet 3DP. By investigating various geometric designs and conducting uniaxial tensile tests on samples printed with a Stratasys Objet500 Connex3 printer, the results reveal that increasing the surface contact area between distinct materials does not necessarily enhance interface strength. The findings highlight the need for further research into multi-material geometric designs and their impact on interface integrity, particularly as interest in PolyJet printing grows in applications like robotics and fluidic circuitry.

2.6. Direct Ink Writing

DIW is AM technique that focuses on extruding viscoelastic inks [166–175], which can be formulated from polymer composites, to create complex geometries with enhanced material properties. This method is particularly effective for producing parts with tailored mechanical characteristics and functionality, suitable for a variety of applications.

The DIW process begins with ink formulation, where a composite ink is created by blending a polymer matrix with reinforcing materials, such as carbon fibers, glass fibers, or ceramic particles [176]. This ink must possess the right viscosity and flow properties to be extruded through a nozzle while maintaining shape fidelity after deposition.

During the printing phase, a syringe or nozzle extrudes the composite ink layer by layer onto a build platform. The printer's movement is controlled by a computer program that follows a pre-defined path, allowing for precise placement of material [177]. As each layer is deposited, it retains its shape due to the viscoelastic properties of the ink, enabling the creation of complex structures, including overhangs and intricate designs.

Post-processing is often required after printing, which may involve curing the printed part through heat or UV light, depending on the type of polymer used. This curing process solidifies the polymer matrix, enhancing the mechanical strength and durability of the final part [88,178]. Additionally, support structures may be incorporated or added during the printing process to ensure stability for more complex geometries. A schematic representation of the DIW process is shown in Figure 9.

DIW enables precise layer-by-layer deposition of functional materials through a positive displacement dispensing system [179]. It excels in printing on flexible substrates, such as polyethylene terephthalate (PET), due to its adaptability to various material viscosities and the importance of controlling process parameters like air pressure and feed rate for achieving high-resolution patterns. One of DIW's advantages is its ability to print conductive inks, crucial for developing sensors and electronic components, and its capability for multi-material printing, allowing for the creation of complex, multifunctional structures. This makes DIW suitable for fabricating stretchable and bendable sensors that can monitor mechanical deformations by varying electrical resistance, while the deposited patterns are typically cured to enhance structural integrity and performance in wearable technologies [180].

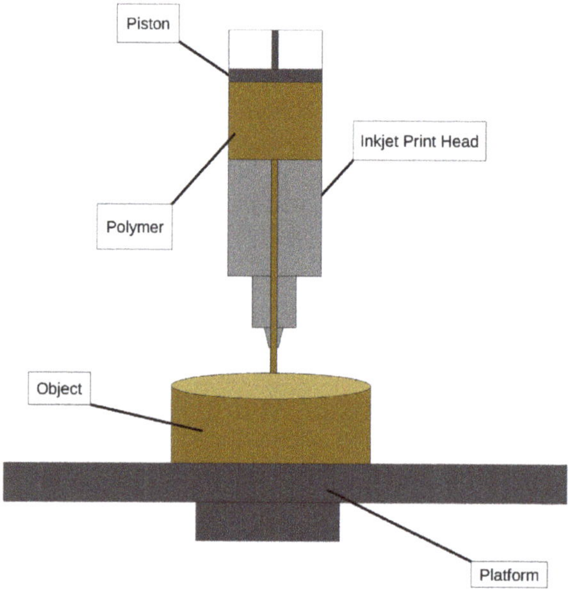

Figure 9. Schematic representation of DIW process.

The emergence of 3DP technology in the 1980s has facilitated the creation of patient-specific products with precise shapes and complexities. Among various techniques, DIW is favored for its affordability, ease of use, and scalability, although the limited variety of printing inks hampers its commercial potential. Injectable hydrogels, known for their quick gelling behavior and shape fidelity, have emerged as promising alternatives for printing inks, made from natural or synthetic polymers to achieve desired properties. Bhardwaj et al. [173] highlight recent advancements in hydrogel inks and their physicochemical aspects for engineered biostructures, and discuss the future prospects and challenges of 4DP in hydrogel-based 3DP applications in healthcare.

DIW advances hydrogel fabrication by enabling precise layer-by-layer deposition of hydrogel inks to create complex three-dimensional structures with tailored properties. Baniasadi et al. [181] explore the diverse applications of DIW in areas such as tissue engineering, soft robotics, and wearable devices, while also examining the various printing techniques and the underlying principles of DIW, including rheological properties and printing parameters. Additionally, they highlight the range of natural and synthetic hydrogel materials used in this process and discuss the latest biomedical applications, particularly in tissue engineering, wound dressings, and drug delivery systems, while outlining future research directions and potential innovations in hydrogel-based manufacturing.

DIW offers a flexible and resource-efficient method for prototyping functional materials and devices with complex geometries. Van et al. [182] focus on the use of graphene nanoplatelets (GNPs) as conductive fillers in printed electronics, addressing the challenges posed by non-spherical colloids that risk nozzle clogging. A workflow was developed to optimize ink rheology and printing parameters, enabling the successful production of filaments ranging from <100 to 1200 μm in width and 30 to 300 μm in height, with conductivities suitable for sensors and electrodes. The predictive models created from this research facilitate high-resolution DIW of platelet-based inks, promoting integrated material and process development for applications in wearable electronics, sensors, RF passives, energy materials, and tissue engineering.

2.7. 4D Printing

This section covers advancements in 4DP, which adds the dimension of time to traditional AM by using materials that change shape in response to external stimuli. Khalid et al. emphasize the potential of shape memory polymers (SMPs) that react to stimuli like heat and humidity for use in various fields. However, challenges like mechanical limitations remain. Qiu et al. explore the benefits of fiber-reinforced polymer composites (FRPCs) in 4DP, enhancing mechanical performance. Yan et al. review SMP composites in 4DP, summarizing advancements and discussing future prospects in biomedical application

Current FDM technology enables the use of multiple polymer filaments, paving the way for complex, responsive structures [63]. Four-dimensional printing represents an advancement beyond traditional 3DP by incorporating the dimension of time. This innovation is made possible through the development of intelligent materials that change shape in response to external stimul. The most promising applications of 4DP are in the creation of smart textiles, which can act as actuators and sensors, allowing for bio-inspired designs. Key areas of potential include smart clothing for extreme environments, auxiliary prosthetics, and orthotic devices that aid muscle recovery.

In recent years, there has been growing interest in AM shape memory polymers (SMPs) and their multifunctional composites, particularly in the realm of four-dimensional (4D) printing, which utilizes time-responsive programmable materials. These stimuli-responsive polymers can return to their original shapes from programmed temporary forms upon exposure to external stimuli such as heat, light, or humidity. The integration of 4DP with shape memory polymer composites (SMPCs) opens up a wide range of engineering applications [183–185], including in automotive, soft robotics, biomedical devices, and wearable electronics. Khalid et al. [186] highlight key 4DP technologies and their functionalities, discuss future opportunities in preprogramming, multi-material printing, and sustainability, and provide illustrative examples of applications, aiming to foster advancements and innovations in the field of 4DP.

Four-dimensional printing technology has gained considerable attention for its capability to reshape 3D-printed structures in response to external stimuli over time. However, challenges such as inadequate mechanical properties, low energy output, and limited design flexibility persist in the 4DP of pure polymers. The advent of fiber-reinforced polymer composites 4DP (FRPCs-4DP) offers promising solutions to these challenges by enhancing mechanical performance and improving actuation capabilities. Qiu et al. [187] explore recent advances in FRPCs-4DP, emphasizing the role of fibers, material compositions, AM techniques, and design strategies, while also outlining the key challenges and future trends for practical applications in this emerging field.

Yan et al. [33] review SMP composites and 4DP technologies, highlighting unique 4D-printed structures and summarizing recent research progress in various fields, particularly biomedical applications. They also discuss the challenges and future prospects for 4D-printed SMPs, serving as a reference for ongoing research and practical applications.

Table 3 provides an overview of the advantages and disadvantages of 3DP methods discussed in the recent literature.

Table 3. Summary of recent studies in AM methods.

Reference	Method	Focus	Advantages	Disadvantages
Franco et al. (2024) [63]	FDM	4DP with responsive structures for smart textiles	Enables complex structures; potential for smart applications	Still evolving; challenges in material properties
Subramani et al. (2024) [64]	FDM	Effect of FDM parameters on mechanical properties of ABS components	Identifies optimal settings for improved mechanical properties	Limited to specific materials and printers
Melentiev et al. (2024) [68]	FDM	Improving adhesion in multi-material components using MPAM	Enhanced structural integrity of metalized plastics	Complex multiprocess setup
Bahrami et al. (2024) [72]	FDM	Enhancing wear resistance in ABS through Fe composite filaments	Improved wear performance with optimized parameters	Limitations in wear resistance of pure FDM parts
Hajjaj (2024) [75]	FDM	Comparison of mechanical properties in zirconia restorations	Insights into material performance for dental applications	FDM-printed parts show inferior mechanical properties
Khan et al. (2024) [76]	FFF	Mechanical properties of lightweight polymer structures	Cost-effective and adaptable for different materials	Process parameters can limit mechanical performance
Kariuki et al. (2024) [77]	FFF	Flexural behavior of carbon fiber-reinforced PA12 parts	Optimized parameters enhance mechanical properties	Requires careful selection of printing parameters
Garcia et al. (2024) [79]	FFF	Comparison of FFF with MIM and PM on stainless steel properties	Superior tribocorrosion resistance in FFF parts	Variability in mechanical properties across methods
Kalinke et al. (2024) [80]	FFF	Sustainable practices in 3DP	Focus on recycling and environmental impact	Challenges in material selection for sustainability
Sun et al. (2024) [97]	SLA	Hydrogel-based electronics for wearable devices	High conductivity and flexibility in applications	Low stretchability in traditional hydrogels
Zhou et al. (2024) [99]	SLA	Producing advanced ceramic objects with complex geometries	High resolution and quality for intricate designs	Thermal debinding can lead to defects
Kulkarni et al. (2024) [101]	SLA	Printing polymer nanocomposites with stimuli-responsive materials	Enhanced mechanical properties with effective particle dispersion	Limited by material formulation options
Curti et al. (2024) [103]	SLA	Personalized medicine through SLA	High resolution suitable for drug formulation	Limited specialized excipients for pharmaceutical SLA
Song et al. (2024) [126]	SLS	Medical engineering applications for implants and prosthetics	Precise production of complex biomedical products	High setup costs and limited material options
Azam et al. (2024) [127]	SLS	Electrically conductive polymer composites	High performance for advanced applications	Process complexity can affect production speed
Han et al. (2024) [128]	SLS	Enhancing properties of PA12 composites with CNTs	Improved mechanical and functional properties	Requires careful control of material interactions
Zhang et al. (2024) [120]	SLS	Impact of process parameters on CF/PEEK composites	Excellent mechanical properties for advanced applications	Complex relationships between parameters can complicate optimization

Table 3. *Cont.*

Reference	Method	Focus	Advantages	Disadvantages
Melentiev et al. (2024) [139]	DLP	LMAM	Produces intricate structures with high resolution and no support structures; ideal for small, precise devices	Limited to specific applications, high dependence on materials
Guo et al. (2024) [137]	DLP	MWCNT-reinforced photosensitive resin	Enhances mechanical and electrical properties; optimized distribution through treatment	Limited research on integration into PR systems
Senthooran et al. (2024) [133]	DLP	Enhancement of mechanical and thermal properties using mica	Improvements in tensile and flexural strength	Material handling and dispersion challenges
Wang et al. (2023) [140]	DLP	Flexible multistage honeycomb structure absorbers	Exceptional EM wave absorption properties; lightweight and flexible	Limited application scope and complexity of design
Alomarah et al. (2024) [143]	MJF	Hybrid auxetic structures in AM	Robust specimens with high dimensional accuracy	Lower print quality with certain techniques like FFF
Tan et al. (2024) [117]	MJF	Simulating fiber-reinforced polymer composites	Improved understanding of pore formation; better material performance predictions	Complexity in modeling and simulation accuracy
Kafi et al. (2024) [142]	MJF	Absorption phenomena in printed polypropylene (PP)	Insights into porosity and mechanical properties	Variability in build orientation effects on performance
Conway et al. (2024) [161]	MJF	Geometric accuracy in surgical guides	High repeatability and accuracy in personalized surgical tools	Time-consuming measurement processes for validation
Patpatiya (2024) [162]	PolyJet	Advanced multi-material structures	Exceptional precision in complex geometries; versatile material options	Challenges with material performance and interfacial bonding
Azpiazu et al. (2024) [163]	PolyJet	Flexural strength in dental prostheses	Significant effects of surface finishing on strength	Thermocycling negatively impacts strength across protocols
Krause et al. (2024) [164]	PolyJet	Microfluidic channels in 3DP	High reproducibility and accuracy for fine features	Limited effective feature sizes for optimal results
Aberdeen et al. (2024) [165]	PolyJet	Bi-material coupons and mechanical failure dynamics	Insights into interface design for multi-material applications	Challenges with interface strength despite geometric improvements
Abas et al. (2024) [179]	DIW	Layer-by-layer deposition of functional materials	Excellent adaptability to flexible substrates and multi-material printing	Limited ink variety restricts commercial applications
Bhardwaj et al. (2024) [173]	DIW	Hydrogel inks in biostructures	Advancements in 4DP for healthcare applications	Challenges in ink consistency and availability
Baniasadi et al. (2024) [181]	DIW	Applications in tissue engineering and robotics	Flexible manufacturing for complex geometries	Limited material choices can restrict applications
Van et al. (2024) [182]	DIW	Conductive fillers in printed electronics	Enables high-resolution printing for sensors and devices	Challenges with material consistency and process optimization
Khalid et al. (2022) [186]	4DP	Shape memory polymers (SMPs)	Responsive structures for various engineering applications	Mechanical property limitations and design flexibility issues
Qiu et al. (2024) [187]	4DP	Fiber-reinforced polymer composites (FRPCs)	Enhanced mechanical performance and actuation capabilities	Challenges in material composition and manufacturing processes

Table 3. *Cont.*

Reference	Method	Focus	Advantages	Disadvantages
Yan et al. (2023) [33]	4DP	SMP composites	Advances in biomedical applications; unique structural designs	Challenges in achieving consistent properties across applications

3. Properties of Polymer Composites in AM Technologies

Polymer composites exhibit unique property combinations. These materials integrate a polymer matrix with reinforcing elements (fibers, particles, nanomaterials), achieving improvements in mechanical, thermal, electrical, and other performance characteristics. Such enhancements are important for applications in aerospace, automotive industries, medicine, and electronics. However, utilizing polymer composites in AM requires careful consideration of their structure, properties, and processing characteristics. Figure 10 illustrates the primary properties of polymer composites used in AM technologies.

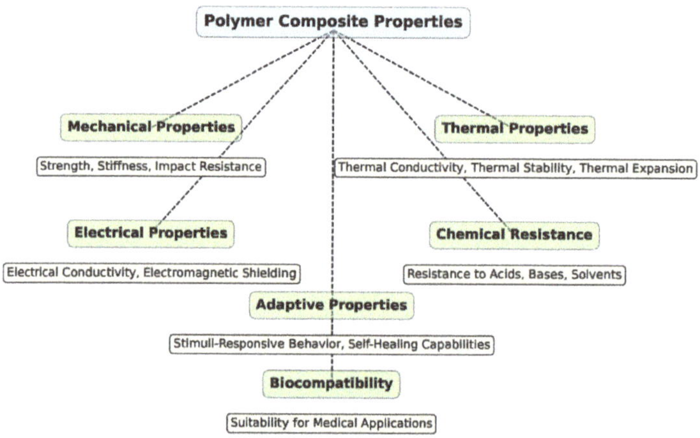

Figure 10. Diagram illustrating key properties of polymer composites in AM.

3.1. Mechanical Properties

One of the key advantages of polymer composites is their ability to provide high mechanical performance with relatively low weight. This is achieved by incorporating reinforcing materials into the polymer matrix, such as carbon or glass fibers. Fiber-reinforced polymer composites enhance strength and stiffness compared with pure polymers. For instance, adding carbon fibers increases tensile strength, making these materials promising for structural components in aerospace and automotive applications, where high strength-to-weight ratios are essential [188]. Additionally, polymer composites exhibit excellent energy absorption characteristics, which makes them resilient to impact loads [189]. In 3DP, impact resistance is particularly important for creating prototypes and end-use parts that are subjected to dynamic stresses. Incorporating high-modulus reinforcing materials increases the stiffness of composites, which means that stiffer parts can be printed without the need for complex metalworking techniques [190]. A key challenge in AM is achieving uniform distribution of reinforcing materials to prevent defects such as delamination or weak zones that can negatively affect the mechanical properties of the printed parts [191].

3.2. Thermal Properties

Thermal resistance is another parameter for polymer composites in AM, especially for high-temperature applications. Adding fillers such as carbon nanotubes, graphene, or metallic particles can improve the thermal conductivity of polymer composites [192]. Aged PLA filaments filled with graphene and carbon nanotubes exhibit improved crystallinity, thermal stability [193], and electrical conductivity but reduced strength and toughness, with annealing treatments enhancing their properties based on the filler type and annealing temperature [194,195]. This is particularly relevant for creating heat-dissipating components, such as heat sinks or electronic housings. By incorporating high-temperature polymers or heat-resistant fillers, composites can withstand higher temperatures compared with standard polymers [196,197]. This makes them suitable for use in high-temperature environments, such as engine components or aerospace structures [87]. Furthermore, composites with low thermal expansion coefficients are more stable against temperature-induced dimensional changes [198], which is important for printing large or precise parts where temperature fluctuations can affect the final dimensions and geometry [199].

3.3. Electrical Properties

Modern polymer composites used in AM can exhibit notable electrical characteristics [200,201]. Electrically conductive polymer composites, combining polymers with metal-like electronic properties, show great potential in additive manufacturing for creating complex designs and rapid production [202], with advancements in various 3DP methods enabling breakthroughs in flexible electronics, energy storage, and other applications [203]. This enables the creation of 3D-printed parts with high electrical conductivity, suitable for sensors [204,205], antennas [206,207], and other functional devices [153,208]. Conductive polymer composites (CPCs) can also provide effective electromagnetic shielding, making them useful for printing enclosures for sensitive electronics [209]. For instance, Maleki et al. [210] created CPCs using material extrusion additive manufacturing by mixing multi-wall carbon nanotubes (MWCNTs) with ABS, resulting in 3D-printed specimens with 26 times higher electrical conductivity, improved electromagnetic interference shielding, and enhanced tensile strength and modulus, though nozzle wear occurred due to the abrasive nature of CNTs [211]. Achieving high electrical properties in AM requires ensuring uniform dispersion of conductive fillers in the matrix and preventing agglomeration, which can adversely affect both electrical properties and mechanical strength [212–214].

3.4. Adaptive Properties and 4DP

Polymer composites can also possess adaptive properties, which are particularly relevant for 4DP, where materials can change their properties or shape in response to external stimuli [215,216]. Polymer composites with shape memory properties can alter their structure or geometry in response to stimuli such as heat, moisture, or other environmental factors. This capability is utilized in 4DP to create products that can change shape over time or in response to operational conditions [217]. Additionally, some polymer composites can include elements that allow the material to self-heal after damage. In AM, such composites can be used to create parts capable of repairing themselves, thus extending their service life [218,219]. Four-dimensional printing with these composites opens new possibilities for creating adaptive and intelligent materials that can be applied in fields such as medicine, robotics, construction, and other advanced areas.

4. Application of ML in 3DP of Polymer Composites

ML has emerged as a transformative technology in the field of AM. By leveraging data-driven approaches, ML techniques offer improvements in various aspects of the printing process, from material optimization to process control and defect detection. This section explores the key applications of ML in 3DP of polymer composites and highlights how these innovations are reshaping the industry. Figure 11 visually organizes the various applications of ML in AM.

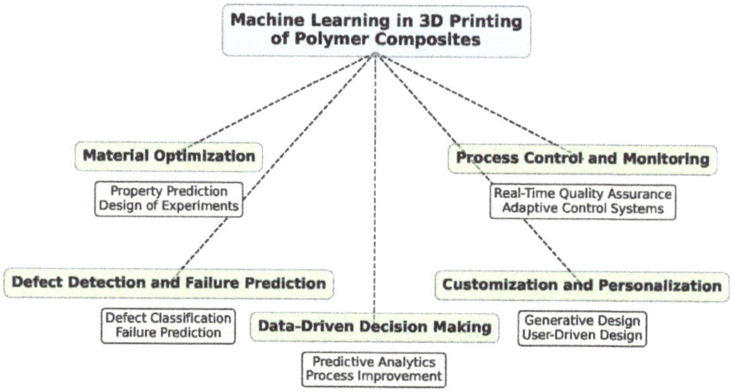

Figure 11. Applications of ML in 3DP of polymer composites.

4.1. Properties Prediction

One of the primary applications of ML in 3DP of polymer composites is material optimization. Traditional methods of developing and refining polymer composite materials can be time-consuming and costly, often requiring extensive experimental trials. ML algorithms, particularly those involving supervised learning and optimization techniques, can expedite this process by analyzing large datasets of material properties and performance metrics.

FDM has enabled personalized drug-loaded formulations tailored to patient needs [220]. However, optimizing fabrication parameters is traditionally time-consuming and requires expert input. To address this, M3DISEEN, a web-based software, was developed [221], utilizing AI and ML techniques (MLTs) to enhance FDM 3DP, including filament production via hot melt extrusion (HME). AI models predict key parameters with high accuracy [222,223], streamlining 3DP for drug development. M3DISEEN is publicly available.

Three-dimensional printing in healthcare enables personalized medicines and devices but is hindered by the lengthy trial-and-error formulation process [224]. Ong et al. [225] combine in-house and literature-mined data on hot melt extrusion (HME) and fuse deposition modeling (FDM) formulations to create a balanced dataset of 1594 formulations, enhancing ML predictive performance. Optimized ML models achieved 84% accuracy in predicting printability and mechanical characteristics, with mean absolute errors of 5.5 °C and 8.4 °C for processing temperatures in HME and FDM, respectively. These models are integrated into the M3DISEEN web application, streamlining the formulation development workflow in pharmaceutical 3DP and improving research throughput [63].

Porous designs, like truss- and sheet-based lattices, offer versatility, but evaluating numerous material-lattice combinations is impractical. Peloquin et al. [226] present a framework for rapidly predicting the mechanical properties of 3D-printed gyroid lattices using base material and porosity data. A kernel ridge regression ML [227] model was trained on experimental data, achieving similar accuracy to numerical simulations but with reduced computation time, advancing ML-driven mechanical property prediction.

AM faces adoption challenges due to inconsistent product properties. Khusheef et al. [228] introduce a novel predictive method using in-process sensing to improve part property prediction in fused deposition modeling (FDM). By integrating Inertial Measurement Unit (IMU) sensors, a thermal camera, and machine settings, the study focuses on predicting key mechanical properties like tensile strength and surface roughness. Utilizing hybrid deep learning models (CNN-LSTM), the best model achieved 99% accuracy in predicting tensile strength. These results highlight the potential of sensor data and advanced modeling to enhance AM reliability and broader industry adoption.

AM of carbon fiber (CF)/epoxy composites is still in early development compared with conventional resin infusion methods. Monticeli et al. [229] predict the flexural strength, modulus, and strain of high-performance 3D-printable CF/epoxy composites using an artificial neural network, analysis of variance, and response surface methodology. The predictions show high reliability with low error, closely matching experimental results. By including different input data, the system can predict various output parameters. Factors such as vacuum pressure, printing speed, curing temperature, and thickness were analyzed, demonstrating efficient fabrication of composite materials with tailored properties.

Malley et al. [230] integrate data analytics with AM to predict the mechanical behavior of samples produced via vat polymerization with varying magnetic particle compositions. A neural network model was developed using mechanical test data from six compositions [231]. The model accurately predicted the mechanical behavior of tested samples and performed well for untested compositions, surpassing traditional data-driven methods. This approach reduces the need for extensive post-manufacturing testing, accelerating product development and improving quality assurance in AM for industrial applications

Griffiths et al. [232] explore the evolution of direct digital and AM from rapid prototyping to rapid production, highlighting its potential for creating personalized, high-quality products with minimal batch sizes. The accessibility of affordable AM machines and open-source software has empowered users, prompting shifts in energy and material consumption patterns. Using a Design of Experiments (DOE) approach, the study optimizes part performance by examining factors such as scrap weight, part weight, energy consumption, and production time. Key findings indicate that optimizing machine parameters can yield desired outcomes, while identical settings across different designs may produce varying results, underscoring the need for design-specific models. The research aims to identify optimal FDM settings for part weight and production time while balancing these with economic factors like energy consumption and scrap weight. Using polylactic acid (PLA) filament for testing, the study analyzes data on weight, build time, and power consumption, employing MiniTab software to visualize parameter interactions through main effects, Pareto, and contour plots. Ultimately, the research contributes valuable datasets for modeling AM processes, facilitating a more accurate assessment of their economic and environmental impacts during the design stage [233].

Table 4 summarizes various studies that utilize ML to predict properties in 3DP.

Table 4. Summary of research on ML applications for predicting properties in 3DP.

Reference	Focus	Data Info	Applied Method
Elbadawi et al. [221]	AI/ML for enhancing FDM 3DP and filament production	N/A	Developed M3DISEEN web-based software
Ong et al. [225]	Balancing dataset for HME and FDM formulations	1594 formulations from in-house and literature data	ML models for predicting printability, mechanical characteristics
Peloquin et al. [226]	Mechanical properties of 3D-printed gyroid lattices	Experimental data for gyroid lattices	Kernel ridge regression ML model
Khusheef et al. [228]	Predicting mechanical properties in FDM	In-process sensing data including IMU and thermal camera	Hybrid deep learning models (CNN-LSTM)
Monticeli et al. [229]	Predicting properties of CF/epoxy composites	Various input parameters: vacuum pressure, printing speed, etc.	Artificial neural network, ANOVA, response surface methodology
Malley et al. [230]	Predicting mechanical behavior in vat polymerization	Mechanical test data from six compositions	Neural network model
Griffiths et al. [232]	Optimizing part production in AM considering environmental impact	Analyzed scrap weight, energy use, production time	Design of Experiments approach

4.2. Process Control and Monitoring

ML enhances process control and monitoring during the 3DP of polymer composites by providing real-time analysis and feedback. This capability helps in maintaining the quality and consistency of printed parts. AI-augmented additive manufacturing (AI2AM) technology was highlighted by Sani et al. [234] and integrates AI-based monitoring and optimization of 3DP parameters to detect and prevent defects, enhance quality and efficiency, and enable more sustainable manufacturing, with a focus on FDM printers and future developments in closed-loop systems.

Real-time defect detection and closed-loop adjustment are essential for ensuring the quality of carbon fiber-reinforced polymer (CFRP) composites in AM. Lu et al. [235] introduce a deep learning-based system for real-time identification and correction of defects in robot-based CFRP AM. The model accurately detects and classifies defects like misalignment and abrasion, while also quantifying their severity through geometric analysis. By integrating this with process parameter adjustments, the system effectively controls defects, achieving what conventional composite fabrication methods cannot.

Narayanan et al. [236] developed a self-monitoring system using real-time camera images, and deep learning detects delamination in FDM 3D-printed parts, while strain measurements predict warping before it occurs. The developed system successfully classifies delamination levels and pre-diagnoses warping, offering potential for automated error detection in various manufacturing processes.

Jin et al. [237] present an automated method for identifying defective 3D-printed polymer parts using images captured during the FFF process. ML (PCA and SVM) and deep learning (CNN) classify parts as good or defective with 98.2% and 99.5% accuracy, respectively, benefiting both manufacturers and hobbyists.

Error detection during extrusion-based AM remains a challenge, with most inspections occurring post-production. Charalampous et al. [238] introduce a vision-based method that compares real-time point cloud data from printed parts to digital 3D models, enabling real-time error detection [239] and performance evaluation to reduce waste and production costs.

An online quality monitoring system using laser scanning detects defects in material extrusion 3DP by comparing surface point clouds with CAD models, and this was investigated by Lin et al. [240]. It reconstructs 3D models of defects, enabling feedback control, and helps reduce material and time waste by determining if the 3D printer should be shut down [241].

The challenge of quality assurance in AM is addressed in [242] by the authors developing an online reinforcement learning (RL) method to detect and mitigate new defects during printing. The method, Continual G-learning, leverages offline knowledge from the literature and online learning during the AM process to minimize required training samples. Applied to a fused filament fabrication (FFF) platform, the method optimally mitigates defects in real time [197,243], demonstrating its effectiveness in both numerical and real-world case studies.

Carrico et al. [244] introduce a new paradigm for manufacturing and controlling soft ionic polymer–metal composite (IPMC) actuators for soft robotics using 3DP. The process creates 3D monolithic IPMC devices with integrated sensors and actuators, and Bayesian optimization is employed to control the actuators, mitigating complex dynamics. The approach improves actuator performance, demonstrated through a modular reconfigurable soft crawling robot, highlighting its potential for more advanced IPMC devices.

Omairi et al. [245] review AI-based predictive models in AM, emphasizing their role in making AM "smart" by improving printability, reducing design complexity, and enhancing real-time control and defect detection. They discuss current trends, research gaps, and opportunities for further collaboration and development in line with Industry 4.0.

Table 5 summarizes the key studies focused on monitoring and adaptive control in AM, detailing their objectives, applied models, and data utilized.

Table 5. Summary of studies on real-time monitoring and adaptive control in AM.

Reference	Focus	Applied Model	Data Info
Lu et al. (2023) [235]	Real-time defect identification in CFRP AM	Deep learning system for defect detection	Utilizes geometric analysis of defect severity based on camera feed images from the printing process.
Narayanan et al. (2019) [236]	Self-monitoring system for FDM	Deep learning for delamination detection	Employs real-time camera images and strain measurements from printed parts to predict warping.
Jin et al. (2020) [237]	Automated defect identification in 3DP	ML (PCA, SVM) and deep learning (CNN)	Utilizes image data captured during the FFF process for classification of parts as good or defective.
Charalampous et al. (2021) [238,239]	Vision-based error detection during extrusion	Comparison of real-time point clouds with digital models	Compares 3D-scanned point cloud data from printed parts against digital models to identify discrepancies.
Lin et al. (2019) [240,241]	Online defect detection via laser scanning	Surface point cloud comparison with CAD models	Involves laser scanning data for 3D reconstruction of defects compared with CAD models for feedback control.
Chung et al. (2022) [197,242,243]	Quality assurance via reinforcement learning	Continual G-learning method	Uses historical data and online learning during AM processes to minimize defects based on previously learned patterns.
Carrico et al. (2019) [244]	Control of soft ionic polymer–metal composite actuators	Bayesian optimization for actuator control	Collects performance data from integrated sensors and actuators to optimize control parameters in real time.
Omairi et al. (2021) [245]	AI-based predictive models in AM	Review of predictive models	Analyzes data from various studies to identify trends and gaps in AI applications for improving AM processes.

4.3. Defect Detection and Failure Prediction

Detecting and predicting defects in printed parts ensure the reliability and performance of polymer composites. ML techniques offer advanced capabilities for identifying and addressing potential issues before they lead to failures. Moreover, predictive models can estimate the likelihood of failure based on historical data and current process conditions. By analyzing factors such as material properties, processing parameters, and environmental conditions, ML can forecast potential issues and recommend preventive measures.

Chen et al. [246] address the challenge of processing large tomography datasets for defect detection in composite materials. Using a micro-CT scan of fiber-reinforced composites, ML models were trained to detect defects. The binarized statistical image features (BSIF) method was applied to compress images without losing defect information. The convolutional neural network (CNN) model achieved high accuracy with a mean square error of 0.001 in fiber orientation prediction, enabling effective defect detection.

Moreover, Chen [247] highlights the unique microstructural signatures in 3D-printed glass fiber-reinforced polymer (GFRP) composites, which can be analyzed using ML to reverse-engineer the tool path. By processing micro-CT images with the BSIF method for data compression, ML models were trained to accurately identify the tool path. This approach poses a potential intellectual property risk for AM, as tool paths could be reconstructed from product microstructures.

FFF faces challenges like inconsistent part quality and print repeatability due to manufacturing defects. Goh et al. [248] developed an on-site monitoring system using computer vision and object detection models to detect and correct such defects in real-time. A camera on the print head captures video, which is processed to detect under-extrusion and over-extrusion anomalies. Various YOLO architectures were tested, with the YOLOv3-Tiny and YOLOv4-Tiny models achieving over 80% accuracy. Optimized models reached 89.8% classification accuracy and 70 frames per second inference speed. A correction algorithm was also implemented, allowing real-time defect detection and correction during printing, advancing FFF process reliability.

A hybrid method combining an artificial neural network (ANN) and micromechanics is developed in [249–251] for predicting failure in IM7/8552 unidirectional composite lamina under triaxial loading. The ANN, trained with data from a finite element method-based representative volume element (RVE) model, achieves over 97.5% accuracy. This approach reveals an elliptical paraboloid 3D failure surface and can refine existing failure criteria.

Wan et al. [252] suggest a data-driven approach incorporating probability, and micromechanical modeling predicts failure in IM7/8552 unidirectional CFRPs under biaxial stress. Using high-fidelity 3D RVE models and ANN training, the method achieves a mean square error of 0.027% and a mean absolute error of 0.78% for regression, and a 98.1% prediction probability for classification. The ANN predictions align well with Tsai-Wu and Hashin failure criteria.

AM of carbon fiber-reinforced polymer (CFRP) composites allows for complex structures but challenges remain in predicting mechanical properties. A data-driven model [253] predicts flexural strength in continuous carbon fiber-reinforced polymers (CCFRPs) fabricated by fused deposition modeling (FDM), considering design factors like fiber layers, fiber rings, and polymer infill patterns. ML validates these predictions against experimental data.

High-fidelity simulations of composite materials are computationally intensive. Sepasdar et al. [254,255] introduce a deep learning framework using two fully convolutional networks to predict post-failure stress distribution and crack patterns in 2D composites based on microstructures. Trained on 4500 synthetic representations, the framework achieves 90% accuracy, aided by a physics-informed loss function.

Table 6 provides a summary of key studies focused on defect classification and failure prediction, detailing the focus, data used, and applied methods for each research effort.

Table 6. Overview of studies focusing on defect classification and failure prediction in polymer composites AM using ML techniques.

Reference	Focus	Data Info	Applied Method
Chen et al. [246]	Defect detection in composite materials using tomography data	Micro-CT scans of fiber-reinforced composites	ML models with binarized statistical image features (BSIF) and CNN
Chen [247]	Tool path analysis in 3D-printed GFRP composites	Micro-CT images for GFRP composites	ML models trained on BSIF-compressed data
Goh et al. [248]	Real-time defect detection in FFF	Video captured from print head	On-site monitoring system using computer vision and YOLO architectures
Chen et al. [249]	Predicting failure in composite lamina under triaxial loading	Data from finite element method-based RVE model	Hybrid method combining ANN and micromechanics
Wan et al. [252]	Predicting failure in CFRPs under biaxial stress	High-fidelity 3D RVE models	Data-driven approach with ANN and micromechanical modeling
Fontes et al. [253]	Predicting flexural strength in CCFRPs fabricated by FDM	Design factors like fiber layers and polymer infill patterns	Data-driven ML model
Sepasdar et al. [254]	Predicting stress distribution and crack patterns in composites	4500 synthetic representations of microstructures	Deep learning framework with fully convolutional networks

4.4. Customization and Personalization

ML enables the customization and personalization of polymer composite parts, making it possible to tailor products to specific user requirements and applications.

Xue et al. [256] propose an optimization framework using a variational autoencoder (VAE) and Bayesian optimization (BayesOpt) to design mechanical metamaterials with specific macroscopic elastic properties. By reducing the design space, this approach efficiently optimizes multi-material 3D-printed samples, validated through experimental testing.

A deep learning approach with high-order Bézier curves and a hybrid neural network–genetic optimization (NN-GO) method is used by Lee et al. [257] to optimize lattice structures for better weight-to-performance ratios. The design shifts material towards weak joint regions, improving modulus and strength, validated through AM and compression testing.

An inverse design method using artificial neural networks and generative adversarial networks (GANs) efficiently designs architectured composite materials. The method by Qian et al. [258] reduces the need for massive labeled training data while maintaining high performance, achieving a reduction in computational resources.

Multi-material inkjet 3DP enables the personalization of medical devices by combining algorithmic design with selective material deposition. He et al. [259] reduce bacterial biofilm formation and allow for user-defined mechanical properties, providing multifunctional customization through generative design and finite element modeling.

A heterogeneous microstructural design methodology is applied in [260] to elasto-electro-active piezoelectric ceramics for sensing and energy harvesting applications. Using a vision transformer-augmented VAE, the study creates a generative neural network to design 3D microstructures with multifunctional properties, optimizing them during the inference phase.

ML is used to discover novel lattice metamaterials that optimize elastic stiffness and wave speed during impact. Garland's et al. [261] AI-driven approach works with minimal simulation calls, overcoming challenges in designing materials for high-performance applications involving complex multi-physics interactions.

Table 7 provides an overview of studies that highlight different approaches to optimization and design using ML.

Table 7. Overview of studies focusing on optimization and design methodologies in AM using ML techniques.

Reference	Focus	Data Info	Applied Method
Xue et al. [256]	Designing mechanical metamaterials with specific elastic properties	Multi-material 3D-printed samples	Optimization framework using variational autoencoder (VAE) and Bayesian optimization (BayesOpt)
Lee et al. [257]	Optimizing lattice structures for weight-to-performance ratios	Lattice structures designed through AM	Deep learning with high-order Bézier curves and hybrid neural network-genetic optimization (NN-GO)
Qian et al. [258]	Inverse design of architectured composite materials	Labeled training data for neural networks	Artificial neural networks and generative adversarial networks (GANs)
He et al. [259]	Personalizing medical devices with multi-material printing	Algorithmic design combined with selective material deposition	Generative design and finite element modeling to reduce bacterial biofilm formation
Hashemi et al. [260]	Designing elasto-electro-active piezoelectric ceramics	Microstructural design methodology for multifunctional properties	Vision transformer-augmented VAE for generative neural network design
Garland et al. [261]	Discovering novel lattice metamaterials	Optimization for elastic stiffness and wave speed in impact scenarios	AI-driven approach with minimal simulation calls

4.5. VAT Photopolymerization and ML

Recent advancements in VAT photopolymerization (VP) enable the creation of complex, customizable materials using techniques like SLA. Sachdeva et al. [262] discuss the evolution, trends, challenges, and future directions of AI in 3DP [46], emphasizing its significance in Industry 4.0.

VP excels at processing polymer composites with high filler content. However, increasing the filler volume raises the suspension viscosity, which conflicts with VP's need for low-viscosity resins. Additionally, factors like filler shape, size, and optical properties affect light interaction. To address these challenges, Nasrin et al. [263] introduce an artificial neural network (ANN)-based classification model to predict the printability of highly filled polymer suspensions in VP. The model, trained on a small dataset, considers both monomodal and bimodal particle distributions and helps map suitable material and process parameters, optimizing printing efficiency and reducing resource usage.

Mechanoluminescent (MechL) materials emit light when subjected to mechanical stimuli, making them promising for structural health monitoring. However, their practical application has been hindered by challenges in producing high-intensity MechL composites and fabricating complex 3D shapes. Jo et al. [264] introduce a novel method for creating $SrAl_2O_4{}^{2+}$, Dy^{3+} particle-based MechL composites using VP 3DP, optimized through ML. A multi-objective Bayesian optimization (MBO) approach with Gaussian process regression (GPR) was employed to fine-tune critical process parameters, including MechL particle content, layer thickness, and cure ratio. This optimization aimed to enhance MechL properties while reducing printing time. The GPR model captured the complex input–output relationships, allowing for the identification of Pareto-optimal solutions that improved the performance of MechL specimens. Additionally, a micromechanical analysis method was developed to examine the influence of MechL particle volume fraction on MechL intensity. The optimized VP process was validated through practical tests on MechL-based stress sensors and mechanical components.

Frumosu et al. [265] focus on enhancing automation in AM by developing an online monitoring system for bottom-up photopolymerization AM (VPP) processes. The system uses sensor data to detect detachment errors in real time, which can lead to wasted material and time if unnoticed. The monitoring procedure involves an offline phase for training a predictive model and an online phase using a control chart to track detachment predictions. This approach improves process efficiency and can be adapted to other AM technologies, contributing to the shift from prototyping to continuous production.

Shan et al. [266] introduce a low-cost smart resin vat for real-time monitoring of VP 3DP to improve quality control, reliability, and minimize waste. Thermistors placed along the vat's edges detect heat changes during polymerization, allowing temperature profiles to reflect the curing patterns. ML algorithms are used to assess printing status, with a Failure Index to detect active or terminated prints. Gaussian process regression predicts the printing area based on temperature data. The system successfully detects printing issues, such as failures and missing features, and can be applied across various VP methods. Limitations and future improvements are discussed.

Cao et al. [267] present a method for predicting the optimal waiting time during bottom-up VP 3DP. The waiting time ensures that the printer's release membrane recovers and the resin becomes stationary between layers, improving print quality. The proposed method, called WTP-VP, uses multilayer perceptrons (MLPs) to predict waiting time based on resin flow and pressure data. This approach reduces waiting time by 47% and overall printing time by 25%, while maintaining surface quality. The method is efficient for real-time predictions in complex topologies and requires fewer data than conventional models.

DLP VP is widely used in AM for creating diverse products layer by layer. A key performance metric is the degree of curing (DoC), which affects material properties like density and elasticity. Current in situ monitoring methods, such as FT-IR, are limited to single-point measurements and can disrupt the process. Zhang et al. [268] introduce a non-invasive, full-field interferometric curing monitoring (ICM) method for real-time tracking of curing dynamics in DLP-VPP. Using a physics-based sensor model and ML, the ICM system estimates refractive index changes to predict DoC, enabling improved process control and print quality.

Table 8 summarizes recent studies that explore the integration of ML models in predicting printability, detecting defects, and optimizing operational parameters in VP-based AM systems.

Table 8. Overview of studies focusing on defect classification, failure prediction, and process optimization in VP-based AM.

Reference	Focus	Data Info	Applied Method
Nasrin et al. [263]	Predicting printability of highly filled polymer suspensions in VP	Small dataset on polymer suspensions with monomodal and bimodal particle distributions	ANN-based classification model for mapping material and process parameters
Jo et al. [264]	Optimizing MechL composites using VP 3DP for structural health monitoring	Data on MechL particle content, layer thickness, and cure ratio	Multi-objective Bayesian optimization with GPR; micromechanical analysis
Frumosu et al. [265]	Online monitoring system for bottom-up photopolymerization AM (VPP) to detect detachment errors	Sensor data from bottom-up VPP processes	Predictive model using a control chart for real-time error detection
Shan et al. [266]	Real-time monitoring system for VP 3DP to improve quality control	Temperature data from thermistors placed along the vat edges	ML algorithms with Gaussian process regression and Failure Index to detect print issues
Cao et al. [267]	Predicting optimal waiting time in bottom-up VP 3DP to improve print quality	Resin flow and pressure data	Multilayer perceptrons (MLPs) for predicting waiting time, and reducing printing and waiting times
Zhang et al. [268]	Real-time tracking of curing dynamics in DLP-VPP using a non-invasive method	Full-field interferometric data on refractive index changes	Physics-based sensor model with ML to estimate degree of curing (DoC)

5. Application of ML in 4DP of Polymer Composites

Four-dimensional printing is an advanced manufacturing technique where 3D-printed objects transform over time in response to external stimuli such as temperature, moisture, light, or magnetic fields. When applied to polymer composites, this technique enables the creation of dynamic structures that can adapt their shape or properties post-fabrication. ML has become instrumental in enhancing the capabilities of 4DP of polymer composites, optimizing both the design and functionality of the materials. Since the number of publications on ML in 4DP is relatively small [269], and even fewer focus on polymer composites, this section will review one article at the intersection of these topics, along with a few related studies.

Wang et al. [270] apply ML to predict the hardness of quaternary polymer blends during 3DP, aiming to reduce development costs and speed up multi-material co-blending technology. Using four polymers (PLA, TPU, PETG, ABS), composite materials of varying hardness were created from random three-material combinations. Hyperparameter optimization of five ML algorithms, using particle swarm and genetic algorithms, produced accurate predictive models. A four-in-one mixing extrusion head was built, validating predictions with real measurements. This approach improves the efficiency of multi-material printing design, reducing time and resource costs, with potential applications in high-cost industries like aerospace and biomedical fields.

Sun et al. [271] integrate active composites and 4DP to enable shape transformation in response to environmental stimuli. The process involves using ML and evolutionary algorithms (EAs) to optimize the design of materials with different expansion properties. A recurrent neural network (RNN) model, trained with finite element simulations, predicts forward shape changes, while the ML-EA approach efficiently solves inverse design problems. Combined with computer vision, this method transforms hand-drawn profiles into 4D-printed active beams that morph into desired shapes. The technique offers an efficient design tool for creating complex 4D-printed structures using grayscale digital light processing (g-DLP) [272].

Hamel et al. [273] explore the design of active composites, materials that respond to environmental stimuli, using a ML approach. By combining the finite element method with an evolutionary algorithm, the paper addresses the challenge of optimizing material distribution within 3D-printed active composites to achieve specific shape changes, a process known as 4DP. The composite structures are divided into voxel units made of either passive or active materials, and the optimization is tested through examples to demonstrate the effectiveness of achieving target shapes.

6. Transformation of Polymer Composites to Ceramics and Other Materials

The transformation of polymer composites into ceramics and other advanced materials typically involves converting polymer precursors into ceramic forms through methods such as pyrolysis, calcination, or thermal treatment. The inherent properties of polymer composites, including their lightweight nature and flexibility, can be harnessed and enhanced during this transformation, resulting in materials with superior mechanical strength, thermal stability, and resistance to harsh environments. Such transformations not only expand the functional capabilities of the original materials but also pave the way for innovative applications across various fields, including aerospace, electronics, and biomedical engineering. This section explore recent studies associated with the conversion of polymer composites into ceramics and other high-performance materials, highlighting their potential impacts on future technological advancements.

Su et al. [274] present a novel precursor-derived SiOC ceramic (PDC-SiOC) architecture for effective terahertz (THz) electromagnetic interference (EMI) shielding and absorption. The bulk SiOC ceramic absorbs over 93% of THz waves between 1.2 and 1.6 THz. A lightweight honeycomb structure, inspired by moth wings, was fabricated using vat photopolymerization 3DP followed by pyrolysis. This architecture demonstrated a maximum shielding effectiveness of 64.1 dB and a transmissivity below 1.4% from 0.2 to 1.6 THz, absorbing 97–99.8% of THz waves in the same range. Additionally, it exhibited good mechanical properties, with compressive and flexural strengths of 1.2 and 16.5 MPa, and thermal stability up to 1100 °C in inert conditions, highlighting its potential for high-efficiency THz EMI shielding applications.

Lyu et al. [275] present the development of diatom frustule-derived porous silica (DFPS) ceramics, which serve as templates for creating Ti3C2Tx/DFPS composites with exceptional electromagnetic interference (EMI) shielding properties. The composites, hot-pressed at 800 °C, achieved a maximum shielding effectiveness (SE) of 43.2 dB in the X-band and a compressive strength of 67.5 MPa. The hierarchical porous structure enhances electromagnetic energy dissipation through scattering and reflection, making these composites promising for delicate electronic components in the aerospace sector.

Wang et al. [276] focus on the fabrication of short carbon fiber-reinforced silicon carbide (Csf/SiC) ceramic matrix composites (CMCs) through material extrusion (ME) 3DP followed by precursor infiltration and pyrolysis (PIP). The study investigates how solid loading and fiber content affect the microstructure and mechanical properties. Optimal compositions yielded high-performance Csf/SiC CMCs with a bending strength of 212.74 MPa and fracture toughness of 5.84 MPa $m^{1/2}$. The findings contribute valuable insights into the 3DP of fiber-reinforced CMCs.

Sarvestani et al. [277] explore the use of polymer-derived ceramics (PDCs) with enhanced toughness and versatility, fabricated through stereolithography (SLA) using a silicon oxycarbide precursor. Triply periodic minimal surface (TPMS) designs are 3D printed and pyrolyzed to produce intricate ceramic structures. The resulting PDCs exhibit a compressive strength of 2.2 MPa and stiffness of 330 MPa, while maintaining a low density of 0.5 g/cm^3. The research highlights the potential of low-cost SLA 3DP for creating customized, bio-inspired ceramic architectures.

Jiang et al. [278] develop ultraviolet (UV)-curable polymer precursors and a two-stage pyrolysis strategy to create polymer-derived ceramics (PDCs) with controllable deformation and complex programmable shapes. Despite a low precursor ceramic yield of 13.5 wt% leading to pyrolysis shrinkage, dense, crack-free SiOC ceramics are achieved. The mechanism of deformation during pyrolysis is analyzed, and the study demonstrates a viable approach for producing programmable PDCs through photopolymerization 4DP.

Zhu et al. [279] optimize the formulation of photosensitive resin for polymer-derived ceramics by incorporating h-BN as a two-dimensional filler. The addition of 1 wt% h-BN enhanced the mechanical properties, achieving a bending strength of 252.4 ± 12.2 MPa and a fracture toughness of 2.7 ± 0.2 MPa·$m^{1/2}$ after pyrolysis. Furthermore, the thermal conductivity of the ceramics increased from 0.44 to 5.34 W·m· K^{-1}. The findings indicate that introducing h-BN effectively improves the thermal, electrical, and mechanical properties of precursor ceramics.

Young et al. [280] address challenges in 3DP polymer-derived ceramics by evaluating various post-processing methods to enhance pyrolysis outcomes. The approaches include UV surface flood curing, solvent soaking, and intermediate heating, aimed at increasing cross-linking and reducing defects. The results show that post-processing improved the pyrolysis survival rate to 97% and the ceramic yield to 53%, enabling the production of larger, complex turbine vanes.

Bobrin et al. [281] introduce a novel method for fabricating nanostructured carbon–ceramic multi-materials through polymerization-induced microphase separation 3DP. By combining inorganic precursors and acrylonitrile within a photocurable resin, nanostructured materials are created, which transform into a carbon–ceramic matrix upon pyrolysis. The study reveals that the initial resin composition influences the microstructure and properties of the resulting materials, allowing for a combination of ceramic and carbon characteristics, including low thermal conductivity and high electrical conductivity.

The study [282] presents an efficient technique for preparing SiC ceramics using selective laser printing combined with precursor impregnation and pyrolysis (PIP) and liquid phase sintering (LPS). A particle gradation technique was utilized to enhance the green body density, resulting in SiC ceramics with a flexural strength of 150 MPa and a relative density of 98.2%. The findings demonstrate a viable strategy for fabricating high-performance SiC ceramics via selective laser printing.

Wang et al. [283] review advancements in ceramic 3DP technology, highlighting its potential to revolutionize the ceramic industry by enabling the direct manufacturing of intricate designs without molds. The review discusses the benefits of advanced ceramics, including high strength and corrosion resistance, and analyzes various ceramic 3DP techniques. It also addresses the limitations and challenges of these technologies, aiming to provide strategies for the development and market implementation of new ceramic 3DP methods.

Table 9 summarizes the key studies that explore approaches to enhancing the properties and functionalities of PDCs, including their fabrication methods, material compositions, and resultant mechanical and thermal characteristics.

Table 9. Summary of recent studies on polymer-derived ceramics.

Reference	Focus	Materials	Methods	Results
Su et al. [274]	THz EMI shielding and absorption	Precursor-derived SiOC ceramic (PDC-SiOC)	Vat photopolymerization 3DP followed by pyrolysis	Absorbs >93% of THz waves (1.2–1.6 THz), SE of 64.1 dB, compressive strength of 1.2 MPa, thermal stability to 1100 °C.
Lyu et al. [275]	EMI shielding properties of composites	Diatom frustule-derived porous silica (DFPS), Ti3C2Tx	Hot-pressing at 800 °C	SE of 43.2 dB in X-band, compressive strength of 67.5 MPa, promising for aerospace applications.
Wang et al. [276]	Short carbon fiber-reinforced SiC CMCs	Short carbon fiber, SiC	Material extrusion 3DP, precursor infiltration and pyrolysis (PIP)	Bending strength of 212.74 MPa, fracture toughness of 5.84 MPa $m^{1/2}$.
Sarvestani et al. [277]	Enhanced toughness and versatility in ceramics	Polymer-derived ceramics (PDCs)	Stereolithography (SLA) using SiOC precursor	Compressive strength of 2.2 MPa, stiffness of 330 MPa, density of 0.5 g/cm^3.
Jiang et al. [278]	Programmable shapes in polymer-derived ceramics	UV-curable polymer precursors	Two-stage pyrolysis strategy	Achieved crack-free SiOC ceramics, despite 59.91% shrinkage; demonstrated programmable shape capability.
Zhu et al. [279]	Optimization of resin formulation for ceramics	Photosensitive resin, h-BN	Incorporation of h-BN in resin formulation	Bending strength of 252.4 ± 12.2 MPa, fracture toughness of 2.7 ± 0.2 MPa·$m^{1/2}$, thermal conductivity improved to 5.34 W·m^{-1}·K^{-1}.
Young et al. [280]	Post-processing methods for improved pyrolysis	Polymer-derived ceramics	Various post-processing techniques	Pyrolysis survival rate of 97%, ceramic yield of 53%, enabling larger turbine vanes production.
Bobrin et al. [281]	Fabrication of nanostructured carbon–ceramic multi-materials	Inorganic precursors, acrylonitrile	Polymerization-induced microphase separation 3DP	Revealed influence of resin composition on microstructure; combined ceramic and carbon properties achieved.
Wang et al. [282]	Efficient SiC ceramics preparation	SiC ceramics	Selective laser printing, precursor impregnation and pyrolysis (PIP), liquid phase sintering (LPS)	Flexural strength of 150 MPa, relative density of 98.2%.

7. Challenges and Limitations

The integration of ML into 3D and 4DP of polymer composites holds promise for advancing the field. However, there are several key challenges and limitations that need to be addressed to realize its full potential. Figure 12 illustrates the challenges and limitations associated with applying ML in 3D and 4DP of polymer composites.

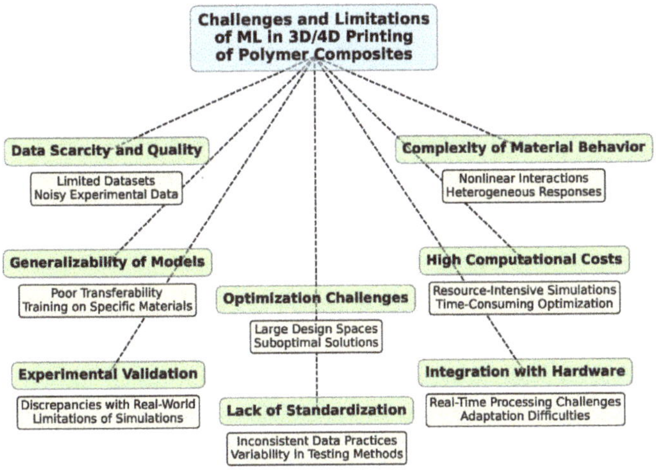

Figure 12. Challenges and limitations associated with applying ML in 3D and 4DP of polymer composites.

One of the foremost challenges is data scarcity and quality [284,285]. ML models typically require large datasets to deliver accurate predictions, but in the field of 3D and 4DP, particularly for polymer composites, such datasets are limited. The data available for material behavior, especially in complex systems involving active composites that respond to environmental stimuli, are often sparse or incomplete. Moreover, experimental data are prone to noise and inconsistencies, making it difficult for ML models to learn effectively and robustly [286].

The complexity of material behavior further complicates the application of ML. Polymer composites, especially when subjected to environmental stimuli in 4DP, exhibit highly nonlinear and heterogeneous behavior [215]. These materials often interact in unpredictable ways when combined, making it difficult for current ML models to accurately predict outcomes [287,288]. For instance, controlling the spatial distribution and phase transitions in multi-material systems, such as mechanical metamaterials or shape-shifting composites, remains a significant challenge due to the complexity of these interactions [289–291].

Another limitation is the generalizability of ML models [292,293]. ML models trained on specific material systems or design configurations often fail to generalize well to new materials or different printing methods. This lack of transferability limits the applicability of ML across diverse materials and printing processes [294]. As a result, models that perform well in controlled experimental settings may not be as effective in real-world industrial applications, where new variables are introduced.

High computational costs also present a barrier to the widespread use of ML in 3D and 4DP [295]. Many predictive models rely on computationally expensive simulations, such as the finite element method (FEM), to generate training data and validate results [296]. This is especially problematic for complex inverse design problems, where the optimization of material distributions or structural configurations requires substantial computational resources. The computational burden makes it difficult to apply ML in real-time applications, limiting its scalability for industrial use.

Optimization challenges are another issue, particularly in the context of designing active composites for 4DP. The design process often involves solving inverse problems with large design spaces, such as voxel-based material distributions [297,298]. Even with advanced optimization techniques, like evolutionary algorithms or Bayesian optimization, navigating these vast design spaces can be difficult, and the risk of converging on suboptimal solutions remains high. The complexity of nonlinear material behaviors further exacerbates these challenges, making it harder to find the optimal design.

Furthermore, experimental validation and real-world implementation of ML-driven models are often difficult to achieve. Discrepancies between simulation predictions and real-world outcomes can arise due to variations in material properties, environmental factors, or the limitations of current printing technologies. In practice, ML models that work well in controlled environments may not perform as expected when scaled to industrial applications, where real-time defect detection and quality control are critical [299].

Another practical limitation lies in integrating ML with printing hardware. The real-time monitoring and control of printing processes using ML models require fast and accurate data processing, which is challenging given the high-speed nature of 3DP systems [46,300]. Additionally, adapting existing hardware, such as multi-material extrusion heads, to accommodate ML-driven optimizations can be technically demanding and expensive.

Lastly, the lack of standardization in material characterization, testing methods, and printing protocols across the AM industry further complicates the application of ML [301]. Without standardized datasets and consistent experimental procedures, it becomes difficult to train and compare ML models across different systems or materials, limiting the broader adoption of ML in the field.

For instance, material extrusion (MEX) lacks standardized testing methods tailored to its unique material and process characteristics. Phillips et al. [302] review current practices for preparing tensile test specimens and propose guidelines for future standards. They emphasize the need to account for slicing parameters, specimen geometry, toolpath optimization [303,304], and material specifications to ensure accurate representation of final part properties. Standardizing these factors could improve comparability between studies and support the development of MEX for advanced applications.

Garcia et al. [305] review current design, material, and process standards for AM [306,307], with a focus on mechanical characterization of polymer-based products. They highlight the reliance on standards from other industries, inconsistencies between documents, and the need for clearer guidance. The work highlights the importance of developing AM-specific standards, particularly for mechanical testing, and addresses the disparity between standards for metallic and polymer materials. This review aims to support both researchers and practitioners in navigating the evolving standardization landscape in AM.

8. Future Directions

As ML continues to evolve, its application in 3D and 4DP of polymer composites presents numerous opportunities for innovation and advancement. To harness the full potential of ML in this field, several key future directions can be explored. Figure 13 illustrates the future directions of ML applications in 3D and 4DP of polymer composites.

One of the most essential areas for future research is the generation of high-quality, diverse datasets [308,309]. This can be achieved through advanced simulation techniques, such as generative adversarial networks (GANs) [310,311], which can synthesize realistic material behavior data. Jabbar et al. [312] review recent advances in using GANs for inverse materials design (IMD), where GANs help discover materials with targeted properties by applying specific constraints. The authors discuss databases, ML criteria, available software tools, and training descriptors for GAN models, highlighting both challenges and future directions in this promising field. On the other hand, Jiang et al. [313] highlight the expanding role of GANs in materials science, covering applications from composition design to microstructure analysis and defect detection. The paper discusses GAN fundamentals, specific use cases, and addresses challenges, underscoring GANs' potential to drive innovative advancements in materials discovery and optimization.

Additionally, efforts to standardize data collection [314] and sharing practices across research institutions and industries can facilitate greater access to valuable datasets [315]. For instance, Shetty et al. [316] developed a pipeline using NLP and trained a specialized language model, MaterialsBERT, to automatically extract material property data from polymer science abstracts, collecting 300,000 records from 130,000 abstracts in 60 h. The data, accessible at polymerscholar.org, offers insights across applications like fuel cells and solar cells, showcasing the potential of automated literature analysis for materials science.

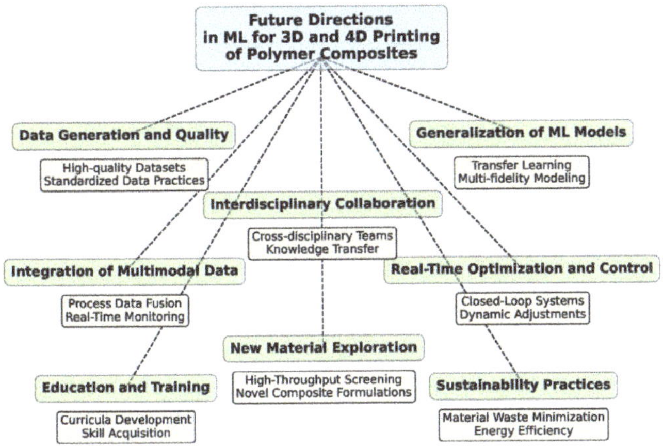

Figure 13. Future directions in ML applications for 3D and 4DP of polymer composites.

Future efforts should also focus on developing ML models that exhibit improved generalization capabilities across various materials and printing conditions. This can be achieved by utilizing transfer learning (TL) approaches [317,318], where models trained on one material system can be adapted for others. Tang et al. [319] review TL in AM modeling, emphasizing its potential to improve model quality despite limited data by reusing models across products. They outline TL methods, current applications, and recommendations for effectively applying TL to enhance AM processes.

Multi-fidelity (MF) modeling [320] techniques can also be explored, allowing for the integration of data from different sources (e.g., experimental and simulation data) to enhance model robustness. Nath et al. [321] introduce an MF modeling approach to predict AM outcomes by combining high-fidelity (HF) and low-fidelity (LF) models with experimental data. Using Bayesian calibration, the method improves LF model predictions, demonstrated here for predicting porosity in laser powder bed fusion.

The incorporation of multimodal data is another promising avenue for future research. Integrating process data (temperature, pressure, speed) with material properties and performance outcomes can provide a more comprehensive understanding of the printing process [322]. Petrich et al. [323] propose a supervised machine learning approach for detecting inter-layer flaws in powder bed fusion additive manufacturing (PBFAM) using in situ multimodal sensor data, with 98.5% accuracy in binary flaw classification. Integrating data from multiple sensors (e.g., imagery, acoustic, and multi-spectral) and scan trajectories, the approach successfully correlates in-process sensor data with post-build CT scans, demonstrating enhanced flaw detection performance by fusing sensor modalities.

Advancements in ML algorithms should be geared towards enabling real-time optimization and control of the 3D and 4DP process. This includes developing closed-loop [324] systems, where feedback from in situ sensors informs adaptive control strategies. By continuously learning from ongoing processes, these systems can adjust printing parameters in real time to mitigate defects and enhance material properties, leading to improved efficiency and product quality. Mercado et al. [325] highlight recent efforts in control system improvements and emphasize the potential advantages of closed-loop control for advancing AM precision and reliability.

Exploration of new material systems is another important direction for future research. The application of ML in discovering and optimizing new polymer composite materials for specific applications is promising. By combining ML with high-throughput experimental methods, researchers can rapidly screen and identify novel composite formulations that meet desired performance criteria [326,327]. Nazir et al. [55] provide a summary of recent advancements in multi-material additive manufacturing (MMAM), exploring its applications, design strategies, and challenges across various industries. They identify limitations in existing processes and software, while also discussing future directions and potential strategies to enhance the functionality and mechanical properties of MMAM-fabricated parts.

Interdisciplinary collaboration can facilitate the transfer of knowledge and technology between academia and industry, driving the adoption of ML in real-world manufacturing settings [328]. Park et al. [329] present a methodology for identifying and prioritizing data analytics (DA) opportunities in AM, highlighting the importance of interdisciplinary collaboration. The framework includes a team of experts, a Data Opportunity Knowledge Base (DOKB), and a prioritization tool utilizing Fuzzy-TOPSIS, resulting in the identification and ranking of 264 DA opportunities for the laser powder bed fusion process, ultimately facilitating ongoing collaboration and knowledge sharing within the AM community.

Furthermore, sustainability is an increasingly important consideration in manufacturing. Future research should explore how ML can contribute to sustainable practices in 3D and 4DP [330–332], such as optimizing material usage to minimize waste [333], improving energy efficiency [334], and developing biodegradable composites [335–337]. Hegab et al. [338] highlight the role of AM in promoting sustainability across various industries, showcasing its benefits in reducing resource depletion, waste, and emissions while improving efficiency in production processes. They discuss the integration of AM within circular economy strategies, identify challenges in its deployment throughout the product life cycle, and emphasize the need for further research on the long-term environmental impacts of AM to encourage its adoption among organizations and policymakers.

As the integration of ML in 3D and 4DP progresses, there will be a growing need for education and training programs that equip researchers and practitioners with the necessary skills and knowledge. Institutions should develop curricula that focus on the intersection of ML, materials science, and AM, ensuring that the next generation of engineers and scientists are prepared to leverage these technologies effectively. Stavropoulos et al. [339] address the lack of expertise hindering the industrial adoption of AM by developing a structured training framework tailored to industry needs. The framework classifies AM into modular educational areas, targeting various professional profiles and emphasizing hands-on practice, while also proposing strategies to enhance accessibility and facilitate the implementation of AM training within the industrial sector.

9. Conclusions

The application of ML in 3D and 4DP of polymer composites represents a shift in the landscape of AM. ML applications range from optimizing the printing process to predicting the performance of materials and enhancing design capabilities. By analyzing vast amounts of data generated during the printing process, ML allows for better control of parameters, improved outcomes, and accelerated innovation. Furthermore, as the complexity of designs and materials increases, the role of ML in facilitating rapid prototyping, quality assurance, and customizability will likely grow, positioning it as a cornerstone of future innovations in additive manufacturing. Based on the literature analysis presented in this manuscript, the following is a summarized version in bullet points:

- The integration of ML in real-time monitoring systems (e.g., AI2AM technology for FDM) improves the quality and consistency of printed polymer composites. This shift to smart manufacturing aligns with Industry 4.0 principles, focusing on defect detection and parameter optimization to prevent errors and enhance efficiency.

- Techniques like the deep learning model developed by Lu et al. for detecting defects in carbon fiber-reinforced polymers (CFRPs) showcase the ability of AI to provide real-time geometric analysis and process adjustments. This automation reduces reliance on traditional methods, enhancing overall manufacturing quality.
- Systems like the self-monitoring approach by Narayanan et al. utilize deep learning to detect delamination and predict warping, demonstrating the capability for early error detection. This proactive management improves automated quality control across various manufacturing sectors.
- Jin et al. [340] achieved up to 99.5% accuracy in classifying 3D-printed parts using ML models. This level of precision benefits both large-scale manufacturers and individual users, emphasizing the competitive advantage gained through intelligent quality assurance.
- By employing laser scanning methods (as demonstrated by Lin et al. [240]) to monitor printed surfaces against CAD models, companies can achieve real-time feedback control that reduces waste and prevents unnecessary production runs, leading to more sustainable manufacturing practices.
- The Continual G-learning method for defect detection exemplifies the potential of reinforcement learning to address emerging defects in real time using historical and real-time data, showcasing an advanced adaptive quality control system that requires minimal training samples.
- Innovations like the multi-material inkjet 3DP method described by He et al. for personalizing medical devices highlight the growing trend towards customization, enabling user-defined mechanical properties and multifunctional device design.
- Xue et al.'s variational autoencoder (VAE) framework [256] for designing mechanical metamaterials demonstrates the ability of ML to efficiently customize 3D-printed parts for specific macroscopic elastic properties, fostering innovation in material science and application design.
- The reduction in printing time (up to 25%) and waiting time (47%) achieved through methods like WTP-VP, as described by Cao et al., signify a positive trend towards more environmentally friendly manufacturing processes by minimizing resource usage.
- Chen et al.'s work [21] on tool path identification in GFRP composites highlights the potential risks to intellectual property as reverse-engineering capabilities using ML can reconstruct manufacturing processes from finished products, necessitating enhanced data protection strategies.
- Recent studies highlight advancements in creating polymer-derived ceramics (PDCs) with superior mechanical properties and functionality. For example, Su et al. [274] developed a precursor-derived SiOC ceramic that has the potential for applications in high-efficiency electromagnetic interference (EMI) shielding, particularly in the aerospace and electronics sectors.

Funding: This research received no external funding.

Institutional Review Board Statement: Not applicable.

Informed Consent Statement: Not applicable.

Data Availability Statement: Data are contained within the article.

Conflicts of Interest: The authors declare no conflicts of interest.

References

1. Osouli-Bostanabad, K.; Masalehdan, T.; Kapsa, R.M.; Quigley, A.; Lalatsa, A.; Bruggeman, K.F.; Franks, S.J.; Williams, R.J.; Nisbet, D.R. Traction of 3D and 4D printing in the healthcare industry: From drug delivery and analysis to regenerative medicine. *ACS Biomater. Sci. Eng.* **2022**, *8*, 2764–2797. [CrossRef] [PubMed]

2. Khan, M.S.; Khan, S.A.; Shabbir, S.; Umar, M.; Mohapatra, S.; Khuroo, T.; Naseef, P.P.; Kuruniyan, M.S.; Iqbal, Z.; Mirza, M.A. Raw materials, technology, healthcare applications, patent repository and clinical trials on 4D printing technology: An updated review. *Pharmaceutics* **2022**, *15*, 116. [CrossRef] [PubMed]
3. Mahmood, A.; Perveen, F.; Chen, S.; Akram, T.; Irfan, A. Polymer Composites in 3D/4D Printing: Materials, Advances, and Prospects. *Molecules* **2024**, *29*, 319. [CrossRef] [PubMed]
4. Ntouanoglou, K.; Stavropoulos, P.; Mourtzis, D. 4D printing prospects for the aerospace industry: A critical review. *Procedia Manuf.* **2018**, *18*, 120–129. [CrossRef]
5. Lakkala, P.; Munnangi, S.R.; Bandari, S.; Repka, M. Additive manufacturing technologies with emphasis on stereolithography 3D printing in pharmaceutical and medical applications: A review. *Int. J. Pharm. X* **2023**, *5*, 100159. [CrossRef]
6. Periyasamy, R.; Hemanth Kumar, M.; Rangappa, S.M.; Siengchin, S. A comprehensive review on natural fillers reinforced polymer composites using fused deposition modeling. *Polym. Compos.* **2023**, *44*, 3715–3747. [CrossRef]
7. Luo, X.; Cheng, H.; Wu, X. Nanomaterials reinforced polymer filament for fused deposition modeling: A state-of-the-art review. *Polymers* **2023**, *15*, 2980. [CrossRef]
8. Verma, N.; Awasthi, P.; Gupta, A.; Banerjee, S.S. Fused deposition modeling of polyolefins: Challenges and opportunities. *Macromol. Mater. Eng.* **2023**, *308*, 2200421. [CrossRef]
9. Yadav, A.; Rohru, P.; Babbar, A.; Kumar, R.; Ranjan, N.; Chohan, J.S.; Kumar, R.; Gupta, M. Fused filament fabrication: A state-of-the-art review of the technology, materials, properties and defects. *Int. J. Interact. Des. Manuf. (IJIDeM)* **2023**, *17*, 2867–2889. [CrossRef]
10. Tao, Y.; Li, P.; Zhang, J.; Wang, S.; Shi, S.Q.; Kong, F. A review of fused filament fabrication of continuous natural fiber reinforced thermoplastic composites: Techniques and materials. *Polym. Compos.* **2023**, *44*, 8200–8222. [CrossRef]
11. Kumar, S.; Singh, R.; Singh, T.; Batish, A. Fused filament fabrication: A comprehensive review. *J. Thermoplast. Compos. Mater.* **2023**, *36*, 794–814. [CrossRef]
12. Credi, C.; Bernasconi, R.; Levi, M.; Magagnin, L. Self-activating metal-polymer composites for the straightforward selective metallization of 3D printed parts by stereolithography. *J. Mater. Res. Technol.* **2023**, *22*, 1855–1867. [CrossRef]
13. Tamburri, E.; Montaina, L.; Pescosolido, F.; Carcione, R.; Battistoni, S. 3D Extrusion and Stereolithography Printing Methods for Producing Multifunctional Polymer Composites. *Macromol. Symp.* **2024**, *413*, 2400030. [CrossRef]
14. Nugraha, A.D.; Kumar, V.V.; Gautama, J.P.; Wiranata, A.; Mangunkusumo, K.G.H.; Rasyid, M.I.; Dzanzani, R.; Muflikhun, M.A. Investigating the characteristics of nano-graphite composites additively manufactured using stereolithography. *Polymers* **2024**, *16*, 1021. [CrossRef]
15. Shen, H.; Wu, W.; Hu, H.; Rui, Z.; Ye, J.; Zhang, C. Preparation of carbon black/graphene nanosheets/PP composites with 3D separated conductive networks based on selective laser sintering. *Polym. Compos.* **2023**, *44*, 3522–3534. [CrossRef]
16. Schappo, H.; Giry, K.; Salmoria, G.; Damia, C.; Hotza, D. Polymer/calcium phosphate biocomposites manufactured by selective laser sintering: An overview. *Prog. Addit. Manuf.* **2023**, *8*, 285–301. [CrossRef]
17. Wu, Z.; Li, Y.; Shi, C.; Lu, Z.; Su, B. A general method for fabricating polymer-based intrinsic superhydrophobic objects by a selective laser sintering 3D printing strategy. *Compos. Part B Eng.* **2023**, *264*, 110910. [CrossRef]
18. Kidalov, S.; Voznyakovskii, A.; Vozniakovskii, A.; Titova, S.; Auchynnikau, Y. The effect of few-layer graphene on the complex of hardness, strength, and thermo physical properties of polymer composite materials produced by digital light processing (DLP) 3D printing. *Materials* **2023**, *16*, 1157. [CrossRef]
19. Huang, W.; Zhang, J.; Singh, V.; Xu, L.; Kabi, P.; Bele, E.; Tiwari, M.K. Digital light 3D printing of a polymer composite featuring robustness, self-healing, recyclability and tailorable mechanical properties. *Addit. Manuf.* **2023**, *61*, 103343. [CrossRef]
20. Yao, J.; Hakkarainen, M. Methacrylated wood flour-reinforced "all-wood" derived resin for digital light processing (DLP) 3D printing. *Compos. Commun.* **2023**, *38*, 101506. [CrossRef]
21. Chen, J.; An, R.; Tey, W.S.; Zeng, Q.; Zhao, L.; Zhou, K. In Situ Filler Addition for Homogeneous Dispersion of Carbon Nanotubes in Multi Jet Fusion–Printed Elastomer Composites. *Adv. Sci.* **2023**, *10*, 2300593. [CrossRef] [PubMed]
22. Chen, M.; Hou, Y.; An, R.; Tey, W.S.; Gao, M.; Chen, J.; Zhao, L.; Zhou, K. Investigation of the mechanical properties of polyimide fiber/polyamide 12 composites printed by Multi Jet Fusion. *Virtual Phys. Prototyp.* **2023**, *18*, e2246032. [CrossRef]
23. Peng, Y.S. *3D Printing of Fibre-Reinforced Polymer Composites via Multi Jet Fusion*; Nanyang Technological University: Singapore, 2023.
24. Kozior, T.; Ehrmann, A. First proof-of-principle of polyjet 3D printing on textile fabrics. *Polymers* **2023**, *15*, 3536. [CrossRef] [PubMed]
25. Andena, L.; Contino, M.; Arioli, M.; De Noni, L.; Volders, T.; Ruffoni, D. Fracture of Polyjet 3D printed materials: A preliminary investigation. In Proceedings of the 9th International Conference on Fracture of Polymers, Composites and Adhesives, Les Diablerets, Switzerland, 24–27 March 2024.
26. Szczygieł, P.; Radoń-Kobus, K.; Madej, M.; Kozior, T. Tribological properties of MED610 medical material used in PolyJet matrix 3D printing technology. *Tribologia* **2023**, *306*, 65–77. [CrossRef]
27. Raj, R.; Dixit, A.R. Direct ink writing of carbon-doped polymeric composite ink: A review on its requirements and applications. *3D Print. Addit. Manuf.* **2023**, *10*, 828–854. [CrossRef]
28. Liu, H.; Mei, D.; Yu, S.; Qian, S.; Wang, Y. Direct ink writing of chopped carbon fibers reinforced polymer-derived SiC composites with low shrinkage and high strength. *J. Eur. Ceram. Soc.* **2023**, *43*, 235–244. [CrossRef]

29. Yang, Z.; Medora, E.; Ren, Z.; Cheng, M.; Namilae, S.; Jiang, Y. Coaxial direct ink writing of ZnO functionalized continuous carbon fiber-reinforced thermosetting composites. *Compos. Sci. Technol.* **2024**, *256*, 110782. [CrossRef]
30. Shinde, S.; Mane, R.; Vardikar, A.; Dhumal, A.; Rajput, A. 4D printing: From emergence to innovation over 3D printing. *Eur. Polym. J.* **2023**, *197*, 112356. [CrossRef]
31. Zeng, C.; Liu, L.; Lin, C.; Xin, X.; Liu, Y.; Leng, J. 4D printed continuous fiber reinforced shape memory polymer composites with enhanced mechanical properties and shape memory effects. *Compos. Part A Appl. Sci. Manuf.* **2024**, *180*, 108085. [CrossRef]
32. Zhao, W.; Yue, C.; Liu, L.; Liu, Y.; Leng, J. Research progress of shape memory polymer and 4D printing in biomedical application. *Adv. Healthc. Mater.* **2023**, *12*, 2201975. [CrossRef]
33. Yan, S.; Zhang, F.; Luo, L.; Wang, L.; Liu, Y.; Leng, J. Shape memory polymer composites: 4d printing, smart structures, and applications. *Research* **2023**, *6*, 0234. [CrossRef] [PubMed]
34. Lalegani Dezaki, M.; Bodaghi, M. Sustainable 4D printing of magneto-electroactive shape memory polymer composites. *Int. J. Adv. Manuf. Technol.* **2023**, *126*, 35–48. [CrossRef]
35. Abdullah, T.; Okay, O. 4D printing of body temperature-responsive hydrogels based on poly (acrylic acid) with shape-memory and self-healing abilities. *ACS Appl. Bio Mater.* **2023**, *6*, 703–711. [CrossRef] [PubMed]
36. Pelluau, T.; Brossier, T.; Habib, M.; Sene, S.; Félix, G.; Larionova, J.; Blanquer, S.; Guari, Y. 4D printing nanocomposite hydrogel based on PNIPAM and Prussian blue nanoparticles using stereolithography. *Macromol. Mater. Eng.* **2024**, *309*, 2300305. [CrossRef]
37. Goyal, R.; Sahu, S.; Mitra, S.; Niranjan, R.; Priyadarshini, R.; Yadav, R.; Lochab, B. Nanocellulose-Reinforced 4D Printed Hydrogels: Thermoresponsive Shape Morphing and Drug Release. *ACS Appl. Polym. Mater.* **2024**, *6*, 1348–1361. [CrossRef]
38. Aufa, A.; Ismail, Z.; Hassan, M.Z. Emerging trends in 4d printing of hydrogels in the biomedical field: A review. *Mater. Today Proc.* 2023, *in press*.
39. Schweizer, K.; Bhandari, S.; Lopez-Anido, R.; Korey, M.; Tekinalp, H. Recycling Large-Format 3D Printed Polymer Composite Formworks Used for Casting Precast Concrete -Technical Feasibility and Challenges. *J. Compos. Constr.* **2024**, *28*, 04024061. [CrossRef]
40. Ejeromedoghene, O.; Omoniyi, A.O.; Akor, E.; Alowakennu, M.; Samson, K.A.; Abesa, S.; Zhang, Z. Progress in stimuli-responsive hydrogel composites for digital technologies. *Appl. Mater. Today* **2024**, *37*, 102088. [CrossRef]
41. Kuang, X.; Yue, L.; Qi, H.J. Introduction to 4D printing: Concepts and material systems. In *Additive Manufacturing Technology: Design, Optimization, and Modeling*; Wiley: Hoboken, NJ, USA, 2023; pp. 1–42.
42. Dixit, G.; Pandey, P.M. Experimental investigations of temperature-sensitive shape memory polymer composites for 4D printing. *J. Thermoplast. Compos. Mater.* **2024**, 08927057241254322. [CrossRef]
43. Kouka, M.A.; Abbassi, F.; Habibi, M.; Chabert, F.; Zghal, A.; Garnier, C. 4D printing of shape memory polymers, blends, and composites and their advanced applications: A comprehensive literature review. *Adv. Eng. Mater.* **2023**, *25*, 2200650. [CrossRef]
44. Ma, T.; Zhang, Y.; Ruan, K.; Guo, H.; He, M.; Shi, X.; Guo, Y.; Kong, J.; Gu, J. Advances in 3D printing for polymer composites: A review. *InfoMat* **2024**, *6*, e12568. [CrossRef]
45. Ghilan, A.; Chiriac, A.; Nita, L.; Rusu, A.; Neamtu, I.; Chiriac, V. Trends in 3D Printing Processes for Biomedical Field: Opportunities and Challenges. *J. Polym. Environ.* **2020**, *28*, 1345–1367. [CrossRef] [PubMed]
46. Ng, W.L.; Goh, G.L.; Goh, G.D.; Ten, J.S.J.; Yeong, W.Y. Progress and opportunities for machine learning in materials and processes of additive manufacturing. *Adv. Mater.* **2024**, *36*, 2310006. [CrossRef] [PubMed]
47. Liu, H.; Ji, X.; Wang, W.; Zhou, L. 3D-Networks Based Polymer Composites for Multifunctional Thermal Management and Electromagnetic Protection: A Mini Review. *Materials* **2024**, *17*, 2400. [CrossRef]
48. Nikooharf, M.H.; Shirinbayan, M.; Arabkoohi, M.; Bahlouli, N.; Fitoussi, J.; Benfriha, K. Machine learning in polymer additive manufacturing: A review. *Int. J. Mater. Form.* **2024**, *17*, 52. [CrossRef]
49. Dananjaya, V.; Marimuthu, S.; Yang, R.C.; Grace, A.; Abeykoon, C. Synthesis, properties, applications, 3D printing and machine learning of graphene quantum dots in polymer nanocomposites. *Prog. Mater. Sci.* **2024**, *144*, 101282. [CrossRef]
50. Oskolkov, A.A.; Bezukladnikov, I.I.; Trushnikov, D.N. Mathematical Model of the Layer-by-Layer FFF/FGF Polymer Extrusion Process for Use in the Algorithm of Numerical Implementation of Real-Time Thermal Cycle Control. *Polymers* **2023**, *15*, 4518. [CrossRef]
51. Behseresht, S.; Park, Y.H.; Love, A.; Pastrana, O.A.V. Application of Numerical Modeling and Finite Element Analysis in Fused Filament Fabrication: A Review. *Materials* **2024**, *17*, 4185. [CrossRef]
52. Bouzaglou, O.; Golan, O.; Lachman, N. Process design and parameters interaction in material extrusion 3D printing: A review. *Polymers* **2023**, *15*, 2280. [CrossRef]
53. Sadaf, M.; Bragaglia, M.; Slemenik Perše, L.; Nanni, F. Advancements in metal additive manufacturing: A comprehensive review of material extrusion with highly filled polymers. *J. Manuf. Mater. Process.* **2024**, *8*, 14. [CrossRef]
54. Bankapalli, N.K.; Gupta, V.; Saxena, P.; Bajpai, A.; Lahoda, C.; Polte, J. Filament fabrication and subsequent additive manufacturing, debinding, and sintering for extrusion-based metal additive manufacturing and their applications: A review. *Compos. Part B Eng.* **2023**, *264*, 110915. [CrossRef]
55. Nazir, A.; Gokcekaya, O.; Billah, K.M.M.; Ertugrul, O.; Jiang, J.; Sun, J.; Hussain, S. Multi-material additive manufacturing: A systematic review of design, properties, applications, challenges, and 3D printing of materials and cellular metamaterials. *Mater. Des.* **2023**, *226*, 111661. [CrossRef]

56. Memarzadeh, A.; Safaei, B.; Tabak, A.; Sahmani, S.; Kizilors, C. Advancements in additive manufacturing of polymer matrix composites: A systematic review of techniques and properties. *Mater. Today Commun.* **2023**, *36*, 106449. [CrossRef]
57. Sun, X.; Mazur, M.; Cheng, C.T. A review of void reduction strategies in material extrusion-based additive manufacturing. *Addit. Manuf.* **2023**, *67*, 103463. [CrossRef]
58. Samykano, M.; Kumaresan, R.; Kananathan, J.; Kadirgama, K.; Pandey, A.K. An overview of fused filament fabrication technology and the advancement in PLA-biocomposites. *Int. J. Adv. Manuf. Technol.* **2024**, *132*, 27–62. [CrossRef]
59. Alqutaibi, A.Y.; Alghauli, M.A.; Aljohani, M.H.A.; Zafar, M.S. Advanced additive manufacturing in implant dentistry: 3D printing technologies, printable materials, current applications and future requirements. *Bioprinting* **2024**, *42*, e00356. [CrossRef]
60. Ali, F.; Kalva, S.N.; Koc, M. Advancements in 3D printing techniques for biomedical applications: A comprehensive review of materials consideration, post processing, applications, and challenges. *Discov. Mater.* **2024**, *4*, 53. [CrossRef]
61. Slepicka, M.; Borrmann, A. Fabrication Information Modeling for Closed-Loop Design and Quality Improvement in Additive Manufacturing for construction. *Autom. Constr.* **2024**, *168*, 105792. [CrossRef]
62. Bănică, C.F.; Sover, A.; Anghel, D.C. Printing the Future Layer by Layer: A Comprehensive Exploration of Additive Manufacturing in the Era of Industry 4.0. *Appl. Sci.* **2024**, *14*, 9919. [CrossRef]
63. Franco Urquiza, E.A. Advances in Additive Manufacturing of Polymer-Fused Deposition Modeling on Textiles: From 3D Printing to Innovative 4D Printing—A Review. *Polymers* **2024**, *16*, 700. [CrossRef]
64. Subramani, R.; Kalidass, A.K.; Muneeswaran, M.D.; Lakshmipathi, B.G. Effect of fused deposition modeling process parameter in influence of mechanical property of acrylonitrile butadiene styrene polymer. *Appl. Chem. Eng.* **2024**, *7*, 3576. [CrossRef]
65. Kut, P.; Pietrucha-Urbanik, K. Bibliometric Analysis of Multi-Criteria Decision-Making (MCDM) Methods in Environmental and Energy Engineering Using CiteSpace Software: Identification of Key Research Trends and Patterns of International Cooperation. *Energies* **2024**, *17*, 3941. [CrossRef]
66. Alamoodi, A.; Al-Samarraay, M.S.; Albahri, O.; Deveci, M.; Albahri, A.; Yussof, S. Evaluation of energy economic optimization models using multi-criteria decision-making approach. *Expert Syst. Appl.* **2024**, *255*, 124842. [CrossRef]
67. Nwanno, C.E.; Thapa, A.; Watt, J.; Simkins Bendayan, D.; Li, W. Field Emission Properties of Cu-Filled Vertically Aligned Carbon Nanotubes Grown Directly on Thin Cu Foils. *Nanomaterials* **2024**, *14*, 988. [CrossRef] [PubMed]
68. Melentiev, R.; Lagerweij, A.; Lubineau, G. Multiprocess additive manufacturing via fused deposition modeling, chemical deposition, and electroplating with tough interfacial adhesion. *Smart Mater. Manuf.* **2024**, *2*, 100043. [CrossRef]
69. Sandhu, H.S.; Alam, A.S.; Jeevan, K.; Yadav, A.K.; Sai, G.; Korapram, V.M. A review of the mechanical properties of 3D printed and electroplated ABS parts. *AIP Conf. Proc.* **2024**, *2962*, 020055.
70. Nizam, M.; Purohit, R.; Taufik, M. Materials for 3D printing in healthcare sector: A review. *Proc. Inst. Mech. Eng. Part H J. Eng. Med.* **2024**, *238*, 939–963. [CrossRef]
71. Singh, S.; Attri, R.K.; Trivedi, S. Optimization of FDM 3D Printing Process Parameters for Improving Wear Characteristics of PLA-nGr Composite using Taguchi DOE. *J. Mater. Eng. Perform.* **2024**, 1–9. [CrossRef]
72. Bahrami, M.H.; Ehteshamfar, M.V.; Adibi, H. The effect, prediction, and optimization of Fe particles on wear behavior of Fe–ABS composites fabricated by fused deposition modeling. *Arab. J. Sci. Eng.* **2024**, *49*, 2001–2016. [CrossRef]
73. Çaparoğlu, Ö.F.; Ok, Y.; Çağlayan Özaydın, N. The Genetic Algorithm-Artificial Neural Networks Integration in the Optimization: An Application for Transportation Systems. In *Intelligent and Fuzzy Systems*; Springer: Cham, Switzerland, 2024; pp. 277–284.
74. Abdel Hamid, E.; Aly, H.; El Naggar, K. Synthesis of nanogeopolymer adsorbent and its application and reusability in the removal of methylene blue from wastewater using response surface methodology (RSM). *Sci. Rep.* **2024**, *14*, 20631. [CrossRef]
75. Hajjaj, M.S.; Alamoudi, R.A.; Babeer, W.A.; Rizg, W.Y.; Basalah, A.A.; Alzahrani, S.J.; Yeslam, H.E. Flexural strength, flexural modulus and microhardness of milled vs. fused deposition modeling printed Zirconia; effect of conventional vs. speed sintering. *BMC Oral Health* **2024**, *24*, 38. [CrossRef]
76. Khan, W.A.; Hassan, M.; Ahmed, I.; Xiao, M.; Faraz, M.I.; Li, K.; Khan, I.; Muhammad, R.; Wu, H.; Hussain, G. Insights into flexural and impact properties of polymer based materials printed through fused filament fabrication: Progress in the last decade. *Int. J. Lightweight Mater. Manuf.* **2024**, *7*, 925–957. [CrossRef]
77. Kariuki, L.W.; Ikua, B.W.; Karanja, S.K.; Ng'ang'a, S.P.; Zeidler, H. Fused filament fabrication of carbon fiber-reinforced polymer composite: Effect of process parameters on flexural properties. *Eng. Rep.* **2024**, *6*, e12807. [CrossRef]
78. Kara, Y.; Lubineau, G. Enhancing structural integrity of the continuous fiber-reinforced 3D printed composites by self-reinforcing nanofiber interleaves. *Compos. Commun.* **2024**, *51*, 102042. [CrossRef]
79. García-Cabezón, C.; Naranjo, J.A.; García-Hernández, C.; Berges, C.; Herranz, G.; Martín-Pedrosa, F. Using fused filament fabrication to improve the tribocorrosion behaviour of 17-4 PH SS in comparison to other metal forming techniques. *Friction* **2024**, *12*, 2325–2343. [CrossRef]
80. Kalinke, C.; Crapnell, R.D.; de Oliveira, P.R.; Janegitz, B.C.; Bonacin, J.A.; Banks, C.E. How to improve sustainability in Fused Filament Fabrication (3D Printing) research? *Glob. Challenges* **2024**, *8*, 2300408. [CrossRef]

81. Nath, P.C.; Sharma, R.; Mahapatra, U.; Mohanta, Y.K.; Rustagi, S.; Sharma, M.; Mahajan, S.; Nayak, P.K.; Sridhar, K. Sustainable production of cellulosic biopolymers for enhanced smart food packaging: An up-to-date review. *Int. J. Biol. Macromol.* **2024**, *273*, 133090. [CrossRef]
82. Celikci, N.; Ziba, C.A.; Dolaz, M.; Tümer, M. Comparison of composite resins containing UV light-sensitive chitosan derivatives in stereolithography (SLA)-3D printers. *Int. J. Biol. Macromol.* **2024**, 136057. [CrossRef]
83. Gruhn, P.; Koske, D.; Storck, J.L.; Ehrmann, A. Three-dimensional printing by vat photopolymerization on textile fabrics: Method and mechanical properties of the textile/polymer composites. *Textiles* **2024**, *4*, 417–425. [CrossRef]
84. Abdul, W.; Durgaiahsangam, H.P.; Jyosyula, S.K.R.; Khan, M.A. 3D Printing of Polymer and Polymer Matrix Composites. In *Polymer Composites: Fundamentals and Applications*; Springer: Singapore, 2024; pp. 281–300.
85. Alshihabi, M.; Kayacan, M.Y. Effect of nanosized carbon nanotubes, Titanium Nitride and cubic Boron Nitride powders on mechanical and thermal properties of SLA 3D printed resin composites. *Polym. Compos.* 2024, *Early View*.
86. Vázquez-Hernández, C.; Ramos-Galicia, L.; Velasco-Santos, C.; Bertolacci, L.; Zahid, M.; Yañez-Limón, J.M.; Perotto, G.; Martinez-Hernandez, A.L. Effect of keratin-rich fibers from rabbit hair in two polymers processed using additive manufacturing: FDM and SLA. *J. Manuf. Process.* **2024**, *120*, 1104–1114. [CrossRef]
87. Wawryniuk, Z.; Brancewicz-Steinmetz, E.; Sawicki, J. Revolutionizing transportation: An overview of 3D printing in aviation, automotive, and space industries. *Int. J. Adv. Manuf. Technol.* **2024**, *134*, 3083–3105. [CrossRef]
88. Nabavi, S.F.; Dalir, H. A review on laser-assisted manufacturing process of thermoset composites: A review of fundamentals, processes, scientific modelling, challenges and prospective. *Opt. Laser Technol.* **2025**, *181*, 111713. [CrossRef]
89. Biswas, A.; Singh, A.K.; Das, D. Stereolithography-Based Polymer Additive Manufacturing Process for Microfluidics Devices: A Review. In *Advances in Additive Manufacturing*; Wiley: Hoboken, NJ, USA, 2024; pp. 237–268.
90. Demoly, F.; André, J.C. 3D stereolithography of polymer matrix composites. In *Additive Manufacturing of Polymer-Based Composite Materials*; Elsevier: Amsterdam, The Netherlands, 2024; pp. 247–280.
91. Modica, F.; Basile, V.; Fassi, I. An Experiment-Based Variable Compensation Method to Improve the Geometric Accuracy of Sub-Mm Features Fabricated by Stereolithography (SLA). *J. Manuf. Mater. Process.* **2024**, *8*, 90. [CrossRef]
92. Rooney, K.; Dong, Y.; Pramanik, A.; Basak, A.K. Additive Manufacturing in Australian Small to Medium Enterprises: Vat Polymerisation Techniques, Case Study and Pathways to Industry 4.0 Competitiveness. *J. Manuf. Mater. Process.* **2023**, *7*, 168. [CrossRef]
93. Ma, R.; Liu, R.; Lu, B. Approaches Used to Design Support Structures for Ceramic Additive Manufacturing: A Review. *Addit. Manuf. Front.* **2024**, *3*, 200136. [CrossRef]
94. Vinay, D.; Keshavamurthy, R.; Tambrallimath, V. Enhanced mechanical properties of metal filled 3D printed polymer composites. *J. Inst. Eng. (India) Ser. D* **2023**, *104*, 181–195. [CrossRef]
95. Shah, M.; Ullah, A.; Azher, K.; Ur Rehman, A.; Akturk, N.; Juan, W.; Tüfekci, C.S.; Salamci, M.U. The influence of nanoparticle dispersions on mechanical and thermal properties of polymer nanocomposites using SLA 3D printing. *Crystals* **2023**, *13*, 285. [CrossRef]
96. Golhin, A.P.; Tonello, R.; Frisvad, J.R.; Grammatikos, S.; Strandlie, A. Surface roughness of as-printed polymers: A comprehensive review. *Int. J. Adv. Manuf. Technol.* **2023**, *127*, 987–1043. [CrossRef]
97. Sun, Y.; Cui, J.; Feng, S.; Cui, J.; Guo, Y.; Liang, C.; Gao, W.; Lu, Z.; Liu, F.; Zhang, B. Projection Stereolithography 3D Printing High-Conductive Hydrogel for Flexible Passive Wireless Sensing. *Adv. Mater.* **2024**, *36*, 2400103. [CrossRef]
98. Carvalho, S.S.; Reis, J.R.; Caldeirinha, R.F. A State-of-the-Art Review on 4D Printed Antennas and Other Adaptable Designs. *IEEE Access* **2024**, *12*, 62861–62881. [CrossRef]
99. Zhou, S.; Liu, G.; Wang, C.; Zhang, Y.; Yan, C.; Shi, Y. Thermal debinding for stereolithography additive manufacturing of advanced ceramic parts: A comprehensive review. *Mater. Des.* **2024**, *238*, 112632. [CrossRef]
100. Meana, V.; Zapico, P.; Cuesta, E.; Giganto, S.; Meana, L.; Martínez-Pellitero, S. Additive Manufacturing of Ceramic Reference Spheres by Stereolithography (SLA). *Appl. Sci.* **2024**, *14*, 7530. [CrossRef]
101. Kulkarni, O.; Enriquez-Cabrera, A.; Yang, X.; Foncy, J.; Nicu, L.; Molnár, G.; Salmon, L. Stereolithography 3D Printing of Stimuli-Responsive Spin Crossover@ Polymer Nanocomposites with Optimized Actuating Properties. *Nanomaterials* **2024**, *14*, 1243. [CrossRef] [PubMed]
102. Ravi, P.; Patel, P. Stereolithography (SLA) in pharmaceuticals. In *Additive Manufacturing in Pharmaceuticals*; Springer: Singapore, 2023; pp. 97–123.
103. Curti, C.; Kirby, D.J.; Russell, C.A. Systematic screening of photopolymer resins for stereolithography (SLA) 3D printing of solid oral dosage forms: Investigation of formulation factors on printability outcomes. *Int. J. Pharm.* **2024**, *653*, 123862. [CrossRef]
104. Wang, J.; Wang, Y.; Wang, R.; Wang, Q.; Wen, M.; Wang, J.; Sheng, L.; Zheng, Y.; Xi, T. A Review on 3D Printing Processes in Pharmaceutical Engineering and Tissue Engineering: Applications, Trends and Challenges. *Adv. Mater. Technol.* **2024**, 2400620. [CrossRef]
105. Nizam, M.; Purohit, R.; Taufik, M. 3D printing in healthcare: A review on drug printing, challenges and future perspectives. *Mater. Today Commun.* **2024**, *40*, 110199. [CrossRef]
106. Ullah, M.; Wahab, A.; Khan, S.U.; Naeem, M.; ur Rehman, K.; Ali, H.; Ullah, A.; Khan, A.; Khan, N.R.; Rizg, W.Y.; et al. 3D printing technology: A new approach for the fabrication of personalized and customized pharmaceuticals. *Eur. Polym. J.* **2023**, *195*, 112240. [CrossRef]

107. Tyagi, N.; Bhardwaj, V.; Sharma, D.; Tomar, R.; Chaudhary, V.; Khanuja, M.; Singh, M.K.; Sharma, G. 3D printing technology in the pharmaceutical and biomedical applications: A critical review. *Biomed. Mater. Devices* **2024**, *2*, 178–190. [CrossRef]
108. Kulkarni, V.R.; Saha, T.; Giri, B.R.; Lu, A.; Das, S.C.; Maniruzzaman, M. Recent Advancements in Pharmaceutical 3D Printing Industry. *J. Drug Deliv. Sci. Technol.* **2024**, *100*, 106072. [CrossRef]
109. Sultana, N.; Ali, A.; Waheed, A.; Aqil, M. 3D Printing in pharmaceutical manufacturing: Current status and future prospects. *Mater. Today Commun.* **2023**, *38*, 107987. [CrossRef]
110. Elbadawi, M.; Basit, A.W.; Gaisford, S. Energy consumption and carbon footprint of 3D printing in pharmaceutical manufacture. *Int. J. Pharm.* **2023**, *639*, 122926. [CrossRef]
111. Feng, S.; Repka, M.A. Future Prospects Including Novel Polymeric Excipients for 3D Printing of Pharmaceutical and Biomedical Applications. In *3D Printing: Emerging Technologies and Functionality of Polymeric Excipients in Drug Product Development*; Springer: Cham, Switzerland, 2023; pp. 273–286.
112. Rouway, M.; Nachtane, M.; Tarfaoui, M.; Sbai, S.J. 3D Printing of a Tidal Turbine Blade Using Two Methods of SLS and FFF of a Reinforced PA12 Composite: A Comparative Study. *Sustain. Mar. Struct.* **2024**, *6*, 1–19. [CrossRef]
113. Tang, H.; Zhang, S.; He, L.; Yang, Z.; Liu, T. 3D printing of high-stiffness and high-strength glass fiber reinforced PEEK composites by selective laser sintering. *Compos. Part A Appl. Sci. Manuf.* **2024**, *187*, 108470. [CrossRef]
114. Nobre, L.; Barros, D.; Bessa, J.; Cunha, F.; Machado, M.; Mendonça, J.P.; Luís, J.; Oliveira, M.; Machado, P.; Fernandes, C.; et al. Enhancing mechanical performance in SLS-printed PA12-slate composites through amino-silane treatment of mineral waste. *Int. J. Adv. Manuf. Technol.* **2024**, *134*, 2979–2992. [CrossRef]
115. Wang, Y.; Ding, Y.; Yu, K.; Dong, G. Innovative polymer-based composite materials in additive manufacturing: A review of methods, materials, and applications. *Polym. Compos.* 2024, *Early View*.
116. Daneshdoost, N.; Peloquin, J.; Gall, K. Structure-performance relationships of multi-material jetting polymeric composites designed at the voxel scale: Distribution and composition effects. *J. Manuf. Process.* **2024**, *131*, 2118–2132. [CrossRef]
117. Tan, P.; Zhou, M.; Tang, C.; Zhou, K. A powder-scale multiphysics framework for powder bed fusion of fiber-reinforced polymer composites. *Adv. Powder Mater.* **2024**, *3*, 100190. [CrossRef]
118. Zhang, G.; Zheng, X.; Wang, Q.; Ni, Y.; Liu, F.; Zhao, K.; Xu, L.R. Limitations of composite strength theory for predicting the ultimate strengths of layered 3D printing polymers. *Compos. Part A Appl. Sci. Manuf.* **2024**, *185*, 108288. [CrossRef]
119. Gajbhiye, T.S.; Waghmare, S.; Dhande, M.; Gondane, R.; Giripunje, M.; Shelare, S.; Belkhode, P. Polymer composite additive manufacturing: Applications, challenges and opportunities. *Mater. Today Proc.* 2024, *in press*.
120. Zhang, S.; Tang, H.; Tang, D.; Liu, T.; Liao, W. Effect of fabrication process on the microstructure and mechanical performance of carbon fiber reinforced PEEK composites via selective laser sintering. *Compos. Sci. Technol.* **2024**, *246*, 110396. [CrossRef]
121. Colucci, G.; Lupone, F.; Bondioli, F.; Messori, M. 3D printing of PBAT-based composites filled with agro-wastes via selective laser sintering. *Eur. Polym. J.* **2024**, *215*, 113197. [CrossRef]
122. Adak, N.C.; Sharia, F.; Lee, W. Laser-Assisted Additive Manufacturing Techniques for Advanced Composites. In *Advances in Additive Manufacturing*; Wiley: Hoboken, NJ, USA, 2024; pp. 217–235.
123. Joshua, R.J.N.; Raj, S.A.; Hameed Sultan, M.T.; Łukaszewicz, A.; Józwik, J.; Oksiuta, Z.; Dziedzic, K.; Tofil, A.; Shahar, F.S. Powder Bed Fusion 3D Printing in Precision Manufacturing for Biomedical Applications: A Comprehensive Review. *Materials* **2024**, *17*, 769. [CrossRef]
124. Yang, Y.; Bharech, S.; Finger, N.; Zhou, X.; Schröder, J.; Xu, B.X. Elasto-plastic residual stress analysis of selective laser sintered porous materials based on 3D-multilayer thermo-structural phase-field simulations. *npj Comput. Mater.* **2024**, *10*, 117. [CrossRef]
125. Bertolini, F.; Mariani, M.; Mercadelli, E.; Baldisserri, C.; Galassi, C.; Capiani, C.; Ardito, R.; Lecis, N. 3D printing of potassium sodium niobate by binder jetting: Printing parameters optimisation and correlation to final porosity. *J. Mater. Res. Technol.* **2024**, *29*, 4597–4606. [CrossRef]
126. Song, Y.; Ghafari, Y.; Asefnejad, A.; Toghraie, D. An overview of selective laser sintering 3D printing technology for biomedical and sports device applications: Processes, materials, and applications. *Opt. Laser Technol.* **2024**, *171*, 110459. [CrossRef]
127. Azam, M.U.; Belyamani, I.; Schiffer, A.; Kumar, S.; Askar, K. Progress in selective laser sintering of multifunctional polymer composites for strain-and self-sensing applications. *J. Mater. Res. Technol.* **2024**, *30*, 9625–9646. [CrossRef]
128. Han, S.; Li, S.; Song, X.; Zhou, Z.; Meng, Q.; Araby, S.; Abdelsalam, A.A. Carbon nanotubes/α-ZrP sheets for high mechanical performance and flame-retarding polyamides using selective laser sintering. *Virtual Phys. Prototyp.* **2024**, *19*, e2368644. [CrossRef]
129. Cai, R.; Luo, X.; Xie, G.; Wang, K.; Peng, Y.; Rao, Y. Effects of the printing parameters on geometric accuracy and mechanical properties of digital light processing printed polymer. *J. Mater. Sci.* **2024**, *59*, 14807–14819. [CrossRef]
130. Cortés, A.; Bañón-Veracruz, M.; Jiménez-Suárez, A.; Campo, M.; Prolongo, M.; Prolongo, S. Enhancing efficiency and sustainability of digital light processing 3D-Printing by novel two-stage processing of carbon nanotube reinforced nanocomposites. *J. Mater. Res. Technol.* **2024**, *29*, 3237–3248. [CrossRef]
131. Cheng, J.; Yu, S.; Wang, R.; Ge, Q. Digital light processing based multimaterial 3D printing: Challenges, solutions and perspectives. *Int. J. Extrem. Manuf.* **2024**, *6*, 042006. [CrossRef]
132. Hussain, M.I.; Xia, M.; Ren, X.; Ge, C.; Jamil, M.; Gupta, M.K. Digital light processing 3D printing of ceramic materials: A review on basic concept, challenges, and applications. *Int. J. Adv. Manuf. Technol.* **2024**, *130*, 2241–2267. [CrossRef]

133. Senthooran, V.; Weng, Z.; Wu, L. Enhancing Mechanical and Thermal Properties of 3D-Printed Samples Using Mica-Epoxy Acrylate Resin Composites—Via Digital Light Processing (DLP). *Polymers* **2024**, *16*, 1148. [CrossRef]
134. Magalhães, F.d.C.; Rubio, J.C.C. Mechanical Properties of Recycled Carbon Fiber-Reinforced Resin Composites 3D Printed via Digital Light Processing. *J. Mater. Eng. Perform.* **2024**, 1–11. [CrossRef]
135. Guessasma, S.; Stephant, N.; Durand, S.; Belhabib, S. Digital Light Processing Route for 3D Printing of Acrylate-Modified PLA/Lignin Blends: Microstructure and Mechanical Performance. *Polymers* **2024**, *16*, 1342. [CrossRef] [PubMed]
136. Lee, D.; Kim, D.H.; Kim, H.; Seung, H.M.; Song, H.C.; Kim, M. Engineering Digital Light Processing Ceramic Composites for Wide-Range Flexible Sensing Arrays. *Compos. Part B Eng.* **2024**, *283*, 111595. [CrossRef]
137. Guo, A.; Li, S.; Wang, S.; Zhai, Z.; Qu, P.; Guo, S.; Kong, H.; Tang, R.; Liu, C.; Han, W.; et al. Digital light processing of multi-walled carbon nanotubes-reinforced photosensitive resin composites: Effects on microstructures and mechanical properties. *J. Mater. Res. Technol.* **2024**, *28*, 434–445. [CrossRef]
138. Swetha, S.; Sahiti, T.J.; Priya, G.S.; Harshitha, K.; Anil, A. Review on digital light processing (DLP) and effect of printing parameters on quality of print. *Interactions* **2024**, *245*, 178. [CrossRef]
139. Melentiev, R.; Harakály, G.; Stögerer, J.; Mitteramskogler, G.; Wagih, A.; Lubineau, G.; Grande, C.A. High-resolution metal 3D printing via digital light processing. *Addit. Manuf.* **2024**, *85*, 104156. [CrossRef]
140. Wang, Y.; Su, R.; Chen, J.; Wang, W.; Zhang, X.; Xu, H.; He, R. 3D Printed Bioinspired Flexible Absorber: Toward High-Performance Electromagnetic Absorption at 75–110 GHz. *ACS Appl. Mater. Interfaces* **2023**, *15*, 53996–54005. [CrossRef]
141. Khorasani, M.; MacDonald, E.; Downing, D.; Ghasemi, A.; Leary, M.; Dash, J.; Sharabian, E.G.; Almalki, A.; Brandt, M.; Bateman, S. Multi Jet Fusion (MJF) of polymeric components: A review of process, properties and opportunities. *Addit. Manuf.* **2024**, *91*, 104331. [CrossRef]
142. Kafi, A.; Khorasani, M.; Downing, D.; Rashed, K.; Leary, M.; Zhang, D.; Bateman, S. Prediction of absorptivity in Multi-Jet Fusion manufactured polypropylene structures through laser flash and corrected porosity method. *Int. J. Adv. Manuf. Technol.* **2024**, *130*, 4041–4052. [CrossRef]
143. Alomarah, A.; Abbas, A.T.; Faisal, B.; Peng, Z.; Ruan, D. The Effects of Manufacturing Techniques on the Mechanical Performance of an Auxetic Structure Manufactured by Fused Filament Fabrication and Multijet Fusion Processes. *Adv. Eng. Mater.* **2024**, *26*, 2302033. [CrossRef]
144. Ráž, K.; Chval, Z.; Kemka, V. Parametric Production of Prostheses Using the Additive Polymer Manufacturing Technology Multi Jet Fusion. *Materials* **2024**, *17*, 2347. [CrossRef]
145. Lupone, F.; Padovano, E.; Lambertini, V.G.; Sampieri, R.; Casamento, F.; Zecchi, S.; Badini, C. Selective Laser Sintering versus Multi Jet Fusion: A Comprehensive Comparison Study Based on the Properties of Glass Beads-Reinforced Polyamide 12. *Adv. Eng. Mater.* **2024**, *26*, 2301345. [CrossRef]
146. Avanzini, A.; Tomasoni, M.; Xu, Z.; Berto, F.; Razavi, N. Fracture assessment of polyamide 12 (PA12) specimens fabricated via Multi Jet FusionTM in the presence of geometrical discontinuities. *Eng. Fract. Mech.* **2024**, *303*, 110118. [CrossRef]
147. Seah, Y.T. *Multi Jet Fusion of ZnO Nanorod-Reinforced PA12 Nanocomposites*; Nanyang Technological University: Singapore, 2024.
148. Tang, C.; Liu, J.; Qiao, J.; Wei, Y.; Shi, C.; Hao, W. The preparation and axial compressive properties of 3D-printed polymer lattice-reinforced cementitious composite columns. *J. Build. Eng.* **2024**, *97*, 110770. [CrossRef]
149. Gao, P.; Liu, R.; Wang, M.; Zhang, H.; Gao, X.; Ma, J. Static and Fatigue Behaviors of Polyamide 12 Hinge Joint Manufactured by Multi-Jet Fusion. *J. Mater. Eng. Perform.* **2024**, 1–12. [CrossRef]
150. Kim, W.; Hwang, J.Y.; Ji, C.H. Large deflection angle resonant 1D scanning micromirror fabricated with multi jet fusion 3D printed parts. *Int. J. Optomechatronics* **2024**, *18*, 2371289. [CrossRef]
151. Bochnia, J.; Kozior, T.; Szot, W.; Rudnik, M.; Zmarzły, P.; Gogolewski, D.; Szczygieł, P.; Musiałek, M. Selected Mechanical and Rheological Properties of Medical Resin MED610 in PolyJet Matrix Three-Dimensional Printing Technology in Quality Aspects. *3D Print. Addit. Manuf.* **2024**, *11*, 299–313. [CrossRef]
152. Nachimuthu, M.; PK, R. Inkjet four-dimensional printing of shape memory polymers: A review. *Rapid Prototyp. J.* **2023**, *29*, 437–446. [CrossRef]
153. Xin, Y.; Zhou, X.; Bark, H.; Lee, P.S. The role of 3D printing technologies in soft grippers. *Adv. Mater.* **2024**, *36*, 2307963. [CrossRef]
154. Tee, Y.L.; Tran, P.; Leary, M.; Pille, P.; Brandt, M. 3D Printing of polymer composites with material jetting: Mechanical and fractographic analysis. *Addit. Manuf.* **2020**, *36*, 101558. [CrossRef]
155. Nguyen, T.T.; Kim, J. 4D-printing—Fused deposition modeling printing and PolyJet printing with shape memory polymers composite. *Fibers Polym.* **2020**, *21*, 2364–2372. [CrossRef]
156. Tee, Y.L.; Peng, C.; Pille, P.; Leary, M.; Tran, P. PolyJet 3D printing of composite materials: Experimental and modelling approach. *JOM* **2020**, *72*, 1105–1117. [CrossRef]
157. Muthuram, N.; Madhav, P.S.; Vasan, D.K.; Mohan, M.E.; Prajeeth, G. A review of recent literatures in poly jet printing process. *Mater. Today Proc.* **2022**, *68*, 1906–1920. [CrossRef]
158. Wu, C.; Do, T.T.; Tran, P. Mechanical properties of polyjet 3d-printed composites inspired by space-filling peano curves. *Polymers* **2021**, *13*, 3516. [CrossRef] [PubMed]
159. Kozior, T.; Mamun, A.; Trabelsi, M.; Sabantina, L. Comparative analysis of polymer composites produced by FFF and PJM 3D printing and electrospinning technologies for possible filter applications. *Coatings* **2022**, *12*, 48. [CrossRef]

160. Palanisamy, C.; Raman, R.; Dhanraj, P.K. Additive manufacturing: A review on mechanical properties of polyjet and FDM printed parts. *Polym. Bull.* **2022**, *79*, 7065–7116. [CrossRef]
161. Conway, C.H.; McGregor, D.J.; Antonsen, T.; Wood, C.; Shao, C.; King, W.P. Geometry repeatability and prediction for personalized medical devices made using multi-jet fusion additive manufacturing. *Addit. Manuf. Lett.* **2024**, *9*, 100200. [CrossRef]
162. Patpatiya, P.; Chaudhary, K.; Shastri, A.; Sharma, S. A review on polyjet 3D printing of polymers and multi-material structures. *Proc. Inst. Mech. Eng. Part C J. Mech. Eng. Sci.* **2022**, *236*, 095440622210795. [CrossRef]
163. Azpiazu-Flores, F.X.; Elfana, A.; Yang, C.C.; Morton, D.; Lin, W.S. Effect of artificial aging and different surface finishing protocols on the flexural strength and surface hardness of a photopolymer for manufacturing monolithic polychromatic complete dentures using PolyJet 3D printing. *J. Prosthodont.* 2024, *Early View*.
164. Krause, M.; Marshall, A.; Catterlin, J.K.; Hornik, T.; Kartalov, E.P. Dimensional Fidelity and Orientation Effects of PolyJet Technology in 3D Printing of Negative Features for Microfluidic Applications. *Micromachines* **2024**, *15*, 389. [CrossRef]
165. Aberdeen, N.J.; Forghani, K.; Sochol, R.D. Brief Paper: Geometric Determinants of Material Jetting-Enabled Bi-Material Interface Integrity Using Polyjet 3D Printing. In Proceedings of the International Manufacturing Science and Engineering Conference, Knoxville, TN, USA, 17–21 June 2024; Volume 88100, p. V001T01A007.
166. Tang, T.; Joralmon, D.; Tiwari, L.; Ravishekar, R.; He, Q.; Pham, L.; Yang, Y.; Li, X. Additive manufacturing of polymer matrix composites via direct ink writing process. In *Additive Manufacturing of Polymer-Based Composite Materials*; Elsevier: Amsterdam, The Netherlands, 2024; pp. 203–245.
167. Chen, C.T.; Chen, Z.L. Direct ink writing of polymer matrix composite with carbon for driving a flexible thermoelectric actuator of shape memory polymer. *Mater. Today Commun.* **2024**, *38*, 108303. [CrossRef]
168. Sole-Gras, M.; Ren, B.; Ryder, B.J.; Ge, J.; Huang, J.; Chai, W.; Yin, J.; Fuchs, G.E.; Wang, G.; Jiang, X.; et al. Vapor-induced phase-separation-enabled versatile direct ink writing. *Nat. Commun.* **2024**, *15*, 3058. [CrossRef]
169. Lin, Z.; Qiu, X.; Cai, Z.; Li, J.; Zhao, Y.; Lin, X.; Zhang, J.; Hu, X.; Bai, H. High internal phase emulsions gel ink for direct-ink-writing 3D printing of liquid metal. *Nat. Commun.* **2024**, *15*, 4806. [CrossRef] [PubMed]
170. Ingram, M.; Campbell, E.; Molotnikov, A.; Feih, S.; Zhong, Y.L. Recent Advances in Multifunctional Polymer/2D Nanocomposite Development for Fused Filament Fabrication and Direct Ink Writing of Electrically and Thermally Conductive Components. *Adv. Nanocomposites* **2024**, *1*, 157–170. [CrossRef]
171. Bao, X.; Meng, J.; Tan, Z.; Zhang, C.; Li, L.; Liu, T. Direct-ink-write 3D printing of highly-stretchable polyaniline gel with hierarchical conducting network for customized wearable strain sensors. *Chem. Eng. J.* **2024**, *491*, 151918. [CrossRef]
172. Zu, W.; Carranza, H.E.; Bartlett, M.D. Enhancing Electrical Conductivity of Stretchable Liquid Metal–Silver Composites through Direct Ink Writing. *ACS Appl. Mater. Interfaces* **2024**, *16*, 23895–23903. [CrossRef]
173. Bhardwaj, D.; Singhmar, R.; Garg, M.; Gupta, D.; Dhiman, A.; Han, S.S.; Agrawal, G. Designing advanced hydrogel inks with direct ink writing based 3D printability for engineered biostructures. *Eur. Polym. J.* **2024**, *205*, 112736. [CrossRef]
174. Li, M.; Huang, S.; Willems, E.; Soete, J.; Inokoshi, M.; Van Meerbeek, B.; Vleugels, J.; Zhang, F. UV-Curing Assisted Direct Ink Writing of Dense, Crack-Free, and High-Performance Zirconia-Based Composites With Aligned Alumina Platelets. *Adv. Mater.* **2024**, *36*, 2306764. [CrossRef]
175. Khecho, A.; Hewlin, R., Jr.; Joyee, E.B. Direct Ink Writing of SiC-Fe3O4-Polymer Nanocomposites for Magnetic Humidity Sensors: Rheology Characterization. In Proceedings of the International Manufacturing Science and Engineering Conference, Knoxville, TN, USA, 17–21 June 2024; Volume 88100, p. V001T02A005.
176. dos Santos, V.I.; Chevalier, J.; Fredel, M.C.; Henriques, B.; Gremillard, L. Ceramics and ceramic composites for biomedical engineering applications via Direct Ink Writing: Overall scenario, advances in the improvement of mechanical and biological properties and innovations. *Mater. Sci. Eng. R Rep.* **2024**, *161*, 100841. [CrossRef]
177. Li, S.; Zhang, H.; Sun, L.; Zhang, X.; Guo, M.; Liu, J.; Wang, W.; Zhao, N. 4D printing of biological macromolecules employing handheld bioprinters for in situ wound healing applications. *Int. J. Biol. Macromol.* **2024**, *280*, 135999. [CrossRef]
178. Xu, P.; Zhu, L.; Zhao, Z.; Yang, Z.; Ning, J.; Xue, P.; Lu, H. Anisotropy behavior of liquid metal elastomer composites with both enhanced thermal conductivity and crack resistance by direct ink writing. *Compos. Part A Appl. Sci. Manuf.* **2024**, *177*, 107890. [CrossRef]
179. Abas, M.; Jan, Z.; Rahman, K. *Direct Ink Writing*; CRC Press: Boca Raton, FL, USA, 2023; pp. 188–209. [CrossRef]
180. Abas, M.; Rahman, K. Fabrication of flex sensors through direct ink write technique and its electrical characterization. *Appl. Phys. A* **2016**, *122*, 972. [CrossRef]
181. Baniasadi, H.; Abidnejad, R.; Fazeli, M.; Lipponen, J.; Niskanen, J.; Kontturi, E.; Seppälä, J.; Rojas, O.J. Innovations in hydrogel-based manufacturing: A comprehensive review of direct ink writing technique for biomedical applications. *Adv. Colloid Interface Sci.* **2024**, *324*, 103095. [CrossRef] [PubMed]
182. van Hazendonk, L.S.; Vonk, C.F.; van Grondelle, W.; Vonk, N.H.; Friedrich, H. Towards a predictive understanding of direct ink writing of graphene-based inks. *Appl. Mater. Today* **2024**, *36*, 102014. [CrossRef]
183. Elumalai, D.; Hosseinnezhad, R.; Bondarenko, V.; Morawiec, J.; Vozniak, I.; Galeski, A. Shape Memory Polymer Foam Based on Nanofibrillar Composites of Polylactide/Polyamide. *Molecules* **2024**, *29*, 5045. [CrossRef]

184. Brossier, T.; Habib, M.; Benkhaled, T.; Volpi, G.; Lapinte, V.; Blanquer, S. 4D printing of hydrogels based on poly(oxazoline) and poly(acrylamide) copolymers by stereolithography. *Mater. Adv.* **2024**, *5*, 2750–2758. [CrossRef]
185. Sindhu, P.M.; Ranjani, M.; Kiran, M.G.; Sachin, M.S. 4D Food Printing: Transforming Culinary Creation with Time-Responsive Technology. *Agrisustain-Int. J.* **2023**, *1*, 34–41. [CrossRef]
186. Khalid, M.Y.; Arif, Z.U.; Noroozi, R.; Zolfagharian, A.; Bodaghi, M. 4D printing of shape memory polymer composites: A review on fabrication techniques, applications, and future perspectives. *J. Manuf. Process.* **2022**, *81*, 759–797. [CrossRef]
187. Qiu, W.; Xu, X.; Dong, K.; Wang, Y.; Xiong, Y. Recent advances in 4D printing of fiber-reinforced polymer composites: A review and outlook. *Compos. Part B Eng.* **2024**, *283*, 111645. [CrossRef]
188. Ahmad, H.; Markina, A.; Porotnikov, M.; Ahmad, F. A review of carbon fiber materials in automotive industry. *IOP Conf. Ser. Mater. Sci. Eng.* **2020**, *971*, 032011. [CrossRef]
189. Wang, W.; Zhang, Y.; Mo, Z.; Chouw, N.; Jayaraman, K.; Xu, Z.d. A critical review on the properties of natural fibre reinforced concrete composites subjected to impact loading. *J. Build. Eng.* **2023**, *77*, 107497. [CrossRef]
190. Sadeghi, B.; Cavaliere, P.D. Reviewing the integrated design approach for augmenting strength and toughness at macro-and micro-scale in high-performance advanced composites. *Materials* **2023**, *16*, 5745. [CrossRef]
191. Sadeghian, H.; Ayatollahi, M.R.; Khosravani, M.R.; Razavi, N. From prototyping to functional parts: A review of mechanical reinforcing procedures for polymeric parts fabricated via material extrusion. *J. Manuf. Process.* **2024**, *130*, 1–34. [CrossRef]
192. Ali, Z.; Yaqoob, S.; Yu, J.; D'Amore, A. Critical review on the characterization, preparation, and enhanced mechanical, thermal, and electrical properties of carbon nanotubes and their hybrid filler polymer composites for various applications. *Compos. Part Open Access* **2024**, *13*, 100434. [CrossRef]
193. Younus, M.M.; Naguib, H.M.; Fekry, M.; Elsawy, M.A. Pushing the limits of PLA by exploring the power of MWCNTs in enhancing thermal, mechanical properties, and weathering resistance. *Sci. Rep.* **2023**, *13*, 16588. [CrossRef] [PubMed]
194. Kotsilkova, R.; Petrova-Doycheva, I.; Menseidov, D.; Ivanov, E.; Paddubskaya, A.; Kuzhir, P. Exploring thermal annealing and graphene-carbon nanotube additives to enhance crystallinity, thermal, electrical and tensile properties of aged poly (lactic) acid-based filament for 3D printing. *Compos. Sci. Technol.* **2019**, *181*, 107712. [CrossRef]
195. De Bortoli, L.; De Farias, R.; Mezalira, D.; Schabbach, L.; Fredel, M. Functionalized carbon nanotubes for 3D-printed PLA-nanocomposites: Effects on thermal and mechanical properties. *Mater. Today Commun.* **2022**, *31*, 103402. [CrossRef]
196. PARIN, F.N.; DEMIRCI, F. Durability of polymer composite materials for high-temperature applications. In *Aging and Durability of FRP Composites and Nanocomposites*; Elsevier: Amsterdam, The Netherlands, 2024; pp. 135–170.
197. Hassine, S.; Chatti, S.; Louhichi, B.; Seibi, A. Experimental Study of the Tensile Behavior of Structures Obtained by FDM 3D Printing Process. *Polymers* **2024**, *16*, 1562. [CrossRef]
198. Li, M.; Hu, Z.; Yan, B.; Wang, J.; Zhang, H.; Ye, F.; Sun, B.; Liu, J.; Li, Y.; Ding, G.; et al. A flexible resistive strain gauge with reduced temperature effect via thermal expansion anisotropic composite substrate. *Microsystems Nanoeng.* **2024**, *10*, 129. [CrossRef]
199. Jia, X.; Luo, J.; Luo, Q.; Li, Q.; Pang, T. Experimental study on the effects of temperature on mechanical properties of 3D printed continuous carbon fiber reinforced polymer (C-CFRP) composites. *Thin-Walled Struct.* **2024**, *205*, 112465. [CrossRef]
200. Tarhini, A.; Tehrani-Bagha, A. Advances in preparation methods and conductivity properties of graphene-based polymer composites. *Appl. Compos. Mater.* **2023**, *30*, 1737–1762. [CrossRef]
201. Wagmare, R.; Harshe, R.; Pednekar, J.; Patro, T.U. Additive manufacturing of continuous fiber-reinforced polymer composites: Current trend and future directions. *Prog. Addit. Manuf.* **2024**, 1–28. [CrossRef]
202. Yan, Y.; Jiang, Y.; Ng, E.L.L.; Zhang, Y.; Owh, C.; Wang, F.; Song, Q.; Feng, T.; Zhang, B.; Li, P.; et al. Progress and opportunities in additive manufacturing of electrically conductive polymer composites. *Mater. Today Adv.* **2023**, *17*, 100333. [CrossRef]
203. Yan, Y.; Han, M.; Jiang, Y.; Ng, E.L.L.; Zhang, Y.; Owh, C.; Song, Q.; Li, P.; Loh, X.J.; Chan, B.Q.Y.; et al. Electrically Conductive Polymers for Additive Manufacturing. *ACS Appl. Mater. Interfaces* **2024**, *16*, 5337–5354. [CrossRef] [PubMed]
204. Bas, J.; Dutta, T.; Llamas Garro, I.; Velázquez-González, J.S.; Dubey, R.; Mishra, S.K. Embedded Sensors with 3D Printing Technology. *Sensors* **2024**, *24*, 1955. [CrossRef]
205. He, X.; Zhang, B.; Liu, Q.; Chen, H.; Cheng, J.; Jian, B.; Yin, H.; Li, H.; Duan, K.; Zhang, J.; et al. Highly conductive and stretchable nanostructured ionogels for 3D printing capacitive sensors with superior performance. *Nat. Commun.* **2024**, *15*, 6431. [CrossRef]
206. Carvalho, S.S.; Reis, J.R.; Mateus, A.; Caldeirinha, R.F. Exploring design approaches for 3D printed antennas. *IEEE Access* **2024**, *12*, 10718–10735. [CrossRef]
207. Dontha, B.; Moulod, M.; Balbaugh, S.; Hoelzle, D.; Li, J.; Miranda, F.A.; Kiourti, A. RF characterization of a photocurable PEDOT: PSS: PEGDA conductive biomaterial for 3D-Printing implantable antennas. *IEEE Trans. Antennas Propag.* **2024**, *72*, 2825–2830. [CrossRef]
208. Saghafi, M.K.; Vasantham, S.K.; Hussain, N.; Mathew, G.; Colombo, F.; Schamberger, B.; Pohl, E.; Marques, G.C.; Breitung, B.; Tanaka, M.; et al. Printed Electronic Devices and Systems for Interfacing with Single Cells up to Organoids. *Adv. Funct. Mater.* **2024**, *34*, 2308613. [CrossRef]
209. Akram, S.; Ashraf, M.; Javid, A.; Abid, H.A.; Ahmad, S.; Nawab, Y.; Rasheed, A.; Xue, Z.; Nosheen, A. Recent advances in electromagnetic interference (EMI) shielding textiles: A comprehensive review. *Synth. Met.* **2023**, *294*, 117305. [CrossRef]
210. Maleki, A.H.; Zolfaghari, A. Investigation of electrical, electromagnetic interference shielding and tensile properties of 3D-printed acrylonitrile butadiene styrene/carbon nanotube composites. *J. Thermoplast. Compos. Mater.* **2024**, *37*, 2409–2424. [CrossRef]

211. Łapińska, A.; Grochowska, N.; Cieplak, K.; Płatek, P.; Wood, P.; Deuszkiewicz, P.; Dużyńska, A.; Sztorch, B.; Głowcka, J.; Przekop, R.; et al. Architecture influence on acoustic performance, EMI shielding, electrical and thermal, properties of 3D printed PLA/graphite/molybdenum disulfide composites. *Mater. Des.* **2024**, *245*, 113241. [CrossRef]
212. Randis, R.; Darmadi, D.B.; Gapsari, F.; Sonief, A.A.a.; Akpan, E.D.; Ebenso, E.E. The potential of nanocomposite-based coatings for corrosion protection of metals: A review. *J. Mol. Liq.* **2023**, *390*, 123067. [CrossRef]
213. Savadatti, M.B.; Kumar, K.; Jaya Christiyan, K.G.; Gajakosh, A.; Thakur, M.; Kumar, R.S.; Paulraj, R.L.; Madhusudhana, H.K. Machine Learning Applications for 3D-Printed Polymers and Their Composites. In *Development, Properties, and Industrial Applications of 3D Printed Polymer Composites*; IGI Global: Hershey, PA, USA, 2023; pp. 239–260. [CrossRef]
214. Ng, W.L.; An, J.; Chua, C.K. Process, material, and regulatory considerations for 3D printed medical devices and tissue constructs. *Engineering* **2024**, *36*, 146–166. [CrossRef]
215. Sajjad, R.; Chauhdary, S.T.; Anwar, M.T.; Zahid, A.; Khosa, A.A.; Imran, M.; Sajjad, M.H. A review of 4D printing–technologies, shape shifting, smart polymer based materials, and biomedical applications. *Adv. Ind. Eng. Polym. Res.* **2024**, *7*, 20–36. [CrossRef]
216. Sun, X.H.; Zhou, K.; Demoly, F.; Zhao, R.; Qi, H. Perspective: Machine Learning in Design for 3D/4D Printing. *J. Appl. Mech.* **2024**, *91*, 030801. [CrossRef]
217. Mahmood, A.; Akram, T.; Shenggui, C.; Chen, H. Revolutionizing manufacturing: A review of 4D printing materials, stimuli, and cutting-edge applications. *Compos. Part B Eng.* **2023**, *266*, 110952. [CrossRef]
218. Ostolaza, M.; Arrizubieta, J.I.; Lamikiz, A.; Plaza, S.; Ortega, N. Latest developments to manufacture metal matrix composites and functionally graded materials through AM: A state-of-the-art review. *Materials* **2023**, *16*, 1746. [CrossRef]
219. Wu, C.; Xu, F.; Wang, H.; Liu, H.; Yan, F.; Ma, C. Manufacturing technologies of polymer composites—A review. *Polymers* **2023**, *15*, 712. [CrossRef]
220. Hamdi, Y.; Lallemand, F.; Benita, S. Drug-loaded nanocarriers for back-of-the-eye diseases-formulation limitations. *J. Drug Deliv. Sci. Technol.* **2015**, *30*, 331–341. [CrossRef]
221. Elbadawi, M.; Castro, B.M.; Gavins, F.K.; Ong, J.J.; Gaisford, S.; Pérez, G.; Basit, A.W.; Cabalar, P.; Goyanes, A. M3DISEEN: A novel machine learning approach for predicting the 3D printability of medicines. *Int. J. Pharm.* **2020**, *590*, 119837. [CrossRef]
222. Saleh, E. 3D and 4D printed polymer composites for electronic applications. In *Processes, Applications, and Challenges*; Elsevier: Amsterdam, The Netherlands, 2020; pp. 505–525. [CrossRef]
223. Zhang, L.; Huang, X.; Cole, T.; Lu, H.; Hang, J.; Li, W.; Tang, S.Y.; Boyer, C.; Davis, T.; Qiao, R. 3D-printed liquid metal polymer composites as NIR-responsive 4D printing soft robot. *Nat. Commun.* **2023**, *14*, 7815. [CrossRef]
224. Peng, H.; Han, B.; Tong, T.; Jin, X.; Peng, Y.; Guo, M.; Li, B.; Ding, J.; Kong, Q.; Wang, Q. 3D printing processes in precise drug delivery for personalized medicine. *Biofabrication* **2024**, *16*, 032001. [CrossRef] [PubMed]
225. Ong, J.J.; Castro, B.M.; Gaisford, S.; Cabalar, P.; Basit, A.W.; Pérez, G.; Goyanes, A. Accelerating 3D printing of pharmaceutical products using machine learning. *Int. J. Pharm. X* **2022**, *4*, 100120. [CrossRef] [PubMed]
226. Peloquin, J.; Kirillova, A.; Rudin, C.; Brinson, L.; Gall, K. Prediction of tensile performance for 3D printed photopolymer gyroid lattices using structural porosity, base material properties, and machine learning. *Mater. Des.* **2023**, *232*, 112126. [CrossRef]
227. Tao, H.; Ahmadianfar, I.; Goliatt, L.; Kazmi, S.S.U.H.; Yassin, M.A.; Oudah, A.Y.; Homod, R.Z.; Togun, H.; Yaseen, Z.M. PM2.5 concentration forecasting: Development of integrated multivariate variational mode decomposition with kernel Ridge regression and weighted mean of vectors optimization. *Atmos. Pollut. Res.* **2024**, *15*, 102125. [CrossRef]
228. Khusheef, A.S.; Shahbazi, M.; Hashemi, R. Predicting 3D printed plastic part properties: A deep learning approach with thermographic and vibration data fusion. *Expert Syst. Appl.* **2024**, *255*, 124605. [CrossRef]
229. Monticeli, F.M.; Neves, R.M.; Ornaghi, H.L., Jr.; Almeida, J.H.S., Jr. Prediction of bending properties for 3D-printed carbon fibre/epoxy composites with several processing parameters using ANN and statistical methods. *Polymers* **2022**, *14*, 3668. [CrossRef]
230. Malley, S.; Reina, C.; Nacy, S.; Gilles, J.; Koohbor, B.; Youssef, G. Predictability of mechanical behavior of additively manufactured particulate composites using machine learning and data-driven approaches. *Comput. Ind.* **2022**, *142*, 103739. [CrossRef]
231. G, K.; Kandasubramanian, B. Exertions of Magnetic Polymer Composites Fabricated via 3D Printing. *Ind. Eng. Chem. Res.* **2022**, *61*, 16895–16909. [CrossRef]
232. Griffiths, C.; Howarth, J.; De Almeida-Rowbotham, G.; Rees, A.; Kerton, R. A design of experiments approach for the optimisation of energy and waste during the production of parts manufactured by 3D printing. *J. Clean. Prod.* **2016**, *139*, 74–85. [CrossRef]
233. Kafle, A.; Luis, E.; Silwal, R.; Pan, H.M.; Shrestha, P.; Bastola, A. 3D/4D Printing of Polymers: Fused Deposition Modelling (FDM), Selective Laser Sintering (SLS), and Stereolithography (SLA). *Polymers* **2021**, *13*, 3101. [CrossRef]
234. Sani, A.R.; Zolfagharian, A.; Kouzani, A.Z. Artificial Intelligence-Augmented Additive Manufacturing: Insights on Closed-Loop 3D Printing. *Adv. Intell. Syst.* **2024**, *6*, 2400102. [CrossRef]
235. Lu, L.; Hou, J.; Yuan, S.; Yao, X.; Li, Y.; Zhu, J. Deep learning-assisted real-time defect detection and closed-loop adjustment for additive manufacturing of continuous fiber-reinforced polymer composites. *Robot.-Comput.-Integr. Manuf.* **2023**, *79*, 102431. [CrossRef]
236. Narayanan, B.N.; Beigh, K.; Loughnane, G.; Powar, N. Support vector machine and convolutional neural network based approaches for defect detection in fused filament fabrication. In Proceedings of the Applications of Machine Learning, San Diego, CA, USA, 11–15 August 2019; Volume 11139, pp. 283–291.

237. Jin, Z.; Zhang, Z.; Gu, G.X. Automated real-time detection and prediction of interlayer imperfections in additive manufacturing processes using artificial intelligence. *Adv. Intell. Syst.* **2020**, *2*, 1900130. [CrossRef]
238. Charalampous, P.; Kostavelis, I.; Kopsacheilis, C.; Tzovaras, D. Vision-based real-time monitoring of extrusion additive manufacturing processes for automatic manufacturing error detection. *Int. J. Adv. Manuf. Technol.* **2021**, *115*, 3859–3872. [CrossRef]
239. Martin Montal, J.; Pernas-Sánchez, J.; Varas, D. Experimental Characterization Framework for SLA Additive Manufacturing Materials. *Polymers* **2021**, *13*, 1147. [CrossRef]
240. Lin, W.; Shen, H.; Fu, J.; Wu, S. Online quality monitoring in material extrusion additive manufacturing processes based on laser scanning technology. *Precis. Eng.* **2019**, *60*, 76–84. [CrossRef]
241. Yao, H.; Yu, M.; Fu, J.; Zhu, M.; Li, Y.; Li, S.; Gan, R.; Zhou, H.; Qi, S. Shape memory polymers enable versatile magneto-active structure with 4D printability, variable stiffness, shape-morphing and effective grasping. *Smart Mater. Struct.* **2023**, *32*, 095005. [CrossRef]
242. Chung, J.; Shen, B.; Law, A.C.C.; Kong, Z.J. Reinforcement learning-based defect mitigation for quality assurance of additive manufacturing. *J. Manuf. Syst.* **2022**, *65*, 822–835. [CrossRef]
243. Aljazara, A.; Abu Tuhaimer, N.; Alawwad, A.; Bani-Hani, K.; Qusef, A.; Alsalhi, N.; Al-dawoodi, A. Quality of 3D Printed Objects Using Fused Deposition Modeling (FDM) Technology in Terms of Dimensional Accuracy. *Int. J. Online Biomed. Eng. (iJOE)* **2023**, *19*, 45–62. [CrossRef]
244. Carrico, J.D.; Hermans, T.; Kim, K.J.; Leang, K.K. 3D-printing and machine learning control of soft ionic polymer-metal composite actuators. *Sci. Rep.* **2019**, *9*, 17482. [CrossRef]
245. Omairi, A.; Ismail, Z.H. Towards machine learning for error compensation in additive manufacturing. *Appl. Sci.* **2021**, *11*, 2375. [CrossRef]
246. Chen, G.L.; Yanamandra, K.; Gupta, N. Artificial neural networks framework for detection of defects in 3D-printed fiber reinforcement composites. *JOM* **2021**, *73*, 2075–2084. [CrossRef]
247. Chen, G.L.; Gupta, N. Image processing and machine learning methods applied to additive manufactured composites for defect detection and toolpath reconstruction. In *Machine Learning Applied to Composite Materials*; Springer: Singapore, 2022; pp. 19–44.
248. Goh, G.D.; Hamzah, N.M.B.; Yeong, W.Y. Anomaly detection in fused filament fabrication using machine learning. *3D Print. Addit. Manuf.* **2023**, *10*, 428–437. [CrossRef] [PubMed]
249. Chen, J.; Wan, L.; Ismail, Y.; Ye, J.; Yang, D. A micromechanics and machine learning coupled approach for failure prediction of unidirectional CFRP composites under triaxial loading: A preliminary study. *Compos. Struct.* **2021**, *267*, 113876. [CrossRef]
250. Shanaswara, A.; Aditya, M.; Rashyid, M.; Herianto, H.; Muflikhun, M. Optimasi Multiobjektif Parameter Proses 3D Printer Jenis Fused Deposition Modelling Menggunakan Grey Relational Analysis-Taguchi. *J. Rekayasa Mesin* **2023**, *14*, 791–810. [CrossRef]
251. Pandžić, A.; Hodzic, D.; Kadric, E. Experimental Investigation on Influence of Infill Density on Tensile Mechanical Properties of Different FDM 3D Printed Materials. *TEM J.* **2021**, *10*, 1195–1201. [CrossRef]
252. Wan, L.; Ullah, Z.; Yang, D.; Falzon, B.G. Probability embedded failure prediction of unidirectional composites under biaxial loadings combining machine learning and micromechanical modelling. *Compos. Struct.* **2023**, *312*, 116837. [CrossRef]
253. Fontes, A.; Shadmehri, F. Data-driven failure prediction of Fiber-Reinforced Polymer composite materials. *Eng. Appl. Artif. Intell.* **2023**, *120*, 105834. [CrossRef]
254. Sepasdar, R.; Karpatne, A.; Shakiba, M. A data-driven approach to full-field nonlinear stress distribution and failure pattern prediction in composites using deep learning. *Comput. Methods Appl. Mech. Eng.* **2022**, *397*, 115126. [CrossRef]
255. Pandžić, A.; Hodzic, D.; Kadric, D.; Kadric, E. The Effect of Ultraviolet Radiation on Mechanical Properties of Fused Deposition Modeling 3D Printed Materials. *TEM J.* **2023**, *12*, 1921–1928. [CrossRef]
256. Xue, T.; Wallin, T.J.; Menguc, Y.; Adriaenssens, S.; Chiaramonte, M. Machine learning generative models for automatic design of multi-material 3D printed composite solids. *Extrem. Mech. Lett.* **2020**, *41*, 100992. [CrossRef]
257. Lee, S.; Zhang, Z.; Gu, G.X. Generative machine learning algorithm for lattice structures with superior mechanical properties. *Mater. Horizons* **2022**, *9*, 952–960. [CrossRef] [PubMed]
258. Qian, C.; Tan, R.K.; Ye, W. Design of architectured composite materials with an efficient, adaptive artificial neural network-based generative design method. *Acta Mater.* **2022**, *225*, 117548. [CrossRef]
259. He, Y.; Abdi, M.; Trindade, G.F.; Begines, B.; Dubern, J.F.; Prina, E.; Hook, A.L.; Choong, G.Y.; Ledesma, J.; Tuck, C.J.; et al. Exploiting generative design for 3D printing of bacterial biofilm resistant composite devices. *Adv. Sci.* **2021**, *8*, 2100249. [CrossRef]
260. Hashemi, M.S.; Nguyen, K.; Kirby, L.; Song, X.; Sheidaei, A. Universal Design Methodology for Printable Microstructural Materials via a New Deep Generative Learning Model: Application to a Piezocomposite. *arXiv* **2024**, arXiv:2402.11102.
261. Garland, A.P.; White, B.C.; Jensen, S.C.; Boyce, B.L. Pragmatic generative optimization of novel structural lattice metamaterials with machine learning. *Mater. Des.* **2021**, *203*, 109632. [CrossRef]
262. Sachdeva, I.; Ramesh, S.; Chadha, U.; Punugoti, H.; Selvaraj, S.K. Computational AI models in VAT photopolymerization: A review, current trends, open issues, and future opportunities. *Neural Comput. Appl.* **2022**, *34*, 17207–17229. [CrossRef]
263. Nasrin, T.; Pourkamali-Anaraki, F.; Hansen, C.J.; Jensen, R.E.; Peterson, A. Predicting Printability of Highly Filled Polymer Suspensions via Vat Photopolymerization: A Classification-Based Machine Learning Approach. Available online: https://papers.ssrn.com/sol3/papers.cfm?abstract_id=4806327 (accessed on 4 November 2024).

264. Jo, J.; Park, K.; Song, H.; Lee, H.; Ryu, S. Innovative 3D printing of mechanoluminescent composites: Vat photopolymerization meets machine learning. *Addit. Manuf.* **2024**, *90*, 104324. [CrossRef]
265. Frumosu, F.D.; Méndez Ribó, M.; Shan, S.; Zhang, Y.; Kulahci, M. Online monitoring for error detection in vat photopolymerization. *Int. J. Comput. Integr. Manuf.* **2023**, *36*, 1313–1330. [CrossRef]
266. Shan, Y.; Krishnakumar, A.; Qin, Z.; Mao, H. Smart Resin Vat: Real-Time Detecting Failures, Defects, and Curing Area in Vat Photopolymerization 3D Printing. In Proceedings of the International Manufacturing Science and Engineering Conference, West Lafayette, IN, USA, 27 June–1 July 2022; Volume 85802, p. V001T01A030.
267. Cao, L.; Lu, L.; Liu, X.; Zhang, J.; Jiang, T.; Tu, C.; Chen, B. Waiting time prediction for bottom-up vat photopolymerization. *Addit. Manuf.* **2023**, *74*, 103693. [CrossRef]
268. Zhang, Y.; Zhang, H.; Zhao, X. In-situ interferometric curing monitoring for digital light processing based vat photopolymerization additive manufacturing. *Addit. Manuf.* **2024**, *81*, 104001. [CrossRef]
269. Alli, Y.A.; Anuar, H.; Manshor, M.R.; Okafor, C.E.; Kamarulzaman, A.F.; Akçakale, N.; Nazeri, F.N.M.; Bodaghi, M.; Suhr, J.; Nasir, N.A.M. Optimization of 4D/3D printing via machine learning: A systematic review. *Hybrid Adv.* **2024**, *6*, 100242. [CrossRef]
270. Wang, Z.; Yang, Y.; Suo, S.; Guo, J.; Rao, W.F. Predicting 4D hardness property from 3D datasets for performance-tunable material extrusion additive manufacturing. *Mater. Today Commun.* **2024**, *38*, 108216. [CrossRef]
271. Sun, X.; Yue, L.; Yu, L.; Shao, H.; Peng, X.; Zhou, K.; Demoly, F.; Zhao, R.; Qi, H.J. Machine Learning-Evolutionary Algorithm Enabled Design for 4D-Printed Active Composite Structures. *Adv. Funct. Mater.* **2022**, *32*, 2109805. [CrossRef]
272. Forte, C.T.; Montgomery, S.M.; Yue, L.; Hamel, C.M.; Qi, H.J. Grayscale digital light processing gradient printing for stress concentration reduction and material toughness enhancement. *J. Appl. Mech.* **2023**, *90*, 071003. [CrossRef]
273. Hamel, C.M.; Roach, D.J.; Long, K.N.; Demoly, F.; Dunn, M.L.; Qi, H.J. Machine-learning based design of active composite structures for 4D printing. *Smart Mater. Struct.* **2019**, *28*, 065005. [CrossRef]
274. Su, R.; Chen, J.; Zhang, X.; Wang, W.; He, R.; Xu, H.; Li, Y. 3D printed SiOC architecture towards terahertz electromagnetic interference shielding and absorption. *Chem. Eng. J.* **2024**, *498*, 155689. [CrossRef]
275. Lyu, S.; Zhao, T.; Wang, Y.; Han, H.; Li, T.; Zhang, C.; Li, D.; Wang, J.K.; Huang, J.; Yu, P.; et al. Ti3C2Tx-coated diatom frustules-derived porous SiO^2 composites with high EMI shielding and mechanical properties. *Ceram. Int.* **2022**, *48*, 22845–22853. [CrossRef]
276. Wang, W.; Gao, X.; Li, Z.; Shen, C.; Ren, X.; He, R. High performance Csf/SiC ceramic matrix composites fabricated by material extrusion 3D printing and precursor infiltration and pyrolysis. *Ceram. Int.* **2024**, *50*, 44511–44520. [CrossRef]
277. Sarvestani, H.Y.; Lacelle, T.; Sohrabi-Kashani, A.; Shashoua, A.; Karamzadeh, V.; Ravanbakhsh, H.; Robitaille, A.; Lavoie, H.; Paquet, C.; Jakubinek, M.; et al. 3D-printed polymer-derived ceramics with tunable cellular architectures. *Appl. Mater. Today* **2024**, *39*, 102327. [CrossRef]
278. Jiang, L.; Long, C.; Xiong, S.; Wang, X.; Mo, Y.; Zeng, J.; Liu, Z.; Liu, C.; Chen, Z. 4D printing of shape-programmable polymer-derived ceramics via two-stage folding-assisted pyrolysis strategy. *Virtual Phys. Prototyp.* **2024**, *19*, e2406408. [CrossRef]
279. Zhu, N.; Hou, Y.; Zhang, L.; Wen, G. Preparation of h-BN/SiCO ceramic matrix composites with high thermal conductivity and strength by vat photopolymerization 3D printing. *J. Eur. Ceram. Soc.* **2024**, *44*, 5885–5895. [CrossRef]
280. Young, J.C.; Brinckmann, S.A.; Fox, A.; Fertig III, R.S.; Lynch, S.P.; Frick, C.P. On enhancing the manufacturability and conversion of additively manufactured polymer-derived ceramics. *J. Eur. Ceram. Soc.* **2024**, *44*, 116691. [CrossRef]
281. Bobrin, V.A.; Hackbarth, H.G.; Bonsu, J.O.; Yao, Y.; Bedford, N.M.; Kundu, D.; Zhang, J.; Corrigan, N.; Boyer, C. Microphase Separation 3D Printing of Binary Inorganic Polymer Precursors to Prepare Nanostructured Carbon-Ceramic Multimaterials. *Adv. Mater. Technol.* **2024**, *9*, 2400337. [CrossRef]
282. Wang, K.; Yin, J.; Chen, X.; Liu, X.; Huang, Z. Microstructure and properties of liquid phase sintered SiC ceramics fabricated via selective laser printing and precursor impregnation and pyrolysis. *Ceram. Int.* **2024**, *50*, 4315–4322. [CrossRef]
283. Wang, Y.; Wu, T.; Huang, G. State-of-the-art research progress and challenge of the printing techniques, potential applications for advanced ceramic materials 3D printing. *Mater. Today Commun.* **2024**, *40*, 110001. [CrossRef]
284. Phan, B.K.; Shen, K.H.; Gurnani, R.; Tran, H.; Lively, R.; Ramprasad, R. Gas permeability, diffusivity, and solubility in polymers: Simulation-experiment data fusion and multi-task machine learning. *npj Comput. Mater.* **2024**, *10*, 186. [CrossRef]
285. Ricci, E.; De Angelis, M.G. A perspective on data-driven screening and discovery of polymer membranes for gas separation, from the molecular structure to the industrial performance. *Rev. Chem. Eng.* **2024**, *40*, 567–600. [CrossRef]
286. Zhang, Q.; Han, S.; Abed, A.M.; Basem, A.; El-Sharkawy, M.R.; Salah, B.; Mahariq, I. Enhancing battery pack safety against cone impact using machine learning techniques and Gaussian noise. *Process. Saf. Environ. Prot.* **2024**, *191*, 448–465. [CrossRef]
287. Hippalgaonkar, K.; Li, Q.; Wang, X.; Fisher III, J.W.; Kirkpatrick, J.; Buonassisi, T. Knowledge-integrated machine learning for materials: Lessons from gameplaying and robotics. *Nat. Rev. Mater.* **2023**, *8*, 241–260. [CrossRef]
288. Osterberg, M.; Henn, K.A.; Farooq, M.; Valle-Delgado, J.J. Biobased nanomaterials—The role of interfacial interactions for advanced materials. *Chem. Rev.* **2023**, *123*, 2200–2241. [CrossRef] [PubMed]
289. Khalid, M.Y.; Arif, Z.U.; Tariq, A.; Hossain, M.; Umer, R.; Bodaghi, M. 3D printing of active mechanical metamaterials: A critical review. *Mater. Des.* **2024**, *246*, 113305. [CrossRef]
290. Zhou, X.; Ren, L.; Song, Z.; Li, G.; Zhang, J.; Li, B.; Wu, Q.; Li, W.; Ren, L.; Liu, Q. Advances in 3D/4D printing of mechanical metamaterials: From manufacturing to applications. *Compos. Part B Eng.* **2023**, *254*, 110585. [CrossRef]

291. Rodríguez, S.E.; Calius, E.; Khatibi, A.; Orifici, A.; Das, R. Mechanical metamaterial systems as transformation mechanisms. *Extrem. Mech. Lett.* **2023**, *61*, 101985. [CrossRef]
292. Chen, Y.; Ebrahimifakhar, A.; Hu, Y.; Yuill, D.P. Generalizability of machine learning-based fault classification for residential air-conditioners. *Energy Build.* **2023**, *295*, 113263. [CrossRef]
293. Li, K.; DeCost, B.; Choudhary, K.; Greenwood, M.; Hattrick-Simpers, J. A critical examination of robustness and generalizability of machine learning prediction of materials properties. *npj Comput. Mater.* **2023**, *9*, 55. [CrossRef]
294. Nasrin, T.; Pourkamali-Anaraki, F.; Peterson, A.M. Application of machine learning in polymer additive manufacturing: A review. *J. Polym. Sci.* **2024**, *62*, 2639–2669. [CrossRef]
295. Manshor, M.R.; Alli, Y.A.; Anuar, H.; Ejeromedoghene, O.; Omotola, E.O.; Suhr, J. 4D printing: Historical evolution, computational insights and emerging applications. *Mater. Sci. Eng. B* **2023**, *295*, 116567. [CrossRef]
296. Nath, D.; Ankit, Neog, D.R.; Gautam, S.S. Application of machine learning and deep learning in finite element analysis: A comprehensive review. *Arch. Comput. Methods Eng.* **2024**, *31*, 2945–2984. [CrossRef]
297. Belkebir, H.; Prod'hon, R.; Foufou, S.; Gomes, S.; Demoly, F. Towards a Multi-view and Multi-representation CAD Models System for Computational Design of Multi-material 4D Printed Structures. In *IFIP International Conference on Product Lifecycle Management*; Springer: Cham, Switzerland, 2023; pp. 287–297.
298. Athinarayanarao, D.; Prod'hon, R.; Chamoret, D.; Qi, H.J.; Bodaghi, M.; André, J.C.; Demoly, F. Computational design for 4D printing of topology optimized multi-material active composites. *npj Comput. Mater.* **2023**, *9*, 1. [CrossRef]
299. Khanam, R.; Hussain, M.; Hill, R.; Allen, P. A comprehensive review of convolutional neural networks for defect detection in industrial applications. *IEEE Access* **2024**, *12*, 94250–94295. [CrossRef]
300. Kumar, S.; Gopi, T.; Harikeerthana, N.; Gupta, M.K.; Gaur, V.; Krolczyk, G.M.; Wu, C. Machine learning techniques in additive manufacturing: A state of the art review on design, processes and production control. *J. Intell. Manuf.* **2023**, *34*, 21–55. [CrossRef]
301. Babu, S.S.; Mourad, A.H.I.; Harib, K.H.; Vijayavenkataraman, S. Recent developments in the application of machine-learning towards accelerated predictive multiscale design and additive manufacturing. *Virtual Phys. Prototyp.* **2023**, *18*, e2141653. [CrossRef]
302. Phillips, C.; Kortschot, M.; Azhari, F. Towards standardizing the preparation of test specimens made with material extrusion: Review of current techniques for tensile testing. *Addit. Manuf.* **2022**, *58*, 103050. [CrossRef]
303. Duarte, G.; Brown, N.C.; Duarte, J.P. Workflow for generating, simulating, and optimizing form and toolpath in 3D concrete printing of vaults. *Autom. Constr.* **2024**, *166*, 105590. [CrossRef]
304. Zhuang, Z.; Xu, F.; Ye, J.; Hu, N.; Jiang, L.; Weng, Y. A comprehensive review of sustainable materials and toolpath optimization in 3D concrete printing. *npj Mater. Sustain.* **2024**, *2*, 12. [CrossRef]
305. García-Domínguez, A.; Claver, J.; Camacho, A.M.; Sebastián, M.A. Analysis of general and specific standardization developments in additive manufacturing from a materials and technological approach. *IEEE Access* **2020**, *8*, 125056–125075. [CrossRef]
306. Chua, C.; Liu, Y.; Williams, R.J.; Chua, C.K.; Sing, S.L. In-process and post-process strategies for part quality assessment in metal powder bed fusion: A review. *J. Manuf. Syst.* **2024**, *73*, 75–105. [CrossRef]
307. Hou, Z.J.; Wang, Q.; Zhao, C.G.; Zheng, J.; Tian, J.M.; Ge, X.H.; Liu, Y.G. Online monitoring technology of metal powder bed fusion processes: A review. *Materials* **2022**, *15*, 7598. [CrossRef]
308. Kim, S.; Schroeder, C.M.; Jackson, N.E. Open macromolecular genome: Generative design of synthetically accessible polymers. *ACS Polym. Au* **2023**, *3*, 318–330. [CrossRef]
309. Jiang, S.; Dieng, A.B.; Webb, M.A. Property-guided generation of complex polymer topologies using variational autoencoders. *npj Comput. Mater.* **2024**, *10*, 139. [CrossRef]
310. Helwing, R.; Mrzljak, S.; Hülsbusch, D.; Walther, F. Cycle-consistent generative adversarial networks for damage evolution analysis in fiber-reinforced polymers based on synthetic damage states. *Compos. Sci. Technol.* **2024**, *254*, 110695. [CrossRef]
311. Cakiroglu, C. Explainable data-driven ensemble learning models for the mechanical properties prediction of concrete confined by aramid fiber-reinforced polymer wraps using generative adversarial networks. *Appl. Sci.* **2023**, *13*, 11991. [CrossRef]
312. Jabbar, R.; Jabbar, R.; Kamoun, S. Recent progress in generative adversarial networks applied to inversely designing inorganic materials: A brief review. *Comput. Mater. Sci.* **2022**, *213*, 111612. [CrossRef]
313. Jiang, Y.; Li, J.; Yang, X.; Yuan, R. Applications of generative adversarial networks in materials science. *Mater. Genome Eng. Adv.* **2024**, *2*, e30. [CrossRef]
314. Karunarathna, I.; Gunasena, P.; Hapuarachchi, T.; Gunathilake, S. The Crucial Role of Data Collection in Research: Techniques, Challenges, and Best Practices. Available online: https://www.researchgate.net/publication/383155720 (accessed on 18 October 2024).
315. McDonough, K.; Battagliarin, G.; Menzies, J.; Bozich, J.; Bergheim, M.; Hidding, B.; Kastner, C.; Koyuncu, B.; Kreutzer, G.; Leijs, H.; et al. Multi-laboratory evaluation of the reproducibility of polymer biodegradation assessments applying standardized and modified respirometry methods. *Sci. Total. Environ.* **2023**, *901*, 166339. [CrossRef]
316. Shetty, P.; Rajan, A.C.; Kuenneth, C.; Gupta, S.; Panchumarti, L.P.; Holm, L.; Zhang, C.; Ramprasad, R. A general-purpose material property data extraction pipeline from large polymer corpora using natural language processing. *npj Comput. Mater.* **2023**, *9*, 52. [CrossRef]

317. Chen, Z.; Li, D.; Liu, J.; Gao, K. Application of Gaussian processes and transfer learning to prediction and analysis of polymer properties. *Comput. Mater. Sci.* **2023**, *216*, 111859. [CrossRef]
318. Shi, J.; Albreiki, F.; Colón, Y.J.; Srivastava, S.; Whitmer, J.K. Transfer Learning Facilitates the Prediction of Polymer–Surface Adhesion Strength. *J. Chem. Theory Comput.* **2023**, *19*, 4631–4640. [CrossRef]
319. Tang, Y.; Dehaghani, M.R.; Wang, G.G. Review of transfer learning in modeling additive manufacturing processes. *Addit. Manuf.* **2023**, *61*, 103357. [CrossRef]
320. Fang, Y.; Lu, X.; Jin, L. Application of Multi-fidelity Surrogate Models to the Noisy Optimization Problems of Carbon Fiber Polymerization Process Parameters. In *International Conference on Bio-Inspired Computing: Theories and Applications*; Springer: Singapore, 2023; pp. 281–295.
321. Nath, P.; Sato, M.; Karve, P.; Mahadevan, S. Multi-fidelity Modeling for Uncertainty Quantification in Laser Powder Bed Fusion Additive Manufacturing. *Integr. Mater. Manuf. Innov.* **2022**, *11*, 256–275. [CrossRef]
322. Quah, T.K.N.; Tay, Y.W.D.; Lim, J.H.; Tan, M.J.; Wong, T.N.; Li, K.H.H. Concrete 3D printing: Process parameters for process control, monitoring and diagnosis in automation and construction. *Mathematics* **2023**, *11*, 1499. [CrossRef]
323. Petrich, J.; Snow, Z.; Corbin, D.; Reutzel, E.W. Multi-modal sensor fusion with machine learning for data-driven process monitoring for additive manufacturing. *Addit. Manuf.* **2021**, *48*, 102364. [CrossRef]
324. Naz, S.A.; Doeven, E.H.; Adams, S.; Kouzani, A.; Guijt, R.M. Closed-loop control systems for pumps used in portable analytical systems. *J. Chromatogr. A* **2023**, *1695*, 463931. [CrossRef]
325. Mercado Rivera, F.J.; Rojas Arciniegas, A.J. Additive manufacturing methods: Techniques, materials, and closed-loop control applications. *Int. J. Adv. Manuf. Technol.* **2020**, *109*, 17–31. [CrossRef]
326. Benavides-Hernández, J.; Dumeignil, F. From Characterization to Discovery: Artificial Intelligence, Machine Learning and High-Throughput Experiments for Heterogeneous Catalyst Design. *ACS Catal.* **2024**, *14*, 11749–11779. [CrossRef]
327. Patel, R.A.; Webb, M.A. Data-driven design of polymer-based biomaterials: High-throughput simulation, experimentation, and machine learning. *ACS Appl. Bio Mater.* **2023**, *7*, 510–527. [CrossRef]
328. Zhang, L.; Banihashemi, S.; Zhu, L.; Molavi, H.; Odacioglu, E.; Shan, M. A Scientometric Analysis of Knowledge Transfer Partnerships in Digital Transformation. *J. Open Innov. Technol. Mark. Complex.* **2024**, *10*, 100325. [CrossRef]
329. Park, H.; Ko, H.; Lee, Y.t.T.; Feng, S.; Witherell, P.; Cho, H. Collaborative knowledge management to identify data analytics opportunities in additive manufacturing. *J. Intell. Manuf.* **2023**, *34*, 541–564. [CrossRef]
330. Khalid, M.Y.; Arif, Z.U.; Noroozi, R.; Hossain, M.; Ramakrishna, S.; Umer, R. 3D/4D printing of cellulose nanocrystals-based biomaterials: Additives for sustainable applications. *Int. J. Biol. Macromol.* **2023**, *251*, 126287. [CrossRef]
331. Farham, B.; Baltazar, L. A Review of Smart Materials in 4D Printing for Hygrothermal Rehabilitation: Innovative Insights for Sustainable Building Stock Management. *Sustainability* **2024**, *16*, 4067. [CrossRef]
332. Ram Kishore, S.; Sridharan, A.; Chadha, U.; Narayanan, D.; Mishra, M.; Selvaraj, S.K.; Patterson, A.E. Natural fiber biocomposites via 4D printing technologies: A review of possibilities for agricultural bio-mulching and related sustainable applications. *Prog. Addit. Manuf.* **2024**, *9*, 37–67. [CrossRef]
333. Rahmatabadi, D.; Khajepour, M.; Bayati, A.; Mirasadi, K.; Yousefi, M.A.; Shegeft, A.; Ghasemi, I.; Baniassadi, M.; Abrinia, K.; Bodaghi, M.; et al. Advancing sustainable shape memory polymers through 4D printing of polylactic acid-polybutylene adipate terephthalate blends. *Eur. Polym. J.* **2024**, *216*, 113289. [CrossRef]
334. Hu, G.; Bodaghi, M. Direct fused deposition modeling 4D printing and programming of thermoresponsive shape memory polymers with autonomous 2D-to-3D shape transformations. *Adv. Eng. Mater.* **2023**, *25*, 2300334. [CrossRef]
335. Bora, L.V.; Vadaliya, K.S.; Bora, N.V. Sustainable feedstocks for 4D printing: Biodegradable polymers and natural resources. *Green Mater.* **2024**, *12*, 192–208. [CrossRef]
336. Vinnikov, S.; Nardid, A.; Gapanyuk, Y. Metagraph Embedding Approach Based on Red-Blue Graphs. In *Advances in Neural Computation, Machine Learning, and Cognitive Research VIII*; Kryzhanovsky, B., Dunin-Barkowski, W., Redko, V., Tiumentsev, Y., Yudin, D., Eds.; Springer: Cham, Switzerland, 2024; pp. 510–522.
337. Shcheglov, G.A.; Mordovskii, A.V. Design optimization of cubic-shaped pressurant tank for CubeSat propulsion system. *Acta Astronaut.* **2024**, *224*, 48–56. [CrossRef]
338. Hegab, H.; Khanna, N.; Monib, N.; Salem, A. Design for sustainable additive manufacturing: A review. *Sustain. Mater. Technol.* **2023**, *35*, e00576. [CrossRef]
339. Stavropoulos, P.; Foteinopoulos, P.; Stavridis, J.; Bikas, H. Increasing the industrial uptake of additive manufacturing processes: A training framework. *Adv. Ind. Manuf. Eng.* **2023**, *6*, 100110. [CrossRef]
340. Qiu, C.; Yang, J. Machine learning applications in composites: Manufacturing, design, and characterization. In *Machine Learning in Materials Informatics: Methods and Applications*; ACS Publications: Washington, DC, USA, 2022; pp. 65–85.

MDPI AG
Grosspeteranlage 5
4052 Basel
Switzerland
Tel.: +41 61 683 77 34

Polymers Editorial Office
E-mail: polymers@mdpi.com
www.mdpi.com/journal/polymers

www.ingramcontent.com/pod-product-compliance
Lightning Source LLC
Chambersburg PA
CBHW061821210926
51850CB00006B/768
* 9 7 8 3 7 2 5 8 4 9 8 5 7 *